Oscillations and Morphogenesis

CELLULAR CLOCKS

Series Editor

Leland N. Edmunds, Jr.

Division of Biological Sciences
State University of New York at Stony Brook
Stony Brook, New York

1. Cell Cycle Clocks, *edited by Leland N. Edmunds, Jr.*
2. Neuronal and Cellular Oscillators, *edited by Jon W. Jacklet*
3. Chronopharmacology: Cellular and Biochemical Interactions, *edited by Björn Lemmer*
4. Molecular Genetics of Biological Rhythms, *edited by Michael W. Young*
5. Oscillations and Morphogenesis, *edited by Ludger Rensing*

Additional Volumes in Preparation

Oscillations and Morphogenesis

Edited by

Ludger Rensing
University of Bremen
Bremen, Germany

Marcel Dekker, Inc. New York • Basel • Hong Kong

Library of Congress Cataloging-in-Publication Data

Oscillations and morphogenesis / edited by Ludger Rensing.
p. cm. -- (Cellular clocks ; v. 5)
Includes bibliographical references and index.
ISBN 0-8247-8765-X (acid-free paper)
1. Morphogenesis. 2. Embryonic periodicity. 3. Biological rhythms. I. Rensing, Ludger. II. Series.
QH491.0577 1993
574.3'32--dc20 92-29094
CIP

This book is printed on acid-free paper.

MARCEL DEKKER, INC.
270 Madison Avenue, New York, New York 10016

Current printing (last digit):
10 9 8 7 6 5 4 3 2 1

PRINTED IN THE UNITED STATES OF AMERICA

Series Introduction

In the earlier part of the twentieth century, it was commonly believed that the evolution of any physiochemical system invariably leads to a state of maximum disorder as a result of the application of the second law of thermodynamics. As a consequence, any coherent behavior, such as self-sustained oscillations, was thought to be ruled out. The development of the concept of open (as opposed to closed) systems, however, typified by living organisms, enabled one to predict in principle the spontaneous appearance (self-organization) of temporal dissipative structures, including autonomously oscillatory systems having steady states sufficiently far from equilibrium. Stable oscillations of the limit cycle type can appear in certain nonlinear systems. Positive or negative feedback, sometimes combined with cross-catalysis or other cross-coupling, is a necessary physical prerequisite.

These theoretical notions are embodied in the biological periodicities prevalent at every level of structural and physiological organization throughout the animal and plant kingdoms: oscillatory behavior is the rule rather than the exception. The period (τ) of known biological oscillators (or 'clocks') ranges from milliseconds, minutes and hours (ultradian, $\tau <$ 20 h), to approximately one day (circadian, 20 h $< \tau <$ 28 h), to longer (infradian, $\tau <$ 28 h) timespans. That organisms can and do measure astronomical time in some manner—as well as the purely private timekeeping reflected in variable-period physiological rhythms—is reflected in the ubiquitous persistent rhythms having solar daily, tidal, lunar monthly, and yearly (circannual) periods; by the *Zeitgedächtnis*, or time sense, of bees and humans; by seasonal photoperiodism, wherein many organisms perform a certain function at a specific time of the year (perhaps by daily measurement of the length of the day or night); and by celestial orientation and navigation, in which the sun, moon, or stars are used as direction givers and for whose daily continuous movement some sort of timing mechanism consequently must compensate. All four types of timekeeping, or functional biochronometry, have external correlates (generated by the movements of the earth, moon, and sun) to which the organism has adapted. The central thesis is that there is a selective premium for each class of temporal adaptation.

An understanding of the physiological and biochemical bases of simpler autogenous oscillators and clocks underlying biological periodicities, such as those characteristic of most cell types, is crucial to the understanding of the more complex timekeeping displayed at the systems, organismal, and even population levels. Accordingly, this series, Cellular Clocks, was initiated with *Cell Cycle Clocks* (ed. L. N. Edmunds, Jr., 1984), which considers the interaction of timers, oscillators, and clocks and the cell division cycle of prokaryotes and eukaryotes. The second volume of the series, *Neuronal and Cellular Oscillators* (ed. J. W. Jacklet, 1989) is particularly appropriate in view of the explosion of interest in the neurosciences and in cellular pacemakers. *Chronopharmacology: Cellular and Biochemical Interactions* (ed. B. Lemmer, 1989) is third in the series and examines the time-dependent effects of drugs, their therapeutic efficacy and toxicity, and their temporal modes of action. The present work, *Oscillations and Morphogenesis*, edited by Ludger Rensing, comprehensively examines at both the theoretical and experimental levels the principles of periodic pattern formation and the periodic timing of morphogenetic processes in plants and animals. A particularly useful overview of this intriguing area is given in the introductory chapter. Soon to follow will be volumes on the molecular genetics of biological rhythms (ed. Michael W. Young) and on quantal oscillators and the evolution of biological timekeeping (ed. Robert R. Klevecz). Future topics are expected to include cellular aspects of sleep-wake cycles and the cellular biochemistry of biological pacemakers.

Leland N. Edmunds, Series Editor

Preface

This book deals with the significance of oscillations in morphogenesis as dissipative structures in self-organizing systems, as signal-transmitting spatial and temporal information in developing living systems, and as time-keeping devices (clocks) serving the coordination or synchronization of system elements. The roles of oscillations in relation to morphogenesis is a rather new topic; thus some of the contributions focus on descriptional evidence. Several chapters are devoted to the experimental analysis of these relations, combined with the application of theoretical approaches such as nonlinear dynamics of self-organized systems developed in the past 25 years. Different nonliving and living systems at different levels of organization are described and analyzed in this book by a wide spectrum of scientists: mathematicians, physicists, chemists, and biologists. There is always some degree of arbitrariness in the selection of topics and contributions; the intention was to document the ubiquitous presence and significance of temporal oscillations as well as of periodic spatial structures and their tight interrelations. Chapter 1 is offered as a guideline to the various roles of oscillations in morphogenesis and the different empirical and theoretical approaches to their analysis. It may also serve to address systems not explicitly analyzed in this volume.

The book is meant to stimulate discussions among colleagues working on different aspects of the intriguing problems of oscillations, morphogenesis, and nonlinear dynamics. It is devoted particularly to overcoming the gap between experimental and theoretical analysis. Hopefully, it may also stimulate graduate students to become interested in these problems.

I am grateful to the Stiftung Volkswagenwerk for their support of part of our experimental and theoretical work.

Ludger Rensing

Contents

Contributors

L. V. Beloussov Department of Embryology, Faculty of Biology, Moscow State University, Moscow, Russia

Silvano Bonotto Professor, Chair of Ecology, Department of Animal Biology, University of Turin, Turin, Italy

Fiona M. K. Corke Department of Cell Biology, John Innes Institute, Norwich, England

Edward C. Cox Professor of Molecular Biology and Edwin Grant Conklin Professor of Biology, Department of Molecular Biology, Princeton University, Princeton, New Jersey

Andreas Deutsch Department of Biology, University of Bremen, Bremen, Germany

Renate Dohmen Physics Department, Institute for Applied Physics, Münster, Germany

Geneviève Dupont Service de Chimie Physique, Faculté des Sciences, Université Libre de Bruxelles, Brussels, Belgium

Leland N. Edmunds, Jr. Division of Biological Sciences, State University of New York, Stony Brook, New York

Peter Fromherz Abteilung Biophysik, Universität Ulm, Ulm, Germany

Albert Goldbeter Service de Chimie Physique, Faculté des Sciences, Université Libre de Bruxelles, Brussels, Belgium

H. Haken Institut für Theoretische Physik und Synergetik, Stuttgart, Germany

Lionel G. Harrison Department of Chemistry, University of British Columbia, Vancouver, British Columbia, Canada

J. Woodland Hastings Professor of Biology, Department of Cellular and Developmental Biology, Harvard University, Cambridge, Massachusetts

Angela J. Hauxwell Department of Applied Genetics, John Innes Institute, Norwich, England

Stéphan Jerebzoff Directeur de Recherche C.N.R.S., Biorhythm Laboratory, Paul-Sabatier University of Toulouse, Toulouse, France

Simonne Jerebzoff-Quintin Chargé de Recherche C.N.R.S., Biorhythm Laboratory, Paul-Sabatier University of Toulouse, Toulouse, France

N. I. Kazakova Department of Embryology, Faculty of Biology, Moscow State University, Moscow, Russia

Ju. A. Labas Institute of Evolutionary Morphology and Ecology of Animals, Russian Academy of Sciences, Moscow, Russia

Thurston C. Lacalli Professor of Biology, University of Saskatchewan, Saskatoon, Saskatchewan, Canada

Jan Löfberg Professor, Department of Zoology, Uppsala University, Uppsala, Sweden

William J. Lucas Professor, Section of Botany, Division of Biological Sciences, University of California, Davis, California

Hans Meinhardt Professor, Max-Planck-Institut für Entwicklungsbiologie, Tübingen, Germany

Karl B. Moritz Professor of Zoology, Zoological Institute, University of Munich, Munich, Germany

Stefan C. Müller Privat-Dozent, Max-Planck-Institut für Ernährungsphysiologie, Dortmund, Germany

Franz-Josef Niedernostheide Physics Department, Institute of Applied Physics, Münster, Germany

Lennart Olsson Department of Zoology, Uppsala University, Uppsala, Sweden

Hans-Georg Purwins Professor, Physics Department, Institute of Applied Physics, Münster, Germany

Ludger Rensing Professor, Department of Zellbiologie, University of Bremen, Bremen, Germany

Christopher A. Richardson Lecturer in Marine Biology, School of Ocean Sciences, Marine Science Laboratories, Menai Bridge, Anglesey, Wales

Till Roenneberg Assistant Professor, Institute of Medical Psychology, University of Munich, Munich, Germany

Helmut W. Sauer Professor, Department of Biology, Texas A&M University, College Station, Texas

David S. Saunders Professor, Institute of Cell, Animal, and Population Biology, Division of Biology, University of Edinburgh, Edinburgh, Scotland

Rodney Arthur Savidge Associate Professor, Tree Biology, Department of Forest Resources, University of New Brunswick, Fredericton, New Brunswick, Canada

James A. Shapiro Professor of Microbiology, Department of Biochemistry and Molecular Biology, University of Chicago, Chicago, Illinois

Florian Siegert Zoological Institute, University of Munich, Munich, Germany

Tetsuo Ueda Professor, Graduate School of Human Informatics, Nagoya University, Nagoya, Japan

Thérèse Vanden Driessche Department of Molecular Biology, Université Libre de Bruxelles, Rhode Saint-Genèse, Belgium

Michael G. Vicker Department of Biology, University of Bremen, Bremen, Germany

Trevor L. Wang Department of Applied Genetics, John Innes Institute, Norwich, England

Cornelis J. Weijer Zoological Institute, University of Munich, Munich, Germany

Heinz Willebrand Physics Department, Institute of Applied Physics, Münster, Germany

1
Introduction and Overview

Ludger Rensing and Andreas Deutsch *University of Bremen, Bremen, Germany*

I. INTRODUCTION

Self-organized order—the appearance of so-called "dissipative structures"—is a characteristic feature of dynamic systems far from thermodynamic equilibrium (Nicolis and Prigogine, 1977). Dissipative structures exist frequently in nonliving systems and characterize the living world. The principles underlying the development of this type of order contrast with the Boltzmann order principle, which applies exclusively to equilibrium systems. Basic principles of self-organization include the amplification of fluctuations and synergetic behavior, such as cooperation and competition among structure elements leading to the selection of ordered structures (Haken, 1978).

Self-organization as such does not require genetic information, as is demonstrated by its presence in nonliving systems. In living systems, however, genetically coded and self-organized processes are intertwined on various levels. It seems plausible that, above all, the specific properties and conditions of basically self-organized processes are encoded by the genome.

The consequences of self-organization are ambiguous for evolutionary development. On the one hand, self-organization provides a great deal of structure formation potential in itself. But on the other hand, it restricts new morphogenetic pathways by its strong dependence on the "ontogenetic constraints" imposed by the boundary conditions set by the already existing organismic structure. Thus self-organization emphasizes internal selection pressures (Bonner, 1982).

Frequently occurring types of self-organized order are periodic in time (oscillations) or space (repetitive structures) or in time and space, as demonstrated by spatiotemporal pattern formation in the well-known Belousov–Zhabotinskii reaction (Fig. 1). Among other well-known periodic

Fig. 1 Concentric rings and spirals in the Belousov–Zhabotinsky reaction: spatiotemporal pattern formation in a chemical reaction. The reactants oscillate between reduced and oxidized state in a petri dish—visible as different colors. (Courtesy of S. C. Müller, Dortmund.)

spatial patterns, which are stationary, are the Liesegang rings (see Chapter 4). In the living world, the predominance of periodic pattern formation is obvious. Examples are provided on different levels of organismic organization, ranging from the arrangement of intracellular structures (Fig. 2) to cellular and multicellular patterning.

Mathematically, states of periodic temporal and spatiotemporal order can be deduced from the same equations by shifting appropriate order parameters (see Chapter 6). This procedure, however, does not elucidate

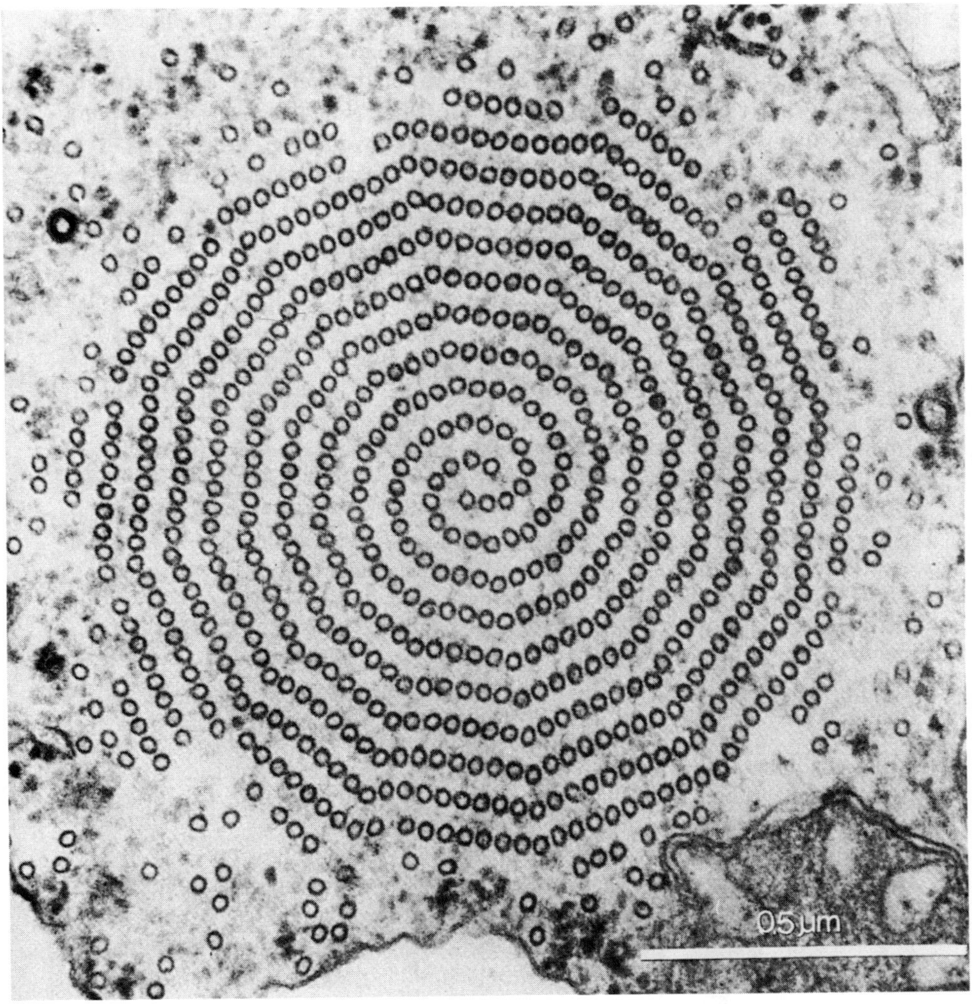

Fig. 2 Electron micrograph of the cross section of a heliozoan axopodium. The axoneme consists of microtubules in a double-spiral arrangement. (Courtesy of De Duve, Die Zelle, Spektrum der Wissenschaft, Heidelberg.)

the particular mechanisms underlying the formation of temporal and spatial order in living and nonliving systems. Furthermore, mathematical analysis cannot assign a functional meaning or such a role to particular principles. A role may appear questionable when assigned to oscillations—a dynamic state of self-organized systems. However, in living systems several roles can be distinguished, because organisms have exploited the principles of self-organization by modifying and using, for example, oscillations as information-transducing signals, which can function as temporal control (clock) devices. Correspondingly, periodic spatial order—as developed, for example, by the segmental organization of insects—may support the mechanical stability.

The aim of this book is to present systems in which the development of periodic temporal and/or spatial order can be observed experimentally. The presented examples here reveal a huge variety of specific mechanisms, leading to rhythmic behavior which, nevertheless, can be reduced to a few principles. The first part of the book is devoted to a general analysis of the functioning of oscillatory systems in morphogenesis, which we are only beginning to understand.

The term "morphogenesis" as used here has a general meaning: the development of spatial order in nonliving and living systems. Most contributions, however, deal with specific aspects of biological morphogenesis. These include the forces and communication systems that determine subcellular organization (e.g., cytoskeletal arrangement), cellular growth, motion, differentiation, and the spatial arrangement of cellular arrays (shape).

Oscillations are involved in each of these processes. In particular, they may establish a basis for temporal and spatial communication between parts of a developing system. Information transfer by oscillations thus provides a means of communication in a morphogenetic field. The second part of this book is concerned with this aspect.

Two main differentiation strategies prevail within biological morphogenesis. In the first, a homogeneous cell aggregate forms first, which subsequently differentiates. In this case, differentiation of a cell occurs according to its position in the aggregate. Alternatively, a reservoir of differentiated cells exists (e.g., in the bone marrow). Subsequently, the cells move from there to their appropriate positions (sorting-out). The essential assumption is the same in both cases: in order to differentiate (or to move) appropriately, the cell must be able to determine its actual position (Wolpert, 1969).

The use of wavelike propagation of oscillating signals for determining spatial information (position) has already been proposed in the phase-shift model (Goodwin and Cohen, 1969). In this model, positional information is encoded by the phase difference between two successive wave trains, each of which arises from particular pacemaker cells. Oscillations may, furthermore, serve as a marker for various cell types. For example, it is known from the cellular slime mold *Dictyostelium discoideum* that prespore and prestalk cells are characterized by different frequencies of cAMP production (Weijer et al., 1984). It is not clear so far, however, whether frequency encoding (Li and Goldbeter, 1990) is widespread within biological morphogenesis.

Many cellular elements can oscillate; not only, for example, enzymatic activities and concentrations of morphogens, but also physical properties, such as changing refractive indices caused by different cellular forms in motile and nonmotile modes. Intracellularly, oscillatory Ca^{2+}-concentration changes, and extracellularly, hormonal (or cyclic nucleotide) changes are well documented (Berridge et al., 1988).

In the third part of this book, emphasis is on a particular aspect of information transfer, the temporal control (timing) of morphogenetic processes. This control may be achieved by a clocklike function of the oscillator: certain phases of the oscillation control, for example, certain

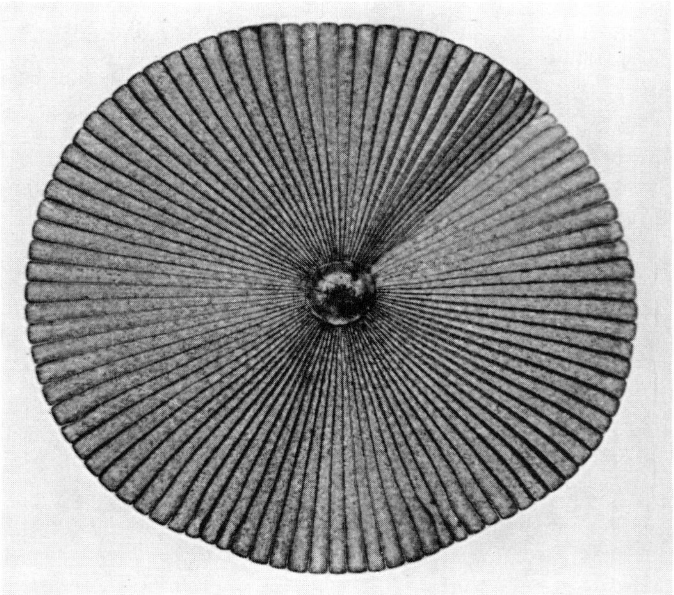

Fig. 3 Cap of the unicellular alga *Acetabularia mediterranea*. Magnification: 25×. (Courtesy of K. Hausmann, Berlin.)

steps in the proliferation or differentiation of cells. In even more refined mechanisms, such as in photoperiodism, a specific coincidence of external and internal oscillatory signals elicits morphogenetic processes.

Periodic morphogenetic structures occurring as a result of oscillatory timing generally develop sequentially diachronous. This phenomenon is particularly widespread in living systems and gives rise to an enormous variety of structures. An alternative possibility for the development of periodic spatial structures is the simultaneous or synchronous formation of structure elements. The fourth part of this book is devoted to examples and models exhibiting this type of periodic order.

The phenomenon of simultaneous pattern formation can be observed in many nonliving and living systems. Development of radial periodic structures in the cap of the giant unicellular alga *Acetabularia* (Fig. 3) provides an example. Cellular extensions emanate from the tip of the alga

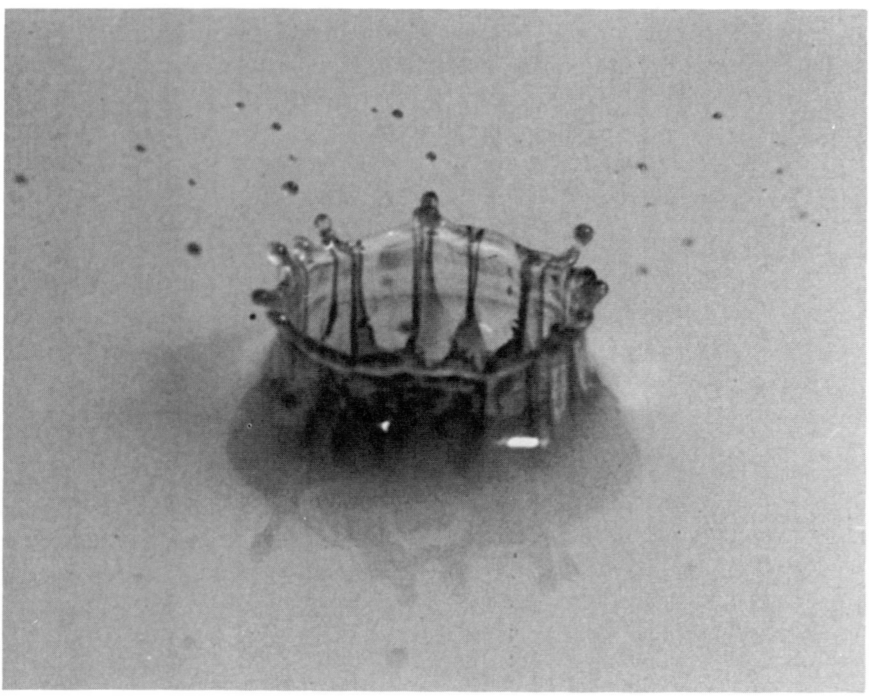

Fig. 4 Water drop falling onto the surface of a fluid and creating a "crater" with a radial symmetrical "crown" (exposure time 1/4000 s). (Courtesy of A. Wilkens, Bremen.)

cell, giving rise to a radial order that is periodic when viewed by an observer moving around the circumference of the cap. Whether oscillations play a role in this type of pattern formation is not clear yet.

A well-known example of simultaneous pattern formation is provided by a drop of water hitting the surface of a fluid. A crater with rather regular tips ("crown") is formed, which releases little droplets (Fig. 4). Subsequently, concentric waves appear on the surface of the fluid.

II. PRINCIPLES OF PERIODIC PATTERN FORMATION

Even though the specific mechanisms involved in the formation of temporal and spatial patterns of cellular arrangements, motion, or intracellular molecular concentrations may be extremely divergent, the principles underlying the formation of these structures are nevertheless similar. In the absence of external periodicities, the formation of spatial periodic patterns requires some sort of spatial interaction between the components constituting the system—be it molecules, cells, or tissues (see Chapter 3). Without spatial interactions only local temporal patterns (e.g., oscillations) can occur. Furthermore, it is clear that spatial periodicity can only be generated sequentially or simultaneously (as demonstrated by the water drop example) (Oster and Murray, 1989).

Mathematical abstraction is extremely valuable for extracting specific principles of pattern formation (see Chapter 6). A mathematical model provides an extremely abstract perspective onto a particular system thereby (hopefully) elucidating the relevant system components. Furthermore, mathematical analysis allows us to compare different systems with regard to the underlying dynamics. Defining a pattern as a more or less orderly heterogeneous structure leads immediately to a number of questions. Which kinds of local dynamics can shift a homogeneous to a heterogeneous state and can, furthermore, give rise to oscillatory behavior? What are the minimum requirements for the system dynamics to create spatial patterns?

Turing was one of the pioneers who worked on these kinds of questions. With the help of a mathematical model, Turing showed that diffusion and interaction of two substances (morphogens) sufficed to create temporal and spatial patterns under appropriate boundary conditions (Turing, 1952). In particular, he analyzed an initially stable homogeneous morphogen distribution along a closed ring of cells and showed that after a small disturbance, wavelike patterns of morphogen concentrations may result. This genius approach was based on spectral analysis and has initiated an intense study of reaction–diffusion systems. Reaction–diffusion models limit the local system dynamics to the kinetics of particular morphogens (chemical concentrations). They only consider spatial interactions caused by diffusion of morphogens. Morphogenetic structures in the perspective of this type of model reflect a chemical prepattern that arises according to the principles sketched above.

Turing was not only the first to present a formal model capable of self-organization, but his model already contained all the essential features that have remained important even in current models of pattern formation. Pattern formation in his system results from amplification of random disturbances of an unstable equilibrium. Furthermore, the two morphogens constitute an antagonistic interaction system. Roughly speaking, the protagonist is responsible for pattern formation at a particular site, while the antagonist limits this pattern in the surroundings (be it spatial or temporal) and thus allows for the establishment of spatial (or temporal) heterogeneities within the system (Fig. 5).

A particular theoretical reaction–diffusion system—the well-known antagonistic activator–inhibitor model introduced by Gierer and Meinhardt—has proven very valuable in modeling many morphogenetic pattern aspects (Gierer and Meinhardt, 1972). In its simplest version a locally self-amplifying substance (the activator) counteracts with a second substance (the inhibitor) that laterally inhibits production of the activator. With these assumptions it is possible to model formation of sequential and simultaneous periodic patterns (see Chapter 2).

A physical example is given by pattern formation in gas discharge systems (see Chapter 5). Here the antagonists are current density and

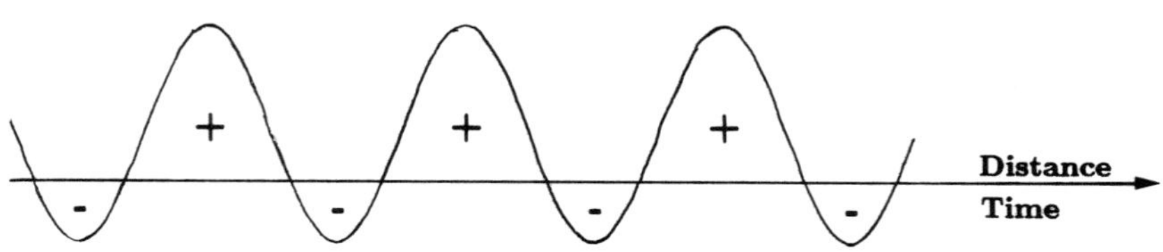

Fig. 5 Schematic view of an antagonistic principle underlying the formation of periodic spatial (or temporal) patterns. Areas where protagonistic or antagonistic effects prevail are indicated by + and − signs, respectively. The protagonistic process often leads to creation of a particular structural element in space (or time), while the antagonistic effect limits the spatial (or temporal) range of the protagonist. Dominance of the antagonistic process is again possible within a certain spatial range (or after a time lag) where (or when) the antagonistic effect is small enough.

resistance, respectively. The importance of various antagonistic schemes leading to the formation of temporal and spatial patterns in chemical reactions is emphasized by Müller (Chapter 4). Other morphogenetic model systems have been analyzed besides reaction–diffusion models. Among them are mechanochemical and mechanical models that emphasize mechanical cell features and mechanical interactions in addition to or instead of chemical kinetics (cf. Perelson et al., 1986). But again, in these models the essential scheme is antagonistic.

A biological example of antagonistic interactions in morphogenesis is provided by the cellular slime mold, *Polysphondylium pallidum*. Here the interplay of chemotaxis and cell adhesion seems to be responsible for pattern formation (see Chapter 26). Also, formation of some membrane patterns seems to be governed by a combination of mainly two effects: the accumulation and dispersion of proteins by electrophoresis (see Chapter 25). Antagonistic systems probably provide the basis for temporal patterns (oscillations) as well (Rensing and Jaeger, 1985).

The relationship between temporal and spatial pattern formation is well documented in so-called "excitable" media (Holden et al., 1991). The components of these systems perform a cycle of different phases of excitability, including a phase of refractoriness, thus establishing a basis for oscillatory behavior. The spreading of oscillations can be modeled by means of an "infection process," for example, by shifting the state of excitability of neighboring components (Markus and Hess, 1990).

Various examples of order development in excitable media are provided in this book. In the case of the aggregation patterns of *Dictyostelium*, a reversible cAMP–receptor desensitization is essential for concentric and spiral aggregation patterns (see Chapter 7). Intracellularly, an example is given by oscillatory Ca^{2+} changes that may be caused by the transient depletion of intracellular Ca^{2+} stores (see Chapter 11). Furthermore, pattern formation in the skin of salamander larvae can be viewed as taking place in an excitable medium (see Chapter 27). The analysis of development in two related salamanders, *Triturus alpestris* and *Ambystoma mexicanum*, provides an example of the concept of heterochrony. Heterochrony accentuates differences of relative timing of developmental processes, which can have dramatic morphogenetic effects. In the terms of excitable media, this corresponds to enlargement or shortening of the excitable phase.

Amplification of random disturbances, antagonistic behavior, and communication among components are basic features of all morphogenetic systems. The examples in this volume demonstrate that an enormous variety of forms may be deduced from the combination of these few fundamental principles.

III. OSCILLATORY SIGNALS IN MORPHOGENETIC FIELDS

The general role of oscillations in information transfer has increasingly gained attention, especially in view of their possible significance in intracellular signaling (Berridge et al., 1988). A frequency-coded information transfer is best known in the case of nerve or muscle action potentials, but may also be relevant in pulsatile hormone release. In Chapter 11, Goldbeter and Dupont summarize evidence for oscillators as signal transmitting channels in morphogenesis and cell communication and present models accounting for oscillatory behavior based on the assumption of excitability (see above). Supportive observations that this type of signaling is also used in developing multicellular systems is provided in Chapter 10 by Beloussov et al., who describe the behavior of hydroid polyps in which oscillations direct morphogenetic cell movements.

Dictyostelium is one of the classical examples of a morphogenetic field of cells organized by pacemaker regions and communicating by way of cAMP oscillations (Gerisch et al., 1975). Especially during the early aggregation phase, but also later (see Chapter 7), the movement of cells toward the pacemaker is periodic (Fig. 6). This tactic movement depends on cells responding to a directed pulse of cAMP, perhaps because cells orient toward that site on their surface which is first touched by cAMP—taxis is not induced by even a steep spatial gradient (Vicker et al., 1984). Each cell, furthermore, produces its own cAMP signal and thus functions as a relay.

What kinds of signals can be received and utilized by cells for morphogenetic purposes? The significance of oscillating temporal signals in morphogenesis through differential taxis, cell cohesion, contraction, and sorting-out is addressed by Vicker (Chapter 8) by way of a specific mechanism depending on temporal signal perception. Another function of oscillating signals in *Dictyostelium* amoebae is their influence on differentiation. In particular, it is known that genes for the cAMP receptor are induced predominantly by periodic cAMP signals (Gerisch, 1987) and that intracellular cAMP changes affect proliferation and differentiation of cells in general. Thus cAMP has multiple functions, including at least differentiation and communication. These functions and, especially, the role of periodic signaling in positional information transfer of differentiating prestalk and prespore cells in the postaggregative phase of *Dictyostelium*, are discussed by Siegert and Weijer (Chapter 7).

Even a single cell may be interpreted as a morphogenetic field. This view is applied to the spatiotemporal organization of *Physarum*—manifested in its cytoskeleton—by Ueda (Chapter 9). The apparent order

Fig. 6 Formation of spirals and concentric rings in the early aggregation phase of the slime mold *Dictyostelium discoideum*. (Courtesy of C. Weijer and F. Siegert, Munich.)

seems to be established by a complex interaction of chemical oscillations of different frequencies which determines, for example, the direction of propagation toward attractive or away from repulsive stimuli.

IV. PERIODIC TIMING OF MORPHOGENETIC PROCESSES

Timing of events is an important task of oscillations in living systems. For example, the well-known periodic annual rings of many woody plants reflect the annual shifts of physical parameters in the habitat of the growing plant. In particular, these (exogenous) changes lead to alternating development of two different forms of xylem cells, visible as successive annual rings (see Chapter 20). But organisms do not rely exclusively on external synchronization devices for their coordination of morphogenetical and physiological processes, which reflect responses to environmental changes. They have gained some independence by developing endogenous timing systems (endogenous oscillators). Especially well known are circadian rhythms with a period length of approximately 24 h, which regulate many physiological and morphogenetical processes (Edmunds, 1988).

The timing of morphogenetic processes, and thereby, in spatially coherent systems, the spatial positioning of a structure, is often determined by a phase-dependent signal from exogenous or endogenous oscillations. These signals act by means of hormones and intracellular second-messenger molecules on gene expression, differentiation, or proliferation. The times of budding, flowering, seed development, or shedding of leaves in many plant species is determined by the photoperiod, temperature, and other annually changing variables—as is the time of sexual reproduction, fur changes, or diapause in animal species.

Such morphogenetic events may occur only once in the lifetime of a member of a species whose individuals undergo just one reproductive or adaptive cycle. They may occur repeatedly within the lifetime of a member of a species whose individuals go through repetitive cycles. In contrast to the annual ring formation in woody plants, for example, there is usually no remaining periodic spatial pattern, since structures as flowers, seeds, leaves, or changes in the ovaries and testes are formed only transiently. Even though the timing mechanisms of morphogenetic processes leading to transient structures and those leading ultimately to the formation of permanent periodic patterns may be basically the same, we distinguish these two types in the following discussion.

A. Single or Repeated Transient Processes

One of the prominent features of living systems is their ability to grow, which is based essentially on replication and mitotic activity at the cellular level. Mitosis implies drastic structural changes within the cell. In the course of the cell cycle, the array of microtubules significantly changes the spatial arrangement: during interphase a more or less radial structure exists that is altered during metaphase into a characteristic spindle structure arising from two organization centers. This may be described as a single morphogenetic process for each individual cell or as a repetitive event in a cell population. Spindle formation and its breakdown are controlled by the cell cycle oscillator. Primarily, this oscillator consists of a rhythmic buildup and breakdown of cyclin, rhythmic activity changes of the cdc2 kinase, and rhythmic phosphorylation/dephosphorylation of proteins (Murray and Kirschner, 1989). The cell cycle oscillator may play a decisive role in the timing of differentiation: At specific cell divisions the further fate of the two daughter cells is determined in *Caenorhabditis* (see Chapter 15).

The cell cycle oscillator itself or the morphogenetic processes it elicits can be further controlled by another oscillator, the circadian clock (for a review, see Edmunds, 1984). Such a temporal control has already been observed in unicellular organisms, such as those analyzed by Edmunds (Chapter 13). In multicellular organisms this control often involves circadian hormonal changes (for a review, see Rensing and Goedeke, 1976). Single waves of mitotic activity were observed in pea cotyledons (see Chapter 14). Many processes of plant and chloroplast development are controlled or influenced by a circadian clock, by exogenous daily light and temperature changes, or both. Some of the genes encoding chloroplast proteins, such as the chlorophyll *a* binding protein, are under circadian control, whereas other genes and their expression are induced by light and elevated temperatures (Otto et al., 1988).

In *Acetabularia*, phytohormones play a particular role in morphogenesis. These hormones are probably distributed according to an apicobasal gradient and are modulated by a circadian rhythm (Chapter 12). In insects, developmental processes, such as the eclosion of adults from the pupal state, are timed or strictly "gated" by a defined phase of the circadian clock (Pittendrigh, 1966). Under natural conditions, the circadian clock is entrained by the light–dark cycle. For *Drosophila* this means that eclosion is strictly confined to dawn. In an arrhythmic mutant (*per*^{0}) of *Drosophila*, the circadian control of eclosion is lacking (Hall and Rosbash, 1987).

Most widespread are transient morphogenetic changes, such as annual flowering in plants as well as gonad development, molting, and diapause in animals. The control of these processes is often exerted by way of photoperiodism (i.e., by the perception of the varying day lengths in the course of a year). A critical daylength triggers flowering, gonad development, or diapause through circadian clock-controlled and light-induced signals. The control mechanism is, apparently, based on either absolute or relative measurements. Thus the organism could measure the absolute duration of the length of the light or dark time by means of an hourglass mechanism. Such a mechanism may be at work, for example, in determining the duration of the cell cycle by the accumulation of cyclin (Murray and Kirschner, 1989). Alternatively, the system may determine phase relations between the light phase of the solar day and the phase of the entrained (endogenous) circadian oscillator. A well-analyzed example of photoperiodic diapause control in insects is presented by Saunders (Chapter 16). He focuses especially on the relations between photoperiodic timing and circadian rhythms.

B. Sequential Formation of Periodic Patterns

If structures develop sequentially and form a coherent, permanent pattern, this pattern can be periodic. The periodic pattern may appear as a result and a permanent record of an oscillatory process eliciting or modifying morphogenetic processes. Morphogenesis in some fungi, such as the formation of stationary periodic patterns by vegetative spores in the ascomycete *Neurospora crassa*, provides a well-studied example: An endogenous circadian oscillator or periodic light or temperature cycles induce rhythmic sporulation. Thus these oscillators serve as triggers of spore differentiation (see Chapter 19). A multitude of other fungi have been analyzed with respect to the involvement of endogenous or exogenous rhythms in their morphogenesis (Chapter 18).

Another example of this type of periodic control—formation of annual rings in woody plants—has already been mentioned. In many woody plants, exogenous annual periodicities, such as changes of temperature, water supply, and day length, control the differentiation of xylem cells by means of shifting the levels of plant hormones. These processes are analyzed in Chapter 20 by Savidge. In mollusk shells, several exogenous periodic and nonperiodic processes as well as endogenous oscillations influence the growth pattern (Chapter 21); endogenous oscillations also seem to play a role in the formation of their pigmentation pattern (Chapter 2) and in the formation of chambers in the spiral-shaped shell of *Nautilus* (Fig. 7). Sequential formation of periodic patterns is possible without considering oscillatory devices. All step-by-step (successive) processes belong to this type of periodic structure formation. Examples are the addition of monomers to polymers or the repeated addition of proteins to larger fibers (collagen, myosin, microtubules). Microtubules can be arranged as complicated multiple spirals by successive addition to the periphery of a bundle (Fig. 2).

At the cellular level, linear rows of cells are sequentially produced, for example, in blue-green bacteria and algae. In two-dimensional layers of bacterial cells additional periodic structures often appear, such as in colonies of *Escherichia coli* (see Chapter 17). In this case the morphogenetic mechanism is not well understood. At least phenomenologically similar is the teloblastic development of segments in arthropod embryos, which first appear as rows of ectoderm and mesoderm cells (Fig. 8).

In plants, an even more widespread occurrence of periodic organization is manifest: The sequential formation of leaves at the stem tip can lead to various arrangements (rows, spirals) of leaves along the stem (Fig. 9). Whether or not the mechanism of the sequential formation of

Fig. 7 Spiral-shaped shell of the cephalopod *Nautilus pompilius*. (Courtesy of H. Stix and H. Landshoff, *Die Muschel.* Belser, Stuttgart, 1978.)

leaf primordia is dependent on the last primordium or, alternatively, is controlled by an independent oscillatory mechanism, is not clear yet.

V. SYNCHRONOUS FORMATION OF PERIODIC PATTERNS

As already mentioned, periodic spatial structures can also be generated simultaneously. A striking example of simultaneous periodic pattern formation is presented by Willebrand et al. in Chapter 5, where they analyze the formation of current density patterns on an electrode surface in a gas

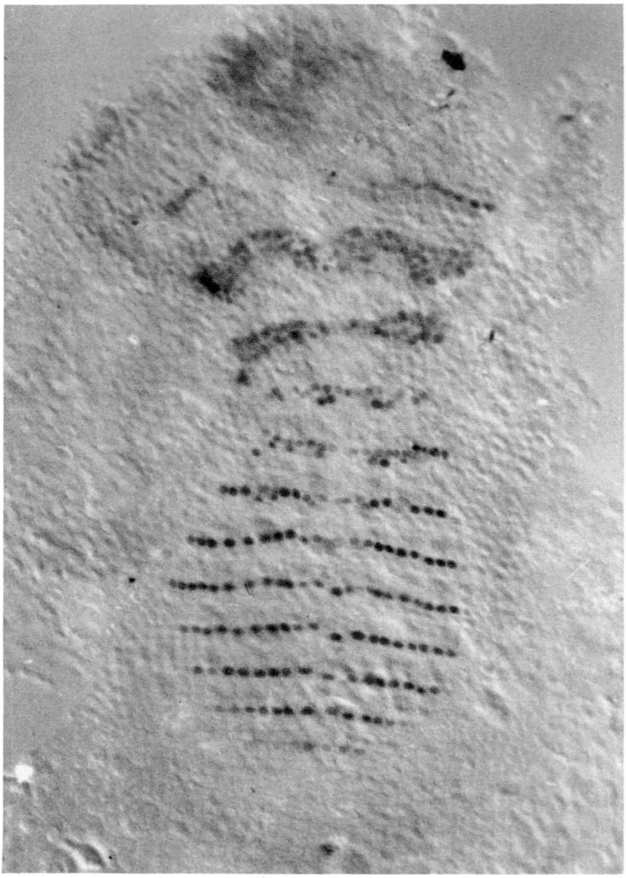

(a)

Fig. 8 Germinal disk of *Orchestia cavimana* (Crustaceae, Amphipoda): (a) stained with ''engrailed'' antibodies; (b) stained in addition with Hoechst blue; both showing the formation of thoracal and abdominal segments. The large stained segments in the upper portion represent simultaneously built anterior segments (antennae, mandible), the smaller ones in the lower portion are sequentially built posterior segments. (Courtesy of G. Scholtz, Berlin.)

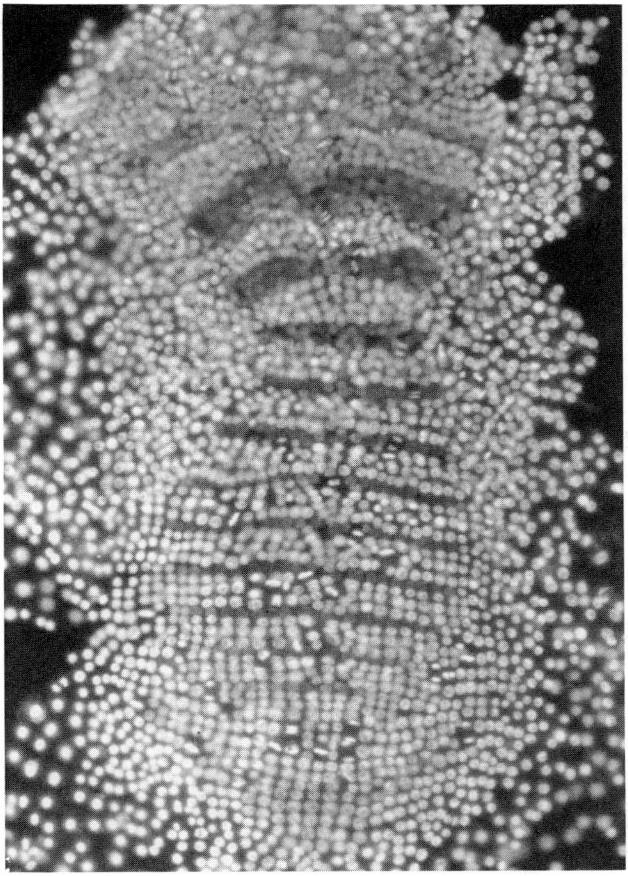

(b)

discharge system. The close principal relationship between sequential and simultaneous pattern formation has been mentioned above. Nevertheless, the mechanisms leading to both types of patterns may be divergent. For example, the ascomycete *Neurospora crassa* expresses sequential as well as synchronous patterns, visible as concentric ring and radial zonation spore patterns, respectively. While the mechanism leading to the concentric ring pattern involves primarily endogenous or exogenous oscillators, the radial zonations probably arise from processes driven by competition and amplification of random patterns in the initial conditions (see Chapter 28). One well-known physical example is the Bernard structure. When a water layer is heated from below, at a critical point circular patterns

Fig. 9 Spiral arrangement of leaves on a stem tip. Electron micrograph of the young inflorescence of *Euphorbia myrsinites*. A sequence of successively developed leafs is indicated by numbers. (Courtesy of R. Rutishauser.)

of water movement can be observed after a state of random fluctuations. Water molecules "cooperate" in a spatially ordered vertical transport of thermal energy (for a review, see Haken, 1978).

Intracellular periodic structures in the alignment of chloroplasts of *Acetabularia* might develop simultaneously as analyzed by Bonotto (Chapter 22). A banding pattern of carbonate deposition develops on the cell surface of *Chara corallina*, a structure based on periodically different states of transport activities of membrane proteins (see Chapter 24). Pattern formation in the distribution of membrane proteins may be a prerequisite for development of cellular structures (see Chapter 25). For example, the emanation of cell extensions during cap formation in *Acetabularia* (Fig. 3) is tentatively deduced from a corresponding radially periodic pattern of calcium or Ca^{2+}-binding proteins along the membrane of the cap "prestructure" (Goodwin and Trainor, 1985).

An important role in intracellular structure formation is played by microtubules. Radial patterns of microtubules are formed during mitosis. Microtubule asters, which are built into a spindle of polar and kinetochore microtubules, develop from both centrosomes. As mentioned above, the formation of this radial periodic structure during mitosis is timed by the cell cycle oscillator by means of phosphorylation changes at the centrosome and, in addition, by a circadian clock (Edmunds, 1988). Microtubules originate mainly from a center (centrosome) near the nucleus and grow with the positive ends toward the periphery of the cell. In heliozoans, but also in other cell types, this radial pattern is periodic in the sense that bundles of microtubules or single microtubules are formed at regular intervals (Fig. 10). The microtubule pattern can lead to a

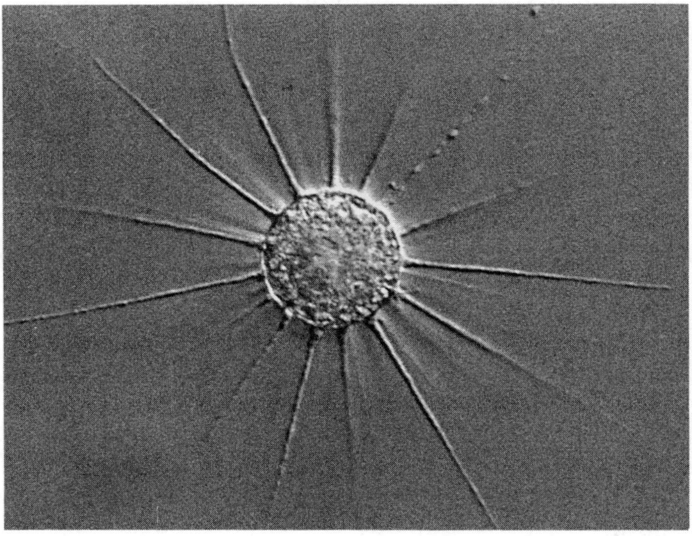

Fig. 10 Radial arrangement of axopodia in the heliozoan *Actinophrys sol*. Each axopodium contains a bundle of microtubuli and grows out from the centrosome. (Courtesy of K. Hausmann, Berlin.)

regular radial distribution of the Golgi complexes and possibly also to a similar arrangement of chloroplasts and plasma membrane components. A centripetal or centrifugal movement of Golgi and endoplasmic reticulum membranes along microtubules may be responsible for the striking periodic spacing of the membranes of these organells.

The development of radial structures may be based on mutual competition: If the available supply of monomers (GTP-tubulin) becomes limiting after a random start of polymerization from the centrosome, a slow growth of some and a rapid degradation of other microtubules is observed, a phenomenon termed "dynamic instability" (Kirschner and Schulze, 1986). In cell-free systems, an oscillatory change between polymerization and depolymerization in microtubule development has been observed (Mandelkow et al., 1989). In the cell, those microtubules are stabilized that maintain a certain distance from neighboring microtubules (similar to the example of water drops at a horizontal bar). Additional protection against breakdown is provided by cap proteins. This protection stabilizes microtubules in defined directions, for example in the direction of movement.

A radial periodic pattern also appears in the fruiting bodies of the true slime mold *Polysphondylium pallidum*, starting from a random pattern. The underlying organizational principle is seen in mutual activation/inhibition processes (see Chapter 26). Another example of radial organization is provided by the "streaming pattern" during the late aggregation phase in the life cycle of *Dictyostelium*. Later in the developmental cycle of this amoeba, a further interesting periodic structure becomes visible if cells were grown with caffeine (a cAMP relay inhibitor) in the medium (see Chapter 7). Possibly, this pattern is also generated simultaneously.

Multicellular periodic structures may develop in populations of moving unicellular organisms, such as in populatons of the dinoflagellate *Gonyaulax polyedra*. The exact mechanism of this type of pattern formation is not known but seems to involve circular movements of cells ("bioconvection") (see Chapter 23). It is interesting to note that circadian oscillations modify this pattern. The migration of differentiated pigment cells leading to formation of simultaneous patterns in salamander larvae is analyzed in Chapter 27. In particular, these larvae provide a beautiful system to study the processes involved in sorting-out mechanisms during embryonic development. One of the best analyzed developmental systems is *Drosophila*. Segment development of *Drosophila* reflects many of the principles typical of the systems presented in this volume. A particularly interesting case of periodic structure formation is the simultaneous development of segments in the early embryo (Fig. 11; see also Chapter 3).

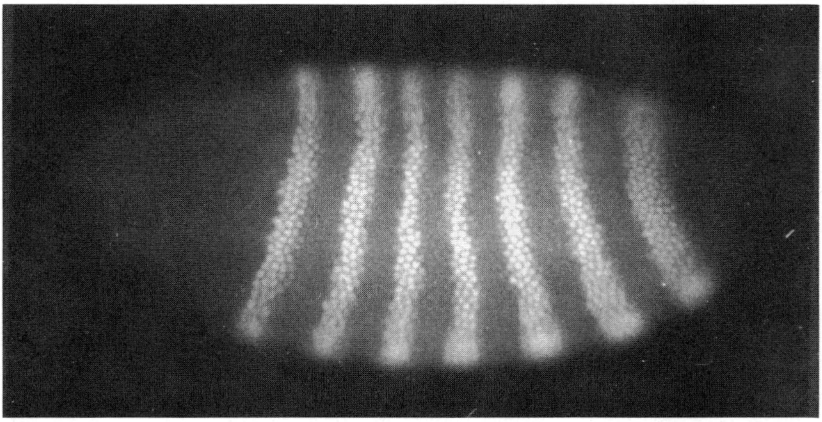

(a)

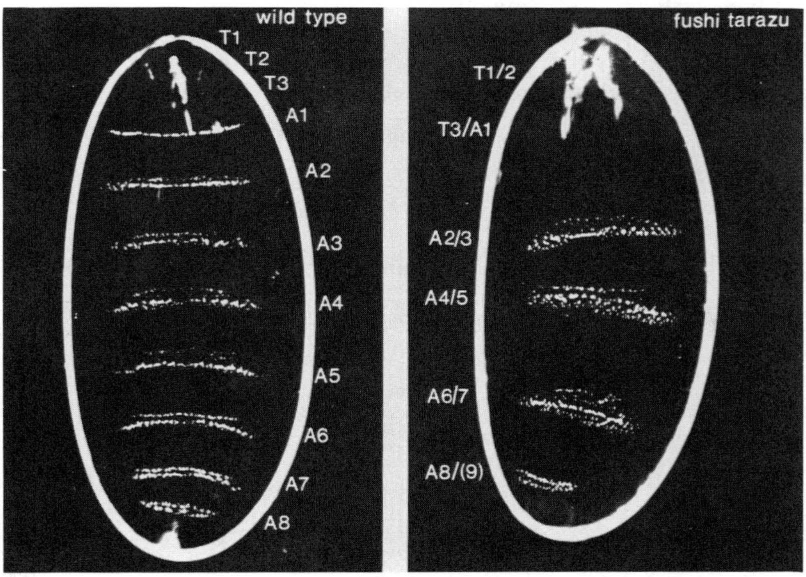

(b)

Fig. 11 Simultaneous formation of periodic structures in the blastoderm stage (a) and larval segments in the wild type and the mutant fushi tarazu (b) of *Drosophila melanogaster*. At the blastoderm stage the expression of the fushi tarazu genes (in the wild type) is marked. The fushi tarazu mutant shows only half of the wild-type number of segments. (Courtesy of W. J. Gehring, Basel.)

From a comparative point of view, a relation between sequential and simultaneous mechanisms may exist, a relation that is already expressed by mathematical models (see Chapters 2 and 6). The rather monotonous segmentation pattern of "ancient" animal groups, such as annelids, onychophores, and millipeds, which develop mainly sequentially, is modified in "modern" arthropods to a more centralized polar organization, as is visible, for example, in insects such as *Drosophila*. In these species the pattern formation is mainly synchronous.

VI. CONCLUSIONS

Confronted with the problem of temporal and spatial order, we have focused on the role of oscillatory systems. The systems presented in this volume demonstrate that oscillations play prominent roles in various kinds of morphogenetic processes. What can be deduced from the examples provided here is that a few principles suffice to create an enormous variety of behavior and forms observed in the living and nonliving world. Even though the principles underlying the development of particular structures may be only a few, the examples demonstrate that the actual mechanisms establishing different forms of spatial and temporal order are often beyond the scope of our understanding.

REFERENCES

Berridge, M. J., Cobbold, P. H., and Cuthbertson, K. S. R. (1988). *Philos. Trans. R. Soc. Lond. B*, *320*: 325.

Bonner, J. T. (ed.) (1982). *Evolution and Development*, Springer-Verlag, New York.

Edmunds, L. N. (1984). *Cell Cycle Clocks*, Marcel Dekker, New York.

Edmunds, L. N. (1988). *Cellular and Molecular Bases of Biological Clocks*, Springer-Verlag, New York.

Gerisch, G. (1987). *Annu. Rev. Biochem.*, *56*: 853.

Gerisch, G., Hülser, D., Malchow, D., and Wick, U. (1975). *Philos. Trans. R. Soc. Lond. B*, *272*: 181.

Gierer, A., and Meinhardt, H. (1972). *Kybernetik*, *12*: 30.

Goodwin, B. C., and Trainor, L. E. H. (1985). *J. Theor. Biol.*, *117*: 177.

Goodwin, B. C., and Cohen, M. H. (1969). *J. Theor. Biol.*, *25*: 49.

Haken, H. (1978). *Synergetics*, Springer-Verlag, New York.

Hall, J. C., and Rosbash, M. (1987). *Trends Genet.*, *3*: 185.

Holden, A. V., Markus, M., and Othmer, H. G. (eds.) (1991). *Nonlinear Wave Processes in Excitable Media*, Plenum Press, New York.

Kirschner, M. W., and Schulze, E. (1986). *J. Cell Sci. Suppl.*, *5*: 293.

Li, Y. X., and Goldbeter, A. (1990). *J. Theor. Biol.*, *146*: 355.
Mandelkow, E. M., Hotani, H., Hess, B., and Müller, S. C. (1989). *Science* (*Washington*, D.C.), *246*: 1291.
Markus, M., and Hess, B. (1990). *Nature*, *347*: 56.
Murray, W., and Kirschner, M. W. (1989). *Science* (*Washington*, D.C.), *246*: 614.
Nicolis, G., and Prigogine, I. (1977). *Self-Organization in Nonequilibrium Systems*, Wiley, New York.
Oster, G. F., and Murray, J. D. (1989). *Exp. Zool.*, *251*: 186.
Otto, B., Grimm, G., Ottersbach, P., and Kloppstech, K. (1988). *Plant Physiol.*, *88*: 21.
Perelson, J., Hyman, A., Maini, P., Murray, J. D., and Oster, G. (1986). *J. Math. Biol.*, *24*: 525.
Pittendrigh, C. S. (1966). *Z. Pflanzenphysiol.*, *54*: 275.
Rensing, L., and Goedeke, K. (1976). *Chronobiologia*, *3*: 53.
Rensing, L., and Jaeger, N. I. (eds.) (1985). *Temporal Order*, Springer-Verlag, New York.
Turing, A. M. (1952). *Philos. Trans. R. Soc. Lond. B*, *237*: 37.
Vicker, M. G., Schill, W., and Drescher, K. (1984). *J. Cell Biol.*, *98*: 2204.
Weijer, C. J., McDonald, S. A., and Durston, A. J. (1984). *Differentiation*, *28*: 9.
Wolpert, L. (1969). *J. Theor. Biol.*, *25*: 1.

I
Principles of Periodic Pattern Formation

2
Oscillations in Biological Pattern Formation

Hans Meinhardt *Max-Planck-Institut für Entwicklungsbiologie, Tübingen, Germany*

I. INTRODUCTION: AUTOCATALYSIS AND LONG-RANGE INHIBITION IN BIOLOGICAL PATTERN FORMATION

A most fascinating aspect of higher organisms is generation of the very complex spatial pattern of cell differentiation in each round of the life cycle. Despite the complexity, the resulting pattern is extremely reproducible. Pattern formation is by no means exclusive to living systems. Sand dunes, all forms of erosion, clouds, and lightning are examples of inorganic pattern formation. Common to all these processes is the fact that small local fluctuations have a strong positive feedback effect on their further growth, while an antagonistic long-range effect limits extension and spread of this self-enhancement. We have proposed that biological pattern formation is based on the same principle: local autocatalysis and long-range inhibition (Gierer and Meinhardt, 1972).

Many molecularly feasible interactions are compatible with the general scheme. For instance, to be called an activator, a substance must have a direct or indirect positive feedback on its own synthesis. This autocatalytic reaction is counteracted by an antagonist, the inhibitor. The antagonistic effect can also result from the depletion of substrate consumed during autocatalytic activator production. Such a reaction is similar to the Brusselator reaction (Lefever, 1968).

A. Formation of Patterns Stable in Time

Formation of stable patterns in space requires that the self-enhancement is a more-or-less local process to create a local instability. In contrast, the inhibition must be of long range to restrict the spread of self-enhancement and to obtain overall stabilization. In addition, the time constant of the inhibition must be shorter than that of the autocatalytic activator; otherwise, oscillations will occur. In terms of the activator–inhibitor scheme, for the generation of stable patterns the inhibitor must have a higher diffusion and a higher decay rate than the activator.

If the range of the activator and the inhibitor is shorter than the size of the field, periodic patterns of high concentrations will emerge. An example will be given in Fig. 1 (see also Chapter 26, this volume). In contrast,

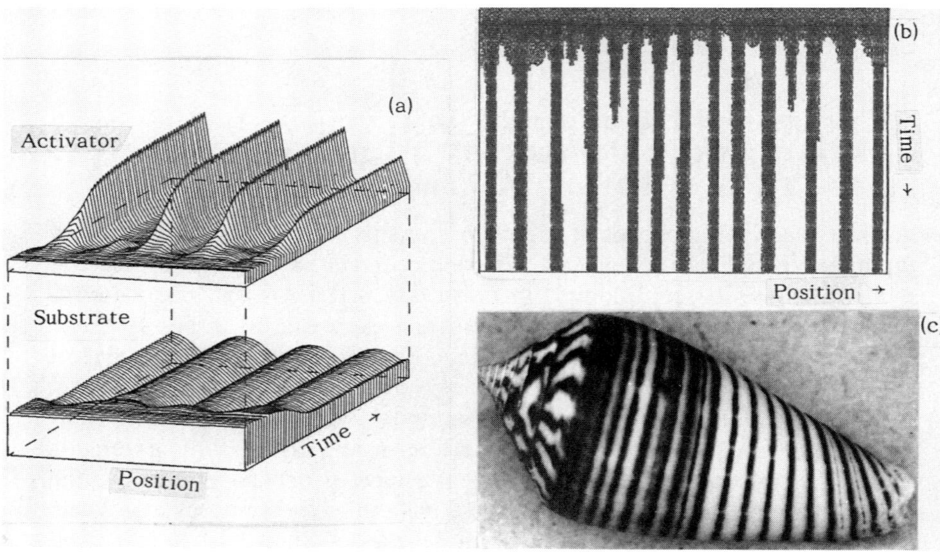

Fig. 1 (a) Formation of stable periodic patterns. Assumed is the interaction of an autocatalytic activator (top) and its rapidly diffusing antagonist, in this example a diffusible substrate that is consumed during activator autocatalysis. Calculation has been made on a linear array of cells. The concentration of the activator and the substrate is plotted as a function of time. The homogeneous steady state is instable. At the surroundings of the emerging activator maxima, the substrate concentration decreases. This limits further increase and spread of activator maxima. (b) Resulting pattern in a space-time plot as produced on shells of snails. Regions in which the activator concentration is above a threshold are drawn in black. Permanent pigment production at particular positions causes stripes parallel to the direction of growth. (c) Natural patterns: shell of *Conus macronatus*. [Computer programs written in Fortran for this and other shell simulations can be found in Meinhardt and Klingler (1986).]

if the range of the activator is comparable with the size of the field, a monotic pattern with a high concentration on one side and a low concentration at the other will be generated. Such a monotonously graded distribution is convenient to act as positional information. The cells obtain a unique signal as to where they are in the tissue. By concentration-dependent gene activation, different genes can be turned on in different regions. The formation of complex structures requires a cascade of pattern-forming reactions that are linked in a hierarchical manner. Detailed theoretical analysis of several developmental processes has been given elsewhere (Meinhardt, 1982, 1986).

B. Oscillating Patterns

As the rule, pattern formation during development of higher organisms requires reactions that are stable in time, at least for a certain interval. In some situations, however, oscillating reactions play an essential role. They can be generated by the same basic mechanisms with other parameters (see Fig. 2). To be in line with the general subject of this book, some examples of the involvement of oscillations in pattern formation are discussed. This includes the pigmentation pattern on shells of mollusks, the involvement of oscillations in chemotaxis, and based on the latter, the generation of filament-like branching structures such as veins of leaves, blood vessels, or nerves.

Oscillations occur if the antagonistic substance adapts too slowly to a change of activator concentration. This is the case if, for instance, the inhibitor has a longer lifetime than that of the activator. Burstlike activations occur. After sufficient accumulation of the inhibitor or after complete depletion of the substrate, the activator production collapses. After a refractory period, the next activation can be triggered from the small activator-independent activator production.

II. THE PIGMENT PATTERN ON THE SHELLS OF MOLLUSKS

The shells of many mollusks are decorated with a relieflike or a pigmentation pattern. These patterns are of great diversity and frequently of great beauty. The formation of these patterns proceeds in most species in a strictly linear manner since new pattern elements are added only along a narrow zone of marginal growth. The second dimension is a protocol of what has happened as a function of time. The shell is, essentially, a space–time plot. Thus the shell patterns provide a unique situation in that the complete history of a very dynamic process is preserved.

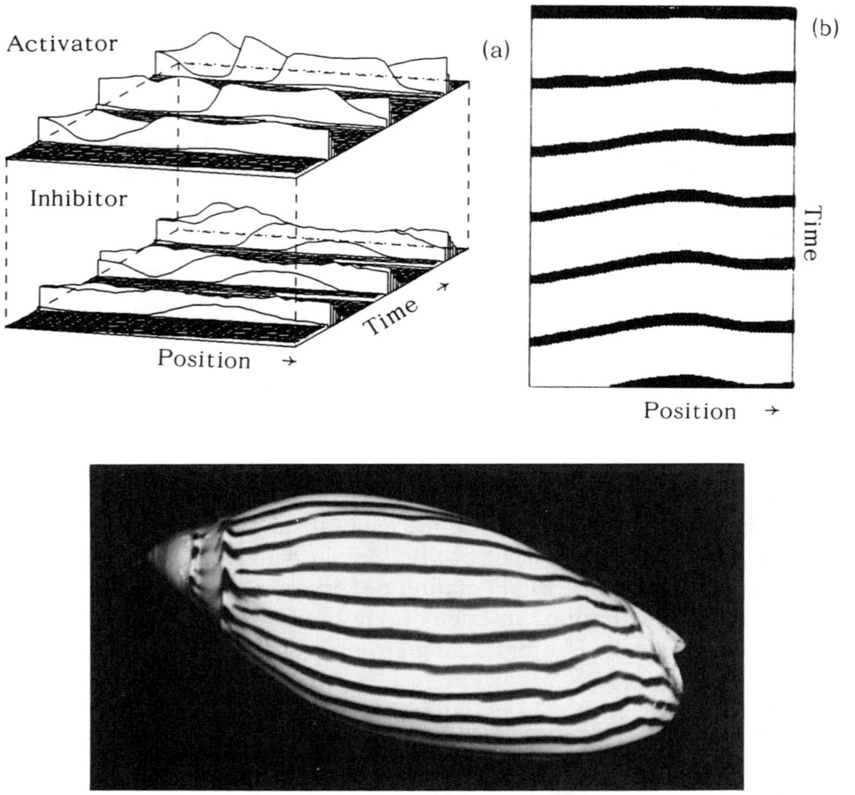

Fig. 2 Oscillating patterns. (a) Assumed is an activator–inhibitor system. Due to a longer time constant of the inhibitor compared with that of the activator, oscillations occur. High diffusion of the activator enforces a near-synchronization between neighboring cells. (b) Applied to shell patterns, this leads to stripes parallel to the growing edge. Nevertheless, despite the fact that the phase differences between neighboring cells are small, a substantial phase difference can accumulate over the entire length of the growing edge. (c) Pattern on the shell of *Amoria ellioti*.

By an application of the theory outlined above to the shell pattern we have shown that many different patterns can be accounted for by the basic model with different parameters (Meinhardt, 1984; Meinhardt and Klingler, 1986, 1987). More complicated patterns result from the superposition of two patterns, one that generates the proper pigment pattern and another that changes the parameters of the pigment-producing reaction in a space- and time-dependent manner. As the rule, the latter pattern is invisible; an example is given in Fig. 5. A model for shell patterning based on similar principles but focusing on the possible role of the nervous system has been proposed by Ermentrout et al. (1986).

Basic elements of shell patterns are lines perpendicular, parallel, and oblique to the growing edge. Keeping in mind the space–time character of the shell pattern, lines perpendicular to the growing edge indicate the formation of a spatial periodic pattern of pigment production along the growing edge that is stable in time (Fig. 1). Lines parallel and oblique to the edge indicate that pigment deposition oscillates. A particular cell produces pigment only during a certain time interval and then enters into an inactive (refractory) period until the next pigment production takes place. A high diffusion of the activator leads to synchronization of the oscillations. Pigment production starts and ceases nearly synchronously along the growing edge. This leads to lines of pigmentation parallel to the shell margin. Figure 2 provides an example together with computer simulations.

A. Oblique Lines Result from Traveling Waves

Oblique lines on the shells originate from traveling waves of pigment production. Such waves arise if pigment-producing cells trigger their neighboring cells such that after a certain delay, these cells also start to produce pigment. In the model, traveling waves of high activator concentration can be formed if the activator has a small diffusion range, but the antagonistic substance is nearly nondiffusible. An activated region can "infect" its neighboring region such that it becomes fully activated. After a certain time delay, it infects its neighboring cell, and so on. In an activated cell, the activation breaks down after a certain delay due to the action of the antagonistic substance. In the time record, such wandering regions of high activation lead to oblique lines of pigmentation.

Characteristic of many such shell patterns is a W-like arrangement of the oblique lines. The ∧-like element of this pattern results from a spontaneous trigger of a cell followed by an "infection" of both neighbors, and so on. In other words, a spontaneous trigger of a cell gives rise to two diverging oblique lines. If two such lines merge, both lines become extinct, since each wave would enter into a region that is refractory due to the passage of the other wave. This leads to the ∨-like element of the pattern. This pattern element is very frequent on the shell shown in Fig. 3.

Fig. 3 Formation of branches. (a) Shell of *Olivia porphyria* L. Oblique lines indicate the formation of traveling waves of pigment production. If two waves meet each other, both become extinct. A branch indicates the sudden formation of a backward wave. (b) Model: branching occurs whenever the number of traveling waves drops below a certain threshold value by a temporary shift of the cells from the oscillatory into a steady-state activator production until a backward wave is initiated. (After Meinhardt and Klingler, 1987.)

B. Traveling Waves with Unusual Properties

Many shell patterns indicate that they result from traveling waves with unusual properties. Some shells show branchings of pigmentation lines (Fig. 3). Keeping in mind the space–time character of the pattern, branches indicate the sudden formation of a backward-oriented wave during normal progression of a wave. The initiation of a backward wave requires a modification of the normal serial trigger mechanism. From a particular moment onward and for a short time interval, the activated cells must remain in the activated state until the refractory period of the neighboring, still oscillating region is over. Then these cells can be reinfected and the backward wave is initiated. Of course, this requires a signal as to where and when this transition from the oscillatory to steady-state activation should occur.

As mentioned, traveling waves normally annihilate each other during a collision. Branching is one of the possible mechanisms to maintain the number of traveling waves. It is reasonable to assume that branches are formed if the number of traveling waves are too low. In the simulation of Fig. 3, this has been achieved by the assumption of a hormonelike, homogeneously distributed substance. It is produced by the activated cells. Its concentration is thus a measure of the total number of activated cells along the growing edge. The hormone has a stabilizing influence on the inhibitor. Thus, with a decreasing number of activated cells the inhibitor lifetime becomes shorter and shorter. When the inhibitor lifetime becomes shorter than that of the activator, the cells shift from the oscillatory into the steady-state mode of activation. After the refractory period of a neighboring, still oscillating cell is over, these neighboring cells can be reinfected and a backward wave initiated. This causes an increase in the number of traveling waves. The hormone concentration and thus the inhibitor lifetime increases and all cells return to the oscillating mode. The formation of the branch is completed.

As shown in Fig. 3, the simulation reproduces many details of the natural pattern. Branchings are initiated at the same time at many different positions. This is indicated by the many V-like structures with the same distance between branch initiation and the tip of the V. In the model, this results from global control of the hormone due to its homogeneous distribution. Therefore, all cells switch simultaneously back and forth between the oscillating and nonoscillating mode of activation.

While the original line remains normal, the branches frequently appear only loosely connected to the original line. In the simulation, this effect results from the passage through the temporary steady-state activation. This is usually lower compared with that achieved during the pulse-like activation in the oscillatory regime. Sometimes a little hook appears close to the branch initiation. This results from an incipient initiation of a wave parallel to the primary wave. Usually this wave does not survive since the medium is not yet exitable enough. If this wave survives, however, a line close to the original line is formed together with the branch.

C. Formation of Crossings by an Additional Diffusible Inhibitor

Pigmentation lines that cross each other can be found on several shells. Figure 4 provides an example. Crossings are unusual for nonlinear waves in exitable media since, as mentioned, usually two colliding waves anni-

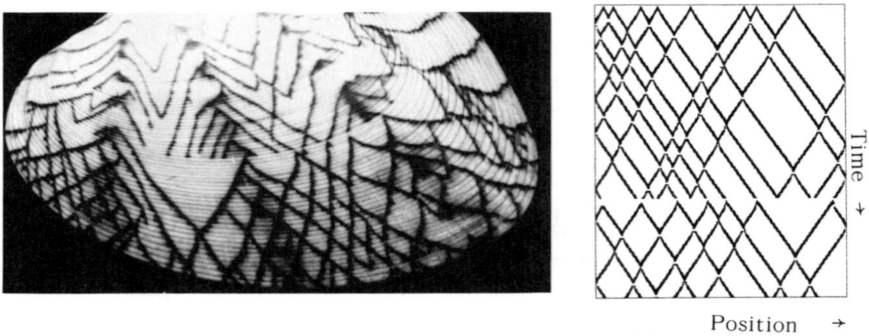

(a) **(b)**

Fig. 4 Crossings of pigmentation lines. (a) Pattern on *Tapes spec*. Crossings indicate that during a collision two waves do not annihilate but penetrate each other. (b) Model: assumed is a three-component system; an activator–substrate mechanism tuned to produce a steady-state activation, and, in addition, a diffusible inhibitor. Traveling waves result from the extinguishing of the activation in each cell by their subsequently triggered neighboring cells. At the point of collision, no new neighbors are available which could extinguish the previous activation. The cells remain in the steady state until two backward waves are initiated. A sudden lowering of the activator concentration can lead either to the termination of a line, to branch formation, or to the continuation of a line. The same features can be seen in the perturbation of the natural pattern. (After Meinhardt and Klinger, 1987.)

hilate each other. Crossings of oblique lines can be regarded as the initiation of two new diverging waves after a collision. As in the case of branch formation, the new waves can result from cells that remain activated for a prolonged period due to a temporary shift into the steady state. In the simulation (Fig. 4), an activator–substrate mechanism has been assumed with parameters such that a cell, once activated, would remain in a steady state. An additional diffusible inhibitor is produced by the activated cells. A normal-appearing traveling wave results since each newly activated cell extinguishes the activation of the preceding cell. However, if two waves collide, the situation is different. At the point of collision, no newly activated cell is available to extinguish the activation of the cells. These cells remain in the steady state until the refractory period of the neighboring region is over. The latter become reinfected and, in turn, extinguish via the diffusible inhibitor the steady-state activation of the cells at the point of collision. Figure 4 shows further that the model can account for regulatory phenomena such as those observed in the shell pattern after a global perturbation.

D. Formation of Droplike, Unpigmented Regions

Oscillations can result not only from a self-enhancing process and its antagonistic reaction but also from two systems that are antagonistic to each other. As an example, the pattern on the shell of *Conus marmoreus* (Fig. 5) will be discussed. It is characterized by a pigmented background with droplike unpigmented regions. The beginnings of these white regions are nearly parallel to the growing edge. This indicates an almost simultaneous breakdown of the pigment production over a larger extension. In the simulation (Fig. 5), two nonlinear systems have been assumed: one that generates pigment as discussed above and a second one that extinguishes the first reaction. Whenever cells were activated for a sufficiently long period, the extinguishing system is triggered. Since it has a faster spreading than the pigment-producing system, the stop of pigment production occurs simultaneously over a larger region. Pigment production survives in regions in which the cells have been activated for too short a period to drive the extinguishing reaction. Due to the spread of activation from these regions, the white regions become pigmented again in the course of time, while in other regions, new pigment-free regions emerge, and so on.

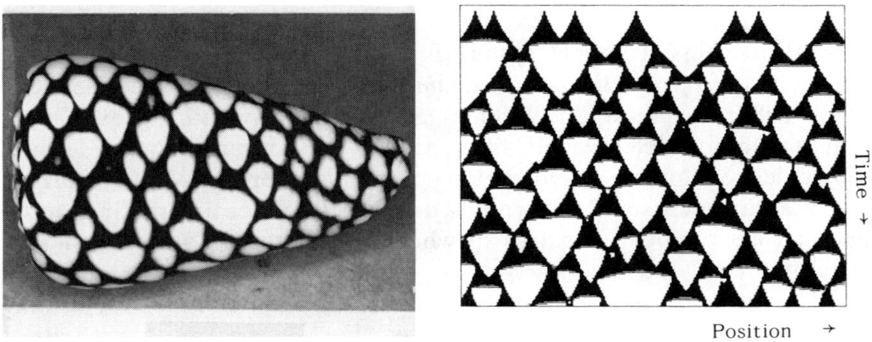

Fig. 5 Unpigmented droplike pattern on a pigmented background. (a) The shell of *Conus marmoreus*. (b) Model: Two pattern-forming reactions are assumed, one that causes pigment deposition and another that extinguishes the pigment-producing reaction. Whenever cells have remained activated for a sufficient time, the extinguishing system is triggered (shaded regions at the top of the white drops). Therefore, pigment production stops almost simultaneously over larger regions which become reactivated in the course of time by the spread of activation from regions in which the activation of the first system has survived.

Other modes of superpositions of two pattern-forming systems exist. For instance, an invisible stable pattern can exist that influences the time constants of the oscillating, pigment-producing pattern-forming system (for a detailed simulation, see Meinhardt and Klingler, 1987; computer programs for these simulations can be found in Meinhardt and Klingler, 1986).

The patterns on the shells of mollusks are presumably not under a strong evolutionary pressure. Many of the most beautiful snails live buried in the sand in the tropical sea. Thus variation could be introduced without endangering the species. The model shows that minor variations and combinations of a basic principle can explain many of the patterns. Especially since the shell patterns are records in time, the condition that the patterns must be stable—at least for a certain period—was no longer required. This has created the opportunity to generate the enormous diversity.

III. USE OF OSCILLATING PATTERN-FORMING SYSTEMS TO ORIENT CHEMOTACTIC CELLS

Many cells are known to react to chemotactic signals. Since the cells are small, the concentration differences across the cell diameter are also small. Thus cells must be able to detect minor concentration differences. A well-investigated model system is the chemotaxis of amoeba of the slime mold *Dictyostelium discoideum* during aggregation (see Gerisch, 1968, 1987; Chapter 8, this volume). Similarly, a navigating growth cone must be able to detect small differences in the concentration of surface molecules to allow an oriented elongation of the nerve toward its correct target area. Such small differences must be amplified internally to produce an unequivocal signal as to where, for instance, filopodia should develop.

The mechanism of autocatalysis and lateral inhibition is able to perform amplifications of small differences. For chemotactical orientation pattern formation has to take place within a cell or within a part of a cell. Therefore, the diffusion range of the activator must be small. This suggests that the self-enhancing process is membrane bound. In contrast, rapid redistribution of the antagonistic substance (e.g., an inhibitor) is essential to suppress secondary maxima that would point in a direction opposite to the one desired. Thus it is advantageous if the antagonistic substance diffuses freely in the cytoplasm.

For a chemotactically sensitive cell it is important that the direction of movement can be permanently readjusted. A stable pattern is inadequate since a once-generated high point on the cortex of a chemotactic cell would be difficult to shift, due to the internal self-amplification. An adjustment in its position and therewith a correction in the orientation of movement would be difficult to achieve. However, as discussed above, if the antagonist has a longer time constant than the activator, the pattern-forming system oscillates. The point at which the activator maximum appears will depend on small external differences (Meinhardt and Gierer, 1974). After an increasing accumulation of an inhibitor or a complete depletion of a substrate, the autocatalysis breaks down. The next activation can appear at a different position of the cell cortex depending on the external signal. The simulation in Fig. 6 shows that minute external differences are sufficient for the formation of a strongly localized signal, and that changes in these external differences are detected unequivocally.

In *Dictyostelium*, the chemoattractant cyclic AMP is released by the amoeba in a pulsatile manner (see Chapters 8 and 11, this volume). The cAMP oscillations have generally been regarded only as a tool for signal relay. In the view of the model, it is also a mechanism to detect small concentration differences. The *Dictyostelium* system has many features expected on the basis of the model. In a particular phase of the oscillation, the release of a cAMP pulse can be triggered by small amounts of externally supplied cAMP (Gerisch and Hess, 1974). Thus cAMP is autocatalytic. The number of cAMP molecules secreted is about 100 times higher than that required for the trigger function. One of the inhibitory elements is a phosphorylation of the cAMP receptor, which thereby becomes nonfunctional (Vaughan and Devreotes, 1988). Still missing is an inhibitory element that can spread within the cell. A possible candidate is the Ca^{2+} ions that enter the cell after the trigger. In the orientation of the growth cone Ca^{2+} ions play an essential role (for a review, see Kater et al., 1988). It is not yet clear whether in *Dictyostelium* the influx of Ca^{2+} ions plays a role in the chemotactic orientation.

Formation of Netlike Structures

A very common structure in almost all higher organisms are filamentlike branching structures. The venation of leaves, the tracheae of insects, the blood, or lymph vessels as well as neurons are examples. We discuss a model for the formation of these structures that is based on the mechanism mentioned above using localized activation and its shift (Meinhardt, 1976). We have seen how a local high activator concentration can be generated and that this signal can be very easily shifted if the system oscillates. A local activator maximum can be used as a signal to change the determination of the particular cell such that it becomes a part of the

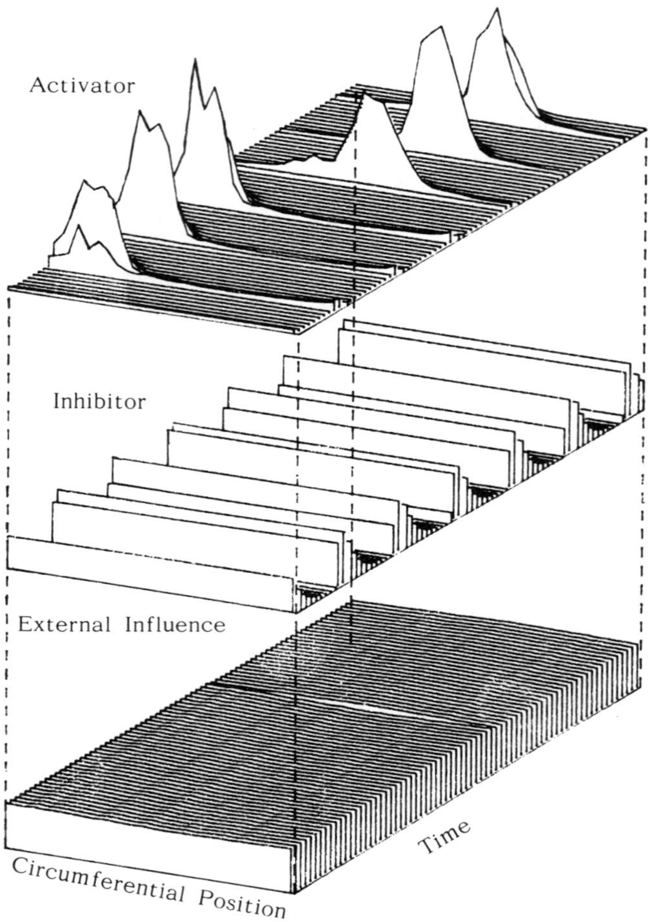

Fig. 6 Use of an oscillating activator–inhibitor system to detect small external differences to orient a chemotactic cell. For computational simplicity simulations were made on the circumference of a circle instead of a sphere. To allow a space-time plot, the circle is cut up. The inhibitor is assumed to be redistributed very quickly within the cell. The left and right elements are adjacent neighbors in reality. Due tó the oscillation, the maximum appears at the position where the external signal (source density) is slightly above average. After a change of the external influence, the activator maximum is formed at the new optimum position. The system can detect and adapt to differences which are so small that they can hardly be seen by eye. This is possible even if, in addition, this signal carries some noise.

vascular system. I assume further that after its determination the cell suppresses the signal that has caused the determination. The signal will then be shifted into a neighboring cell, which will also be determined. A repetition of this process—determination/shift of the signal/determination—leads to a filament of determined cells behind a wandering activator maximum.

The simulation in Fig. 7 is based on interaction of four substances. An activator–inhibitor system generates local high concentrations. A third substance functions as a switch. It is a measure whether the cell is differentiated or not. The switchlike behavior can be realized by a gene whose gene product has a saturating positive feedback on its own production. Exposure of a cell to a sufficiently high activator concentration leads to the activation of this gene. Due to the feedback, the gene remains in the on-state even if the activator concentration decreases. To achieve shift of the activator maximum away from a cell after its differentiation has been induced, it is assumed that (1) all cells produce a substrate, (2) the differ-

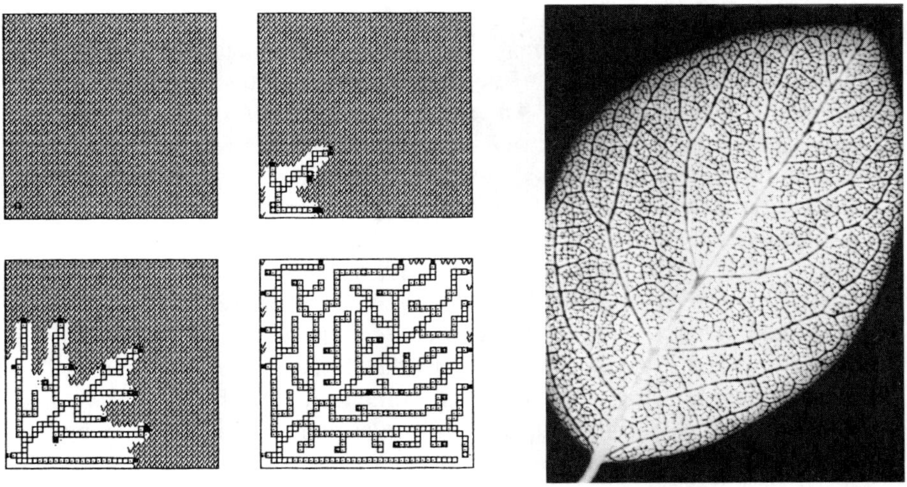

(a) **(b)**

Fig. 7 Stages in the formation of a netlike structure. An interaction of four substances can create a very complex netlike structure. (a) Two substances are used to form local activator maxima (■) (see Fig. 1). The corresponding cells differentiate (□) and remove a substrate (wave lines) from the surrounding cells. The activator maximum moves toward higher substrate concentrations (i.e., away from cells already differentiated), leaving behind a trail of newly differentiated cells. (b) For comparison, venation of a leaf. (After Meinhardt, 1976, 1982.)

entiated cells remove the substrate, and (3) the activator–inhibitor system depends on this substrate. Due to the depletion of the substrate in the differentiated cells, neighboring cells contain more of this substrate and are thus more efficient in activator production. The highest substrate concentration in the neighborhood of differentiated cells is usually present at the tip of a growing filament since at the side more cells contribute to substrate removal. Therefore, in the absence of other constraints, the elongation of a filament will take place at the tip and the filament will become straight. For the reasons outlined above, the shift of the activator maximum into the neighboring cell is greatly facilitated if after differentiation, the activation for a cell is shifted to an oscillating mode.

Branches are formed whenever activator maxima become sufficiently remote from each other during elongation of the filaments. Then the inhibitor concentration can become so low locally that a new activator maximum is triggered along an existing vein. This maximum is shifted again into a neighboring cell and a branch is initiated. Since the activator maximum attempts to escape from the inhibitory influence of the filament as fast as possible, branches will usually occur under 90°.

This model of vein formation is also able to account for pattern regeneration. After removal of veins at a particular region, the substrate is no longer reduced there. An increase in the substrate concentration results, which attracts activator maxima on filaments near the border of the injured region. New filaments grow into the damaged region. The resulting pattern is similar but not identical to the original one.

Although it may be surprising that interaction of only four substances can generate such complicated structures, it is not the simplest possible model. There are two inhibitions involved: one that keeps the activator signal localized and another that causes its shift. Both inhibitions can be mediated by one substance, but this has the consequence that an existing filament is so inhibitory that no lateral branches can be formed. Branches can occur only by a split of the activator maximum at the tip of a growing filament. The venation of leaves of the evolutionary older *Ginkgo* tree as well as those of some lower ferns show this pattern. Thus the model suggests that a separation of the two inhibitions took place during the evolution of leaves.

Leaf venation is frequently regarded as an example for a fractal process in which similar structures are formed at very different scales. According to the model, there is only one process that works only at a single scale. However, since the leaf grows, the spaces between the veins enlarge, and from a certain size onward, new filaments grow into these regions, in a similar way as in the regeneration, as mentioned above.

As mentioned, in the modeling of cell determination or differentiation, an autocatalytic input of a gene product on the activation of the corresponding gene has been assumed. Meanwhile, the prediction that autocatalytic regulatory mechanisms are involved in the activation of genes required for determination (Meinhardt, 1976, 1978) has found much support by experimental observations (see Serfling, 1989).

IV. CONCLUSIONS

By self-enhancement and long-range inhibition many patterns can be generated that are important elements in the development of higher organisms. Depending on diffusion, production, and decay rates of the substances involved, stable monotonic and periodic patterns as well as oscillating patterns and traveling waves can be generated. Coupling of several such reactions allows the generation of more complex patterns in a reliable and reproducible way. By a change of parameters under evolutionary pressure the pattern generated can be adapted to the requirements of a particular developmental situation. As exemplified for the shell patterns, in the absence of such evolutionary pressure an overwhelming diversity of patterns can be formed.

REFERENCES

Ermentrout, B., Campbell, J., and Oster, G. (1986). *The Veliger*, *28*: 369.
Gerisch, G. (1968). *Top. Dev. Biol.*, *3*: 157.
Gerisch, G. (1987). *Annu. Rev. Biochem.*, *56*: 853.
Gerisch, G., and Hess, B. (1974). *Proc. Natl. Acad. Sci.* (*USA*), *71*: 2118.
Gierer, A., and Meinhardt, H. (1972). *Kybernetik*, *12*: 30.
Kater, S. B., Mattson, M. P., Cohan, C., and Connor, J. (1988). *Trends Neurosci.*, *11*: 315.
Lefever, R. (1968). *J. Chem. Phys.*, *49*: 4977.
Meinhardt, H. (1976). *Differentiation*, *6*: 117.
Meinhardt, H. (1978). *J. Theor. Biol.*, *74*: 307.
Meinhardt, H. (1982). *Models of Biological Pattern Formation*, Academic Press, London.
Meinhardt, H. (1984). *J. Embryol. Exp. Morphol.*, *83*(Suppl.): 289.
Meinhardt, H. (1986). *J. Cell Sci. Suppl.*, *4*: 357.
Meinhardt, H., and Gierer, A. (1974). *J. Cell Sci.*, *15*: 321.
Meinhardt, H., and Klingler, M. (1986). *Lect. Notes Biomath.*, *71*: 184.
Meinhardt, H., and Klingler, M. (1987). *J. Theor. Biol.*, *126*: 63.
Serfling, E. (1989). *Trends Genet.*, *5*: 131.
Vaughan, R. A., and Devreotes, P. N. (1988). *J. Biol. Chem.*, *23*: 14538.

3
Controversy over Concepts: Long-Range Communication Versus "No Crosstalk"

Lionel G. Harrison *University of British Columbia, Vancouver, British Columbia, Canada*

Thurston C. Lacalli *University of Saskatchewan, Saskatoon, Saskatchewan, Canada*

I. CYCLIC PROCESSES AND SPATIAL PATTERNS

Building a spatially organized structure always involves cyclically repetitive processes: the artisan's laying of bricks into a wall; the enzyme molecule's processing of 100 substrate molecules every second; the cycles of cell division, whether leading to a structure of increased size or repatterning of a nonfeeding embryo. One could even include the formation of populations and ecosystems, involving the life cycles and reproductive processes of the organisms. Life is a hierarchy of cyclic processes, from the molecular scale to the geographical.

Not all such cycles have the character of oscillation; that is, the occurrence of the cyclic repeats at precisely regular intervals of time. If an enzymic reaction is diffusion limited, its frequency of, say, 100 s^{-1} is a statistical mean of a Poisson distribution. In any given second the number of molecules processed will be 100 $\pm$ 20 at 95% confidence. This raises an issue which, in our view, requires much more attention that it has yet received in developmental biology: error reduction. Clearly, 20% random error is intolerable in many of the time-order or space-order aspects of development and morphogenesis.

There are two obvious mechanisms for error reduction. The first is participation of a very large number of molecules in the controlling events. Suppose, for instance, that we envisage a DNA-binding molecule, such as a homeobox protein, playing an important role in patterning by exerting concentration-dependent control of the activity of a gene. This is in fact well established in the *Drosophila* blastoderm for the control of patterned expression of genes in the pair-rule class by the DNA-binding *bicoid* protein (Driever and Nüsslein-Vollhard, 1988a,b). The mechanism of this control is not definitely known (Lacalli and Harrison, 1991), although there is a superabundance of experimental evidence establishing a hierarchy of three classes of gene (for reviews, see Akam, 1987, and Ingham, 1988). But through whatever intermediate stages and by whatever mechanisms, it appears that a single matched pair of genes in a diploid nucleus can respond differently to *bicoid* concentration levels differing by something on the order of 1%. If the error reducer is the sheer number of *bicoid* molecules interacting with gene control sequences in a Poissonian-statistical manner, there must be about 40,000 such molecules for the system to work properly 19 times out of 20.

The second mechanism for error reduction is also likely to use large numbers of molecules but involves deterministic organization of their dynamics rather than mere Poissonian decrease of fractional error by increase of numbers. Where this leads to cyclic processes on the spatial scale of the organized dynamics, the cycles are likely to become equally spaced in time as a result of the organization; that is, they become oscillations. The cell cycle is the prime example. If morphogenesis requires spatially patterned differential rates of cell proliferation (e.g., morphogenesis of multilobed glands; for examples, see Bernfield and Banerjee, 1972; Bernfield et al., 1984), it seems likely that a cell must receive information on what its division rate is to be from extracellular signals. These can be read much more precisely by the averaging abilities of thousands of receptors on a single cell surface than by one or two copies of a gene in a syncytium.

Error reduction of this kind provides an evolutionary perspective on the usefulness of cellular structure in the control needed for development of higher organisms. But it makes it all the more remarkable that the crucial initial steps in formation of the body plan of *Drosophila* occur while the multinucleate blastoderm (interphase 14, about 8000 nuclei) is still syncytial and just about to cellularize. Why couldn't nature wait another 20 minutes? And *Drosophila* is not the only organism to form a complex pattern without cellularization. The whorls of *Acetabularia* are spatially periodic patterns of many parts formed within a very large single cell. Work on these by one of us (e.g., Harrison et al., 1988) has provided evidence both for the involvement of dynamic organization and for hierarchical ordering of patterning processes. Frankel (1989) has assembled extensive evidence for cellular-scale dynamic controls in the ciliates.

Such instances raise two questions on generalities. The first is indicated by the title of this chapter. When a complex pattern of many repeated parts arises hierarchically from a simpler pattern such as a gradient, does this occur by local reading of the gradient (e.g., by each nucleus in the *Drosophila* blastoderm independently of all the others), or is there long-range interaction involved at the level of the complex pattern? We believe the latter to be, a priori, the more promising possibility. It shows that nature did not need to wait another 20 minutes. The power of numbers to overcome random errors need not be present at every local site if it is achieved by continual dynamic interaction of all the molecules throughout the pattern-forming region. In that case the second question is whether dynamics for patterns periodic in space are closely related to those for activity periodic in time. Here we indicate that the same reaction–diffusion dynamics which can generate a pattern as a number of repeats each some tens of micrometers long can also produce oscillations such as those seen on the cellular scale in the cell cycle.

II. CHEMICAL MECHANISMS FOR TIME AND SPACE ORDER

Reaction–diffusion theory involving a pair of mutually interacting and diffusible substances was first proposed by Turing (1952). His differential rate equations contain both time and space variables and are capable of representing both spatial patterning and oscillations in the concentrations of those two substances. The Turing equations also represent a common property of most reaction–diffusion models which have been proposed subsequently (i.e., the dynamic behavior in the early stages of emergence of pattern out of uniformity). This is shown by the mathematical procedure of linearization about the spatially uniform steady state, and we have earlier discussed the results of that procedure for two well-known reaction–diffusion models (Lacalli and Harrison, 1979). We outline linear Turing analysis in Section IV. But the closeness of the relation between mechanisms for time order and space order is best shown first in terms of the generalized chemical equations for two of the later models.

A. Relation of the Brusselator and the Sel'kov Model

In the same year, Prigogine and Lefever (1968) published a two-substance dynamic model that has subsequently been used mainly in connection with spatial patterning; and Sel'kov (1968) published a model for self-oscillations in glycolysis. The former authors stated in their paper: "As we shall see, there is a remarkable parallelism between the problems of structural dissipative order and time order." In relation to cellular oscillators, it has later been acknowledged (e.g., by Klevecz, 1984) that "most recent analyses of this problem have focused on nonlinear oscillations of the general form described by Prigogine and his coworkers."

The nonlinear property referred to arises from a part of the reaction dynamics in which a substance self-enhances its own rate of formation in proportion to the *square* of its concentration. If one reads the cited papers of Sel'kov and of Prigogine and Lefever side by side, it is not immediately obvious that in this respect they contain the same dynamic postulate; but they do. The Brusselator scheme shows it explicitly as the third step in a four-step mechanism for conversion of reactants A and B into products D and E via two intermediates X and Y.

	Rate constant	
$A \longrightarrow X$	a	(1a)
$B + X \longrightarrow Y + D$	b	(1b)
$Y + 2X \longrightarrow 3X$	c	(1c)
$X \longrightarrow E$	d	(1d)

The kinetic equations for this model, including diffusion of X and Y, are

$$\frac{\partial X}{\partial t} = aA - bBX + cX^2Y - dX + D_x\nabla^2 X \tag{2a}$$

$$\frac{\partial Y}{\partial t} = bBX - cX^2Y + D_y\nabla^2 Y \tag{2b}$$

This mechanism is very good at producing regular spatial periodicity in one or two dimensions (Lacalli, 1981; Harrison, 1982; Harrison and Lyons, 1991). This property appears to be related to the two equal and opposite terms in the growth rates of X and Y, which include the nonlinear term.

In their earliest accounts of this model, the Prigogine group indicated equation (1c) as being unrealistic because it is termolecular; and their comment is echoed frequently as a criticism of the model even today. There is in fact no problem of unreality if equation (1c) is regarded not as a single step but as an abbreviation of two or three steps of a type that is quite common in catalyst activation in liquid solutions. Termolecular steps are taught as unlikely mainly in those courses in elementary chemical kinetics that are based almost entirely on gas-phase reactions. In liquid solution, activation of a catalyst by attachment of two ligands gives the same dynamics.

In the Sel'kov model for glycolytic oscillations an enzyme E acts as an ATPase. Hydrolysis of ATP (Sel'kov's S_1, our Y) is catalyzed by E when it has been activated by a number γ of product ADP molecules (our X). We shall use $\gamma = 2$. Hunding (1989, 1991) has used similar models with $\gamma = 3$ extensively in relation to *Drosophila* patterning. If the allosterically activated complex EX_2 is formed by a rapid equilibration with X,

$$E + 2X = EX_2 \qquad \text{equilibrium constant } K \tag{3}$$

the equilibration together with the catalytic activity

$$Y + EX_2 \longrightarrow EX_2 + X \qquad \text{rate constant } k \tag{4}$$

are together kinetically equivalent to equation (1c). In the equilibration (3), it is not necessary to envisage a termolecular step. The two X ligands could be added one at a time successively [Harrison, 1981, equations (24)]. Equations (3) and (4) together give the kinetic equation

$$\frac{\partial X}{\partial t} = k(EX_2)Y = kKEX^2Y = \frac{kKE_0X^2Y}{1 + KX^2} = \frac{cX^2Y}{1 + KX^2} \tag{5}$$

where E_0 is the total concentration of enzyme from which KX^2E has been converted into the active complex with X. Here the simplifying assumption has been made that any part of the total enzyme supply bound up in a complex YEX_2 in the course of reaction (4) is small compared to KX^2E.

The remaining features of the Sel'kov model are a constant rate of formation of Y out of some precursor (unnamed in Sel'kov's equations; we call it A) and a first-order decay of X, which is equation (1d) of the Brusselator model. For the initial step,

$$A \longrightarrow Y \qquad \text{rate constant } a \tag{6}$$

The kinetic equations are then

$$\frac{\partial X}{\partial t} = \frac{cX^2Y}{1 + KX^2} - dX \tag{7a}$$

$$\frac{\partial Y}{\partial t} = aA - \frac{cX^2Y}{1 + KX^2} \tag{7b}$$

B. Oscillations, Patterns, and Gradients

Equations (7) are written without diffusion terms because they were originally proposed as a model for temporal oscillations. If diffusion terms are added, they clearly have a strong resemblance to the Brusselator equations (2a) and (2b). Both are capable of producing spatial patterning. The Sel'kov model actually corresponds even more closely to one of the variants of the Brusselator devised by Tyson et al. (Tyson and Light, 1973; Tyson and Kauffman, 1975) which has been used extensively by one of us (e.g., Lacalli, 1981). In the work of our two groups, the LGH group has most often used the original Brusselator, and the TCL group that particular one of Tyson's models, in which the first step is production of Y instead of X. We consider that the term Brusselator should be used for a class of models, including Sel'kov's with the added diffusion terms, all of which have the particular pattern-forming behavior associated with the X^2Y kinetics and the equal and oppositely signed nonlinear terms in growth rates of X and Y.

When does a two-morphogen reaction–diffusion model produce patterns, and when does it generate oscillations? This is not just a matter of diffusion *versus* stirring for the morphogens X and Y. Both kinds of behavior can be found with the diffusion terms in the equations. For the kind of dynamics required, it is necessary that the parameters of the model fall into certain numerical ranges. This can be seen best by linear Turing analysis, as described in Section IV. There we show that quite moderate changes in parameter values could change the behavior of a model from producing a pattern with a spatial periodicity of a few tens of micrometers to generating oscillations with periods from infinity down to 2 to 3 min, covering the range of cell cycles during active development.

Brusselator-class models are set up explicitly as generalized reaction mechanisms containing putative chemical substances. These include reactants such as the A and B in equations (1) and, in the Sel'kov model, enzyme E, the total concentration E_0 of which becomes part of the rate constant c in equation (5). These inputs to the X, Y patterning system might themselves be patterned. The models therefore provide a rather clear way of representing the hierarchical linking of developmental processes. Other classes of reaction–diffusion model are also capable of representing hierarchies. In fact, the model of Gierer and Meinhardt (1972) was specifically set up to show how a patterning mechanism could make use of a preexisting gradient. But the Brusselator type of model can assist significantly in attempts to relate dynamic features to types of chemical reaction step.

Long-range gradients of unknown chemical substances have long been postulated on the basis of indirect evidence from a morphological study of development, in examples as diverse as *Hydra* (Wilby and Webster, 1970), segments of the insect cuticle (Locke, 1967), and the chick limb bud (Saunders and Gasseling, 1968; for a review, see Wolpert et al., 1975). The concept that cells possess information on their positions in a developing organism is more than a century old: for plants, Vöchting (1877); for animals, Driesch (1893, 1908). Wolpert (1970) associated positional information with the local reading by cells of concentration levels in the putative gradients. He referred to the "French flag problem": how three concentration ranges in a gradient can specify three different types of cellular differentiation, symbolized by the blue, white, and red regions of the tricolor (Fig. 1C).

C. Gradients and Periodicity: The American Flag Problem

In the *Drosophila* blastoderm, the segmentation pattern first arises as a pattern of 14 alternating stripes of expression of the pair-rule genes. This is one more stripe than the American flag, and a much more severe demand, quantitatively, on a patterning mechanism than Wolpert's French flag problem (Fig. 1E). The gradient reading proceeds via an intermediate set of genes, the gap genes (Fig. 1D), and mechanisms have been suggested wherein this involves ratios of concentrations in two gap gene gradients (e.g., Pankratz et al., 1989). Nevertheless, the pair-rule expression in each nucleus is ultimately dependent on local concentration in the monotonic gradient at the top of the hierarchy (*bicoid*, Fig. 1A,B). Over the entire length of the egg, this value (which we shall call A) drops by a factor of about 50; over the stripe-forming region, by a factor of about 10. If error limits are Poissonian, varying with $A^{-1/2}$, they should increase by a factor of about 3 from the most anterior stripe to the most posterior (Fig. 1F). There is no hint of this in published pictures that show every nucleus (e.g., the cover of the second edition of Gilbert, 1988). Irregularities in the edges of the stripes are very similar, from the most anterior to the most posterior.

In this instance there is no indication in the evidence of any linkage of the patterning phenomenon to the nuclear division cycle, except in a negative sense. Stripe formation occurs during interphase 14, which is a long pause in a previously rapid sequence of divisions. We seek here, however, to raise the issue of error limits more generally, and this might include multicellular instances in which a gradient controls the frequency of cell division. For Brusselator-class models for the timing mechanism in each cell, with the controlling gradient as input A in equation (1a), the oscillation frequency is likely to go with A^2 [Section IV, equation (19)]. In that case the same A gradient that gives a factor of 3 increase in the reading error of A from the high to the low end would give a factor of 10 increase when A^2 is being read.

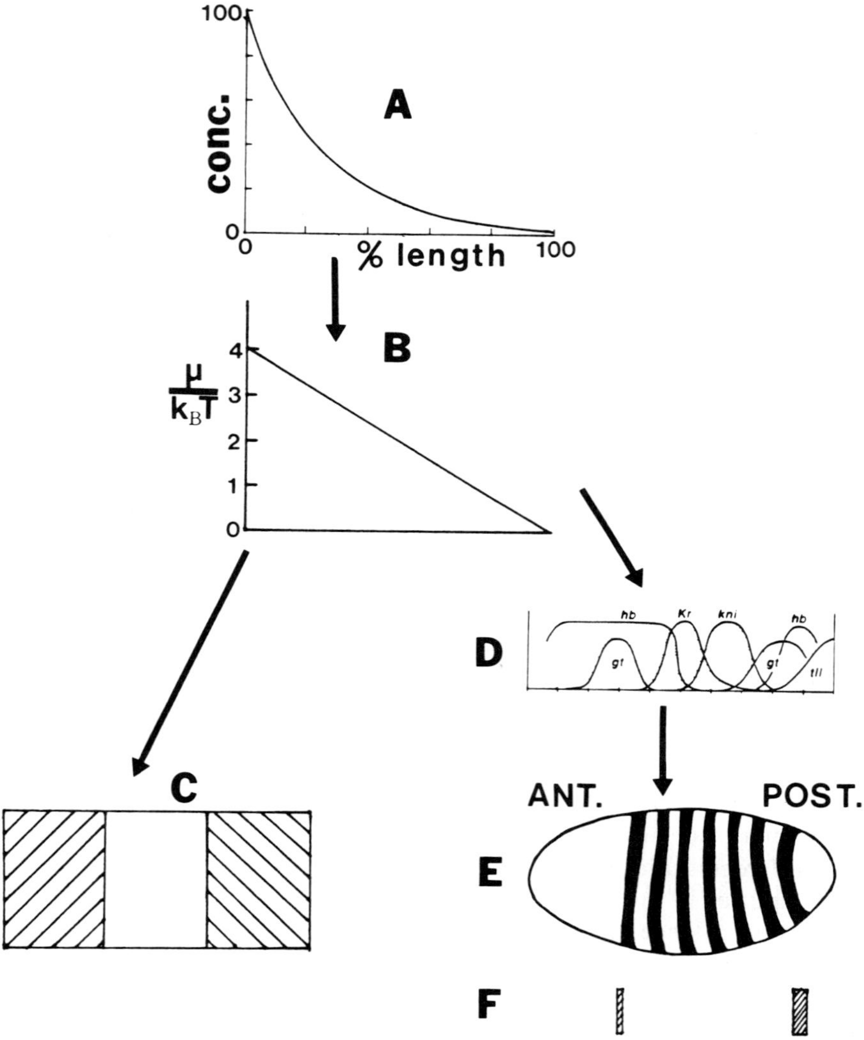
100
A
conc.
0
0
% length
100
B
4
3
2
1
0
μ
$k_B T$
D
hb
Kr
kni
hb
gt
gt
tll
C
ANT.
POST.
E
F

Another well-known feature of the American flag is that it also has a pattern of stars, in which two spatial dimensions are treated equally with regard to spacing. The tendency of a Brusselator-class model in two dimensions is to do this, producing a spotted rather than a striped pattern (Lacalli, 1981). We have discussed elsewhere two ways in which biological patterning can be asymmetrized between two spatial dimensions, to give stripes: the use of one-dimensional gradients (Lacalli et al., 1988), and the use of reaction–diffusion mechanisms different from the Brusselator, having within their nonlinear dynamics an intrinsic tendency to form stripes (Lacalli, 1990; Lyons and Harrison, 1991; Harrison and Lyons, 1991).

III. CONCLUSIONS: A CHARACTER OF LONG-RANGE COMMUNICATION

In well-regulated biological development, which leads to the formation of properly shaped and properly functioning organisms, error distributions are clearly visible when pattern formation is observed to the normal limits of resolution of optical microscopy. Our main purpose here is to point out

---

Fig. 1 (A) *Bicoid* gradient along the anteroposterior direction of the *Drosophila* egg, modified from Driever and Nusslein-Volhard (1988a) by allowance for background, as concentration (percent of maximum) versus distance. (B) Same as logarithm of concentration versus distance, indicated as chemical potential (μ, arbitrary zero) versus distance; ideal solution assumed. For 1% of egg length, the change in free energy of the protein is only $0.04k_BT$ (i.e., 0.04 times thermal noise). This illustrates the need for many *bicoid* protein molecules to be involved in conveying positional information locally if the information is to rise above thermal noise. (C) French flag problem. This does not impose severe requirements on the informative power of the gradient, because each of the three types of behavior corresponds to a broad range of concentration or free energy in the gradient. (D) Spatial distributions of several gap genes in *Drosophila*, an intermediate stage in reading of information from the gradient to the segmentation-patterning system. (E) American flag problem, a much more severe test of precision in reading a gradient. This is the distribution of the segmentation gene product *fushi tarazu* (black stripes: the protein) at interphase 14 in the *Drosophila* egg. Alternating white stripes are occupied by another protein, *even-skipped*. (F) At the left, an indication of the breadth of irregularities in the edge of a stripe. At the right, an indication of how this should have increased from the first to the last stripe, based on Poisson statistics and the gradient in (A). The breadth of the irregularities in fact shows no change from first to last stripe, pointing to operation of long-range organization rather than local gradient reading.

that the character of these distributions is important, and may yield significant evidence to decide between rival theories of morphogenesis. This field of work is almost untouched. In current literature, especially on the *Drosophila* segmentation problem, there is a tendency to focus on the evidence that each stripe rapidly acquires an individual character and can be modified by local influences. We believe that the evidence from error distributions points to a character of uniformity in stripe formation throughout the pattern and hence to a character of long-range communication in the patterning mechanism at the level in the hierarchy at which complex pattern is being formed and not only at the higher level of gradients. Similar considerations may readily be applicable to both time order and space order in morphogenesis.

IV. APPENDIX: LINEAR TURING ANALYSIS

For two mutually interacting substances (morphogens) X and Y, the linear equations of Turing (1952) refer to a spatially uniform steady state X_0, Y_0 and to the departures of concentrations from that state,

$$U = X - X_0 \quad \text{and} \quad V = Y - Y_0 \tag{8}$$

The Turing equations are

$$\frac{\partial U}{\partial t} = k_1 U + k_2 V + D_x \nabla^2 U \tag{9a}$$

$$\frac{\partial V}{\partial t} = k_3 U + k_4 V + D_y \nabla^2 V \tag{9b}$$

Particular solutions exist in the form

$$U = a_0 e^{k_g t} \cos \frac{2\pi s}{\lambda} \tag{10}$$

and similarly for V, the U and V patterns being of the same wavelength or pattern spacing λ and in phase with each other. These solutions are for one spatial dimension s. a_0 is an initial amplitude. The constant k_g can be positive, negative, or complex, according to the values of the six parameters k_1 to k_4, D_x, and D_y. Positive k_g means growth of amplitude. Where this exists only over a narrow range of wavelength, the model has a spatial bandpass character which allows it to take all the short and long wavelengths present in thermal noise and selectively amplify one. This is the patterning behavior. Negative k_g indicates relaxation back to an unpatterned state. Complex k_g indicates temporal oscillations. The frequency can be obtained from the imaginary component. Both patterned and unpatterned concentrations can oscillate. The frequency for unpatterned oscillation can be found by setting $\lambda = \infty$.

$$2k_g = k_1 + k_4 - \left(\frac{2\pi}{\lambda}\right)^2 (D_x + D_y) + (b^2 + 4k_2k_3)^{1/2} \quad (11)$$

where

$$b = k_4 - k_1 + \left(\frac{2\pi}{\lambda}\right)^2 (D_x - D_y).$$

The wavelength for maximum positive value of k_g is, in an approximation for $D_y \gg D_x$,

$$\lambda_{\max} = \frac{2\pi(D_xD_y)^{1/4}}{(-k_2k_3)^{1/4}} \quad (12)$$

When k_g is complex, its imaginary part is the last term in equation (11); and for an unpatterned system, the oscillation frequency is found from this as

$$\nu = -i\frac{1}{4\pi}[(k_1 - k_4)^2 + 4k_2k_3]^{1/2} \quad (13)$$

where i is the square root of -1. It is a condition for the Turing model to produce a pattern that one of k_2 and k_3 must be positive and the other negative. Hence in equation (12) the quantity $-k_2k_3$ is positive, and in equation (13) k_2k_3 is negative, giving the possibility that the quantity of which a square root is taken is negative, which indicates oscillation.

For complex k_g, the real part may be positive (growing oscillations) or negative (damped oscillations). The real part is zero, corresponding to stable oscillations, for $k_4 = -k_1$. In that condition,

$$\nu = -i\frac{1}{2\pi}(k_1^2 + k_2k_3)^{1/2} \quad (14)$$

Using the parameter space described in Lacalli and Harrison (1979), we find that the available range for k_1^2 is from $-k_2k_3$ down to a small fraction of that value. This corresponds to a frequency from zero to a maximum close to

$$\nu_{\max} = \frac{1}{2\pi}(-k_2k_3)^{1/2} \quad (15)$$

This can be related to the chemical wavelength in the pattern-forming region through equation (12), yielding

$$\nu_{max} = \frac{2\pi(D_x D_y)^{1/2}}{\lambda_{max}^2} \tag{16}$$

If we use the probable values $D_x = 10^{-9}$ cm^2 s^{-1} and $D_y = 10^{-7}$ cm^2 s^{-1} (Lacalli and Harrison, 1987) and $\lambda_{max} = 30$ μm, a common order of magnitude for many morphogenetic spacings in both animal and plant kingdoms, we obtain $\nu_{max} = \pi/450$ s^{-1} (i.e., the period of oscillation is from infinity down to 2 or 3 min). It is probable that faster oscillations could be stabilized by the nonlinearities in chemically realistic models such as the Brusselator class. But this estimate is adequate to show that the same general type of model that will account for the common range of spacings in periodic structures arising during morphogenesis will also describe oscillations covering all of the very wide cell-cycle range of periods.

For any two-morphogen reaction–diffusion model, linearization to the Turing model and analysis of parameter spaces arising from the constants k_1 to k_4 is an important procedure. We find it continually necessary in our work. But it involves one difficulty. The relation of these four constants to the rate constants of the putative chemical steps in the mechanism (a, b, c, and d for the Brusselator) is not at all obvious. For the standard Brusselator, equations (1a) to (1d):

$$k_1 = bB - d \qquad k_2 = \frac{a^2A^2c}{d^2} \qquad k_3 = -bB \qquad k_4 = \frac{-a^2A^2c}{d^2} \tag{17}$$

With the approximation that d is small compared to bB, which is good for most of the pattern-forming range of the Brusselator, $k_3 = -k_1$; and along the stable oscillation line for which $k_1 = -k_4$ we have, consequently,

$$-k_2k_3 = A^4 \frac{a^4c^2}{d^4} \tag{18}$$

and from equation (15),

$$\nu_{max} = \frac{1}{2\pi} \frac{a^2c}{d^2} A^2 \tag{19}$$

This is the A^2 dependence for reading of an A gradient by a local oscillation frequency referred to in the discussion above of the American flag problem.

Linear Turing analysis makes the condition for stable oscillation seem rather restrictive: exact equality of k_1 and $-k_4$. If k_1 is smaller, oscillations decay and the model forms stable, nonoscillating patterns. If

k_1 becomes larger, oscillations become unstable. For realistic nonlinear models such as the Brusselator class, the effect of the nonlinearities in curbing indefinite exponential growth should extend the stable oscillation region into the range $k_1 > -k_4$. A few computations (for which we thank M. J. Lyons) have shown us that behavior for the Brusselator. Changes of a few percent in parameter values lead from pattern to oscillation.

ACKNOWLEDGMENTS

We thank Michael J. Lyons and Bernard C. Lakowski for much discussion of pattern formation, some Brusselator computations, and an analysis of published data on the *bicoid* gradient in *Drosophila*; and NSERC Canada for financial support.

REFERENCES

Akam, M. E. (1987). *Development*, *101*: 1.

Bernfield, M., and Banerjee, S. D. (1972). *J. Cell Biol.*, *52*: 664.

Bernfield, M., Banerjee, S. D., Koda, J. E., and Rapraeger, A. C. (1984). *Proc. 42nd Symp. Soc. Dev. Biol.*, Irvine, Calif., Alan R. Liss, New York, pp. 545–572.

Driesch, H. (1893). *Z. Wiss. Zool.*, *55*: 34.

Driesch, H. (1908). *Science and Philosophy of the Organism*, Vol. 1, Adam and Charles Black, London.

Driever, W., and Nüsslein-Vollhard, C. (1988a). *Cell*, *54*: 83.

Driever, W., and Nüsslein-Vollhard, C. (1988b). *Cell*, *54*: 95.

Frankel, J. (1989). *Pattern Formation: Ciliate Studies and Models*, Oxford University Press, New York.

Gierer, A., and Meinhardt, H. (1972). *Kybernetik*, *12*: 30.

Gilbert, S. F. (1988). *Developmental Biology*, 2nd ed., Sinauer Associates, Sunderland, Mass.

Harrison, L. G. (1981). *Chem. Soc. Rev.*, *10*: 491.

Harrison, L. G. (1982). *Proc. 40th Symp. Soc. Dev. Biol.*, Boulder, Colo., Alan R. Liss, New York, pp. 3–33.

Harrison, L. G., Graham, K. T., and Lakowski, B. C. (1988). *Development*, *104*: 255.

Harrison, L. G., and Lyons, M. J. (1991). *Proc. Winter Workshop, Centre de Physique*, Les Houches, France (in press, Nova Science Publishers, New York).

Hunding, A. (1989). In *Cell to Cell Signalling: From Experiment to Theoretical Models* (A. Goldbeter, ed.), Academic Press, London, pp. 229–236.

Hunding, A. (1991). In *Complexity, Chaos and Biological Evolution* (E. Mosekilde, ed.), Plenum Press, New York (in press).

Ingham, P. W. (1988). *Nature*, *335*: 25.

Klevecz, R. R. (1984). In *Cell Cycle Clocks* (L. N. Edmunds, ed.), Marcel Dekker, New York, p. 47.
Lacalli, T. C. (1981). *Philos. Trans. R. Soc. Lond. B*, *294*: 547.
Lacalli, T. C. (1990). *J. Theor. Biol.*, *144*: 171.
Lacalli, T. C., and Harrison, L. G. (1979). *J. Theor. Biol.*, *76*: 419.
Lacalli, T. C., and Harrison, L. G. (1987). *Can. J. Bot.*, *65*: 1308.
Lacalli, T. C., and Harrison, L. G. (1991). *Seminars in Developmental Biology*, Saunders Scientific Publications, Philadelphia (in press).
Lacalli, T. C., Wilkinson, D. A., and Harrison, L. G. (1988). *Development*, *104*: 105.
Locke, M. (1967). *Adv. Morphogen.*, *6*: 33.
Lyons, M. J., and Harrison, L. G. (1991). *Chem. Phys. Lett.*, *183*: 158.
Pankratz, M. J., Gaul, U., Hoch, M., Seifert, E., Nauber, U., Gerwin, N., Rothe, M., Bronner, G., Forsbach, V., Goerlich, K., and Jäckle, H. (1990). *Proc. 48th Symp. Soc. Dev. Biol.*, Berkeley, Calif., pp. 17–29.
Prigogine, I., and Lefever, R. (1968). *J. Chem. Phys.*, *48*: 1695.
Saunders, J. W., and Gasseling, M. T. (1968). In *Epithelial–Mesenchymal Interactions* (R. Fleischmajer and R. E. Billingham, eds.), Williams & Wilkins, Baltimore, p. 78.
Sel'kov, E. E. (1968). *Eur. J. Biochem.*, *4*: 79.
Turing, A. M. (1952). *Philos. Trans. R. Soc. Lond. B*, *237*: 37.
Tyson, J. J., and Kauffman, S. (1975). *J. Math. Biol.*, *1*: 280.
Tyson, J. J., and Light, J. C. (1973). *J. Chem. Phys.*, *59*: 4164.
Vöchting, H. (1877). *Pfluegers Arch. Gesamte Physiol. Menschen Tiere*, *15*: 153.
Wilby, O. K., and Webster, G. (1970). *J. Embryol. Exp. Morphol.*, *24*: 595.
Wolpert, L. (1970). In *Towards a Theoretical Biology* (C. H. Waddington, ed.), Vol. 3, Edinburgh University Press, Edinburgh, p. 198.
Wolpert, L., Lewis, J., and Summerbell, D. (1975). *Cell Patterning*, CIBA Foundation Symposium 29 (new series), Elsevier/Excerpta Medica/North-Holland, Amsterdam, pp. 95–130.

4
Spatial Patterns in Chemical Reactions

Stefan C. Müller *Max-Planck-Institut für Ernährungsphysiologie, Dortmund, Germany*

I. INTRODUCTION

Macroscopic spatial structures and their dynamic evolution are frequently observed in open chemical and biochemical systems in which the reactions occur far from thermodynamic equilibrium. Reports on beautiful patterns can be found throughout the literature over the past century. But only in recent years the study of nonequilibrium dynamics in chemical systems has become a subject of intensive research. In particular, the discovery of the oscillating Belousov–Zhabotinsky (BZ) reaction (Belousov, 1959; Zhabotinsky, 1964), and subsequently, of traveling chemical waves in this famous system by Busse (1969) and Zaikin and Zhabotinsky (1970), triggered unprecedented activity in this field (Field and Burger, 1985). Nowadays, with the maturation of nonequilibrium thermodynamics and rapid progress in understanding nonlinear system dynamics (Nicolis and Prigogine, 1977; Haken, 1983), the phenomenon of spatio-temporal organization in initially homogeneous reactions is generally accepted, although a long period of hesitation had to be overcome. It is obvious now that in many cases of chemical pattern formation one finds useful models for biological organization on a higher level of complexity, for instance in morphogenesis and differentiation processes.

Early examples of spatially inhomogeneous reactions are documented in Runge's work around 1855. After impregnating a piece of filter paper with one chemical solution, he placed a sequence of drops of an appropriately chosen second reactant on this paper at regular intervals. Runge thus discovered a large number of pairs of solutions for which the procedure results, upon spreading of the drops into the surrounding area, in quite spectacular and colorful spatial structures, which he called "paintings"

(Fig. 1). Whereas Runge's methodological contribution to the development of chromatography is generally accepted, his more artistic approach to produce the "paintings" was not appreciated by the scientists. Nevertheless, it should be emphasized that he was the first to observe ring-shaped regions of precipitated reaction product, many years before Liesegang, to whom the first systematic study of periodic precipitation is credited, rediscovered this phenomenon in 1896.

Another early observation was reported by Lippmann (1873). He described an electrochemical–mechanical oscillator, the "beating mercury heart." A mercury drop, placed in a watchglass and submerged in a

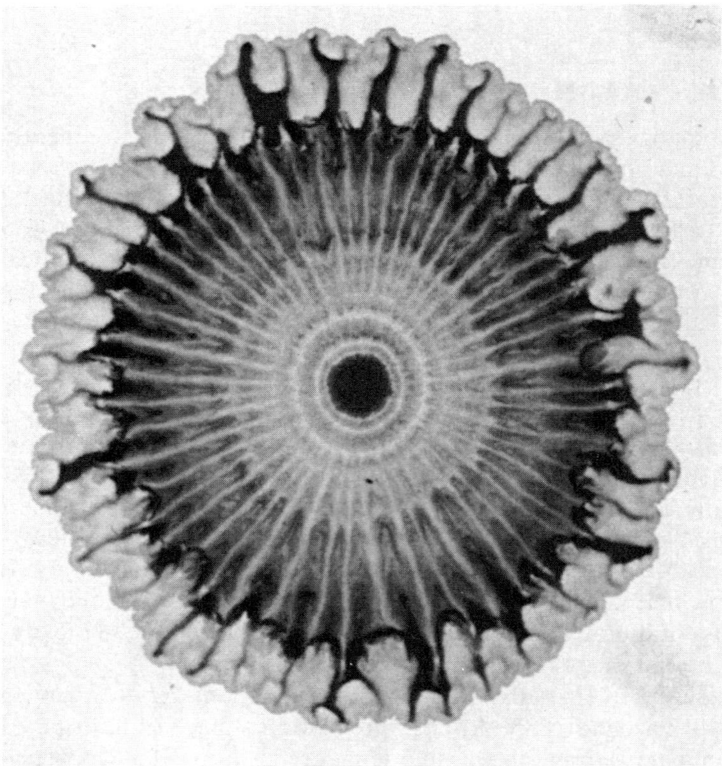

Fig. 1 "Painting" according to a recipe from Runge (1855), produced by repetitively placing drops of one chemical solution on a filter paper impregnated with a second solution, which undergoes a reaction with the first one. Diameter ≈ 10 cm. (Courtesy of L. Kuhnert, Berlin.)

pool of acidic solution of an oxidizing reagent, starts to pulsate with a frequency of about 1 Hz as soon as an iron needle touches the drop laterally. Depending on the details of the experimental setup the periodic geometrical distortions are isotropic or anisotropic (e.g., triangular) (Avnir, 1989). Following these and other sporadic reports, patterns in heterogeneous systems received considerable attention during the early decades of our century, especially the Liesegang phenomenon and electrochemical oscillations. This is well documented in a book by Hedges and Myers (1926).

Today, homogeneous liquid-phase reactions belong to the most frequently investigated systems, and in addition to classical systems such as the BZ reaction, many new oscillators have been found by the use of flow reactors (CSTR) (Epstein, 1984). For spatial patterns caused by the nonlinear coupling of complex reaction kinetics with diffusion, the BZ reaction still plays a predominant role, especially for traveling waves in an excitable medium. But also in this respect, other reactions become increasingly important in the study of continuously fed, spatially extended systems, in which the influx of substrates is controllable by the use of gel reactors (Noszticzius et al., 1987). A recent example is the study of the stationary Turing structures, observed for the first time in the chlorite–iodide system (Castets et al., 1990).

In this chapter a brief description of precipitation patterns is presented, followed by a more comprehensive discussion of the current knowledge of spiral-shaped traveling waves in an excitable BZ medium and their close analogy to aggregation patterns in the cellular slime mold *Dictyostelium discoideum*. More complex patterns were found recently to occur in excitable media under externally imposed gradients, and these are included on a qualitative basis. Furthermore, the role of additional transport processes other than diffusion, such as chemotaxis and hydrodynamic convection, is pointed out.

II. PRECIPITATION PATTERNS

The periodic precipitation process known as Liesegang ring formation has been investigated over the past 90 years. In the first experiments a drop of silver nitrate was placed on a gelatin layer containing potassium dichromate (Liesegang, 1896). The subsequent process of diffusion leads to the deposition of silver dichromate in concentric rings as shown in Fig. 2a. More generally, if a soluble electrolyte is placed in contact with a second electrolyte and, on interdiffusion, both react to form a poorly soluble salt, a metastable solution can be produced. For many combinations of electrolytes at suitable concentrations the subsequent precipita-

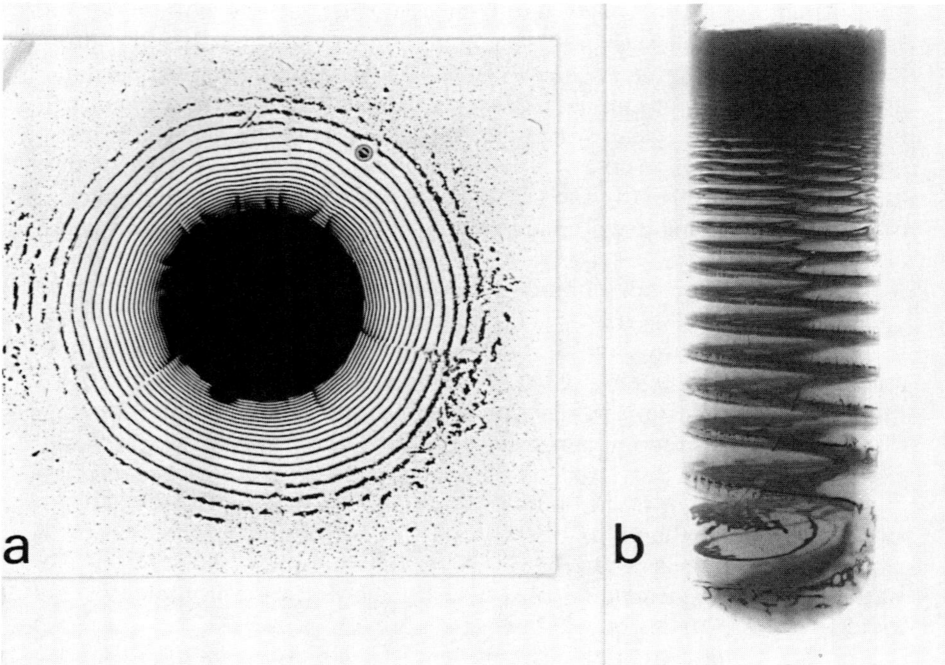

Fig. 2 Periodic precipitation patterns (Liesegang bands). (a) Ring-shaped bands of $Ag_2Cr_2O_7$ precipitate in a thin layer of gelatin gel. Initial reactants: $AgNO_3$ and $K_2Cr_2O_7$. Diameter of pattern: 3.5 cm. (b) Helicoidal band of PbI_2 precipitate in a tubular container with agar gel. Initial reactants: KI and $Pb(NO_3)_2$. Inner tube diameter: 13 mm.

tion process does not occur homogeneously in the volume. Instead, well-separated bands of precipitate appear parallel to the diffusion front. They develop sequentially in a period of hours to several days. Usually, a gel-forming material is added to the solutions to prevent sedimentation and convection, but structures can also be observed without a gel in very thin layers (films) or capillary tubes. Excellent reviews (Henisch, 1988) and bibliographies (Stern, 1967) exist concerning this subject.

Since the discovery of Liesegang patterns phenomenological descriptions have been reported in abundance, while the quantification of spatial concentration distributions is still quite rare. An early study yielded concentration data after performing a chemical analysis of thin cuts of the gel column in which the pattern of interest evolved. Other techniques for the investigation of local properties, such as light scattering or light transmission of a focused laser beam, were applied (Kai et al., 1982). For this purpose, a one-dimensional system was prepared in a test tube in which well-separated bands of $Mg(OH)_2$ appear sequentially. The following properties were determined locally along the tube: the moving front of a color change of a suitable indicator, corresponding to a specific value of OH^- concentration; the front of a slight increase of turbidity following a certain distance apart and caused by growth of colloidal particles; the onset of localized, pronounced changes in refractive index, indicating a focusing of material into a narrow spatial region; and finally, the formation of visible bands of $Mg(OH)_2$ at these locations. These findings provide strong evidence in favor of a mechanism for pattern formation acting after the nucleation stage of the new phase.

Such experiments and others led to a theoretical discussion concerning the validity of various physicochemical processes which have been proposed to explain band formation. Their mathematical formulation is based on a set of reaction–diffusion equations. Most frequently discussed are the traditional supersaturation theory, originally proposed by Ostwald (see Keller and Rubinow, 1982, and references therein), and the more recent hypothesis of a chemical instability due to the competitive growth of colloidal particles coupled with diffusion, first put forward by Ross and co-workers (see, e.g., Venzl and Ross, 1982). A major difference between these two approaches is that the former theory postulates pattern formation to occur at the nucleation stage, while the latter assumes uniform nucleation and subsequent patterning due to the autocatalytic nature of colloidal particle growth, meaning that larger particles grow at the expense of smaller ones.

Recent investigations indicate that a common basis may be found by combining elements of both approaches (LeVan and Ross, 1987). The model by Chernavskii et al. (1991) includes the description of a propagating supersaturation wave and the kinetics of precipitate nucleation and growth. It gives a clue for explaining not only simple sets of bands but also such complex patterns as the helicoidal precipitation band shown in Fig. 2b. The helix shape is created by the superposition of two motions: the one along the axis of the test tube with the reaction front, and the other, circular motion in the plane normal to the axis due to the asymmetrical distribution of the already formed precipitate.

III. CHEMICAL WAVES

A. Rotating Spirals

Chemical gradients traveling through a thin solution layer have been described in various reaction mixtures (Winfree, 1972; DeKepper et al., 1982; Nagy-Ungvarai et al., 1989). Quantitative studies, however, are restricted primarily to the classical Belousov–Zhabotinsky (BZ) reaction, in which malonic acid is oxidized and decarboxylyzed by bromate in the presence of a catalyst, usually the redox couple ferroin (red)/ferriin (blue). The propagation of such waves of chemical activity is one of the most striking examples of pattern formation due to the coupling of reaction and diffusion. Wave patterns may evolve in an oscillating or in a quiescent, excitable medium. Here we are concerned only with the so-called trigger waves in an excitable reaction mixture.

The most common geometries of two-dimensional wave fronts are a circle, a set of concentric circles (''target'' patterns), or rotating spirals (see Fig. 3). Circular waves are triggered by, for example, a hot platinum wire or evolve around impurities. Because of the excitability of the surrounding regions this local perturbation then travels outward, while excited solution volumes gradually return—through a refractory period of time—back to a quiescent and newly excitable state.

Spiral-shaped waves in an excitable medium can be produced by disrupting the front of an expanding circular wave (Winfree, 1972;

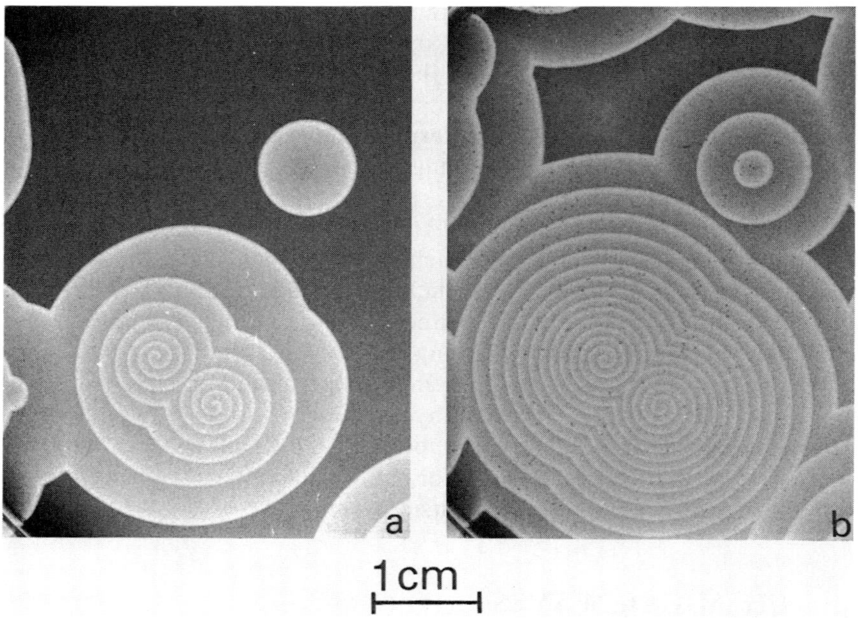

Fig. 3 Target and spiral patterns of chemical waves in a thin layer (thickness 0.7 mm) of an excitable Belousov–Zhabotinsky solution at an early (a) and a later stage (b). The reaction mixture contains malonic acid, $NaBrO_3$, H_2SO_4, and ferroin. Initial concentrations as given in Müller et al. (1985).

Müller et al., 1985). At the open wave ends spiral structures evolve with highly regular geometry (Fig. 3). To obtain quantitative information about concentration distributions, spiral waves were observed with a computerized two-dimensional spectrophotometer (Müller et al., 1986). Figure 4 gives an example for the visualization of the steep concentration gradients that are characteristic for the excited wavefronts.

The singular properties of the core of a spiral wave were analyzed in some detail. In a standard BZ system, a sequence of spiral images covering one spiral revolution were digitally overlayed. As shown in Fig. 5a, the overlay visualizes a circular area in which the maximum degree of excitation is never reached—the spiral core (outer diameter 0.7 mm). The spiral tip rotates on a circle around the rotation axis, which is located at the center of the orange "hole." This center is a singular site (diameter

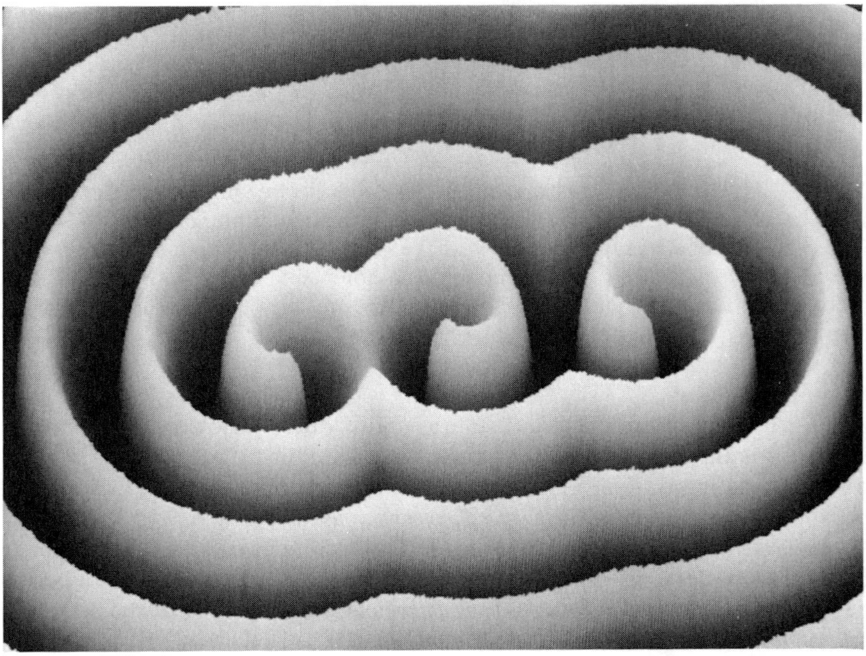

Fig. 4 Digital image of a spiral wave pattern in the BZ reaction (preparation as in Fig. 3), recorded in light absorption at 490 μm with a computerized two-dimensional spectrophotometer. The spatial distribution of the gray level represents the oxidized state of the catalyst and is enhanced in three-dimensional perspective.

(a)

Fig. 5 (a) Digital overlay of six images of a spiral wave produced in a reaction mixture as in Fig. 3, covering one spiral revolution (rotation period ≈ 18 s). (b) Three-dimensional perspective view of the cores of four counterrotating spiral waves, derived from an overlay image of a composite spiral pattern [compare picture (a)]. Each of the funnel-shaped structures merges into the rotation axis of a spiral wave. (Co-authors: T. Plesser, B. Hess)

≤ 10 μm) where the chemical state remains quasistationary, while at all points outside the chemical composition varies in time between the two extremes of maximum and minimum excitation. In a three-dimensional perspective representation a spiral overlay as in picture (a) results in a "tornado"-like structure of the core. This is illustrated in Fig. 5b for a composite pattern of four counterrotating spirals. Inside each of the core regions the maximum of excitation decreases rapidly toward the singular site at the rotation center.

Until recently, most studies were devoted to the analysis of spiral waves which maintain a regular Archimedean shape during rotation. In actual works, some emphasis is also given to the analysis of spirals with non-Archimedean shape forming in systems with low excitability (Plesser et al., 1990; Skinner and Swinney, 1991). These evolve readily in solu-

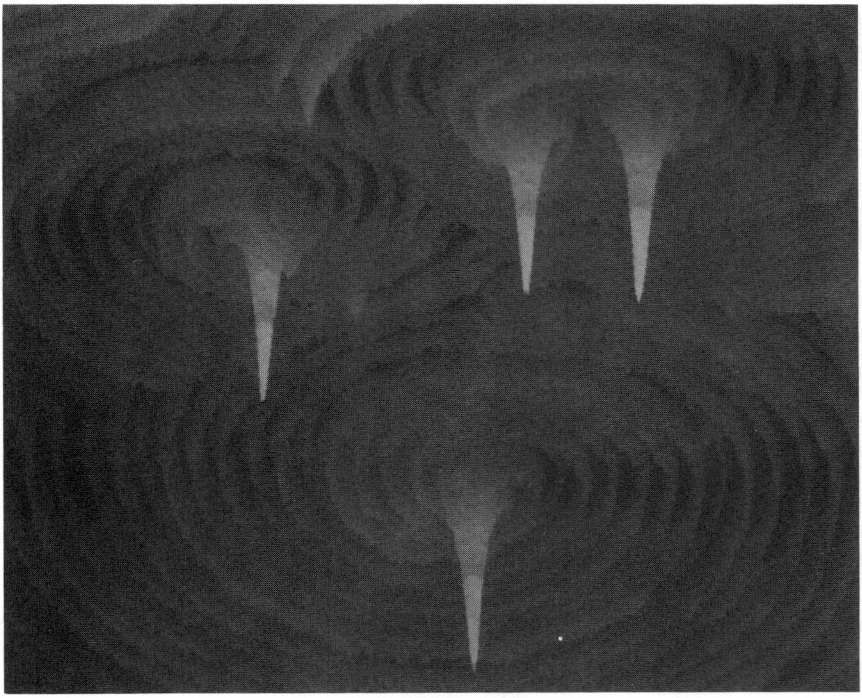

(b)

tions with a sulfuric acid concentration lower than that frequently used ($\approx$0.35 *M*, e.g., in Fig. 3). An example is shown in Fig. 6a (initially 0.19 *M* sulfuric acid). In an overlay of a sequence of spectrophotometric images a short interval of the motion of the spiral tip can be traced (Fig. 6b). In this case it does not perform a steady rotation around a spatially stable center, but it follows a compound motion that can be described as a repeated alternation between an almost straight path and a loop around a temporary rotation center (see the dotted curve in Fig. 6a).

Thus the overall trace of the spiral tip resembles a prolate epicycloid. Here the remarkable stability of the spiral center, as observed in Archimedean-shaped spirals (Figs. 3 to 5), is only temporarily maintained. After one turn around such a center the spiral tip moves on to an adjacent one, and for each turn this dynamic behavior produces a singular site in the solution layer, at which one cycle of the periodic redox transition is omitted.

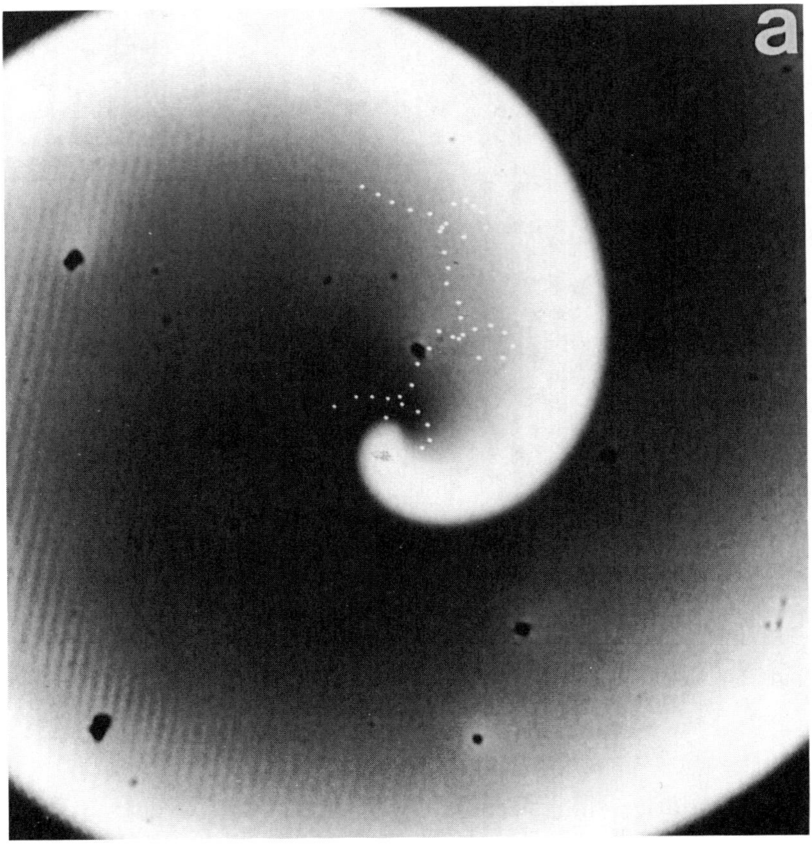

Fig. 6 Under low-excitability conditions (decrease of sulfuric acid concentration) the spiral tip follows a compound motion. (a) Snapshot of a spiral wave, bright dots indicating the path of the spiral tip in 6-s steps. (b) Overlay of consecutive spiral images (6-s time intervals) showing one loop of the compound motion around a temporary spiral center inside the dark spot. Observation area: 11.4 by 11.4 mm^2. (From Plesser et al., 1990.)

B. Complex Front Geometries

With recent advances in the use of gels and more sophisticated reactor design it has become feasible to produce quite easily more complex wave patterns under externally imposed constraints (e.g., gradients of concentration or electric fields). Without going further into this newly developing direction, I will illustrate only two phenomena: the spontaneous distortion of a wavefront (Fig. 7a) and the formation of a helix in a capillary

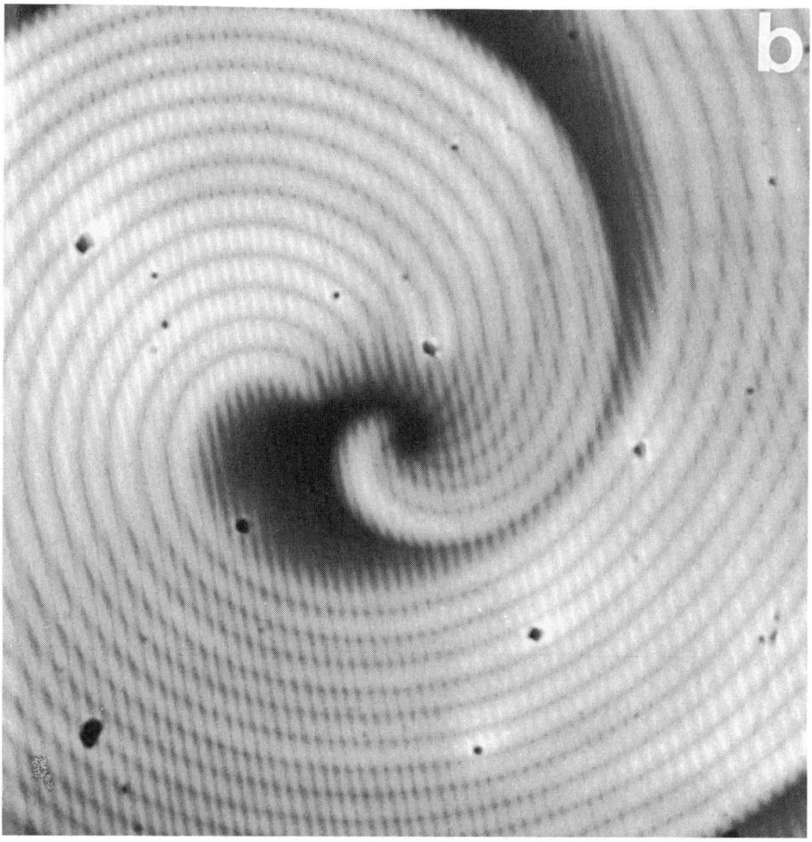

tube (Fig. 7b). Both are observed in a smooth gradient of excitability (Yamaguchi and Müller, 1991).

The distortion shown in Fig. 7a is followed by a fragmentation into three front pieces—thus a novel structure appears which evolves perpendicularly to the direction of the macroscopic gradient. Other phenomena occurring under anisotropic initial conditions are, for instance, the drift of spiral waves, interaction of adjacent spiral cores, or the formation of staircase-shaped front geometries.

In three dimensions various topologically interesting structures of scroll waves can be observed (Winfree, 1987; Pertsov et al., 1990). There is a prediction that a twisted scroll wave can be stable only under certain anisotropic conditions (Panfilov et al., 1984). The helicoidal wave in Fig. 7b, observed in a smooth concentration gradient, appears to verify this prediction. The gradient decreases the rotation speed of a scroll wave

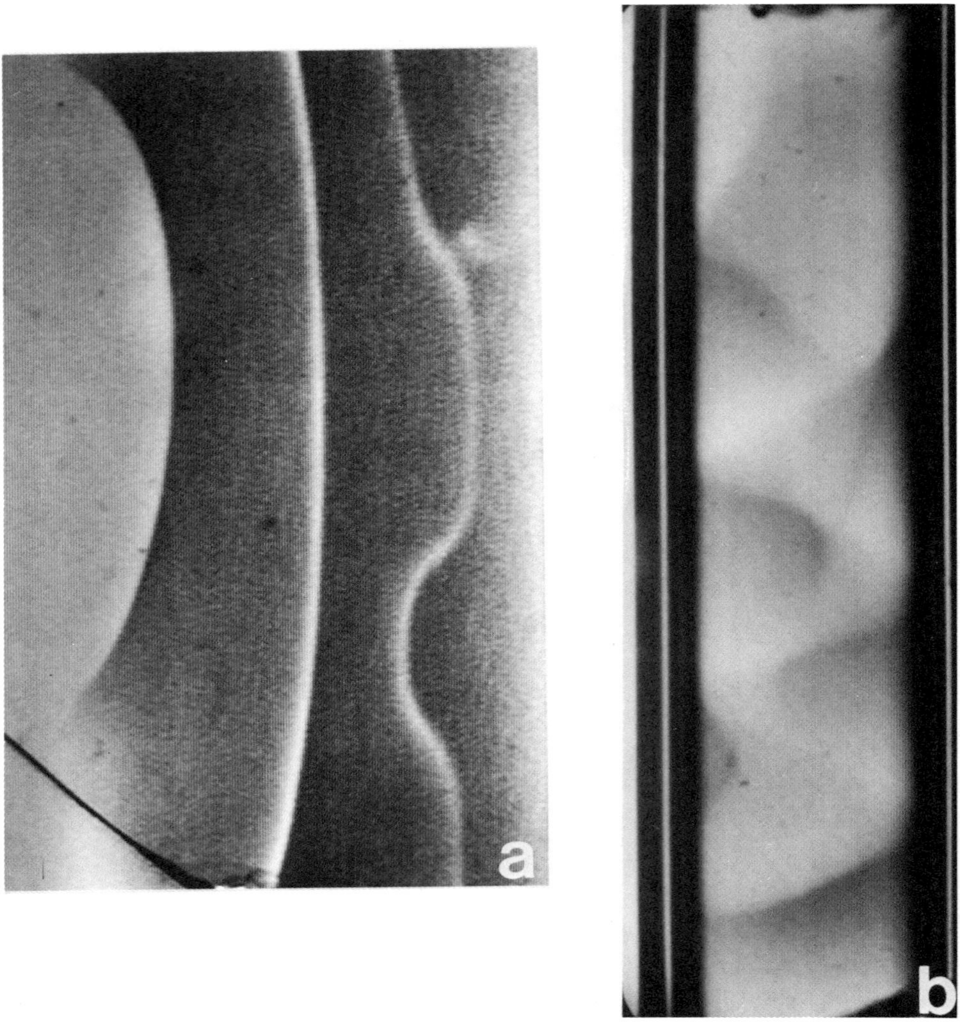

Fig. 7 (a) Top view of a ferroin-immobilized two-dimensional silica gel. Under externally imposed concentration gradients the stability of a propagating wavefront in the BZ solution is broken. From left to right: malonic acid decreases, H_2SO_4 increases. (b) Chemical helix observed in silica gel in a capillary tube (diameter 1.1 mm). The tube was immersed in reacting solution containing 0.33 *M* of malonic acid, $NaBrO_3$, and H_2SO_4. (From Yamaguchi and Müller, 1991.)

with increasing distance from the open end and should cause a winding into a twisted structure. This structure is currently being analyzed in terms of its precise topological nature.

C. Reaction–Diffusion Modeling

In the recent literature an increasing number of theoretical approaches is presented to explain patterns in excitable media, especially the formation of spiral waves. In many cases, simplified model schemes of excitable kinetics are coupled with the diffusion of relevant system variables (e.g., by Meron and Pelcé, 1988, and Barkley et al., 1990). More abstract models, for instance on the basis of cellular automata (Markus and Hess, 1990), are also discussed. There are only a few models which start from the real reaction kinetics and diffusion properties of the investigated system and thus lead to predictions that can be verified directly by experimentation. For the BZ spirals a reaction–diffusion approach based on the Oregonator model was proposed by Keener and Tyson (1986). After supplementing the resulting rate equations by diffusion terms, they derived the following reduced, dimensionless reaction–diffusion equations for a fast activator and a slow recovery variable:

$$\begin{aligned} \varepsilon \frac{\partial u}{\partial t} &= \varepsilon^2 \Delta u + f(u,v) \\ \frac{\partial v}{\partial t} &= \varepsilon \Delta v + g(u,v) \end{aligned} \tag{1}$$

where u is proportional to the concentration of the autocatalytic variable ($HBrO_2$), v is proportional to the oxidized form of the catalyst ferroin, Δ is the Laplacian operator, $f(u,v)$ and $g(u,v)$ describe the kinetics of the reaction, and ε is a time scale factor (the spatial scale has been chosen such that the diffusion coefficients of u and v become identical to ε).

The functions f and g, which include nonlinear terms, have such a form that the main properties of an excitable system emerge: a resting stationary state, short and fast excursion to large u after a sufficiently strong perturbation, and recovery, through a refractory period, characterized by slow variation of v.

Because $\varepsilon \ll 1$, the system can be treated by singular perturbation arguments. After changing variables into a moving coordinate system, one obtains the following two results for the wave solutions of the equation system. The first is the relationship between the normal velocity N and the curvature K of a wavefront (cf. Zykov and Morozova, 1979):

$$N = c - DK \tag{2}$$

Here c denotes the velocity of plane waves and D represents the diffusion coefficient of the autocatalytic species. The relationship predicts a minimal radius

$$R_{\text{crit}} = \frac{D}{c} \tag{3}$$

below which propagation of circular waves will not take place. Furthermore, treatment of equation (1) results in a dispersion relation for wave propagation

$$c = \sigma(\omega) \tag{4}$$

as described in detail by Keener and Tyson (1986). This relation expresses the dependence of the speed of plane waves c on the frequency ω of the wave trains. It shows a decrease of the velocity of the wave propagation when the frequency of wave initiation is increased. The validity of both relationships has been proven experimentally (Pagola et al., 1988; Foerster et al., 1989).

In the framework of their model, Keener and Tyson (1986) predict spiral solutions, the shape of which is close to that of an Archimedean spiral. Since the curvature of a spiral increases considerably toward its rotation center, the size of the critical radius (about 20 μm in a standard BZ system) is of particular importance for their arguments. Critical curvature is assumed to govern the motion of the spiral tip around the singular core region into which it cannot penetrate. There are other models in the recent literature, which may not follow precisely the same picture, but nevertheless criticality of curvature effects also constitutes a major argument (Davydov et al., 1989).

During compound tip motion (Fig. 6) the curvature of the tip region increases while it performs a loop and decreases when it passes to the next loop. The distance to the relaxing back of the precursor wave changes with the same rhythm. Therefore, the actual dynamic behavior can be regarded as a delicate balance between the influence of curvature and dispersion on the local propagation velocity.

IV. WAVES AND CHEMOTAXIS IN CELL AGGREGATION

The aggregation patterns formed in the excitable system *Dictyostelium discoideum* have been investigated over a long time as a model case for biological self-organization (Gerisch, 1971; Klein et al., 1988). There is a remarkable similarity to the trigger waves found in the BZ reaction. The aggregation patterns are usually detected by dark-field illumination of macroscopic areas, which takes advantage of the difference between the light-scattering behavior of moving, slightly elongated cells and that of resting circular cells. Thus the regions of moving cells appear as bright bands (circles or spirals, Fig. 8a). By combining this detection optics with digital recording of video images and computerized evaluation techniques, it was shown that the spiral shape is close to that of an Archimedean spiral. An overlay of image sequences revealed that there exists a core region in spiral patterns of approximately 300 μm radius around which the dark-field waves rotate (Fig. 8b). The relationship between normal propagation velocity and local curvature of a wavefront fulfills a linear law, as in the BZ reaction (Foerster et al., 1990).

A reaction–diffusion model for spirals was proposed on the basis of the Martiel–Goldbeter kinetic scheme (Tyson et al., 1989). Its structure is closely analogous to equations (1). In this system the two variables are the autocatalytically produced signal transmitter molecule cAMP and the membrane receptor to which the secreted cAMP is bound, thus stimulating further cAMP production. While in the BZ system both species are allowed to diffuse, in the cell aggregation a diffusion term is implemented only for extracellular cAMP.

For the slime mold system a direct comparison between theory and experiment is complicated because the detection of cAMP is very difficult and available only for a few frozen-in snapshots. The commonly used dark-field data supply only indirect evidence for comparison with theory.

To improve the experimental knowledge about the dynamic behavior, a newly developed method for measuring chemotactic velocity was used (Miike et al., 1986). The method is based on mutual correlation analysis of temporal sequences of microscopic digital images, recorded during the aggregation phases of the cell population. It uses the fact that cell motion causes a temporal change of the gray level at each pixel of a digital image. The local velocity at a certain pixel site is estimated by computing the mutual correlation between the temporal brightness change of a selected centered pixel and that of neighboring pixels. From this, the direction and speed of motion are determined. In this way, periodic cell motion in a spiral arm due to the repetitive passage of cAMP pulses could already be quantified in some detail (Steinbock et al., 1991).

The method is currently applied to analyze the properties of the core of spiral aggregation. The cell distribution in this region is shown in the microscopic photograph of Fig. 9a. In the center part of this rather complex cell distribution there is a more or less circular region of lower cell density. A quantitative study yields that velocity vectors are distributed

(a)

Fig. 8 (a) Spirals in the *Dictyostelium discoideum* system observed in dark-field illumination. Bright areas indicate pronounced cell motion toward the aggregation center, which is the core region of the spiral. Image area 19 by 19 mm^2. (b) Overlay of eight contour maps of bright areas, taken at 6-s intervals, showing a small spot (diameter $\approx$ 0.5 mm) where no pronounced cell motion takes place. (From Foerster et al., 1990.)

like a vortex in this core region (Fig. 9b). A temporal sequence of such vector fields reveals that a wave of excitation circulates around the low-density region. Cell motion is directed roughly tangentially to the boundary of a core of 300 μm diameter. Cells inside this boundary tend to move outward, thus depleting the core region.

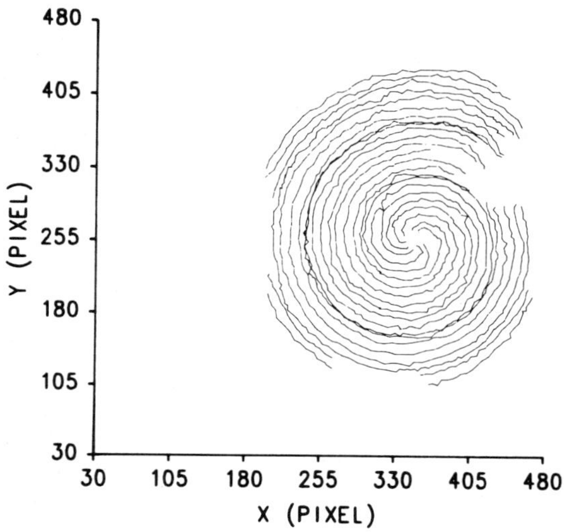

(b)

V. INFLUENCE OF CONVECTION

Beyond the coupling of reaction with diffusion there are frequently additional transport processes involved in the formation of chemical spatial patterns. Chemotaxis was just mentioned as an example. In many observations hydrodynamic convection of the liquid system has to be taken into account, driven either by surface tension or density gradients caused by changes in temperature or chemical composition. This opens a wide field of complex phenomena due to reaction–convection coupling which calls for enhanced future activities to improve our mechanistic understanding.

One particularly interesting consequence of reaction–convection coupling is the turbulent decomposition of wavefronts (Fig. 10). It was observed that under suitable initial conditions a periodic train of chemical waves induces, after a sufficiently long induction time, convective flow which changes its direction periodically, following the imposed rhythm of the traveling fronts (Miike et al., 1988). This leads to an alternation between a sharpening and a flattening of wave profiles (Fig. 10a,b). If the amplitude of flow oscillation exceeds the velocity of wave propagation, a strong distortion and finally a complete disruption of the fronts follow (Fig. 10c,d). Thus the periodicity of the original pattern causes—by liquid motion—transition to an aperiodic state.

A pattern of biological interest was recently discovered in a suspension of oscillating microtubuli (Mandelkov et al., 1989; see Fig. 11).

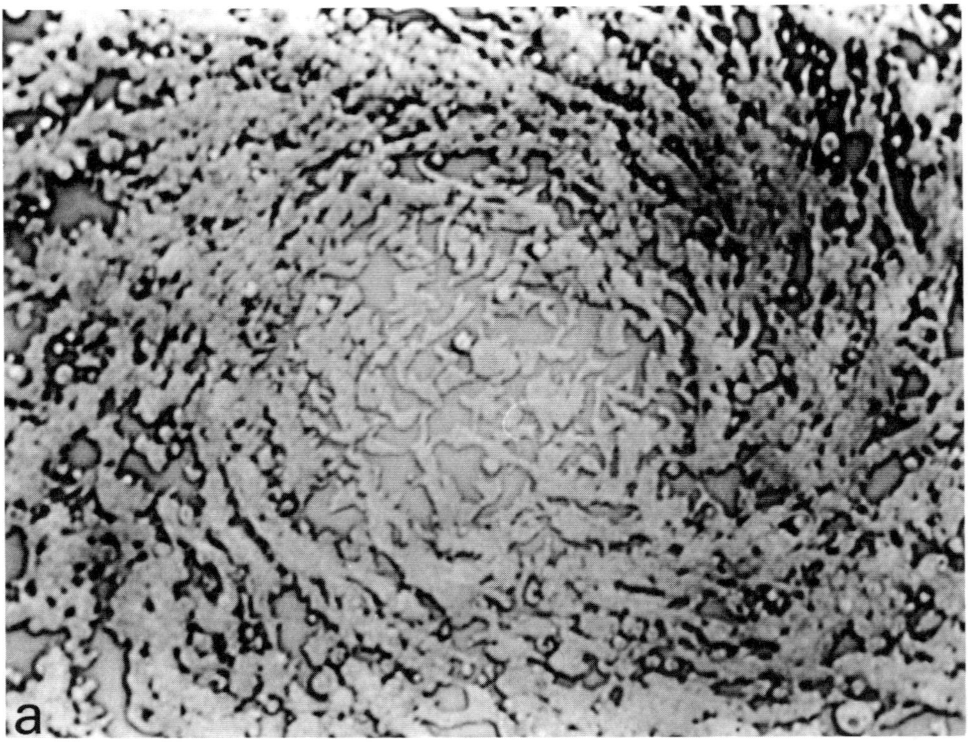

Fig. 9 (a) Microscopic photograph in transmitted light of a spiral vortex of aggregating *Dictyostelium* amoebae. Image area: 0.39 by 0.32 mm^2. (b) Cell velocity field in the same area, obtained by mutual correlation analysis of pixel brightness. Each of the 10 × 10 vectors represents averaged magnitude and direction of cell motion. (Courtesy of Steinbock and Müller, unpublished data.)

Waves of microtubule assembly travel on the background of a cellular network-type pattern. It probably represents a set of convective cells created in a layer by an instability of the liquid–gas interface of the suspension, similar to repetitively appearing structures observed earlier in oscillating glycolysis (Boiteux and Hess, 1980).

VI. A BRIEF OUTLOOK

Various examples of spatial patterns in chemical and biochemical reactions were presented, which demonstrate the rich and complex, regular or irregular features of spatiotemporal ordering processes acting under far-

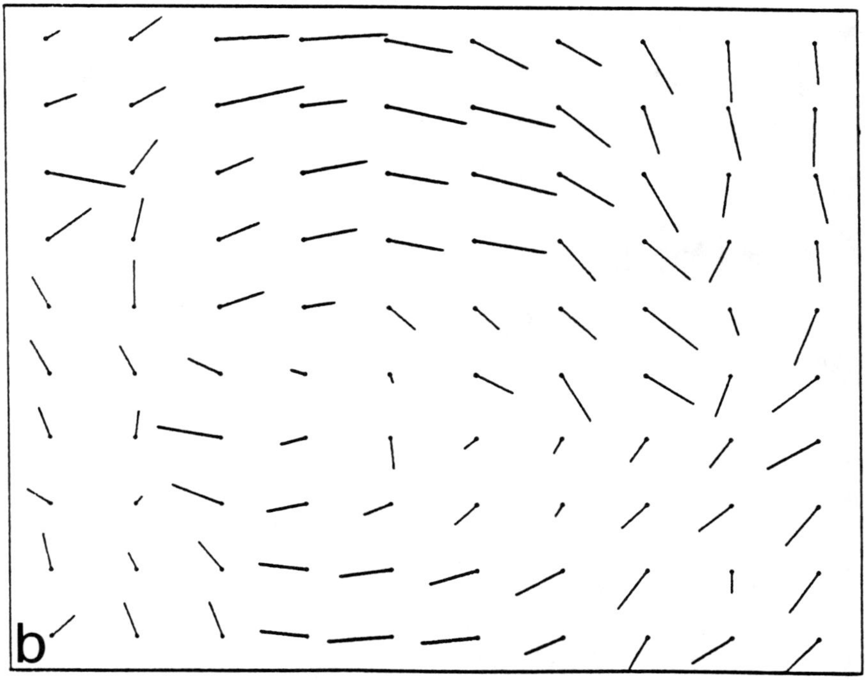

from-equilibrium conditions. The investigations described here show that we now have tools at hand which allow for a comprehensive quantitative description of the evolution of such pattern-forming systems. The recent improvement of experimental approaches contributes significantly to the elucidation of the underlying mechanisms on theoretical grounds. Much work has been done already on the modeling of spatial patterns on the basis of reaction–diffusion coupling, but further work is needed to develop extended models that take into account, for instance, chemotaxis or convective liquid motion.

In this field of self-organization it is useful to emphasize the structural analogies which are found in systems of apparently quite different nature. This is of particular interest for our understanding of spatial order in living organisms. Periodic spatial patterns in chemical reactions may be taken as useful models for morphogenetic development. In excitable media it is noteworthy that beyond the close analogy between chemical spirals and spirals in cell aggregation, the occurrence of rotating vortices has been reported in a number of other biological systems, recently, for example, in isolated ventricular muscle (Davidenko et al., 1991) and in

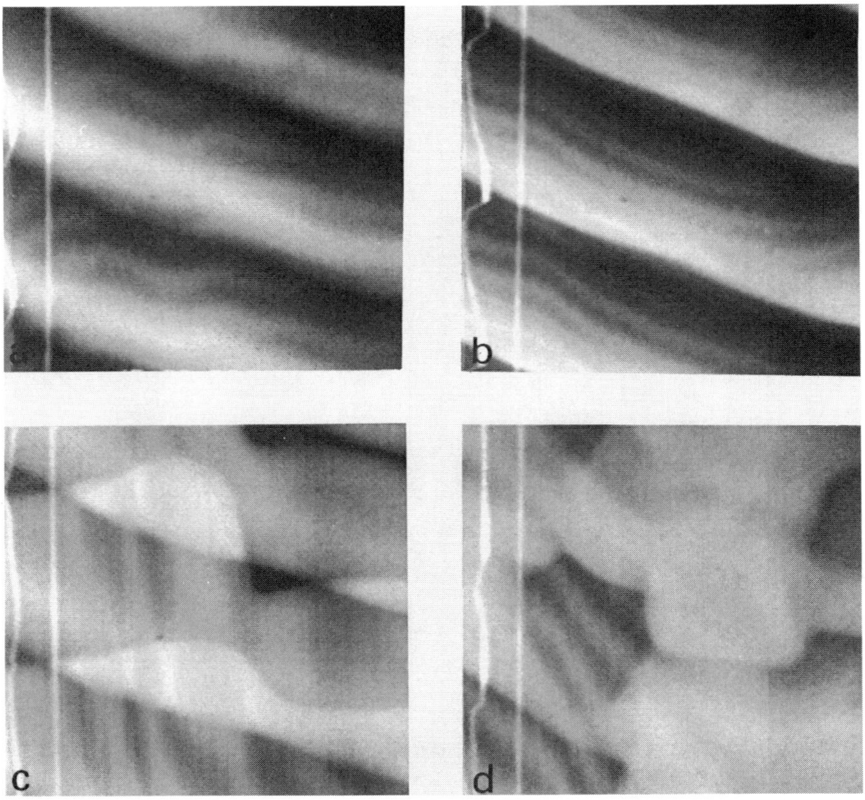

Fig. 10 Influence of hydrodynamic flow on chemical wave trains: chemically induced flow oscillations lead to flattening (a) and sharpening (b) of wave profiles. High amplitudes of flow oscillations result in the breaking of front geometry (c), and finally, in a turbulent decomposition of the wave pattern (d). (From Miike et al., 1988.)

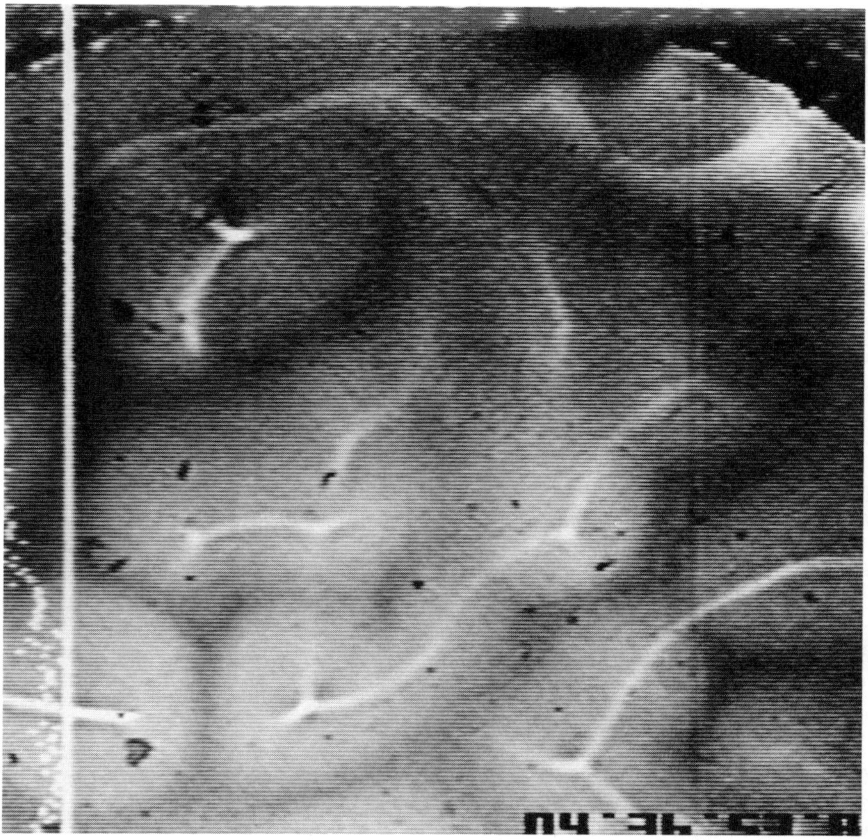

Fig. 11 Network of polygons observed in a thin layer of a microtubule system undergoing periodical oscillations in the degree of assembly. Observation area: 5 by 5 mm^2. (From Mandelkow et al., 1989.)

patterns of calcium activity in *Xenopus* oocytes (Lechleiter et al., 1991). Model systems such as the BZ reaction will certainly play an important role in future research on excitable systems, in general.

REFERENCES

Avnir, D. (1989). *J. Chem. Ed.*, *66*: 211.

Barkley, D., Kness, M., and Tuckerman, L. S. (1990). *Phys. Rev. A*, *42*: 2489.

Belousov, B. P. (1959). *Collection of Abstracts on Radiation Medicine*, Medgiz, Moscow, p. 145.

Boiteux, A., and Hess, B. (1980). *Ber. Bunsen-Ges. Phys. Chem.*, *84*: 392.

Busse, H. G. (1969). *J. Phys. Chem.*, *73*: 750.

Castets, V., Dulos, E., Boissonade, J., and DeKepper, P. (1990). *Phys. Rev. Lett.*, *64*: 2953.

Chernavskii, D. S., Polezhaev, A. A., and Müller, S. C. (1991). *Physica D*, *54*: 160.

Davidenko, J. M., Kent, P., and Jalife, J. (1991). *Physica D*, *49*: 182.

Davydov, V. A., Zykov, V. S., and Mikhailov, A. S. (1989). In *Nonlinear Waves in Active Media* (J. Engelbrecht, ed.), Springer-Verlag, Berlin.

DeKepper, P., Epstein, I. R., Kustin, K., and Orbán, M. (1982). *J. Phys. Chem.*, *86*: 170.

Epstein, I. R. (1984). *J. Phys. Chem.*, *88*: 187.

Field, R., and Burger, M. (eds.) (1985). *Oscillations and Traveling Waves in Chemical Systems*, Wiley, New York.

Foerster, P., Müller, S. C., and Hess, B. (1989). *Proc. Natl. Acad. Sci. USA*, *86*: 6831.

Foerster, P., Müller, S. C., and Hess, B. (1990). *Development*, *109*: 11.

Gerisch, G. (1971). *Naturwissenschaften*, *58*: 430.

Haken, H. (1983). *Synergetics*, 3rd ed., Springer-Verlag, Berlin.

Hedges, E. S., and Myers, J. E. (1926). *The Problem of Physico-Chemical Periodicity*, Arnold, London.

Henisch, H. K. (1988). *Crystals in Gels and Liesegang Rings*, Cambridge University Press, Cambridge.

Kai, S., Müller, S. C., and Ross, J. (1982). *J. Chem. Phys.*, *76*: 1392.

Keener, J. P., and Tyson, J. J. (1986). *Physica D*, *21*: 307.

Keller, J. B., and Rubinow, S. I. (1982). *J. Chem. Phys.*, *74*: 5000.

Klein, P. S., Sun, T. J., Saxe, Ch. L., III, Kimmel, A. R., Johnson, R. L., Johnson, R. L., and Devreotes, P. N. (1988). *Science*, *241*: 1467.

Lechleiter, J., Girard, S., Peralta, E., and Clapham, D. (1991). *Science*, *252*: 123.

LeVan, M. E., and Ross, J. (1987). *J. Phys. Chem.*, *91*: 6300.

Liesegang, E. S. (1896). *Phot. Archiv.*, *21*: 321.

Lippmann, G. (1873). *Ann. Phys. 2nd Ser.*, *149*: 544.

Mandelkow, E., Mandelkow, E.-M., Hotani, H., Hess, B., and Müller, S. C. (1989). *Science*, *246*: 1291.

Markus, M., and Hess, B. (1990). *Nature*, *347*: 56.
Meron, E., and Pelcé, P. (1988). *Phys. Rev. Lett.*, *60*: 1880.
Miike, H., Kurihara, Y., Hashimoto, H., and Koga, K. (1986). *Trans. IEICE Japan E*, *69*: 877.
Miike, H., Müller, S. C., and Hess, B. (1988). *Phys. Rev. Lett.*, *61*: 2109.
Müller, S. C., Plesser, Th., and Hess, B. (1985). *Science*, *230*: 661.
Müller, S. C., Plesser, Th., and Hess, B. (1986). *Naturwissenschaften*, *7*: 165.
Nagy-Ungvarai, Zs., Müller, S. C., and Hess, B. (1989). *Chem. Phys. Lett.*, *156*: 433.
Nicolis, G., and Prigogine, I. (1977). *Self-Organization in Nonequilibrium Systems*, Wiley, New York.
Noszticzius, Z., Horsthemke, W., McCormick, W. D., Swinney, H. L., and Tam, W. Y. (1987). *Nature*, *329*: 619.
Pagola, A., Vidal, C., and Ross, J. (1988). *J. Phys. Chem.*, *92*: 163.
Panfilov, A. V., Rudenko, A. N., and Pertsov, A. M. (1984). In *Self-Organization: Autowaves and Structures Far from Equilibrium* (V. I. Krinsky, ed.), Springer-Verlag, Heidelberg, p. 103.
Pertsov, A. M., Aliev, R. R., and Krinsky, V. I. (1990). *Nature*, *345*: 419.
Plesser, Th., Müller, S. C., and Hess, B. (1990). *J. Phys. Chem.*, *94*: 7501.
Runge, F. F. (1855). *Der Bildungstrieb der Stoffe veranschaulicht in selbständig gewachsenen Bildern*, Selbstverlag, Oranienburg, Germany.
Skinner, G. S., and Swinney, H. L. (1991). *Physica D*, *48*: 1.
Steinbock, O., Hashimoto, H., and Müller, S. C. (1991). *Physica D*, *49*: 233.
Stern, K. H. (1967). *A Bibliography of Liesegang Rings*, 2nd ed., U.S. Government Printing Office, Washington, D.C.
Tyson, J. J., Alexander, K. A., Manoranjan, V. S., and Murray, J. D. (1989). *Physica D*, *34*: 193.
Venzl, G., and Ross, J. (1982). *J. Chem. Phys.*, *77*: 1308.
Winfree, A. T. (1972). *Science*, *175*: 634.
Winfree, A. T. (1987). *When Time Breaks Down*, Princeton University Press, Princeton, N.J.
Yamaguchi, T., and Müller, S. C. (1991). *Physica D*, *49*: 40.
Zaikin, N., and Zhabotinskii, A. N. (1970). *Nature*, *225*: 535.
Zhabotinskii, A. M. (1964). *Biofizika*, *9*: 306.
Zykov, V. S., and Morozova, G. L. (1979). *Biofizika*, *23*: 717.

5
Stationary and Dynamic Patterns of Current Density in Gas Discharge Systems

Heinz Willebrand, Franz-Josef Niedernostheide, Renate Dohmen, and Hans-Georg Purwins *Institute of Applied Physics, Münster, Germany*

I. INTRODUCTION

In many branches of biology, chemistry, and physics nonlinear effects are responsible for the appearance of spatially inhomogeneous (e.g., periodic or solitary) structures. Well-known examples are biological morphogenesis (Gierer and Meinhard, 1972; Murray, 1989); the Belousov–Zhabotinski reaction (Kuramoto, 1984), in which the competition of two chemical components leads to dynamical patterns; and current filaments or electric field domains in semiconductors (Kerner and Sinkevich, 1982; Jäger et al., 1986; Peinke et al., 1989). Also long known are electrical and thermal effects in gas discharges, which cause patterns consisting of inhomogeneous light-intensity distributions. Such patterns occur, for example, as standing or moving striations and as alternating bright and dark regions perpendicular to the axis of the discharge column (Nedopasov, 1968; van den Heuvel, 1988). Solitary current filaments were observed in different systems. Müller (1988) investigated a dc gas discharge between a hollow steel cathode and a spherical steel anode and found luminous spots appearing at the surface of the anode. In an ac-driven gas discharge system consisting of two dielectrically covered electrodes, Boyers and Tiller (1973, 1982) observed filamentary and striped light distributions comparable to magnetic bubble domains. Similar structures were found by Bragin et al. (1984).

In many cases theoretical considerations of the underlying microscopic processes lead to a mathematical description in terms of reaction–

diffusion equations. Such approaches were applied, for example, by Kerner and Osipov (1989) to describe inhomogeneous current and temperature distributions in semiconductors and gas discharges. A more phenomenological model based on the idea of a system containing competing components has been derived by Radehaus and Purwins to explain pattern formation in two-layer systems with certain nonlinear electrical properties (Radehaus et al., 1987; Purwins et al., 1987; Purwins and Radehaus, 1988; Willebrand et al., 1990, 1991).

In the following we report on experimental results obtained with two different gas discharge systems. First, results concerning stationary inhomogeneous structures of the light intensity in the positive column of a dc- and an ac-driven system are presented. The dc gas discharge system consists of a high ohmic semiconductor and a metal electrode between which the discharge takes place. The ac gas discharge system is similar to the arrangement investigated by Boyers and Tiller. In a subsequent paragraph the dynamical behavior of these systems is described. Regular and irregular oscillations of the total current can be related directly to the behavior of filaments (e.g., to splitting processes). A phenomenological model to explain the qualitative behavior of the dc gas discharge system is presented in a final section.

II. EXPERIMENTAL INVESTIGATIONS CONCERNING THE STATIONARY BEHAVIOR OF SPECIAL GAS DISCHARGE SYSTEMS

A. Experimental Setup

The experimental setup of the dc-driven gas discharge system is shown in Fig. 1. One of the electrodes consists of copper, the other of doped silicon material of p- or n-type and a resistivity on the order of 1 kΩ·cm. The semiconductor is ohmically contacted by a thin aluminum layer. The gas itself is a highly nonlinear material that has an S-shaped characteristic. The length of the discharge gap is denoted by l, its width by a, the lateral extension of the semiconductor electrode by b, and the thickness of the gap by d (Fig. 1a,b). Typical dimensions are l = 40 mm, a = 3 mm, b = 10 mm, and d = 0.3 mm. To ensure discharge only between the faceplates of the electrodes, the entire electrode system is mounted in two glass plates. The width of the gas discharge gap can be varied by moving the copper electrode. A water-cooling system serves to cool the arrangement and to maintain an approximately constant temperature. The experimental investigations take place in a vacuum chamber that contains a gas

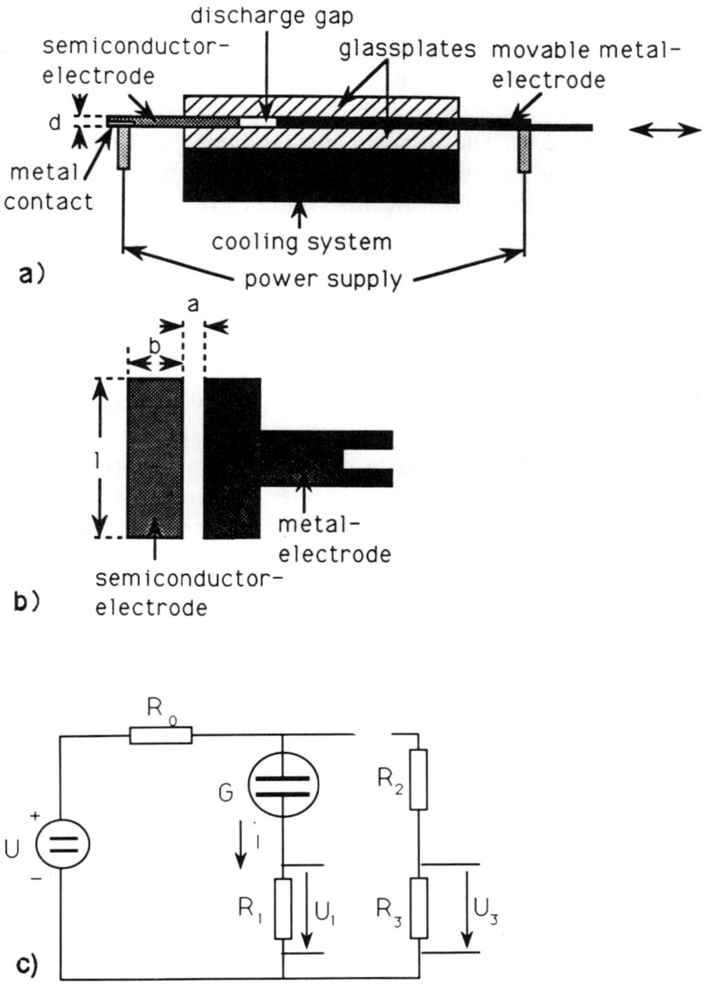

Fig. 1 Experimental setup of the dc gas discharge system: (a) schematic cross section; (b) top view; (c) measurement circuit. $R_0 = 0$ to 160 kΩ; $R_1 = 50\ \Omega$; $R_2 = 1\ \mathrm{M}\Omega$; $R_3 = 1\ \mathrm{k}\Omega$; $I = U_1/R_1$; $U_V = U_3(R_2 + R_3)/(R_3 - R_1)I$.

mixture consisting of an inert gas (He or Ar) and several volume percent points of air. The gas pressure is varied between 10 and 250 hPa. The electrodes of the gas discharge system G are connected to a high-voltage dc power supply via a load resistor R_0 which serves to limit the current (Fig. 1c).

The ac gas discharge system consists of two dielectrically covered electrodes, one of which is of copper, while the other is made of a transparent indium tin oxide (ITO) covered glass plate. The dielectric layers are formed by glass plates with a thickness of 50 to 170 μm and a dielectric constant $\varepsilon_r = 2.4$. The experiment has been performed with helium, argon, nitrogen, or air, respectively, at pressures between 50 and 1000 hPa; the experiments discussed in this chapter are carried out with helium, but the qualitative behavior for all gases investigated is the same. The thickness of the gas gap typically comprises the range from 50 to 600 μm. The schematical setup and the electric circuit are shown in Fig. 2.

B. Stationary Patterns in the DC-Driven System

The global $I(U_V)$ characteristic of the gas discharge device and the relative radiation density distribution of the gas discharge slit are chosen to characterize the stationary current density patterns. Here I represents the total current through and U_V the voltage drop over the entire device (Fig. 1c). For measurement of the relative radiation density distribution a vidicon camera is used which takes up the entire discharge gap. By electronic

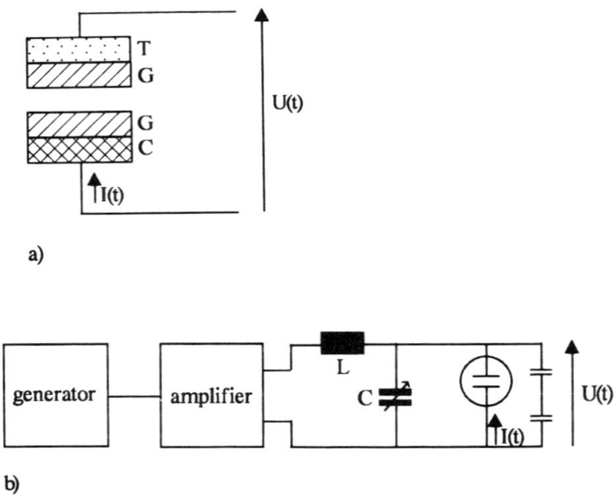

Fig. 2 (a) Schematical setup of the ac-driven system. The gas discharge takes place between the two dielectric glass plates (G) connected to a copper (C) and a transparent ITO covered electrode (T), respectively. (b) Electrical circuit of the ac-driven system, $L = 0.1$ to 2 mH, $C = 50$ to 1000 nF; the voltage $U(t) = U_{SS} \sin 2\pi ft$ is measured by the capacitive divider, the current $I(t)$ by a current probe. The capacity C is adjusted such that the circuit operates in resonance.

means one video line parallel to the electrodes is selected. The amplitude of this video line is an appropriate measure for the radiation density distribution. The behavior of the system is investigated in dependence on various system parameters, such as the gas pressure p, the applied voltage U, and the width of the discharge gap. In the experiments described in the sequel the polarity of the electrode system is chosen such that the copper electrode works as the cathode and the semiconductor electrode as the anode.

By increasing the applied voltage U a current density filament appears when a certain breakdown voltage is reached. The breakdown of the voltage across the discharge gap is due to the Townsend mechanism, which means that electron generation processes at the cathode are responsible for maintenance of the discharge. The most important electron generation or secondary processes are photoemission and electron emission, due to the bombardment of positive ions on the surface of the cathode. In addition to the generation of charge carriers, recombination takes place. The recombination of charge carriers can in general be observed in the volume (volume processes) or on the wall (wall processes) of the discharge tube. In the discharge cell under discussion the wall processes are at least as important as the volume processes. For a detailed description of the physics of glow discharges, see, for example, Kapzow (1955), Wiesemann (1976), and von Engel (1965).

In the environment of a current filament a large voltage drop at the semiconductor electrode appears, and consequently the voltage drop over the discharge gap is reduced, resulting in a lateral inhomogeneous voltage distribution which stabilizes the filament. An enlargement of U leads first to a small broadening of the filament; then a splitting process takes place that results in the generation of a second filament. A further increase of U leads to the generation of further filaments by a splitting of existing filaments. Obviously, to allow for a larger current flow, the discharge system prefers to generate new filaments instead of changing the size of existing current paths. In this sense the filaments may be said to show solitary properties. Figure 3a shows photographs of the discharge slit depicting the successive appearance of current filaments and revealing their solitary character by constant width and radiation intensity. We point at the typical funnel-shaped positive column of the discharge and the negative glow near the cathode which is placed at the bottom of each picture.

The $I(U_V)$ characteristic of the gas discharge device corresponding to this scenario is represented in Fig. 3b. The appearance of the first filament at the breakdown voltage is accompanied by a discontinuity in the characteristic. This typical behavior is determined by the gas properties

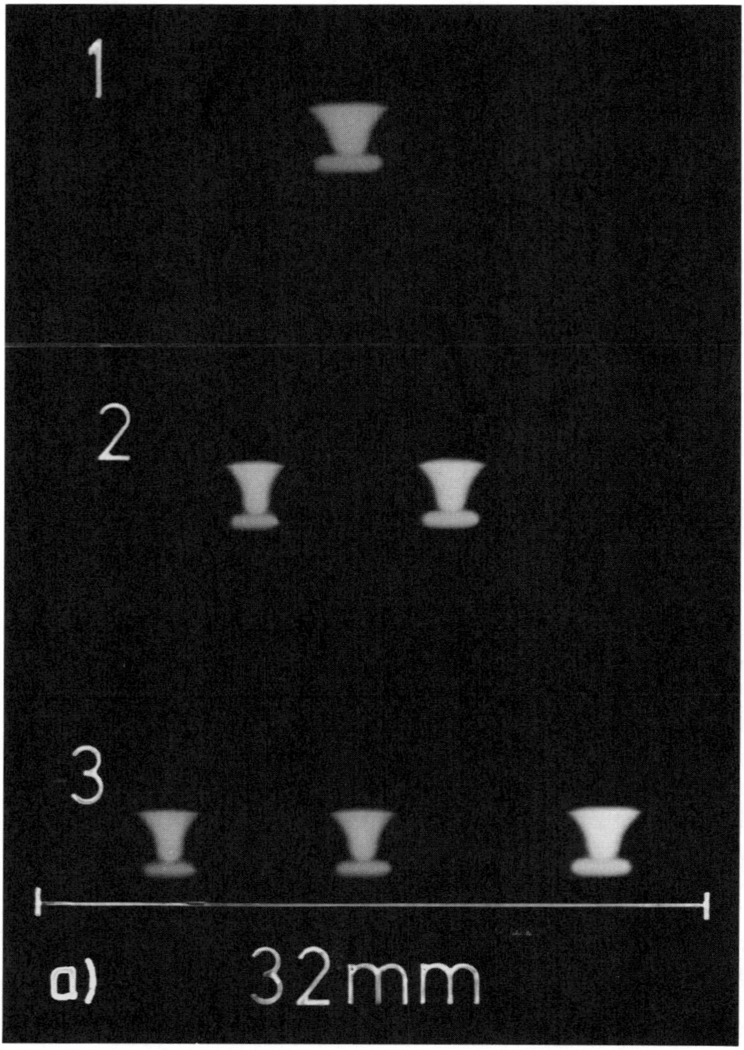

(a)

Fig. 3 (a) Stationary filament patterns when increasing the applied voltage U. (b) Corresponding $I(U_V)$ characteristic; the numbers correspond to situations labeled by the numbers in part (a). The parameters are $R_0 = 164$ kΩ, $\rho = 1.7$ kΩ·cm, $p = 200$ hPa, $l = 32$ mm, $b = 10$ mm, $a = 0.3$ mm, $d = 3.4$ mm. The gas mixture is argon and 10 vol % of air.

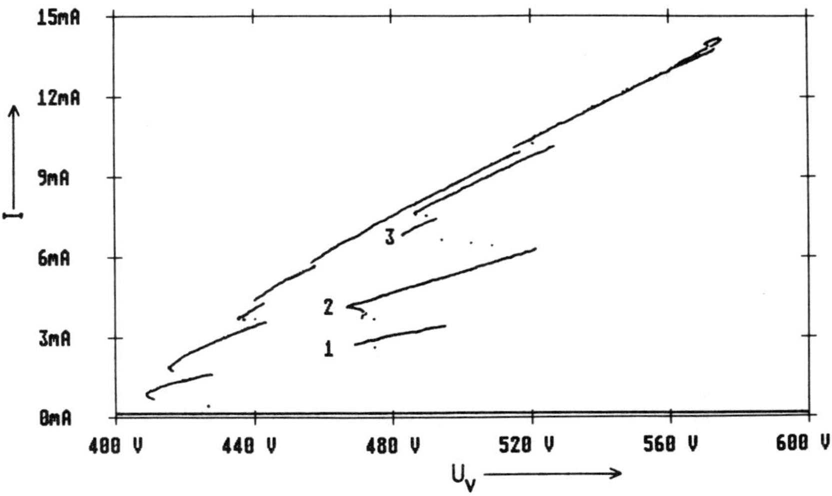

(b)

and reflects the transition from the spatially homogeneous high-ohmic state into a low-ohmic state. The generation of further filaments is based on the destabilization of existing filaments, which in general takes place at a lower voltage than the breakdown voltage and leads to the above-mentioned splitting process and the appearance of another discontinuity in the $I(U_V)$ characteristic. With decreasing U the shape of the characteristic exhibits a hysteresis. This behavior indicates the multistability of the discharge system, which is also reflected in the fact that the general shape of the characteristic can be reproduced but not its details. This nonreproducibility is also valid for the current filament patterns; in particular, the position in the discharge gap at a certain value of the externally applied voltage must not be exactly the same when repeating the experiment. Also, in the descending branch of the characteristic discontinuities show up that are associated with the successive disappearance of filaments.

Carrying out the experiment with a smaller load resistor R_0 the generation of even more than one filament at the first jump in the $I(U_V)$ characteristic can be registrated. This behavior underlines the solitary character of the filaments in the following way. A smaller shunt resistor allows for a larger total current I when the voltage drop across the discharge gap breaks down. The system does not react with the generation of a single broad filament but creates several filaments, all of them having nearly the same shape and carrying almost the same current I. The increase in applied voltage leads to the generation of further filaments in the same manner as described above.

Figure 4 shows the total current I immediately after a jump as a function of the number of filaments for different values of R_0 and reflects a nearly linear relationship. Particularly, the current flow through a single filament depends on the chosen value of the gas pressure. With increasing gas pressure the current through a single filament becomes larger. This behavior has been verified in a gas pressure range from 120 to 200 hPa.

For appropriately chosen parameters of the gas discharge system, another kind of behavior can be observed: the transition from a homogeneously lightening discharge into a spatially periodic discharge pattern transversal to the current flow. This bifurcation phenomenon can be observed by decreasing the applied voltage U or by increasing the gas pressure p, the width a of the discharge gap, or the value of the load resistor R_0. In Fig. 5a the radiation density distribution transversal to the current flow is represented for the case of decreasing voltage U. At a voltage of U = 1280 V a nearly homogeneous structure appears. The fluctuations of the radiation density are caused by unavoidable material inhomogenities at the semiconductor electrode surface. For decreasing U a spatially periodic structure arises starting from the boundary of the discharge gap and propagating with a further decrease of U. The particular role of the boundary is probably caused by a gradient in the discharge gap width or by a material inhomogeneity of the semiconductor electrode. At U = 1038 V the periodic structure grows up also in the center of the

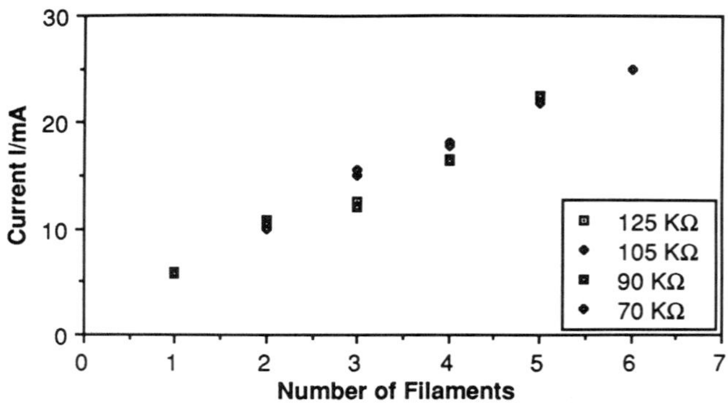

Fig. 4 Total current I immediately after a jump as a function of the number of filaments for different values of R_0. Parameters: ρ = 0.9 kΩ·cm, p = 150 hPa, l = 45 mm, b = 12 mm, a = 3 mm, d = 0.3 mm. Gas mixture: 135 hPa argon and 15 hPa air.

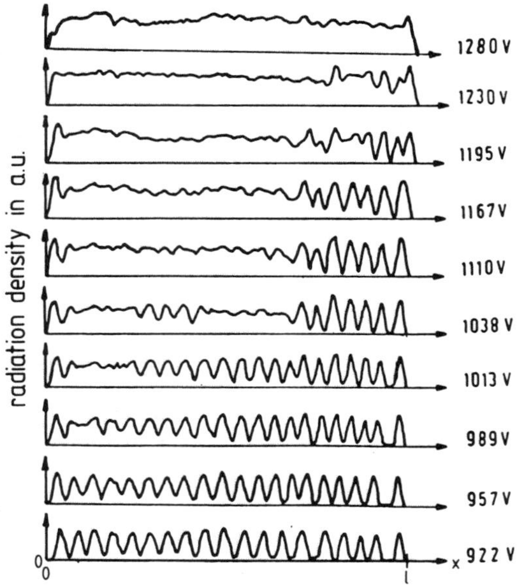

a)

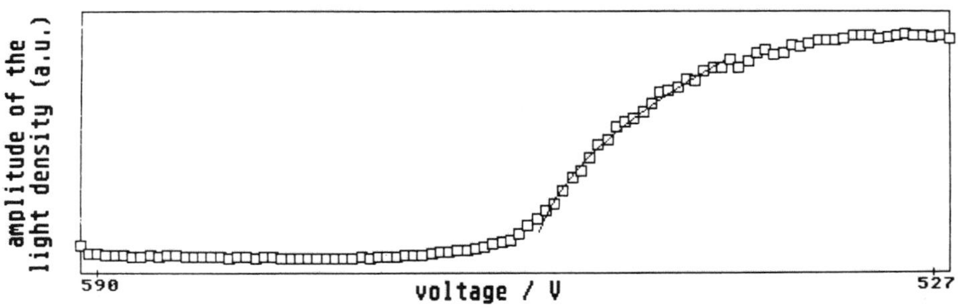

b)

Fig. 5 (a) Radiation density distribution as a function of the space coordinate x for decreasing voltage U. (b) Amplitude of the spatially periodic structure versus the applied voltage. Parameters: $\rho = 0.9\ \text{k}\Omega\cdot\text{cm}$, $p = 90$ hPa, $l = 45$ mm, $b = 12$ mm, $a = 4.6$ mm, $d = 0.3$ mm, $R_0 = 20\ \text{k}\Omega$. Gas mixture: 80 hPa helium and 10 hPa air.

discharge gap. As this structure has the same wavelength as the structure appearing at the boundary of the discharge gap, it can be concluded that the destabilization of the homogeneous discharge is caused by a certain wavelength, which is intrinsic for a given parameter set. This behavior is typical for the well-known Turing bifurcation (Turing, 1952). With decreasing bifurcation parameter, in this case the external voltage, the amplitude grows in a square root–like manner near the bifurcation point (Fig. 5b).

C. Stationary Patterns in the AC-Driven System

Applying a sinoidal driving voltage $U(t) = U_{SS} \sin 2\pi ft$ with frequency f to the ac system described above, the total current $I(t)$ responds with the signal shown in Fig. 6. For a short time interval the applied voltage exceeds the breakdown voltage, resulting in a glow discharge accompanied by a rapid current rise; time-resolved measurements reveal that the light radiation takes place simultaneously with this current rise.

Depending on the pressure p, the electrode distance d, and the amplitude of the applied voltage U_{SS} various stationary structures can be observed when the electrodes are sufficiently large. Choosing d and p in an

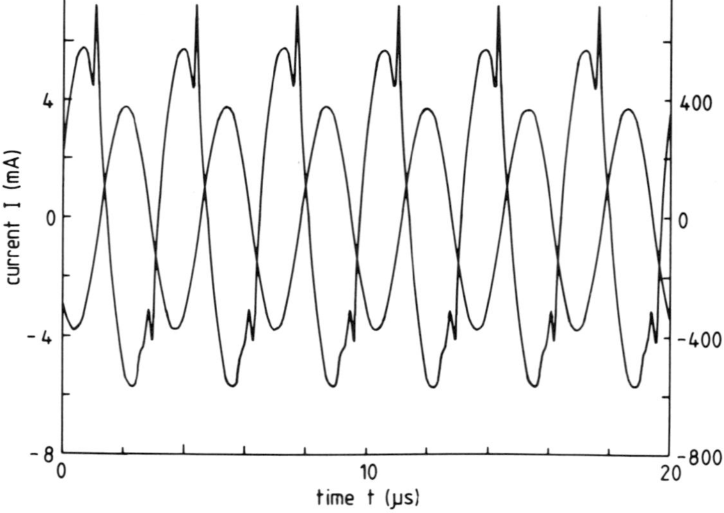

Fig. 6 Typical current response $I(t)$ to the driving voltage $U(t)$. The pressure and electrode distance are $p = 100$ hPa and $d = 150$ μm, respectively.

appropriate way and increasing the voltage beyond the breakdown voltage, the generation of current filaments is observed and stationary light-intensity patterns in form of points appear as described by Boyers and Tiller (1973, 1982) (Fig. 7a). The arrangement of the arising filaments depends on the shape of the electrodes: for example, circular electrodes cause a hexagonal order, while rectangular ones create a cubic order of the discharge channels. Further increase of the voltage leads to the coalescence of the filaments, resulting in discharge stripes (Fig. 7b). Finally, the discharge extends over the entire area of the electrodes. The same scenario takes place when the electrode distance or the pressure is varied.

The inhomogeneous light patterns are determined essentially by the dielectric glass plates. In the environment of a current filament the glass plates cause a large voltage drop such that the voltage drop over the discharge gap is reduced below the breakdown voltage. In a certain distance of the filament, however, the voltage over the gap may again be large enough to generate another filament. Therefore, in this system the dielectric layers act as resistivity layers just as the semiconductor electrode in the dc-driven system.

III. DYNAMIC BEHAVIOR OF THE GAS DISCHARGE SYSTEMS

According to Section II, the generation of multiple filamentary structures in the dc gas discharge system is, except for the first filament, connected with splitting processes. These splitting processes are associated with temporal oscillations of the total current accompanied by spatiotemporal oscillations of the radiation density distribution in the positive column of the gas discharge. The spatiotemporal behavior of the radiation density distribution in the positive column of the glow discharge can be recorded with a streak camera.

A. Measurement Setup

The incident light is projected onto a slit via a relay lens and the resulting slit image is focused on the photocathode of the streak tube where the photons generate electrons (Fig. 8). The electrons are accelerated between the photocathode and a mesh electrode by a strong electrostatic field. With the help of a deflection field, synchronized with the arriving electrons, the electron beam is swept perpendicular to the slit length. After an amplification by means of a micro channel plate (MCP), the electrons generate an optical streak image on a phosphor screen. Therefore, the time axis of the incident light corresponds to the vertical axis on

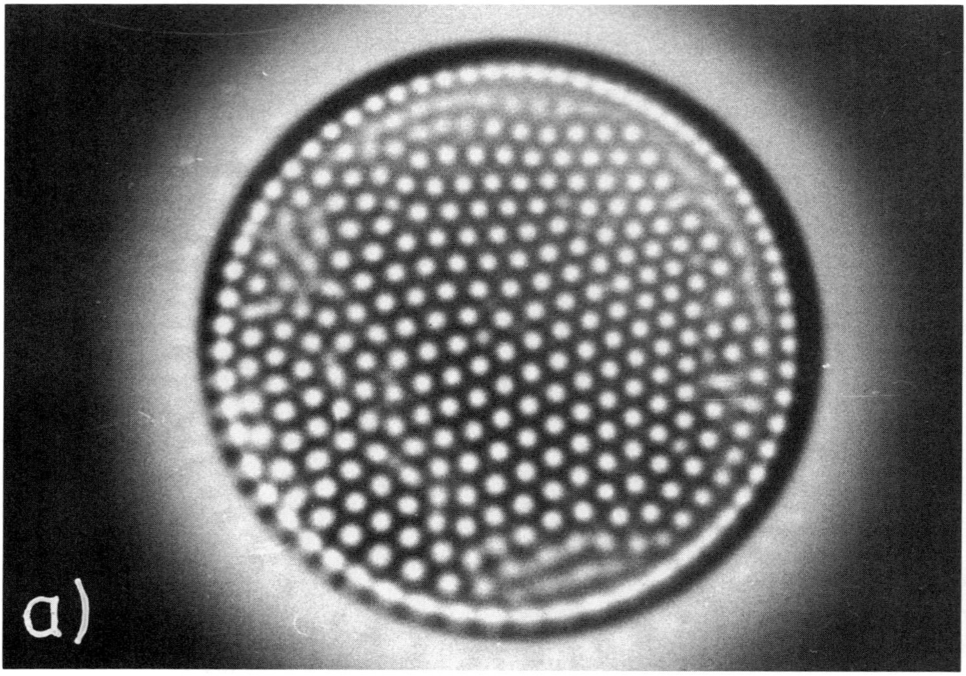

Fig. 7 Stationary radiation density distribution of the two-dimensional ac-driven system; the spots are corresponding to current density filaments. The radius of the electrode is $r = 1.2$ mm, the thickness of the dielectric glass plate is 150 μm, the other parameters are $f = 300$ kHz, $p = 1000$ hPa, $d = 300$ μm. (a) $U_{SS} = 900$ V; (b) $U_{SS} = 1050$ V.

the phosphor screen, the horizontal axis gives the spatial resolution, and the intensity of the incident light can be determined from the brightness of the streak camera record. For a detailed information of streak camera operation, see, for example, Tsuchiya (1984).

B. Dynamic Patterns in the DC-Driven System

Starting from a stationary state with one filament, the applied voltage U is increased up to the critical point where the splitting process starts and the current consists of a dc part superimposed by a regularly oscillating ac component, which has an amplitude in the range of a few percentage of the dc component and a frequency in the range of a few hundred hertz. A slight increase in U causes a change in the frequency and leads to the

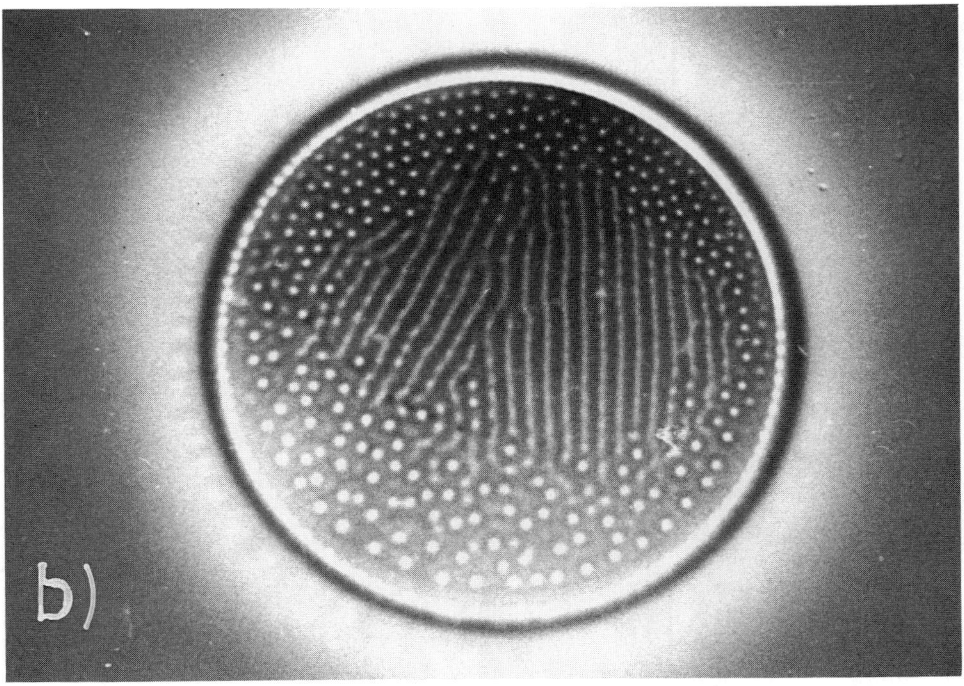

generation of harmonics, while the amplitude of the oscillation slowly increases. By further enlargement of U the oscillation becomes irregular; finally, the second filament gets separated and the total current I becomes stationary again. A similar scenario can be observed for further generation processes of filaments. In Fig. 9 the time series of the described scenario is presented. For $U = 1120$ V a nearly sinoidal oscillation of 233 Hz is measured. The slight increase of U to 1125 V results in a period doubling. Up to $U = 1150$ V the above-mentioned shift in the frequency appears while the amplitude of the ac component remains nearly constant. Beyond $U = 1190$ V two stationary filaments are stable.

The temporal oscillations of the total current are accompanied by spatiotemporal oscillations of the radiation density distribution in the positive column of the splitting filament. Figure 10a shows a streak camera record of a typical radiation density distribution as a function of space and time. The picture shows regular oscillations of the radiation density in the form of a splitting and extinction of filamentary structures.

To find the correlation between this spatiotemporal signal and the temporal behavior of the total current I, the integral of the radiation

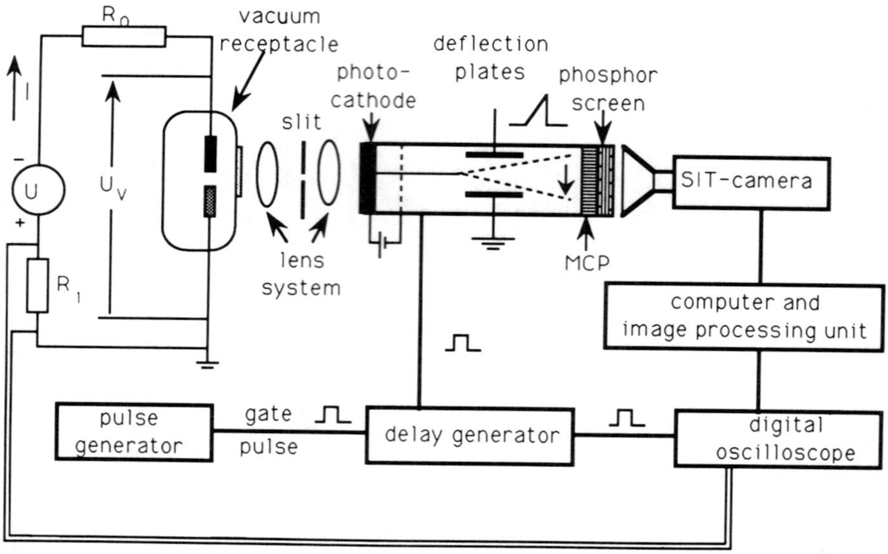

Fig. 8 Experimental setup for the measurement of spatiotemporal structures by means of a streak camera.

density in the area between the two vertical solid lines in Fig. 10a has been determined. In Fig. 11 the time series of the calculated integral and of the current I are displayed, respectively. Both signals exhibit regular pulse-like oscillations consisting of a long and a short pulse with widths of $\Delta t_1 = (2.18 \pm 0.02)$ ms and $\Delta t_2 = (0.55 \pm 0.01)$ ms, respectively. For this reason it may be concluded that the splitting and extinction of filamentary structures are responsible for temporal oscillations of the total current. By increasing the applied voltage U the spatiotemporal behavior of the filament changes analogously with the time signal of the total current.

The splitting processes extend over a wide range of parameters. Different forms of oscillations are found for gas pressures between 30 and

---

Fig. 9 Typical time series of the total current I measured during the splitting of a filament for various values of the applied voltage U. The parameters are $R_0 = 150$ kΩ, $\rho = 1.7$ kΩ·cm, $p = 200$ hPa, $d = 3.4$ mm. The gas mixture is argon and 10 vol % of air. I_{dc} and $I_{ac,SS}$ mean the direct current and the peak-to-peak amplitude of the alternating current, respectively.

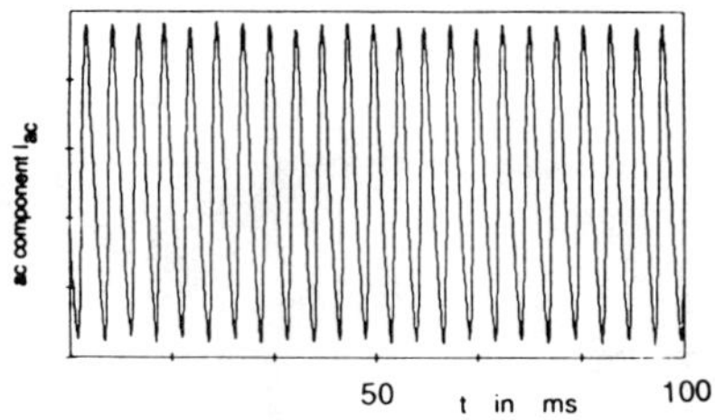

$U = 1120\ V$ $I_{dc} = 2.50\ mA$ $I_{ac,ss} = 12\ \mu A$

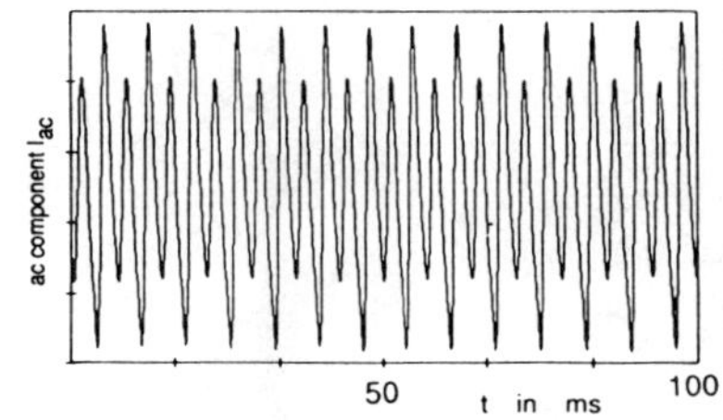

$U = 1125\ V$ $I_{dc} = 2.52\ mA$ $I_{ac,ss} = 12\ \mu A$

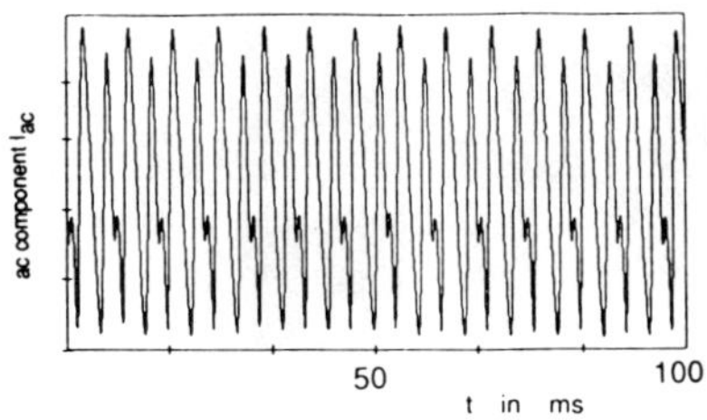

$U = 1150\ V$ $I_{dc} = 2.62\ mA$ $I_{ac,ss} = 15\ \mu A$

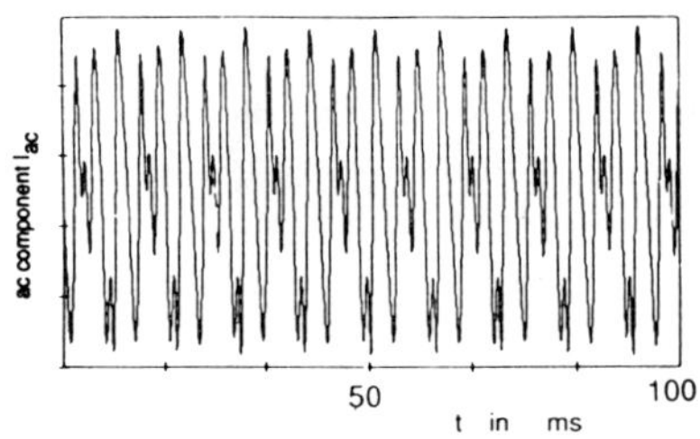

$U = 1170\ V$ $I_{dc} = 2.71\ mA$ $I_{ac,ss} = 16\ \mu A$

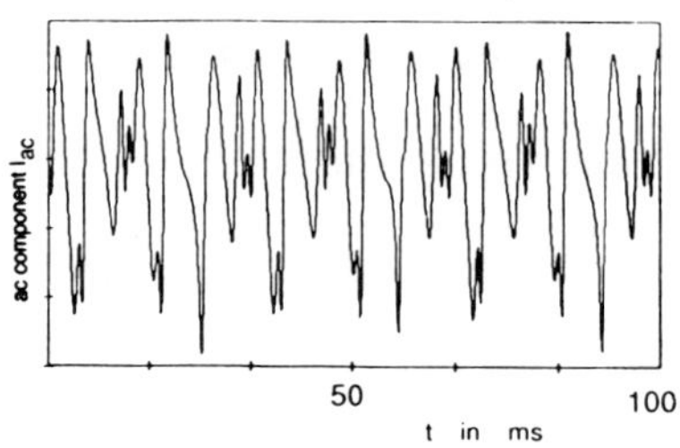

$U = 1190\ V$ $I_{dc} = 2.80\ mA$ $I_{ac,ss} = 19\ \mu A$

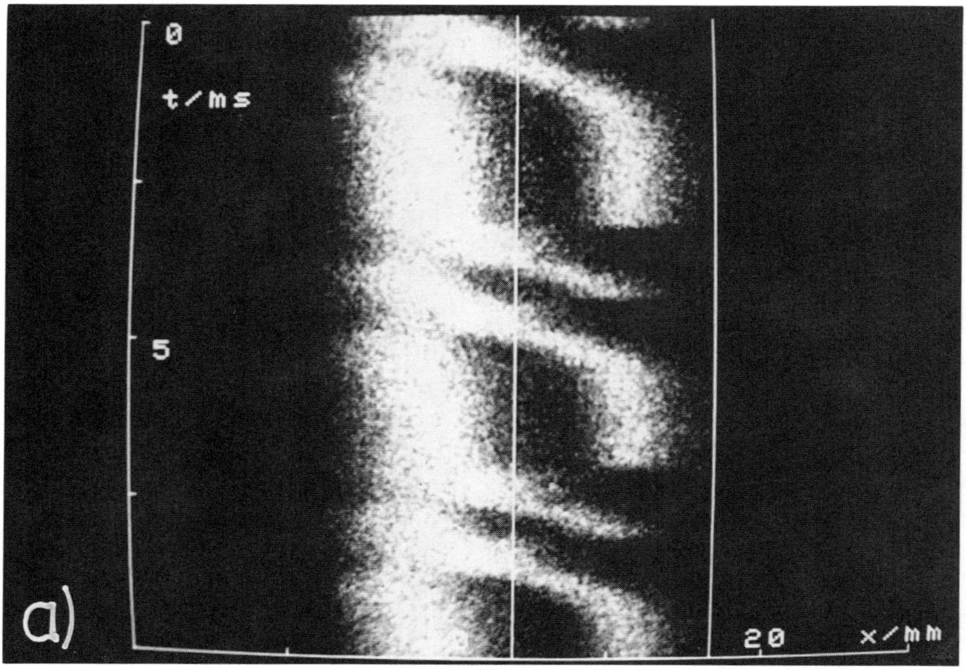

Fig. 10 Streak camera record of splitting filaments. The parameters are: (a) R_0 = 200 kΩ, ρ = 1.7 kΩ·cm, p = 30 hPa, a = 3.5 mm, b = 8 mm, l = 30 mm, U = 900 V; (b) R_0 = 74 kΩ, ρ = 1.7 kΩ·cm, p = 180 hPa, a = 3.5 mm, b = 8 mm, l = 30 mm, U = 1300 V. The gas mixture is argon and 10 vol % of air.

300 hPa and electrode distances between 2 and 6 mm. The common feature of these oscillations is that they are always connected with the splitting and extinction of filamentary structures. The process of splitting is reversible under a decrease in the applied voltage until a second filament is able to exist independently from the original one, at which moment the oscillations vanish. However, when the division process has finished, a decrease in the voltage does not result in a recombination of the two filaments but in a successive extinction, which is not accompanied by spatiotemporal oscillations of the radiation density. In a similar way, spatiotemporal oscillations of the light intensity distribution occur when further filaments are created by the splitting of already existing filaments. In this case the splitting filament exhibits oscillations while the other filaments are stationary. Such a scenario is represented in Fig. 10b.

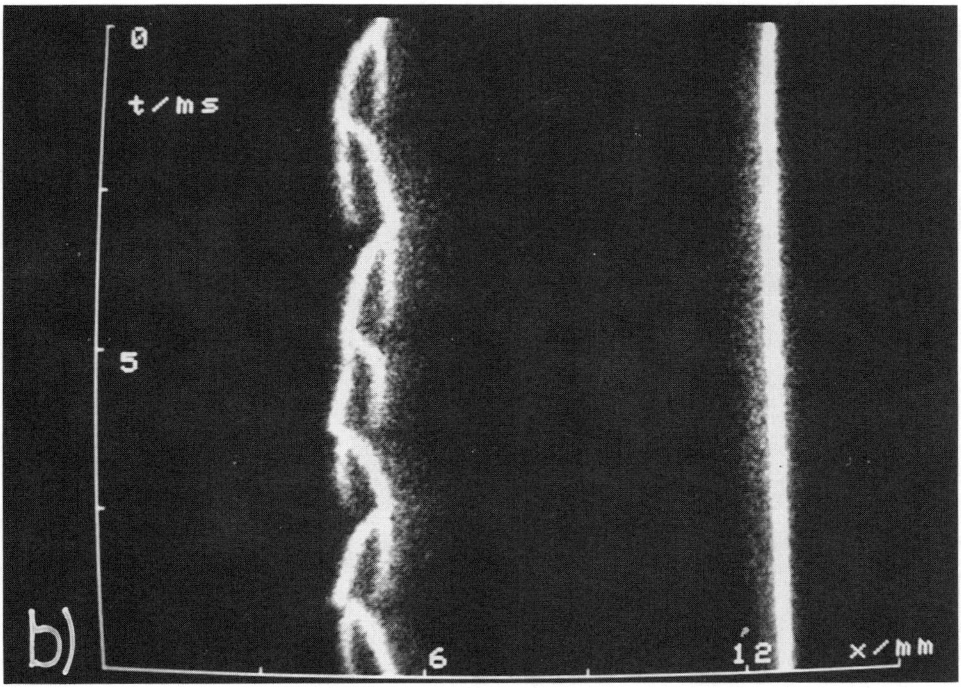

Another kind of spatiotemporal behavior can be found by choosing the metal electrode as anode and the semiconductor electrode as cathode. There are no stationary filamentary structures but instead, an irregular behavior in that solitary filaments may move, appear, and disappear. This is connected with irregular oscillations of the total current. The corresponding streak camera record is shown in Fig. 12.

C. Dynamic Patterns in the AC-Driven System

In analogy to the dc-driven system the dynamic behavior during the filament generation process has been investigated in the ac-driven system. For this purpose a quasi-one-dimensional electrode has been prepared by etching the ITO layer in an appropriate manner. The streak camera system has been used as described above to get spatiotemporally resolved measurements of the light-intensity distribution depending on the ampli-

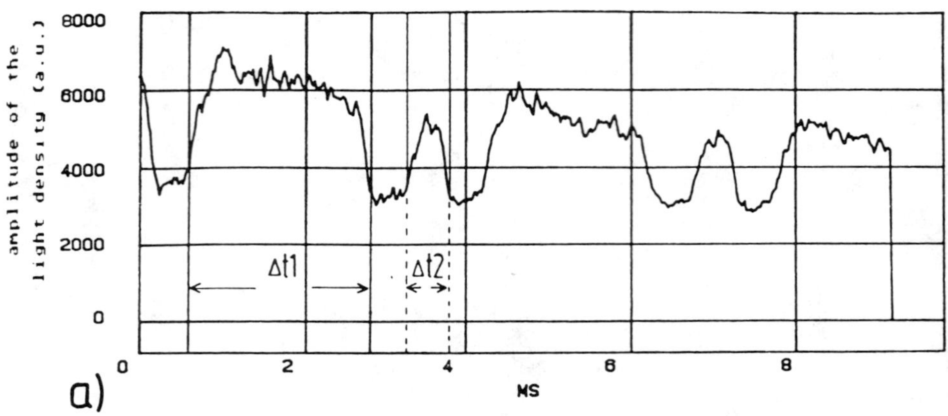

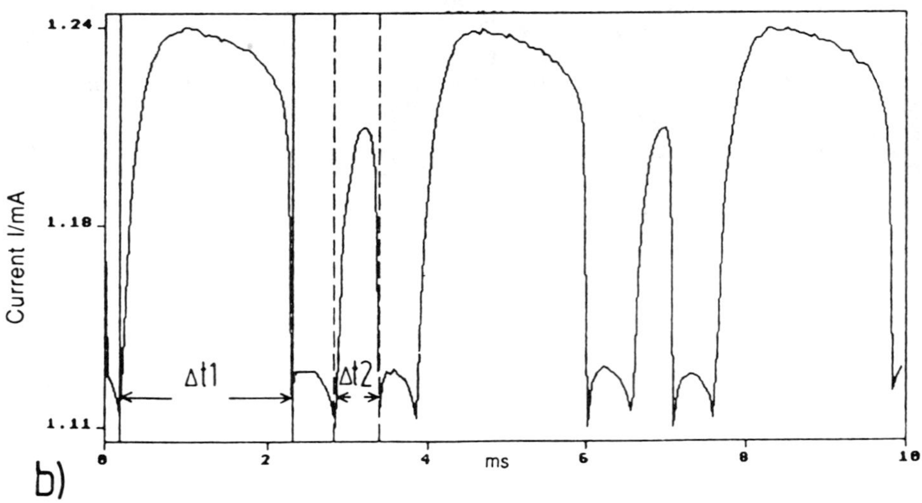

Fig. 11 (a) Total light intensity of the area marked in Fig. 10a as a function of the time with time intervals $\Delta t_1 = 2.17$ ms, $\Delta t_2 = 0.54$ ms; (b) corresponding time series of the total current with time intervals $\Delta t_1 = 2.19$ ms, $\Delta t_2 = 0.55$ ms.

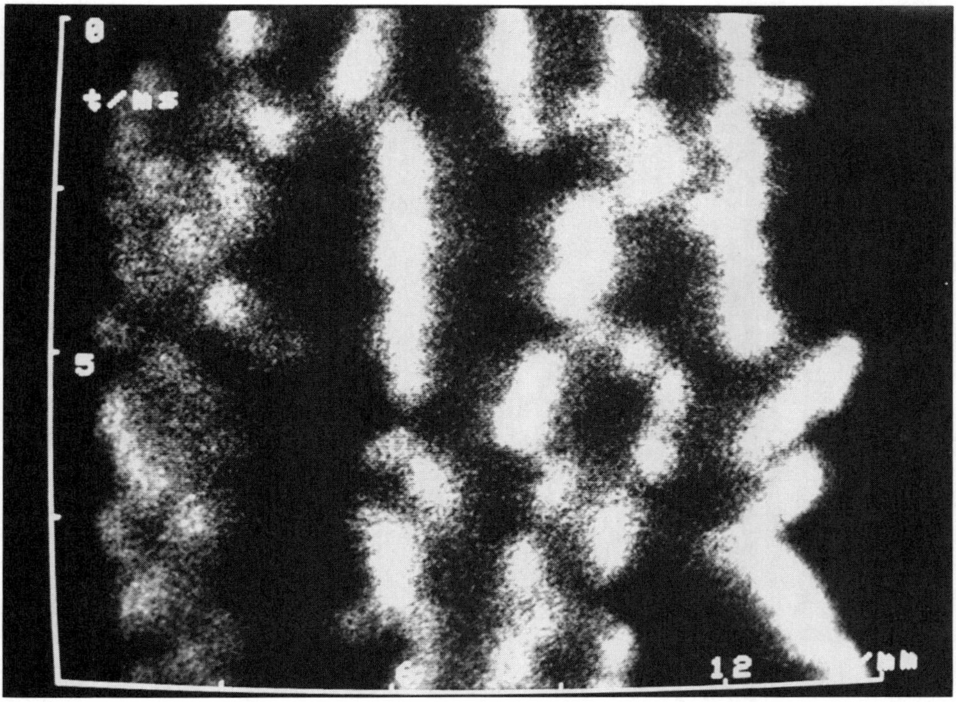

Fig. 12 Streak camera record showing spatiotemporally irregular motion of filamentary structures. Parameters: R_0 = 74 kΩ, ρ = 0.9 kΩ·cm, p = 80 hPa, a = 2.3 mm, b = 8 mm, l = 15 mm, U = 1300 V. The gas mixture is argon and 10 vol % of air.

tude U_{SS} of the applied voltage. When the breakdown voltage is reached, one stationary filament of a typical size arises. An increase in the voltage leads to spatiotemporal instabilities of this filament, and at a certain voltage, a second stable filament appears. Whenever a new filament is generated, such instabilities emerge. There are different forms of spatiotemporal oscillations, depending on the pressure and the electrode distance. In Fig. 13a, for example, a transition from four to five filaments is shown, which can be interpreted as follows. Two filaments coalesce, resulting in a four-filament state, and after a short time another filament splits into two filaments, the system reaches the initial five-filament state.

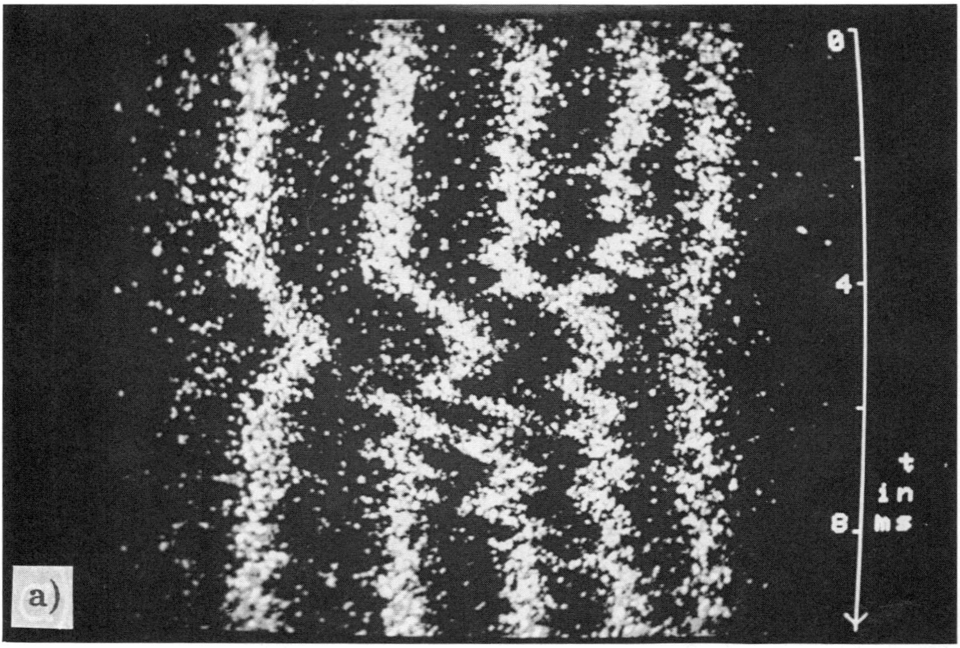

Fig. 13 Spatiotemporally resolved measurements of the light-density distributions in a quasi-one-dimensional ac-driven gas discharge system. The horizontal and vertical axis correspond to the space and time axis, respectively. The bright regions mark positions of filaments. The area of the electrode is $A = 0.5 \times 10$ mm^2, the thickness of the glassplates is 150 μm, and the other parameters are: (a) $U_{SS} = 1020$ V, $p = 420$ hPa, $f = 456$ kHz, $d = 566$ μm; (b) $U_{SS} = 670$ V, $p = 360$ hPa, $f = 115$ kHz, $d = 480$ μm; (c) $U_{SS} = 1072$ V, $p = 420$ hPa, $f = 456$ kHz, $d = 566$ μm.

Another form of oscillations is shown in Fig. 13b. The transition from two to three filaments consists of the spontaneous generation and a spatial oscillation of one filament. The spatial amplitude of this filament grows in the course of time, and finally, the adjacent filament extincts, whereas the oscillating filament moves to the position of the dying filament and the process is repeated. A third type of oscillation emerges when almost the whole electrode is filled with filaments (Fig. 13c). Several filaments travel through the discharge gap.

All these processes occur on a time scale much larger than the driving force but comparable with that of the oscillations in the dc system.

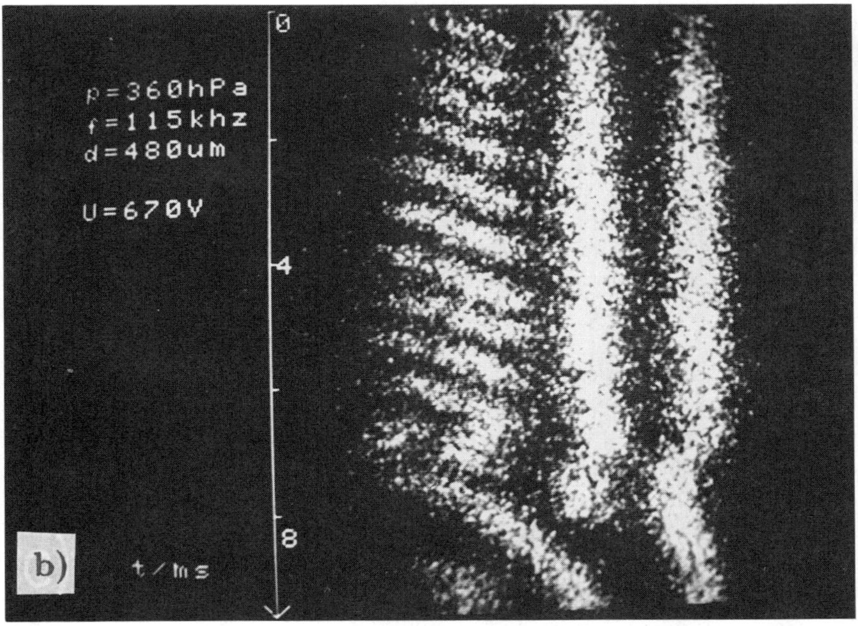

b)

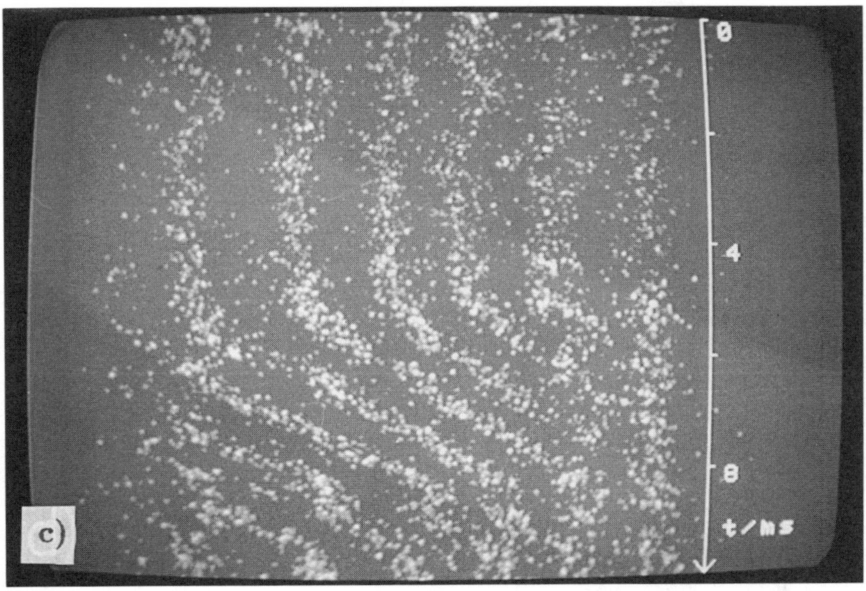

c)

IV. A PHENOMENOLOGICAL MODEL

To describe the qualitative behavior of the systems investigated, a phenomenological model is useful. In the following we restrict our considerations to the dc-driven system. Let us divide the system in two parts with different electrical properties: A nonlinear layer (N) exhibiting S-type negative differential resistivity is represented by the gas discharge gap, and a linear layer (L) is formed by the high ohmic semiconductor electrode (Fig. 14). The thickness of the nonlinear layer is assumed to be small compared to the thickness of the linear layer. The entire device is provided with ohmic contacts and connected to a voltage source via a load resistor R_0.

Using a suitable potential ansatz in the linear layer and taking into account diffusion processes of charge carriers in the nonlinear layer, we can deduce a two-component system of reaction diffusion equations of the activator–inhibitor type (cf. Radehaus et al., 1987). The two components are represented by the current density $j(x,y,t)$ and the potential $U_N(x,y,t)$ in the interface Ω between the layers. The current density j acts as activator based on the negative differential resistivity of the nonlinear $j(E)$ characteristic of the gas discharge, while U_N represents the inhibitor and stabilizes current filaments. The mathematical description of this model leads to the following system of coupled differential equations:

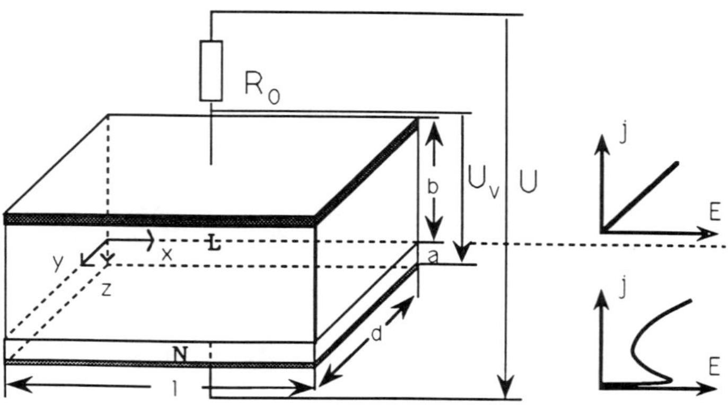

Fig. 14 Schematical illustration of the two-layer model. The linear (or at least monotonically increasing) $j(E)$ characteristic refers to the layer L while the second layer N is characterized by a bistable S-shaped $j(E)$ characteristic.

$$\frac{\partial v}{\partial \tau} = \sigma\, \Delta v + f(v) - w + \mu(\nabla v)(\nabla w) + \kappa_1 - \frac{\kappa_2}{|\Omega|}\int v\, d\Omega$$

$$\delta \frac{\partial w}{\partial \tau} = \Delta w + v - w \tag{1}$$

$$v \sim \text{const} + j \qquad w \sim \text{const} - U_N$$

where σ denotes the ratio of the diffusion constants and $1/\delta$ is the ratio of the time constants of the activator v and the inhibitor w. The parameter μ is essentially proportional to the ambipolar mobility of the charge carriers, κ_1 to the externally applied voltage U, and κ_2 to the load resistor R_0. The function $f(v)$ is determined by the S-shaped $j(U)$ characteristic of the nonlinear layer. In analogy to the experimental results described above, we now present some analytical and numerical results of the system of equations (1).

From linear stability analysis it is known that a stationary solution (v_0, w_0) is stable if the eigenvalues of the linear matrix of Eqs. (1) have negative real parts and that a bifurcation occurs if by changing a parameter one of the eigenvalues crosses the imaginary axis. To calculate the eigenvalues we develop v and w with respect to the eigenfunctions of the Laplace operator with suitable boundary conditions. In a one-dimensional case with Neumann boundary conditions we obtain the eigenfunctions $\varphi_0 = 1/l$ and $\varphi_n(x) = 2/l \cos(k_n x)$, $k_n = n\pi/l$ and $n \geq 1$, where l is the length of the system. In this orthonormal basis equations (1) can be written as an infinite system of nonlinear differential equations determining the temporal development of the amplitudes (v_n, w_n) referring to the functions φ_n. If the linearized (2×2)-matrix to each mode k_n is denoted with C_n the eigenvalues $\omega(k_n)$ of C_n are given by

$$\omega^2(k_n) - \omega(k_n)\,\mathrm{Tr}(C_n) + \mathrm{Det}(C_n) = 0$$

The system is stable if $\mathrm{Det}(C_n) > 0$ and $\mathrm{Tr}(C_n) < 0$ and may become linearly unstable in two different ways, $\mathrm{Tr}(C_n) = 0$ or $\mathrm{Det}(C_n) = 0$, called hard-mode and soft-mode instability, respectively. In the following we restrict ourselves to the soft-mode instability in which a variation of the external voltage, of the load resistor or of the electrode distance causes the real part of one of the eigenvalues to become positive at a critical point called Turing point. This eigenvalue refers to a critical mode k_c which is given by

$$k_c = (\sigma^{-1/2} - 1)^{1/2}$$

In order to investigate the development of the arising spatially periodic structure near the instability point the center manifold approach can be used (Carr, 1981; Radehaus et al., 1992). As a result, we obtain that the amplitude of the linearly unstable mode grows proportional to the square root of the bifurcation parameter. Depending on the system parameters, the bifurcating branch is stable. This case is called supercritical and leads

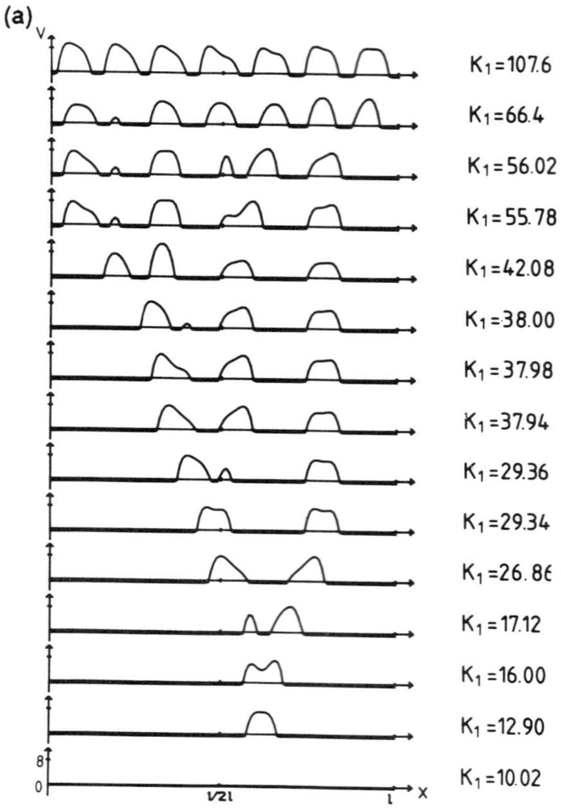

Fig. 15 (a) Numerically calculated stationary spatial current density distributions corresponding to certain values of the global current J in (b); (b) numerically calculated global current–voltage characteristic $J(T)$ with $T = \kappa_1 - (\kappa_2/l)J$ and $J = \int_0^1 v(x)\, dx$. The shape of the characteristic is similar to that of the experimental $I(U)$ characteristic in Fig. 3b. Parameters: $\sigma = 0.1$, $\mu = 0.000035$, $\delta = 1$, $\kappa_2 = 10$,

$$f(v) = \begin{cases} -11.5v^3 + 8v & v \leq v_f \\ 0.6v + \theta & v \geq v_f \end{cases}$$

with v_f and θ such that $f(v)$ is continuously differentiable.

to an amplitude curve, which is similar to that of the gas discharge system (cf. Fig. 5b). Using Dirichlet instead of Neumann boundary conditions, the critical mode propagates from the boundaries. The amplitude of the periodic structure decays with growing distance from the boundaries, and a further increase in the bifurcation parameter results in a spatially varying wavenumber (Ahlers, 1989).

In a qualitative manner the model also describes the appearance of solitary current filaments. Figure 15a shows numerically calculated current distributions in dependence on the external voltage κ_1. As in the gas discharge system the successive generation of filaments is accompanied by discontinuities in the global current voltage characteristic (Fig. 15b).

In the model two different kinds of filament generation processes have been found. The first is connected strongly to the Turing instability, as discussed above: At a certain voltage the Turing point is reached and a spatially periodic structure with a wavevector k_c destabilizes the homogeneous distribution. Fluctuations may favor some regions in which the activator develops stronger than in the rest of the system, resulting in a high-current region. The large inhibitor diffusion prevents a spreading of the filament to the neighboring region. In the vicinity of the Turing point (i.e., near the breakdown voltage) the distributions of v and w lateral to the filament walls are modulated by a periodic structure with wavenumber k_c and an amplitude that is strongly decaying with growing distance from the filament. Therefore, if the external voltage is increased up to the critical point, the second filament appears at the largest maximum of this modulation (i.e., near the first one in a distance determined by the wavevector k_c). If the parameter μ is sufficiently large, the generation

(b)

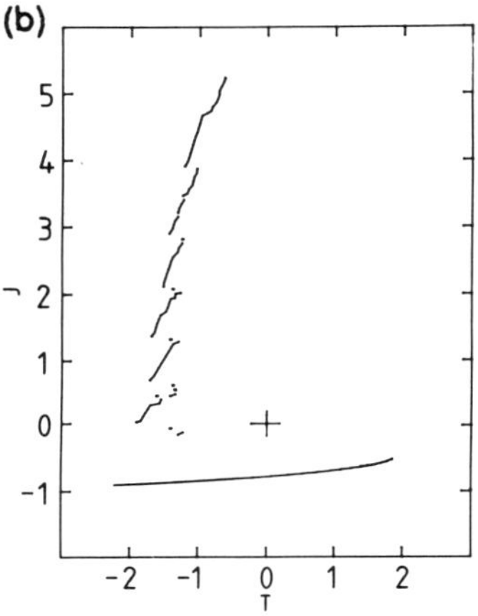

process changes. Then at the walls of the current filament additional terms have to be taken into account, resulting in a rise of the walls, and when the bifurcation parameter exceeds a critical value in a splitting process of the filament. In contrast to the experimental behavior, no current oscillations take place during the splitting process.

The ratio $1/\delta$ of the time constants of the activator and inhibitor essentially determines the dynamic behavior of the system. All considerations until now dealt with $\delta \leq 1$. In the case $\delta > 1$, the inhibitor is not able to follow the activator immediately, which results in different kinds of dynamic structures. In addition to breathing and traveling filaments, we observe swaying (i.e., in phase oscillations of the filament walls), and spatiotemporally irregular current distributions in which filaments appear by division and extinguish or coalesce in a random way (Fig. 16), which is similar to the experimental result in Fig. 12.

V. CONCLUSION

The gas discharge systems described above show a rich variety of stationary and dynamical current density patterns. Because of the linear relation between current and radiation density the appearance of complex inhomogeneous structures can be easily detected in contrast to other physical systems. The solitary properties of the filaments are reflected in the fact that for a given set of parameters, all filaments have nearly the same width, and even the current flow through a single filament is approximately the same. The generation of filaments can be reduced to the division of already existing filaments accompanied by temporal oscillations in the total current and spatiotemporal oscillations in the radiation density. Special features of the current signal (e.g., certain components in the power spectra) can be directly related to the branching of filamentary structures, especially in the dc system. The generation of new filaments is connected with well-pronounced jumps in the global $I(U_V)$ characteristic. Based on these facts it may be concluded that the current signal itself contains sufficient information to reconstruct the spatial distribution of current density in the device. The main properties of the dc system are well described by a phenomenological model, which leads to a reaction–diffusion system of the current density and the voltage drop across the gas gap. Numerically calculated current density distributions show the same solitary behavior of the filaments as well as comparable global current–voltage characteristics. Spatially periodic patterns observed in the gas discharge system can be related to the Turing instability, the treatment of which in terms of the center manifold theory leads to the same dependence of the growing amplitude on the bifurcation parameter.

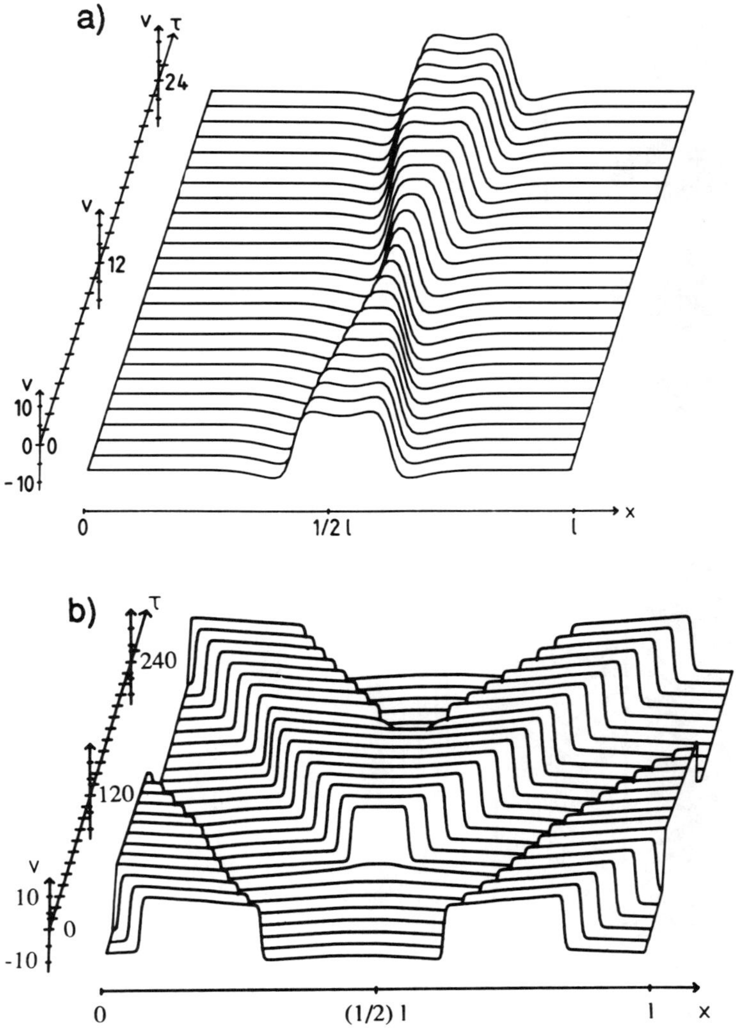

Fig. 16 Time evolution of the spatial distribution of v for the case of: (a) a breathing filament (parameters: $\sigma = 0.1$, $\mu = 0$, $\kappa_1 = -0.5$, $\kappa_2 = 0$, $\delta = 5$, $f(v) = -0.01v^3 + 1.37v$); (b) self-created traveling pulses (parameters: $\sigma = 10$, $\mu = 0$, $\kappa_1 = -4$, $\kappa_2 = 4$, $\delta = 100$, $f(v) = -0.02v^3 + 1.5v$); (c) a swaying filament [parameters: $\sigma = 0.1$, $\mu = 0.05$, $\kappa_1 = -2.15$, $\kappa_2 = 10$, $\delta = 3$, $f(v)$ as in Fig. 15]; (d) irregular behavior of fronts [parameters as in (b), but $\delta = 3$]. In (c) and (d) the filled areas mark that part of the distribution which exceeds a certain threshold.

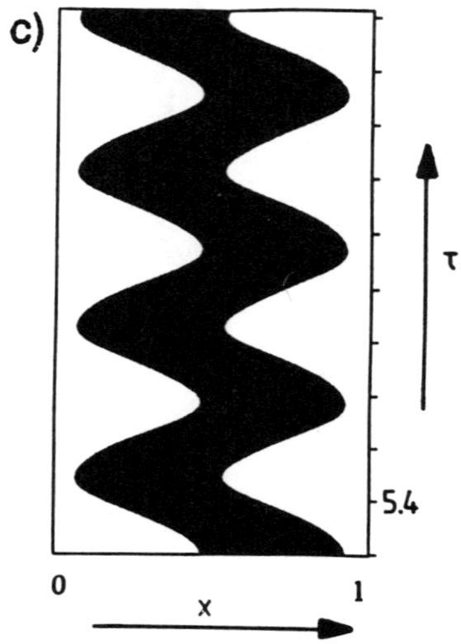

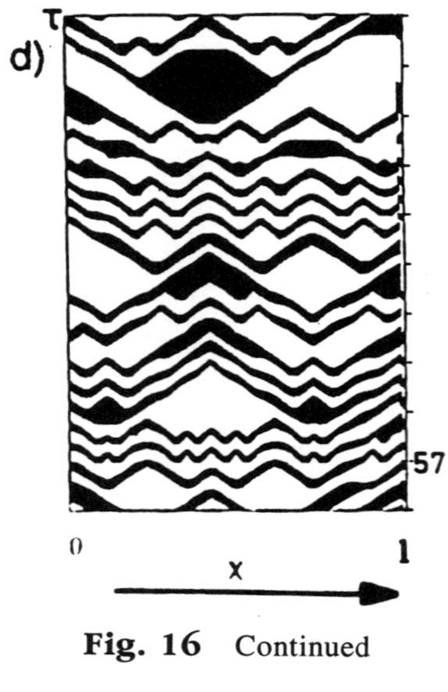

Fig. 16 Continued

REFERENCES

Ahlers, G. (1989). In *Complex Systems*, SFI Studies in the Science of Complexity, Addison-Wesley, London, pp. 175–224.

Boyers, D. G., and Tiller, W. A. (1973). *J. Appl. Phys.*, *44*: 3102.

Boyers, D. G., and Tiller, W. A. (1982). *Appl. Phys. Lett.*, *41*: 28.

Bragin, V. E., Matyukhin, V. D., and Muravev'ev, V. F. (1984). *Sov. Phys. Tech. Phys.*, *29*(7): 780.

Carr, J. (1981). *Applications of Centre Manifold Theory*, Springer-Verlag, Berlin.

Gierer, A., and Meinhardt, H. (1972). *Kybernetik*, *12*: 30.

Jäger, D., Baumann, H., and Symanczyk, R. (1986). *Phys. Lett. A*, *177*: 141.

Kapzow, N. A. (1955). *Elektrische Vorgänge in Gasen und im Vakuum*, VEB Deutscher Verlag der Wissenschaften, Berlin.

Kerner, B. S., and Osipov, V. V. (1989). *Sov. Phys. Usp.*, *32*: 101.

Kerner, B. S., and Sinkevich, V. F. (1982). *JETP Lett.*, *36*: 436.

Kuramoto, Y. (1984). *Chemical Oscillations, Waves, and Turbulence*, Springer-Verlag, Berlin.

Müller, K. G. (1988). *Phys. Rev. A*, *37*: 4836.

Murray, J. D. (1989). *Mathematical Biology*, Springer-Verlag, Berlin.

Nedopasov, A. V. (1968). *Sov. Phys. Usp.*, *11*: 174.

Peinke, J., Rau, U., Clauss, W., Richter, R., and Parisi, J. (1989). *Europhys. Lett.*, *9*: 743.

Purwins, H.-G., and Radehaus, Ch. (1988). In *Neural and Synergetic Computers* (H. Haken, ed.), Springer-Verlag, Berlin.

Purwins, H.-G., Klempt, G., and Berkemeier, J. (1987). *Festkörperprobleme*, *27*: 27.

Radehaus, Ch., Kardell, K., Baumann, H., Jäger, D., and Purwins, H.-G. (1987). *Z. Phys. B*, *65*: 515.

Radehaus, Ch., Willebrand, H., Dohmen, R., Niedernostheide, F.-J., Bengel, G., and Purwins, H.-G. *J. Phys. Rev. A*, *45*: 2546 (1992).

Tsuchiya, Y. (1984). *IEEE J. Quantum Electron.*, *QE-20*: 1516.

Turing, A. M. (1952). *Philos. Trans. R. Soc. B*, *237*: 37.

van den Heuvel, F. C. (1988). *Philips Tech. Rev.*, *44*: 89.

von Engel, A. (1965). *Ionized Gases*, Clarendon Press, Oxford.

Wiesemann, K. (1976). *Einführung in die Gaselektronik*, Teubner Studienbücher, Stuttgart.

Willebrand, H., Radehaus, Ch., Niedernostheide, F.-J., Dohmen, R., and Purwins, H.-G. (1990). *Phys. Lett. A*, *149*: 131.

Willebrand, H., Niedernostheide, F.-J., Ammelt, E., Dohmen, R., and Purwins, H.-G. (1991). *Phys. Lett. A*, *153*: 437 (1991).

6
Oscillations and Morphogenesis: A Unifying Mathematical Approach

H. Haken *Institut für Theoretische Physik und Synergetik, Stuttgart, Germany*

I. INTRODUCTION

Oscillations are observed in a great variety of fields, ranging from physics to biology to economics. Although the individual mechanisms causing oscillations may be quite different in nature, there is a unifying mathematical description of their occurrence that we want to outline below. When a system in equilibrium is perturbed in some way and then the perturbation is cut down, the system may relax toward its equilibrium state by means of damped oscillations. Far more interesting is the case in which undamped oscillations occur where in all systems treated in the natural sciences an energy flux into the system must be provided to overcome damping. Under specific circumstances, undamped autonomous oscillations may be entertained by a system. They must be distinguished from forced oscillations where the driving force acting on the system is itself oscillatory in nature.

In this chapter we are concerned with undamped autonomous oscillations. In biological systems, oscillations or, more loosely speaking, rhythms can be observed over a large range of frequencies (e.g., in the millisecond regions of nerve pulses over the circadian rhythms up to rhythms with periods of a year or longer). Study of the biological causes of such rhythms is left to researchers in biology and medicine (cf. Rensing et al., 1987, for definitive references); in this chapter we are concerned with the mathematical aspects. As will transpire from our considerations, oscillations may occur under quite general conditions. This has the advantage that oscillations can be treated from a unifying point of view; on the other hand, the mere occurrence of oscillations or rhythms does not allow much of a conclusion on the detailed underlying mechanisms, be they physical, chemical, or biological.

A second large field of study concerns morphogenesis. Formation of patterns can be observed in both the inanimate and animate world. In the inanimate world we may think of the growth of crystals, where the growth of snow crystals is a particularly striking example, the formation of dendrites, and so on. In biology we are concerned primarily with the growth of plants and animals in which, in one way or another, the genetic information is transformed into specific structures, which then perform specific functions. According to a generally accepted argument, the information contained in DNA is not sufficient to provide the information required to build a complete body. For this reason the concept of self-organization has been invoked. We shall come to this point below. A basic model for the causes of morphogenesis by self-organization was put forward by Turing (1952) in his fundamental paper. He considered a model consisting of two cells. In each cell a rather simple chemical reaction is going on. Initially, both cells undergo the same reaction processes. But then a coupling (i.e., an exchange of matter between these two cells) is introduced. By means of this coupling the spatial symmetry of the two cells can be broken and one cell produces a higher amount of certain molecules than the other one. If one assumes that genes can be switched on by means of a specific minimal amount of some molecules, one may easily come to the conclusion that by such a mechanism cell differentiation may be caused. Such a mechanism may also lead to the formation of differentiated cells in *extended tissues*. The basic idea is that at first molecules of two specific kinds, activators and inhibitors, are generated, migrate in the tissue, and interact with each other in a certain nonlinear fashion. As discussed below, spatial patterns of concentrations of molecules can be produced by such a mechanism. If one then assumes that at high concentrations of the activator molecules, genes can be switched on, it appears that now the formation of organs becomes understandable. Examples to which these concepts have been applied include *Hydra*, with the formation of its head and foot and *Drosophila*. Again it will not be our concern to elucidate the detailed chemical mechanisms which are still under detailed study, but rather to illustrate the underlying mathematical concepts.

II. UNIFIED MATHEMATICAL APPROACH

In the interdisciplinary field of synergetics (Haken 1977, 1987) a unified mathematical approach for the study of oscillations and morphogenesis has been developed over the last decades. As a starting point we assume that there exists a medium, a biological tissue, and so on, which is in a spatially homogeneous and quiescent state. This state is achieved by setting a specific set of control parameters to some values that guarantee the stability of that state. When we change the size of a control parameter, say of a chemical concentration, the system may become unstable. The old homogeneous state will then be replaced by a new state which may be inhomogeneous in space (e.g., periodic) or which may exhibit temporal oscillations. Thus the generality of our approach is achieved by studying the transition from a homogeneous quiescent state to an inhomogeneous or/and oscillatory state. Once such a new state is established, one may extrapolate the results and find such states also being stable at higher amplitudes (until a new point of instability is reached).

As we shall see below, the study of oscillations and morphogenesis may be reduced to a few concepts: namely, the instability of a system when one or several control parameters are changed, the occurrence of a few macroscopic degrees of freedom, the so-called order parameters, and the slaving principle by which the behavior of the individual parts of a system is determined by the order parameters. As a conclusion we may state that the phenomenon of morphogenesis as well as the occurrence of oscillations is governed by few degrees of freedom: the order parameters.

For the mathematical treatment we start from a so-called state vector $\mathbf{q}$. This vector lumps together all the relevant variables of the system. These variables may be chosen at different levels: microscopic, mesoscopic, or macroscopic. In what follows we shall start from the microscopic or mesoscopic level and then derive properties of the system at the macroscopic level. Thus our approach can be considered as a treatment of emergent properties. These emergent properties will be oscillations or morphogenesis (i.e., temporal or spatiotemporal structures).

A precise definition of the microscopic or mesoscopic level may not always be simple, especially not in biological systems, and is partly a matter of definition. However, to illustrate this distinction, let us consider a fluid. At the microscopic level it consists of the individual atoms or molecules, which are described by their positions and velocities and which obey the equations of classical mechanics. At the mesoscopic level we may divide the fluid into elements which are large compared to the atomic dimensions and still small compared to length scales at which macroscopic patterns emerge. In each volume element the fluid may be described by its density, its temperature, and perhaps by a mean overall velocity. Similarly, in biological systems the microscopic level may be that of individual biomolecules and the mesoscopic level that of concentrations of those molecules. However, when we think of population dynamics, the microscopic level may consist of individual animals and the mesoscopic level of the density of animals in a certain area where the animals may be bacteria as well as, say, foxes or hares.

After these remarks we introduce the state vector $\mathbf{q}$ where we shall be concerned mainly with the state vector at the *mesoscopic level.*

$$\mathbf{q}(\mathbf{x},t) = (q_1(\mathbf{x},t), q_2(\mathbf{x},t), \ldots, q_N(\mathbf{x},t)) \tag{1}$$

where the individual components of $\mathbf{q}$ are functions of space $\mathbf{x}$ and time t. In a great variety of systems, including practically all the model systems of biology, we may assume that $\mathbf{q}$ undergoes a temporal evolution which is described by equations of the form

$$\dot{\mathbf{q}} = \mathbf{N}(\mathbf{q},\alpha) + \mathbf{F}(t) \tag{2}$$

The dot above $\mathbf{q}$ on the left-hand side of (2) denotes the derivative of $\mathbf{q}$ with respect to time t; $\mathbf{N}$ is a nonlinear function of $\mathbf{q}$ which contains one or a set of control parameters α and which may contain spatial derivatives ∇. $\mathbf{F}(t)$ are fluctuating forces that represent the action of external or internal fluctuations on the state vector $\mathbf{q}$. An important special case of (2) is represented by reaction–diffusion equations, where $\mathbf{N}$ may be split into a part describing reactions, $\mathbf{R}$, and diffusion $D\,\Delta\mathbf{q}$, where D is the diffusion matrix. In this representation $\mathbf{F}$ represents the spontaneous generation of certain kinds of molecules described by the state vector $\mathbf{q}$. Thus the reaction–diffusion equations read

$$\dot{\mathbf{q}} = \mathbf{R}(\mathbf{q},\alpha) + D\,\Delta\mathbf{q} + \mathbf{F}(t) \tag{3}$$

Of course, equations (2) and (3) describe a huge variety of different processes and it will be a hopeless task to solve these equations in general.

However, the following strategy has proved to be useful and adequate. We assume that for a certain set of control parameters α, some solution $\mathbf{q}_0$ of (2) has been found. In a variety of cases this may be a solution that is time and space independent. We denote this solution by $\mathbf{q}_0$. We now assume that the control parameter α is changed and we study the stability of the solution $\mathbf{q}_0$. In general, in physics the change of α is caused by *external* means; for instance, a temperature difference between the upper and lower surface of a liquid in a vessel may be established. In biology, α may depend on *external* or *internal* causes. For instance, α may be the temperature to which a biological system is subjected, but it may equally well be a concentration of chemicals, for instance of hormones, which are produced in the body itself. To study the (linear) stability of the solution $\mathbf{q}_0$, we make, as usual, the hypothesis

$$\mathbf{q} = \mathbf{q}_0 + \mathbf{w} \tag{4}$$

Inserting (4) into (2), and keeping only the linear parts on both sides of (2), we obtain

$$\dot{\mathbf{w}} = L\mathbf{w} \tag{5}$$

L is a matrix,

$$L = (L_{\mu\nu}) \tag{6}$$

with components

$$L_{\mu\nu} = \frac{\partial N_\mu}{\partial q_\nu} \tag{7}$$

This definition may also be applied if (2) depends on spatial derivatives as in (3). To solve (5), we make the hypothesis

$$\mathbf{w} = e^{\lambda t}\mathbf{v}(\mathbf{x}) \tag{8}$$

which reduces (5) to a time-independent eigenvalue equation of the form

$$\lambda\mathbf{v}(\mathbf{x}) = L\mathbf{v}(\mathbf{x}) \tag{9}$$

In the case of spatially extended media, where diffusion according to (3) may take place, equation (9) is of the form of a partial differential equation. Depending on the boundary conditions, an infinite set of eigenfunctions $\mathbf{v}$ may be found together with their corresponding eigenvalues λ. We distinguish the eigenvalues and eigenfunctions by an index μ. We may distinguish two cases; namely, for the so-called unstable modes, we have

$$\operatorname{Re} \lambda_\mu \geq 0 \tag{10}$$

In this case we shall put the index $\mu = u$ (u = "unstable"). For the other case, we have

$$\operatorname{Re} \lambda_\mu < 0 \tag{11}$$

where we put $\mu = s$ (s = "stable"). A typical spatial dependence of $\mathbf{v}$ may be in the form of sine functions, such as

$$\mathbf{v}(\mathbf{x}) = \mathbf{v}_0 \sin k_x x \sin k_y y \tag{12}$$

but the general form of $\mathbf{v}$ depends on the specific boundary conditions. At least in general, the eigenvalues λ may be complex, so that we put

$$\lambda = \lambda' + i\lambda'' \tag{13}$$

where λ' and λ'' are real quantities.

To study the behavior of the system close to its instability points, we make the hypothesis

$$\mathbf{q} = \mathbf{q}_0 + \sum_u \xi_u(t)\mathbf{v}_u(\mathbf{x}) + \sum_s \xi_s(t)\mathbf{v}_s(t) \tag{14}$$

Equation (14) is still quite general and even valid outside the instability region. However, this form is particularly well suited for the region close to the instability point defined by $\alpha = \alpha_c$. α_c is the critical value of the control parameter α, where Re λ becomes positive. Inserting (14) into (2) and projecting the resulting equations on the eigenfunctions $\mathbf{v}_u$ or $\mathbf{v}_s$, we obtain the equations

$$\dot{\xi}_u = \lambda_u \xi_u + \hat{N}_u(\xi_u, \xi_s) + \hat{F}_u(t) \tag{15}$$

and

$$\dot{\xi}_s = \lambda_s \xi_s + \hat{N}_s(\xi_u, \xi_s) + \hat{F}_s(t) \tag{16}$$

Note that $\hat{N}_u$ and $\hat{N}_s$ are nonlinear functions starting with quadratic or bilinear terms.

At this stage, equations (15) and (16) are entirely equivalent to the original equations (2), and their solution is as difficult. However, we now utilize the fact that the system is close to its instability point. In such a case, λ'_u is a small quantity, whereas $|\lambda'_s|$ is much bigger than λ'_u. This allows one to apply the slaving principle of synergetics, according to which the stable amplitudes ξ_s can be explicitly expressed by the unstable ones, ξ_u, and especially at the same time, so that we obtain

$$\xi_s(t) = f_s(\xi_u(t), t) \tag{17}$$

The explicit time dependence of f_s stems from the fluctuating forces and vanishes entirely if these forces are absent.

It is beyond the scope of this chapter to represent the rather lengthy procedure of how to construct f_s (cf. Haken, 1987). It may suffice to say that in the lowest approximations f_s can be well approximated by a polynomial of low degree in ξ_u. Inserting (17) into (15), we may eliminate the variables ξ_s entirely and end up with equations of the form

$$\dot{\xi}_u = \lambda_u \xi_u + \hat{N}_u(\xi_u, f_s(\xi_u)) + \hat{F}_u(t) \tag{18}$$

There are some subtleties if equations (3) and (15) are stochastic equations of the Îto or Stratonovich type, but we do not have the space here to dwell on these problems and refer the reader to the literature (Haken, 1987).

The importance of the slaving principle as expressed by (17) consists of the following: As could be shown in a great variety of specific cases, the number of order parameters ξ_u is small, whereas the number of stable modes ξ_s is, in general, very large. Thus via (17) and (18) an enormous reduction of the degrees of freedom is achieved, or, in other words, the entire dynamics of a complex system close to its instability points is governed by a few collective variables: namely, the order parameters ξ_u. In this way it becomes possible to discuss complex systems in terms of few adequate variables.

A side remark may be of importance here: In a number of cases, especially in biology, it may not be possible to establish the state vector and the equations of motion at the microscopic or mesoscopic level. Close to instability points, however, the occurrence of macroscopic variables, namely of the order parameters ξ_u, is still guaranteed. In this way one finds a sound basis for modeling the behavior of biological systems at a macroscopic scale via order parameters ξ_u and their corresponding equations (18). But in contrast to the procedure outlined above, we now have to resort to a phenomenological reasoning in order to establish the explicit form of $\hat{N}_u$.

In many cases close to instability points, the order parameters ξ_u may be considered as small quantities. This allows one to expand the right-hand side of (18) into a power series of ξ_u whereby it is sufficient to retain the leading terms: that is, in general, terms up to the third power:

$$\dot{\xi}_u = \lambda_u \xi_u + \sum B_{uu'u''}\xi_{u'}\xi_{u''} + \sum C_{uu'u''u'''}\xi_{u'}\xi_{u''}\xi_{u'''} + \hat{F}_u(t) \tag{19}$$

Equations (19) describe both oscillations and morphogenesis at the level of the order parameters. How this can be worked out is illustrated in the next sections.

III. SOME EXAMPLES

In the simplest case there is only one order parameter ξ_u, which is real and λ' is also real. In this case equations (19) reduce to

$$\dot{\xi}_u = \lambda' \xi_u - \beta \xi_u^3 + F(t) \tag{20}$$

For simplicity, we omitted the quadratic term, ξ_u^2. The right-hand side of (20) can be written in the form of a derivative of a potential V

$$\dot{\xi}_u = -\frac{\partial V}{\partial \xi_u} + F(t) \tag{21}$$

where the potential is given by

$$V = -\frac{1}{2}\lambda'\xi_u^2 + \frac{\beta}{4}\xi_u^4 \tag{22}$$

The potential function (22) is plotted in Fig. 1. As may be seen, its qualitative behavior changes dramatically when λ' passes from negative to positive values. When λ' is negative, the system is stable and relaxes to an equilibrium point $\xi_u = 0$. On the other hand, for positive λ' the system may now acquire two equivalent stability points. If fluctuating forces $F(t)$ are taken into account, the behavior of the order parameters ξ_u can be described as a ball moving in a potential landscape (Fig. 1) and being subject to random pushes. Without fluctuating forces, the time-independent solution ξ_u can be written as

$$\text{for } \lambda' \le 0\text{: } \xi_u = 0, \quad \text{stable} \tag{23}$$

$$\text{for } \lambda' \ge 0\text{: } \begin{cases} \pm\sqrt{\dfrac{\lambda'}{\beta}}, & \text{stable} \\ 0, & \text{unstable} \end{cases} \tag{24}$$

The behavior of ξ_u as a function of the control parameter λ' is plotted in

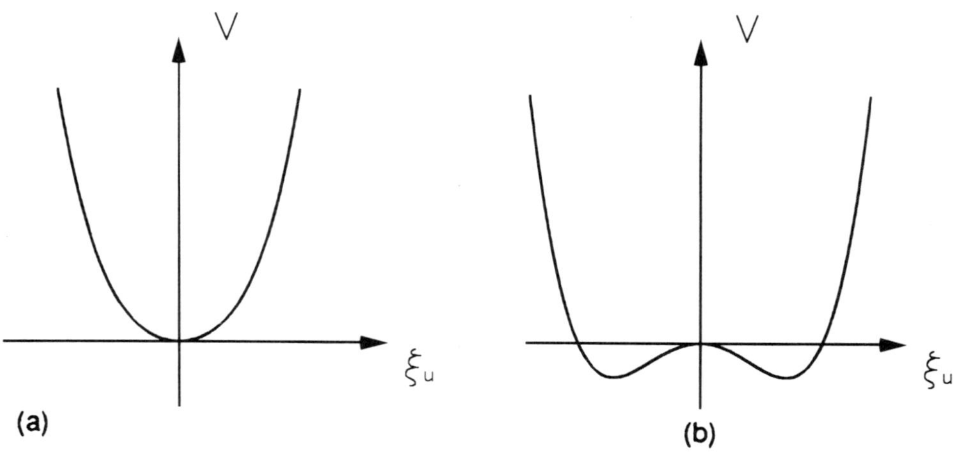

Fig. 1 (a) Potential V as a function of ξ_u for $\lambda' < 0$. Equation (18) can be interpreted as the equation for a particle with coordinate ξ_u moving in the potential V in an overdamped motion. As is obvious, the stable position is at $\xi_u = 0$. (b) Potential function for $\lambda' > 0$. Note the qualitative change compared to the left-hand side of this figure.

Fig. 2. This figure represents a typical bifurcation diagram. Note that our present approach is considerably more general than the conventional bifurcation analysis, because it includes transients, in which ξ_u is truly time dependent and may be subject to fluctuating forces $F(t)$. The relaxation behavior of $|\xi_u|$ for $\lambda' > 0$ in the absence of fluctuations is shown in Fig. 3. In order to treat the case with fluctuations properly, a convenient means is to proceed to the Fokker–Planck equation, although we shall not dwell on this aspect here (Risken, 1988).

Let us apply these results to *pattern formation* (*morphogenesis*). As may be shown, the term in (14) in the first and second sums are of a different size. In general, the pattern is described by the leading part represented by the first sum, which may be called the mode skeleton. In case of a single order parameter, the mode skeleton reduces to a single term, so that

$$\mathbf{q} = \mathbf{q}_0 + \xi_u(t) v_u(\mathbf{x}) \tag{25}$$

Figure 4 illustrates the growth of a pattern beyond the instability point, where $v_u(x) = \sin kx$. Equation (25) shows quite clearly the mechanism

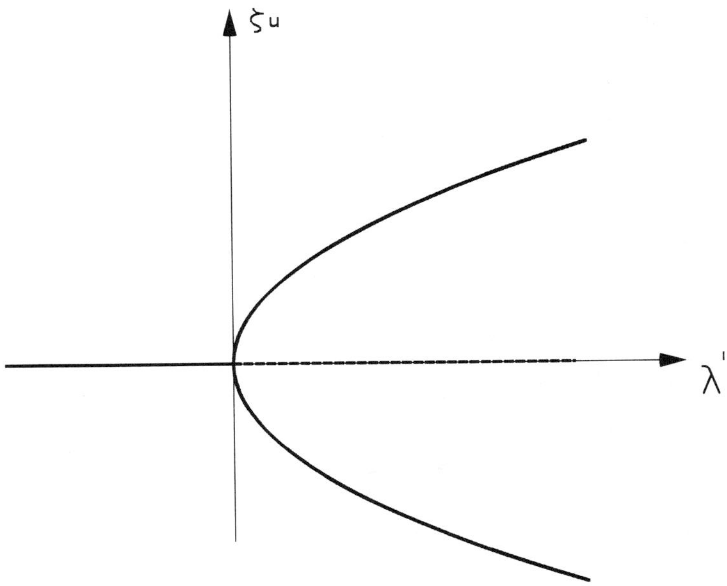

Fig. 2 Bifurcation diagram. ξ_u is plotted versus the control parameter λ'. The branching (bifurcation) is quite obvious.

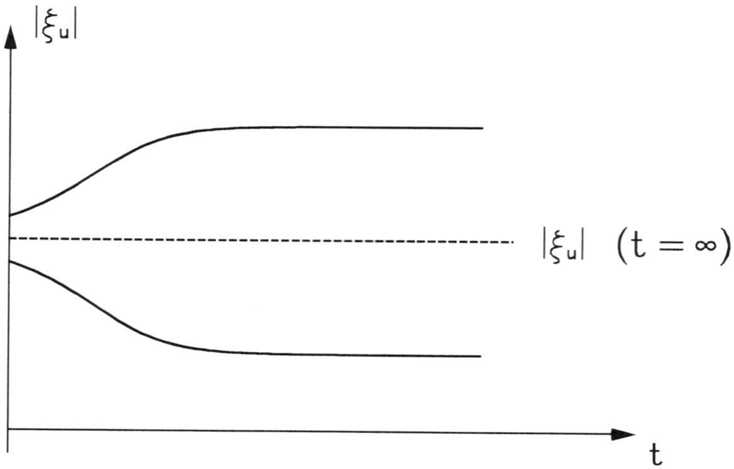

Fig. 3 $|\xi_u|$ as a function of time for two different initial conditions. The solution converges toward the stationary solution given in Eq. (24).

of the entire procedure. While the linearized equations (5) provide us with the spatial form of $v_u(\mathbf{x})$, the order parameter equation (20) provides us with the time-dependent amplitude $\xi_u(t)$. This also holds in the case of several order parameters, in which case the full mode skeleton [i.e., the first sum in (14)] comes into play. The resulting pattern is then a superposition of the elementary patterns $\mathbf{v}_u$ which are fixed by the linearized equations (5) and the boundary conditions. Which linear combination is realized depends on equation (18) or (19), initial conditions, and fluctuations. A number of general remarks may be necessary:

1. Because equations (18) and (19) are nonlinear, there may be different sets of possible stable time-independent solutions, as illustrated by the special case of a single order parameter by means of Fig. 2, where the stable solution may be positive or negative according to (24).

2. Depending on the coefficients B, C, and λ_u, there may be a competition or cooperation of modes. In the first case only one mode survives, so that $\xi_u \neq 0$ for only one index u, whereas in the other case, several nonvanishing ξ_u occur which stabilize each other. Examples of the results on the spatial patterns are shown in Figs. 5 and 6. Whether competition or cooperation occurs can also depend on specific initial conditions in addition to the coefficients mentioned above. As one may show in detail, the same coefficients B and C can be produced by a variety of different nonlinear functions $\mathbf{N}$ in equation (2). Thus, in general, an occurring pattern does not allow us to deduce the underlying microscopic or meso-

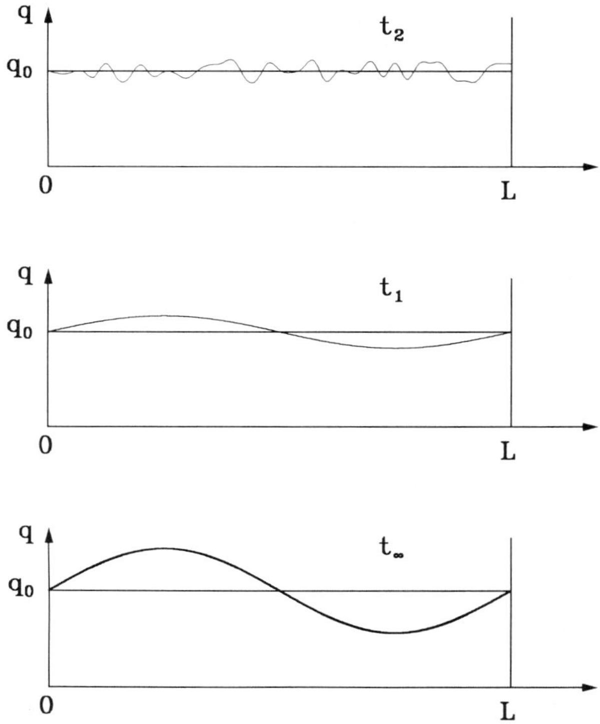

Fig. 4 From top to bottom: time evolution of the variable q from the initial state for $t = 0$ via an intermediate state for $t = t_1$ to its final state. If q describes a concentration of a chemical, its inhomogeneous distribution is clearly seen. The reaction diffusion equations leading to such a kind of behavior may be considered as models for the formation of stripes on zebras, to mention just one example.

scopic mechanism in detail. A pattern can give us at best some hints on the kind of the underlying processes.

Now let us turn to *oscillations*. Let us consider the case in which the order parameter ξ_u and the eigenvalue λ are complex. The simplest order parameter equation reads

$$\dot{\xi}_u = (\lambda' + i\lambda'')\xi_u - \mu|\xi_u|^2\xi_u + F(t) \tag{26}$$

By means of the hypothesis

$$\xi_u = e^{i\lambda'' t}\nu_u(t) \tag{27}$$

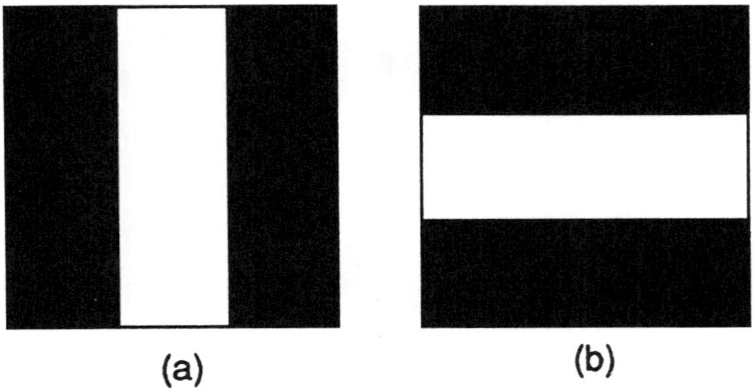

Fig. 5 Spatially inhomogeneous patterns in a square. Their final formation depends on the initial conditions. Here the case of competition (i.e., of a single-order parameter) is considered. In (a) one order parameter has won the competition; in (b), another.

we may transform equation (26) into the form (20), where η_u replaces ξ_u. Equation (26) thus describes an oscillation with the frequency λ'', but the kind of solution remains the same, as for $\lambda' < 0$ v relaxes to zero, whereas for $\lambda' > 0$ it approaches a stable value η_0 [compare (24)]. In the latter case, an autonomous oscillation occurs.

Let us briefly discuss some important features of equations (19) in the case of several order parameters, where at least part of the λ_u's are complex. Depending on the size of the coefficients λ_u, B, and C, quite

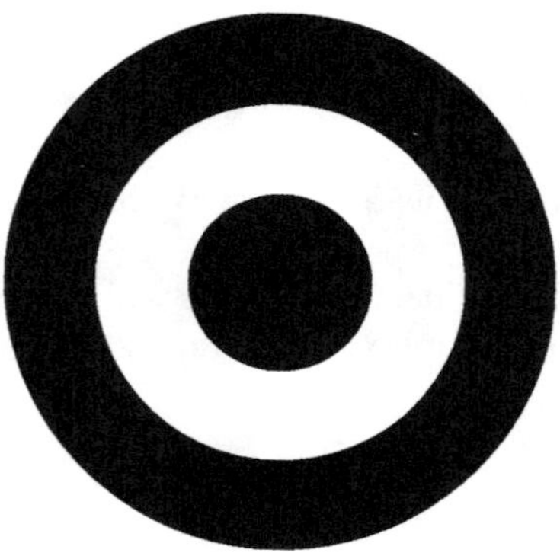

Fig. 6 Circular pattern of chemical concentrations.

different phenomena may occur, among which the following are particularly well known and important:

1. *Free-running oscillators.* In this case the interaction between different mode amplitudes ξ_u is such that the nature of the oscillatory motion according to (27) is not changed in principle, but where the interaction between the modes may lead to frequency shifts and to new values of amplitudes η_u. $\mathbf{q}$ is then a superposition of oscillations with different frequencies.

2. *Frequency locking.* If the difference between the frequencies λ_u'' and $\lambda_{u'}''$ for some u, u' becomes smaller than a critical value, the oscillation may occur at the *same frequency*, so that only one frequency is observed. In addition, the phases between the modes become locked. Frequency-locking phenomena may occur between two mode amplitudes as well as between several or many. A number of physiological systems (heartbeat, breathing, etc.) may show frequency locking, where $\lambda_1'' = (m/n)\lambda_2''$, with m and n integers (e.g., 3 and 4). In other cases of mode locking, the frequencies may have a constant spacing so that $\lambda_u'' = \lambda_0 + u\omega$, $u = 0, \pm 1, \pm 2, \ldots$. If the mode amplitudes are connected with spatial modes, which represent running waves, regular pulses may appear (Fig. 7).

3. *Quasi-periodic motion.* In other cases, the interaction between the modes may lead to amplitudes ξ_u which can be represented as a superposition of the following form:

$$\mathbf{q} = \mathbf{q}_0 + \sum \mathbf{q}_{m_1,\ldots,m_N} e^{i(m_1\omega_1 + \cdots + m_N\omega_N)t}$$

In such a case one speaks of a quasi-periodic motion.

IV. CHAOS

In a number of cases, order parameter equations can give rise to *deterministic chaos*. Two by now well-known examples are provided by the Rössler system and the Lorenz system. The equations for the state vector $\mathbf{q} = (q_1,q_2,q_3)$ of the Rössler system read

$$\begin{aligned}\dot{q}_1 &= -q_2 - q_3\\ \dot{q}_2 &= q_1 + aq_2\\ \dot{q}_3 &= b + q_3(q_1 - c)\end{aligned} \tag{28}$$

whereas the equations of the Lorenz system are given by

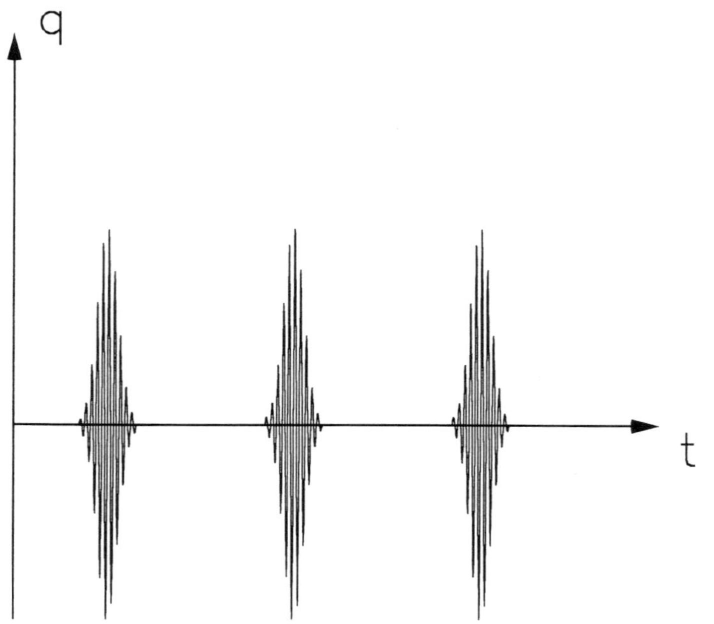

Fig. 7 Example of a pulse solution.

$$\begin{aligned}\dot{q}_1 &= \sigma(q_2 - q_1)\\ \dot{q}_2 &= q_1(r - q_3) - q_2\\ \dot{q}_3 &= q_1 q_2 - b q_3\end{aligned} \tag{29}$$

Quite generally, the behavior of $\mathbf{q}$ or ξ can be visualized as follows: The solution described by the variables $\xi_1, \ldots, \xi_N$, where each ξ_j is a function of time, can be represented as a point in phase space that is spanned by the components ξ_j. In the course of time, this point runs through a so-called trajectory. In the case of the order parameter equation (20), such a trajectory approaches one fixed point ($\lambda' < 0$) or two fixed points ($\lambda' > 0$) (Figs. 8 and 9). In the case of equation (26), the trajectories approach a so-called limit circle or lie on a limit circle provided that $\lambda' > 0$ (Fig. 10); otherwise, they spiral down to a so-called focus (Fig. 11). Quasi-periodic motion can be visualized by means of a trajectory lying on a torus (or a doughnut) (Fig. 12). *Chaotic motion* can be characterized as follows: There is a region in phase space, the chaotic attractor, to which all nearlying trajectories are attracted. When the trajectory has entered the chaotic attractor, it moves around within this area forever

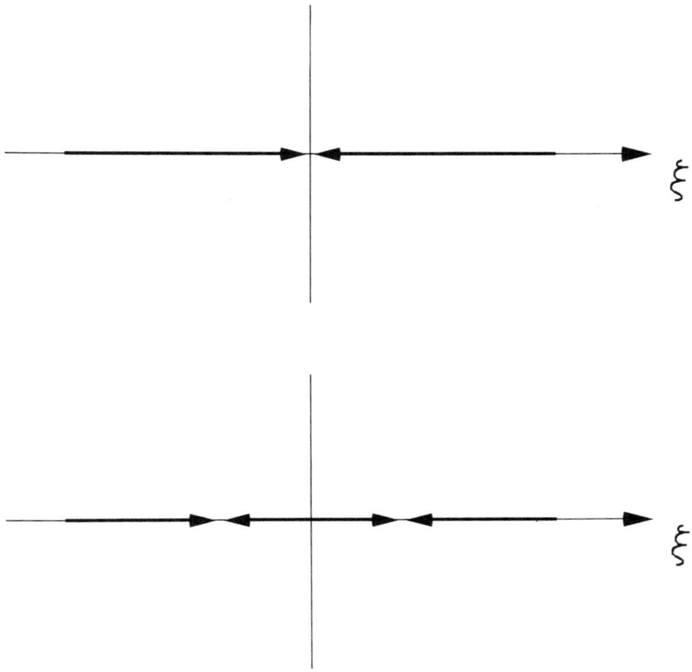

Fig. 8 (a) Trajectories move to a single fixed point; (b) trajectories move to two stable fixed points.

(Fig. 13). If we consider two trajectories, which at a certain instant are closely together, their distance D increases exponentially, $D = D_0 \exp(\lambda t)$, where λ is called the Lyapunov exponent. For the occurrence of chaotic motion at least one Lyapunov exponent must be bigger than zero. As may be shown, for the occurrence of chaotic motion at least three order parameters are needed. The trajectories do not fill the phase space fully, but only with a fractional dimension. There are several definitions of how to measure such a dimension d. One possibility is

$$d = \frac{\ln N}{\ln r} \tag{30}$$

where r is the radius of a sphere in an m-dimensional space and N the number of points enclosed in that sphere. In this definition it is assumed that $r \to 0$. Equation (30) is obtained by generalizing the definition of a

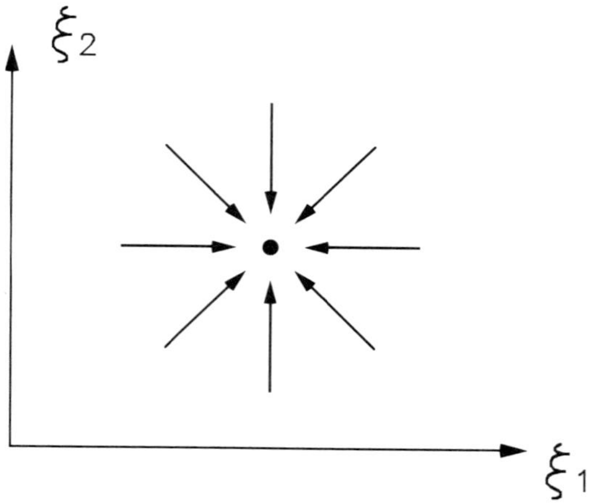

Fig. 9 Trajectories in the neighborhood of a stable fixed point in a two-dimensional example.

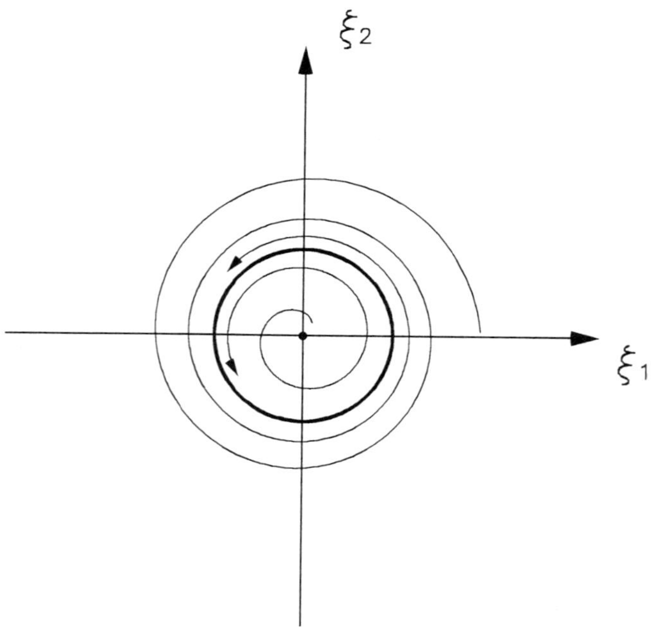

Fig. 10 Limit circle in a plane.

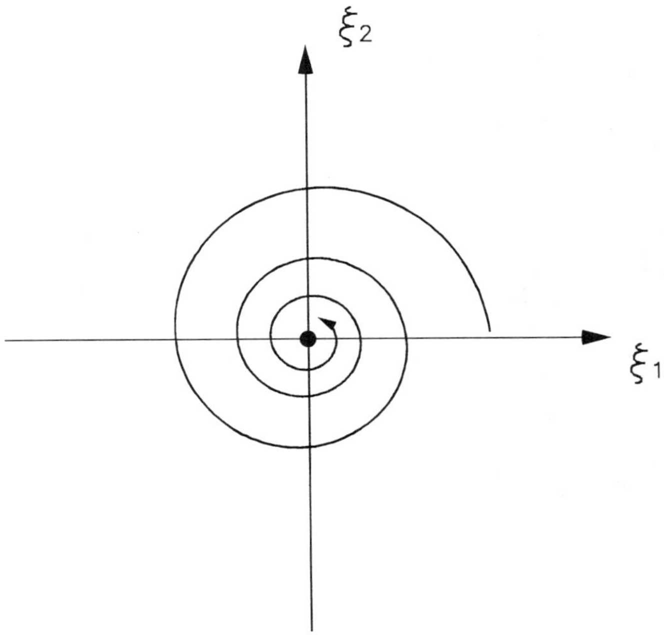

Fig. 11 Stable focus.

dimension along a line, within a circle, or within a sphere, where $N \propto r^d$ for $d = 1, 2, 3$, respectively.

Another important concept is that of the Poincaré cross section, which is explained in Fig. 14. In this way an originally continuous motion can be described by a sequence of discrete points, x_n. When we plot the point x_{N+1} versus its previous point, x_N, we obtain a discrete map. A well-known example, the logistic map, is shown in Fig. 15. For a height h of this parabola bigger than h_c, the sequence of points x_N becomes chaotic. The transition to chaos shows remarkable regularities.

V. MORE COMPLICATED SPATIOTEMPORAL PATTERNS

If the eigenvalue spectrum of equation (9) is *continuous* and crosses the imaginary axes, equations (15) and (16) are no more sufficient. In such a case, the linear part $(\lambda'\xi_u)$ in equation (15) must be replaced by an operator acting on the spatial variables, where ξ_u becomes a slowly varying function of space in addition to its dependence on time (Haken, 1987). A

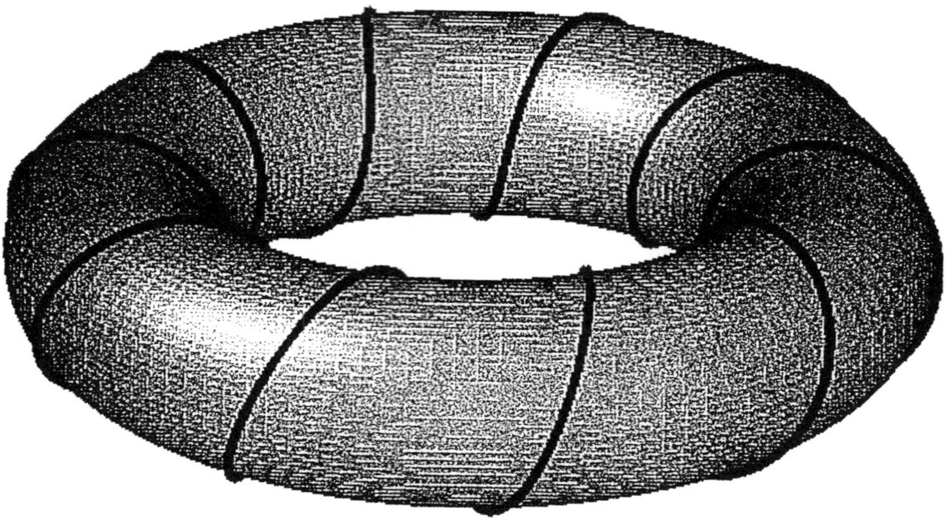

Fig. 12 Visualization of trajectories on a torus.

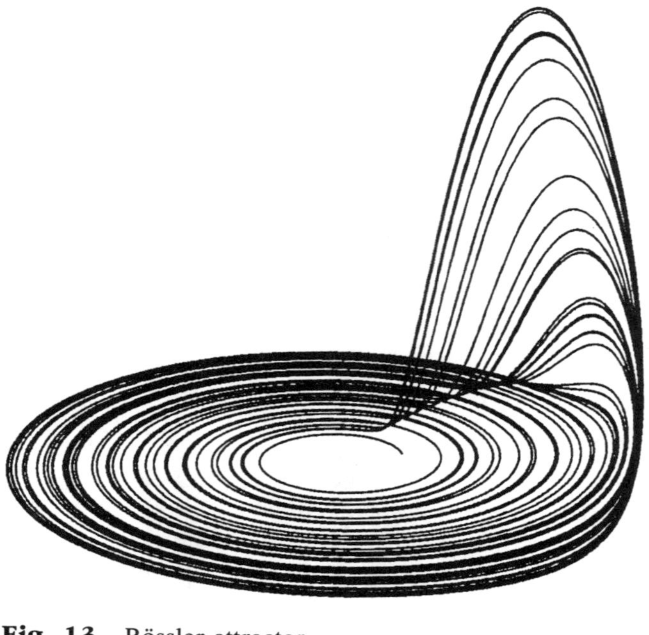

Fig. 13 Rössler attractor.

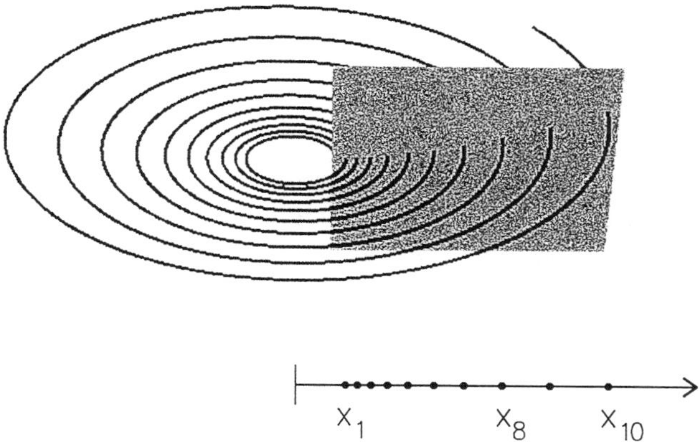

Fig. 14 When the trajectories (e.g., in three-dimensional space) are cut by a plane, the plane is hit in a sequence of points. This sequence is plotted in the lower part of this figure, where an appropriate coordinate connecting the individual points of the upper part is introduced.

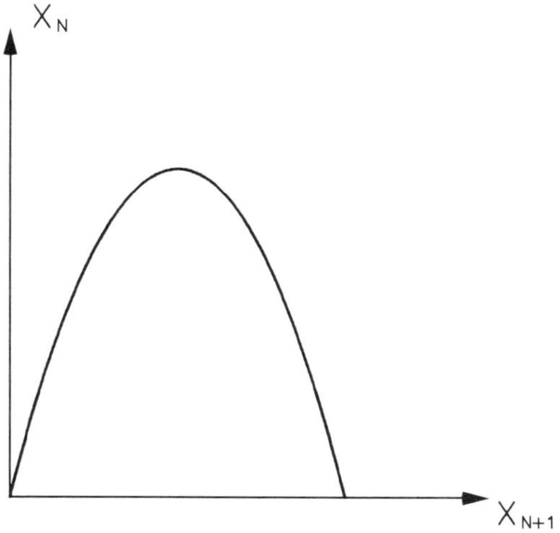

Fig. 15 Logistic map, where x_{N+1} is plotted versus x_N. Depending on the height of the parabola, different kinds of behavior of x as a function of N are possible (e.g., a fixed point, periodic behavior with higher periods 2, 4, 8, . . . , and chaos).

typical dependence of λ' on spatial derivatives reads $\lambda' = (\lambda_0 + a\Delta)$, where Δ is the Laplace operator. As shown, in such a case a number of new phenomena occur: in particular, concentric and spiral waves. Such spiral waves can be found in certain chemical reactions (e.g., the Belousov–Zhabotinski reaction), in the propagation of concentration waves of cAMP molecules during the first stage of assembly of the cells of aggregating slime mold (*Dictyostelium discoideum*), and in excitation waves of the heart. Phenomena of this sort are also reported of the EEG potentials in epileptic seizures. While such spiral waves can be found rather easily in numerical simulations, their analytic treatment has been confined to a few papers only; I mention in particular the book by Kuramoto (1984).

REFERENCES

Haken, H. (1977). *Synergetics, An Introduction*, Springer-Verlag, Berlin.

Haken, H. (1987). *Advanced Synergetics*. Springer-Verlag, New York.

Kuramoto, Y. (1986). *Chemical Oscillations, Waves, and Turbulence*, Springer Series in Synergetics, Springer-Verlag, New York.

Rensing, L., Mackey, M., and an der Heiden, U. (1987). *Temporal Disorder in Human Oscillatory Systems*, Springer-Verlag, Heidelberg.

Risken, H. (1988). *The Fokker–Planck Equation*, 2nd ed., Springer-Verlag, New York.

Turing, A. M. (1952). *Philos. Trans. R. Soc. B*, *234*: 37.

II
Oscillatory Signals in Morphogenetic Fields

7
The Role of Periodic Signals in the Morphogenesis of *Dictyostelium discoideum*

Florian Siegert and Cornelis J. Weijer *Zoological Institute, University of Munich, Munich, Germany*

I. INTRODUCTION

Morphogenesis, the process by which cells differentiate and are organized in space and time, is a central problem in biology. There are two totally different possible mechanisms to set up a spatial pattern: (1) Cells differentiate in situ and the problem reduces to the generation of a spatial morphogen gradient. (2) Cells differentiate position independent and then move to their proper destination, a process called cell sorting. Morphogenesis in the cellular slime mold *Dictyostelium discoideum* is a prime example of the latter mechanism. Single cells collect in an aggregate by chemotaxis to periodic signals. In the aggregate the cells differentiate in a random fashion and then sort out to form a simple spatial pattern in the multicellular slug stage. The slug is a three-dimensional excitable medium and is organized by a variety of twisted scroll and helicoidal waves. The cellular slime mold therefore provides a suitable system for the study of the relationship between cellular communication, biological oscillations, and differentiation.

II. THE LIFE CYCLE

Amoebae of *Dictyostelium discoideum* live in the soil and feed on bacteria. They multiply as single cells. The developmental cycle is initiated when the food source is exhausted. Some cells on the substratum start to secrete periodic cyclic adenosine monophosphate (cAMP) signals. If surrounding cells detect the cAMP signal with their cell-surface cAMP receptors, they respond with the production of cAMP and chemotactic movement toward the autonomous oscillator (Durston, 1974a,b). This leads to periodic cell movement toward the aggregation center. Streams are formed by cell–cell contacts, and after 4 h all cells of a field are gathered in a multicellular aggregate (mound) consisting of 10^3 to 10^5 cells. The mound transforms either directly into the fruiting body, which consists of two cell types, the stalk cells and spores, or alternatively, into a migratory slug. External signals such as light, temperature, or humidity direct slug migration and regulate culmination.

In the mound stage, the cells begin to differentiate in at least three cell types—prestalk, prespore, and anterior-like cells—which are initially found distributed at random. They then sort out to form a relatively simple pattern with prestalk cells in the anterior one-quarter of the slug and prespore cells in the back three-quarters of the slug. The anterior-like cells are distributed at random in the prespore zone; they resemble prestalk cells but will sort out from them (Sternfeld and David, 1981). Morphogenesis from the tipped aggregate stage onward is guided by the tip, a prominent structure, which forms during the mound stage. The tip shows all the properties of a classical organizer (Spemann, 1938; Raper, 1940): It determines the polarity of the prespore–prestalk pattern and coordinates slug movement and fruiting body formation.

The cellular communication system during aggregation is already well characterized. Still largely unknown are the exact mechanisms by which cells in the later stages of the development communicate. There is, however, increasing evidence that periodic signals secreted by the tip and chemotaxis to these signals organize cell movement and differentiation in postaggregative development. In this chapter we summarize the evidence that all aspects of morphogenesis can be accounted for by a mechanism involving periodic signals and chemotactic cell sorting. We start with a short summary of what is known about these signals during early aggregation and then summarize the evidence for their involvement in slug morphogenesis.

III. EARLY AGGREGATION

The aggregation process is brought about by three cellular competences:

1. Periodic production and secretion of cAMP into the medium by the cells in the aggregation center (pacemaker).
2. Detection of this signal by surrounding cells via their cell-surface cAMP receptors, followed by amplification of the cAMP signal through the activation of adenylate cyclase (relay response). This

response shows adaptation, which ensures outward propagation of cyclic AMP waves. Extracellular cAMP is continuously destroyed by membrane-bound and secreted phosphodiesterases.

3. Chemotaxis toward increasing cAMP concentrations. Cells respond to temporal increases in cAMP and therefore cells aggregate unidirectionally in the direction of the aggregation center (Devreotes, 1989).

The pulsatile nature of the cAMP signal can be visualized using dark-field optics. Aggregating amoebae on nonnutrient agar show propagating waves of light and dark bands which form concentric rings and/or spirals (Fig. 1). The light bands are composed of elongated inwardly moving cells, while the amoebae in the dark bands are not moving directionally

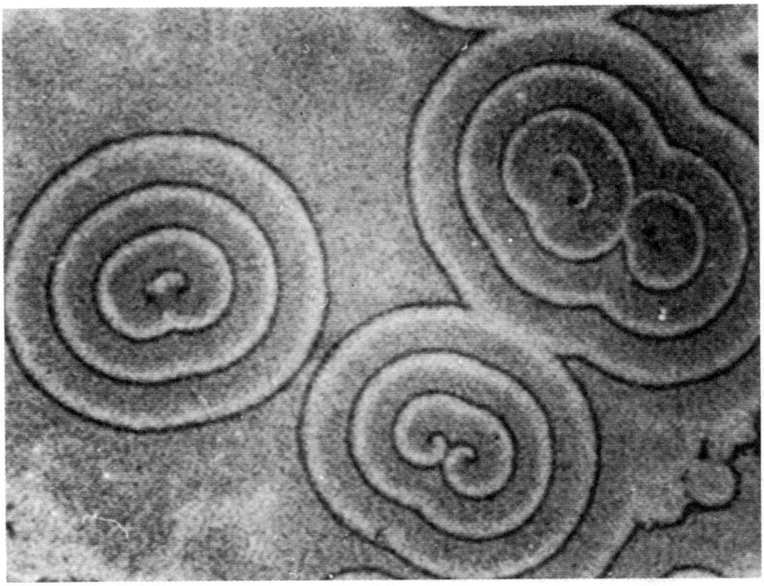

Fig. 1 Aggregation waves seen in strain AX-2 on agar plates containing 2 m*M* caffeine. The cAMP waves are seen as optical density waves in a dark field. The waves move from the center outward. The waves' leading edge is seen as a small dark band where the cells are rounded up (cringe response) at the detection of cAMP. This band is followed by a lighter band of cells actively engaged in chemotaxis. A darker gray band follows where the cells sit around and stick out pseudopods in order to try to detect new signals. Several centers are visible, all are scroll waves.

(Alcantara and Monk, 1974). The relay of the cAMP signal is visible as an outward-moving ring or spiral wave.

Different strains and mutants show different dark-field patterns. In aggregation fields of the wildtype strain NC4, concentric rings are predominant (Durston, 1974a), while in the axenic strain AX-2 concentric rings are rare and inevitably convert to spirals during aggregation. Centers emitting concentric rings consist of cells that autonomously secrete periodic cAMP signals. Spiral centers are set up by local differences in cAMP concentration, which break the symmetry of cAMP diffusion and are self-preserving (i.e., they do not require autonomous signaling) (Durston, 1974b). Once initiated, the signal is relayed around a core of cells. The period of the signal and the wave propagation speed determine the diameter of the core. Spiral centers often oscillate with frequencies higher than those of concentric ring centers, because the core evolves to a minimum size that is defined by the minimal refractory period of the cells. This agrees with the continuous decrease in period length between successive dark-field spiral waves.

Another observation shows that early morphogenesis is influenced by the frequency of the oscillator. Caffeine, a specific inhibitor of the cAMP relay (Brenner and Thoms, 1984), reduces the wave propagation speed and oscillation frequency (Siegert and Weijer, 1989). At concentrations above 2 m*M* caffeine, one observes a widening of the circumference of the center loop during the aggregation process (Fig. 2). This process continues until the loop breaks up to form several small mounds at the end of aggregation, which shows that the cells are still able to oscillate autonomously.

During aggregation the frequency of cAMP waves increases, which results in a decrease of the outward-moving velocity of successive waves. This negative correlation between oscillation frequency and wave propagation speed, the so-called dispersion relation, is also observed in the Belousov–Zhabotinski reaction (Pagola et al., 1988; Dockery et al., 1988). The explanation for this phenomenon is that when frequency increases, successive waves travel through a not yet completely recovered medium, and therefore the medium takes a longer time to become excited, which results in a reduction of wave propagation speed. This implies that diffusion of the signal from cell to cell is not the rate-limiting step in wave propagation but is determined by the kinetics of the underlying chemical reaction.

To obtain stable spirals there has to be an influence of curvature of the spiral wavefront on wave propagation velocity (i.e., the wavefront close to the aggregation center), showing that high curvature should propagate more slowly than the wavefront farther away from the center (describing

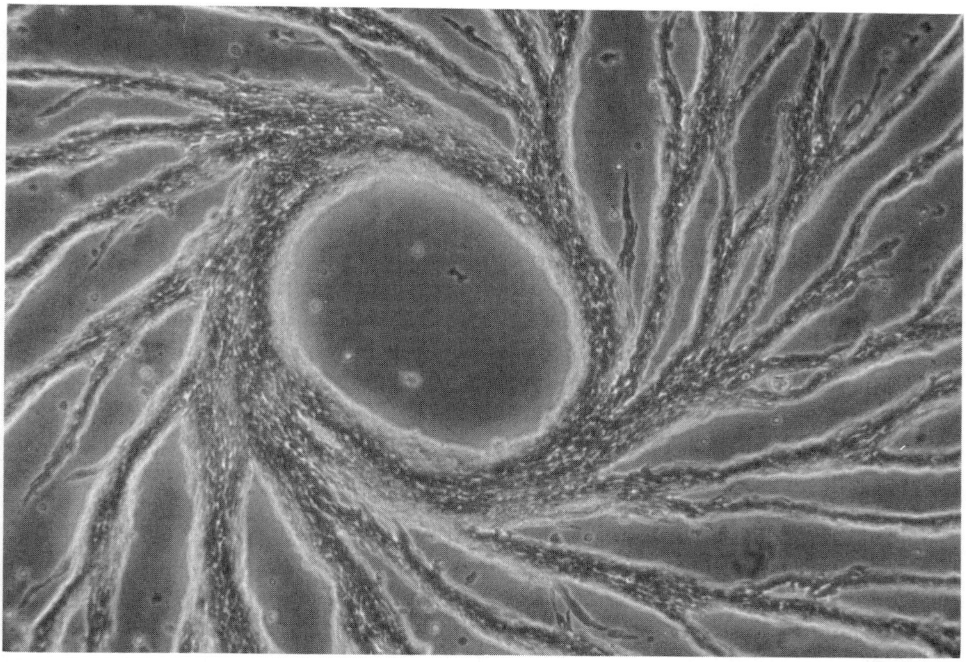

Fig. 2 Circular aggregation center of AX-2 cells aggregating on 2 m*M* caffeine. The cells rotate around in a loop while other cells are still flowing in from aggregation streams. Cell movement is clockwise and wave propagation is counterclockwise.

a greater radius of rotation). This curvature relationship was derived from theoretical model calculations (Keener and Tyson, 1986). For negative curvatures (waves curving away from the direction of propagation), the wave propagation velocity decreases, while for positive curvatures (waves curving in the direction of propagation as in the cusp of two colliding waves) it increases. This implies a minimum radius for the excitation center, below which wave propagation decays. Curvature was measured in spiral waves of both the Belousov–Zhabotinski reaction (Foerster et al., 1988) and in *Dictyostelium* and is in good agreement with theory. The minimum radius for aggregation centers on 2 m*M* caffeine was calculated to be 130 μm (Foerster et al., 1990). Our own calculations, based on signal propagation velocity and frequency, resulted in a radius of 100 to 200 μm.

IV. LATE AGGREGATION

Dark-field waves can only be seen for about 20 periods during early aggregation and disappear when cells form aggregation streams. The cAMP signaling system is difficult to study later in development since cell behavior is not visible, due to firm cell–cell contacts. A few reports based mainly on the movement of neutral red-stained cells have been published (Durston and Vork, 1979; Clark and Steck, 1979; Takeuchi et al., 1988). We have developed a method whereby we label cells fluorescently (with the cell type nonspecific label rhodamine–dextran) at the beginning of development and study the movement and behavior of individual cells in streams and slugs. Once loaded into the cells the rhodamine–dextran stays evenly dispersed in the cytoplasm until final differentiation. Digital image processing allows us to follow the tracks of several single cells simultaneously and to analyze velocity, frequency, direction of movement, and changes in cell shape (Siegert and Weijer, 1991).

Figure 3 shows that cell movement is periodic during early aggregation, in streams, and in slugs. Just before making cell–cell contacts, single cells of the axenic strain AX-2 move with an average period of 4.95 min, cells in streams with an average period of 3.3 min (Fig. 4). This confirms our earlier observation (Siegert and Weijer, 1989) that the frequency of cAMP signal propagation decreases during aggregation and decreases further after the disappearance of dark-field waves. As can be seen in Fig. 3, individual cells do move periodically in streams, although the periodicity is not as clear as in the dark-field waves. This must be attributed to the fact that we are now looking to the response of an individual cell and not to a population response as in the case of dark-field waves.

V. POSTAGGREGATIVE DEVELOPMENT

A. Evidence for Periodic Signals in Slugs

The main objective of this chapter is to summarize the evidence for periodic signaling and chemotaxis in the slug and culmination stages of development. During later development all morphogenetic movements are organized by the tip, a distinct morphological structure, which is present from late aggregation onward. The tip behaves as an organizer because it will induce a secondary axis by taking over part of the tissue behind it to form a new secondary slug when it is grafted into the side of another slug (Rubin and Robertson, 1975). Transplantation experiments have shown that the tip inhibits the formation of new tips (Durston, 1976). If a tip is transplanted into a slug containing a tip, its success of

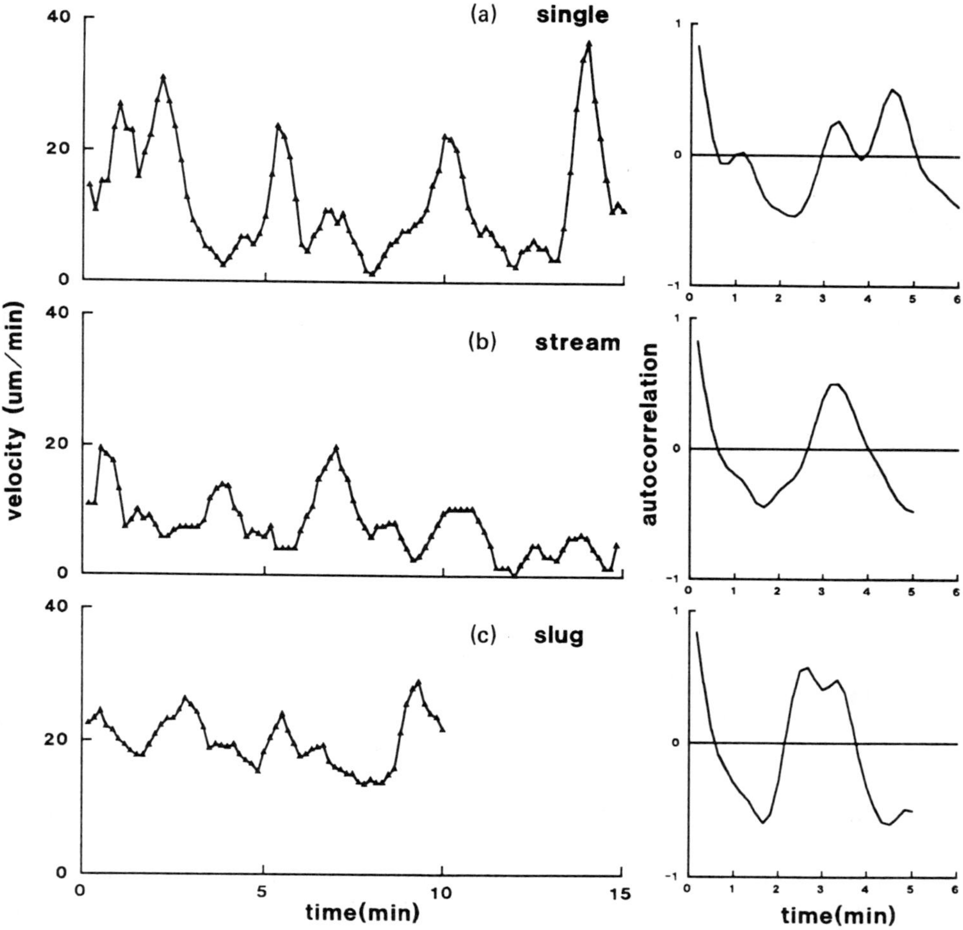

Fig. 3 Rate of movement of single cells at different developmental stages. The rate of movement of fluorescently labeled cells was measured by calculating the displacement of the cells' center of mass between two successive measurements (10 s). (a) Single cell in the early aggregation stage; (b) cell in an aggregation stream; (c) cell in the prespore zone of a slug. Also shown are the corresponding autocorrelation periodograms showing periods of movement between 3 and 2.5 min.

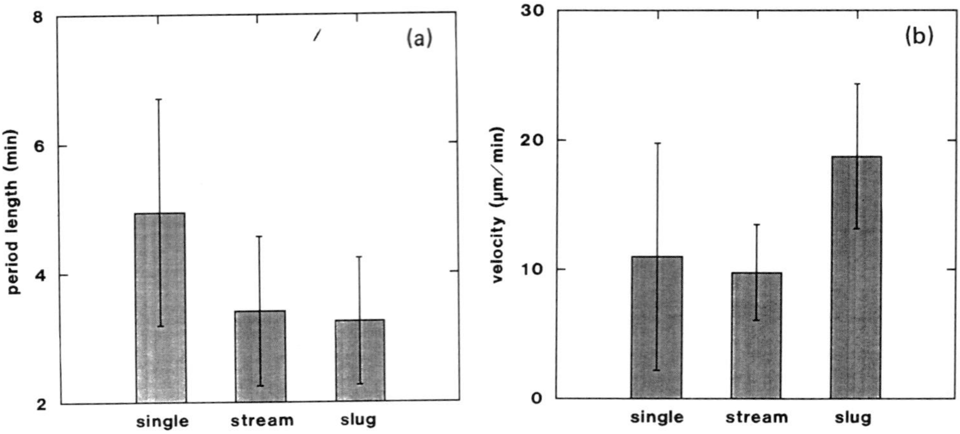

Fig. 4 Period and velocity of cell movement at different developmental stages: (a) mean period of cell movement for single cells at the stage before entering streams, in aggregation streams and in the prespore zones of slugs; (b) mean velocity of cell movement in single cells, cells in aggregation streams, and cells in the prespore zones of slugs.

forming a new tip is greater the larger the distance between the original tip and the grafted tip. If the slug is decapitated before tip transplantation, the rate of success of new tip formation increases dramatically. Tissue from the region of the tip is much more effective in inducing secondary tips in host slugs as tissue from the prespore zone of a slug, a phenomenon called tip activation (MacWilliams, 1982). Both these properties, tip activation and tip inhibition, can readily be explained by waves and oscillations. The tip inhibition signal is a periodic signal suppressing the emergence of autonomous centers, while tip activation is correlated with the excitability of the cells tested. The fact that prestalk cells form tips much more readily than prespore cells shows that prestalk cells are more excitable than prespore cells.

If the tip (less than 10% of the slug) is removed and placed on agar, it continues to move. The prespore piece stops immediately, rounds up, and forms a mound. The anterior-like cells in the prespore piece sort out and form a new tip. During this time some of the prespore cells redifferentiate to prestalk and anterior-like cells until a normal proportioned slug is formed and morphogenesis continues (Raper, 1940). This suggests that the tip coordinates cell differentiation and slug movement. These properties of the tip can be explained by assuming that it is a pacemaker for cAMP signals. Circumstantial evidence for periodic cAMP pulses as signal comes from experiments in which dissected slug tips have been shown to attract aggregation competent amoebae in a periodic fashion (Rubin and Robertson, 1975).

B. Relay Inhibitor Caffeine Removes Tip Inhibition

Further evidence for the involvement of periodic cAMP signals in pattern formation comes from experiments where slugs are treated with cAMP relay inhibitors. Neutral red-stained slugs placed on agar containing 5 m*M* caffeine stop migrating immediately. The anterior-like cells start to form local aggregation centers, precursor structures for new tips. The formation of these structures is completed after 2 h; however, further development into tips is blocked. After removal of caffeine these foci transform almost immediately into tips, which organize the surrounding tissue to slugs (Fig. 5). These experiments indicate that cAMP relay is the tip inhibition signal and that reduction of the signal amplitude and oscillation frequency by caffeine removes tip inhibition. The reduced oscillation frequency might be responsible for the inability of the aggregated anterior-like cells to complete tip formation.

C. Cell Sorting in Slugs Involves Chemotaxis to cAMP

There is good evidence that chemotaxis to cAMP is involved in cell sorting in slugs. If slugs, stained with the prestalk and anterior-like cell specific vital dye neutral red, are mixed on agar plates, the prestalk cells will sort out to the top of the aggregate, form a new tip, and then the aggregate forms a slug. If the same experiment is performed on plates containing cAMP, the prestalk cells will sort to the periphery of the aggregate, showing that prestalk cells sort preferentially to cAMP. It has also been shown that movement of the cells is directed, suggesting that the mechanism of movement is chemotaxis to cAMP (Matsukama and Durston, 1979). If dissociated neutral red-stained slug cells are mixed and reaggregated in suspension, the prestalk cells will sort to the center of the aggregate. If cAMP at 10^{-6} *M* is added to the buffer, the prestalk cells will sort to the periphery of the aggregate, showing that cAMP at this concentration interferes with the aggregate's own signaling system (Sternfeld and David, 1981).

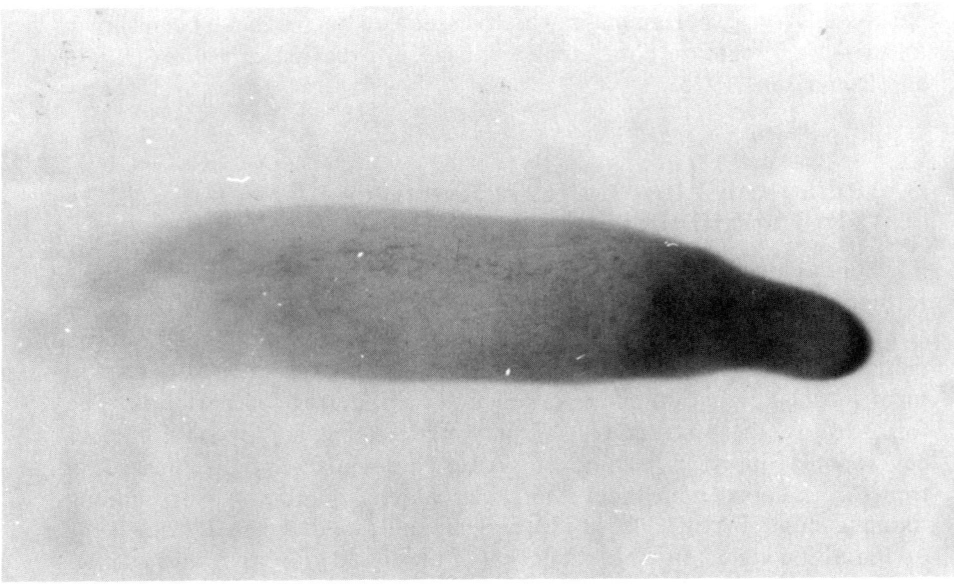

(a)

Fig. 5 Disappearance of tip inhibition in slugs on the relay inhibitor caffeine. (a) Photograph of a neutral red-stained slug. (b) Photograph of a slug after incubation on agar containing 5 m*M* caffeine for 2 h. The anterior-like cells in the prestalk zone have sorted out to form aggregation centers in the prespore zone of the slug. Development of the centers is arrested at a stage before tip formation. (c) Slug 2 h after being removed from caffeine containing agar to normal agar. The cells in the aggregation centers in the prespore zone have developed into tips characterized by small circumference. Each center normally forms a tip, except the original prestalk cells. Most likely they have irreversibly differentiated to stalk cells.

D. Cell Movement in Slugs Is Directed and Periodic

To test the hypothesis that the tip is a pacemaker and secretes periodic cAMP signals, it will be necessary to measure these signals in slugs. Until now it has not been possible to measure cAMP directly in three-dimensional structures as done for two-dimensional aggregation fields (Tomchik and Devreotes, 1981). Since the well-characterized dark-field waves are no longer visible in slugs, we used an indirect approach and investigated the cell movement response in slugs. By digital image processing we measured the velocity and periodicity of individual cells and the tracks

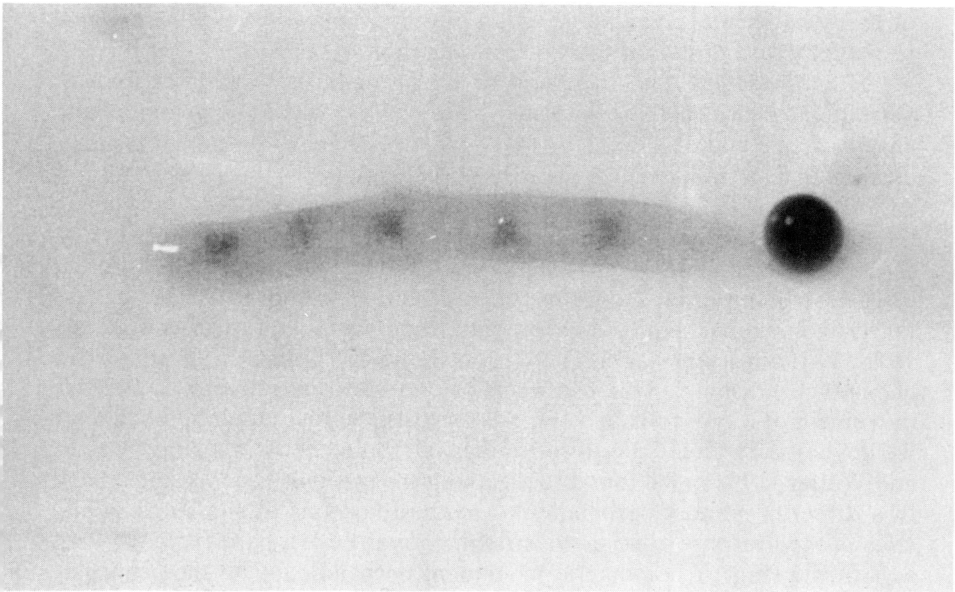

(b)

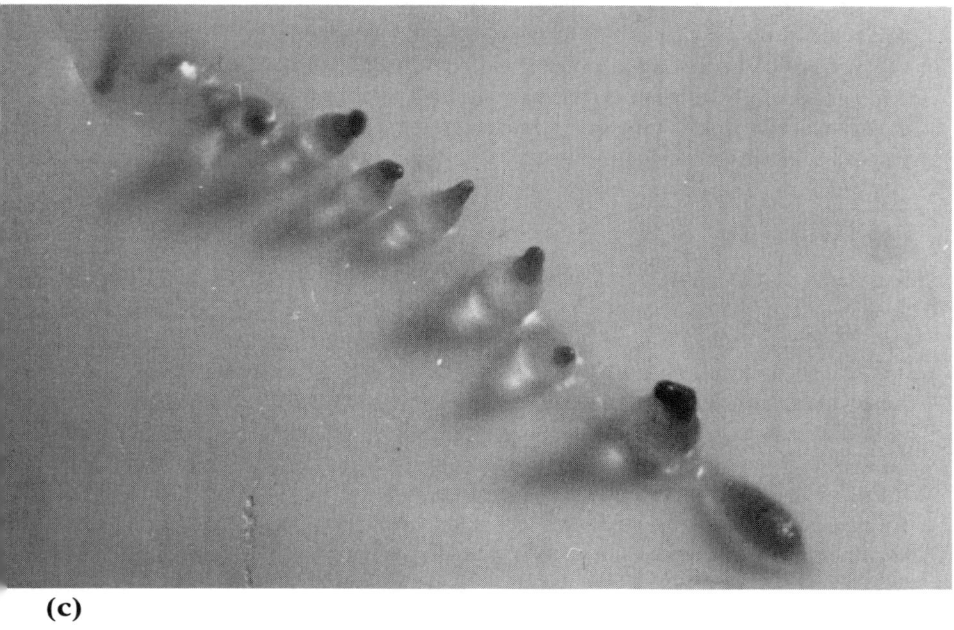

(c)

of many cells in a given slug. From these parameters we deduce signal periodicity and direction of wave propagation.

It appears that the movement of single cells in slugs is at least as periodic as that of cells in aggregation streams. Cells in slugs move with an average period of 3 min (Fig. 4). The periodicity of the signal is not as clear as in dark-field waves, for the same reasons as mentioned for aggregation streams (Fig. 3). Furthermore, it can be seen that the cells change their shape in a characteristic fashion during the periodic movement (Fig. 6). At the times of slow movement they stick out pseudopods in different directions before they decide in which direction to continue. This behavior is well known from cells moving chemotactically (Gerisch et al., 1975; Varnum-Finney et al., 1987) and shows that cells in slugs move in a chemotactic fashion. This technique allows us to investigate the precise movement of a few cells in a slug. To investigate the population behavior we developed a technique to follow many labeled cells in a slug (Siegert and Weijer, 1991). We found that cells in the prestalk zone of slugs move in a direction almost perpendicular to the long axis of the slugs, while cells in the prespore zone move straightforward (i.e., parallel to the long axis of the slug). Chemotactic movement perpendicular to the propagating wavefront in the prestalk zone implies that the wavefront must extend along the long axis of the tip. Therefore, the signal in the slug tip is a (twisted) scroll wave which decomposes in a wave traveling perpendicular to the long axis of the slug in the prespore zone (Fig. 7). An average slug length of 1 mm, a period length of 3 min, and a cAMP wave propagation speed of 200 μm/min (the lowest dark-field wave propagation velocity measured) imply that only one or two cAMP waves travel through a slug at a given point in time.

E. Model for Wave Propagation in Slugs

The decomposition of scroll waves into twisted scroll waves and then into planar waves was recently also observed in a three-dimensional Belousov–Zhabotinski reaction system along a concentration gradient of substrate (Yamaguchi and Müller, 1991). The change in the geometry of the propagating waves in *Dictyostelium* is in our opinion probably caused by a difference in the oscillatory properties of the prestalk and prespore cells. We have shown previously that one can separate aggregation stage cells that will sort to the tip of the slug from those that will sort to the back of the slug. The cells that will sort to the front have higher intrinsic oscillation frequencies than cells that will sort to the back of a slug (Weijer et al., 1984). Thus prestalk cells in the tip could dominate the remaining cells in the slug by their higher oscillation frequency. This

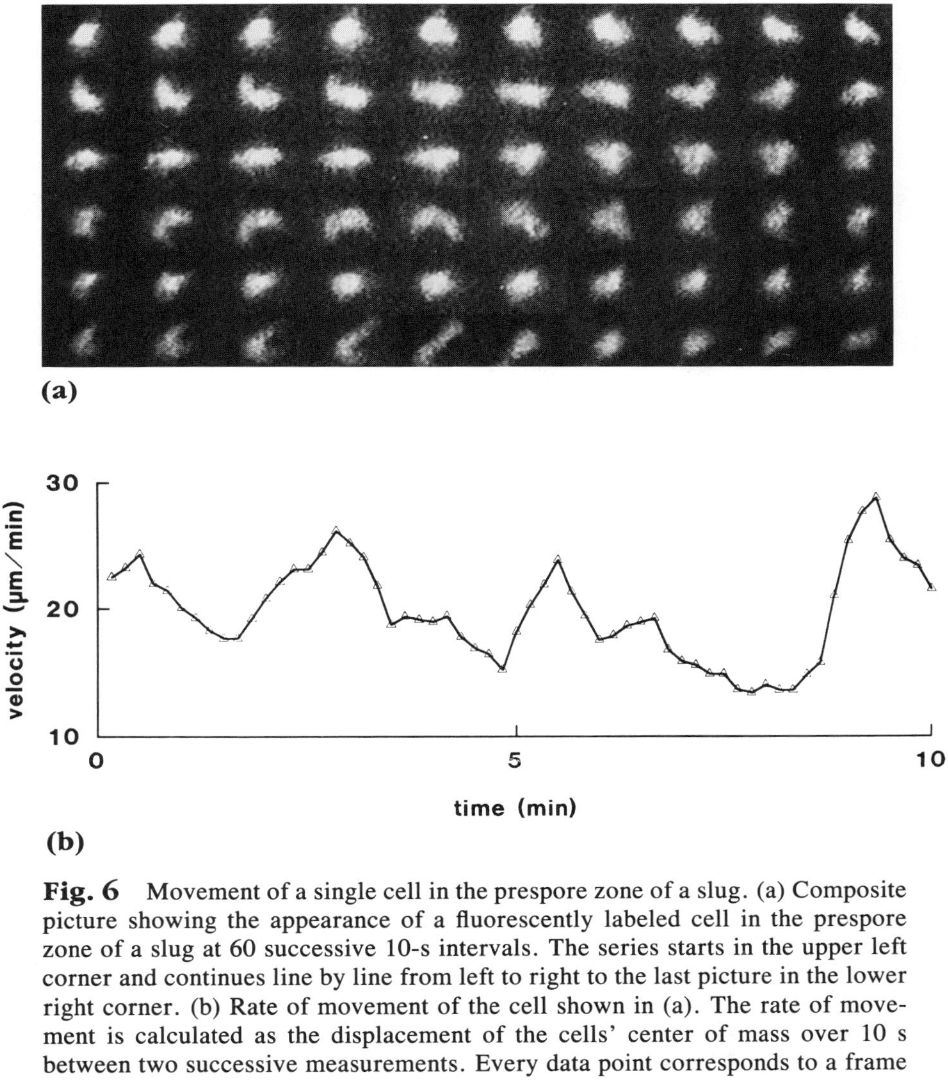

Fig. 6 Movement of a single cell in the prespore zone of a slug. (a) Composite picture showing the appearance of a fluorescently labeled cell in the prespore zone of a slug at 60 successive 10-s intervals. The series starts in the upper left corner and continues line by line from left to right to the last picture in the lower right corner. (b) Rate of movement of the cell shown in (a). The rate of movement is calculated as the displacement of the cells' center of mass over 10 s between two successive measurements. Every data point corresponds to a frame in (a).

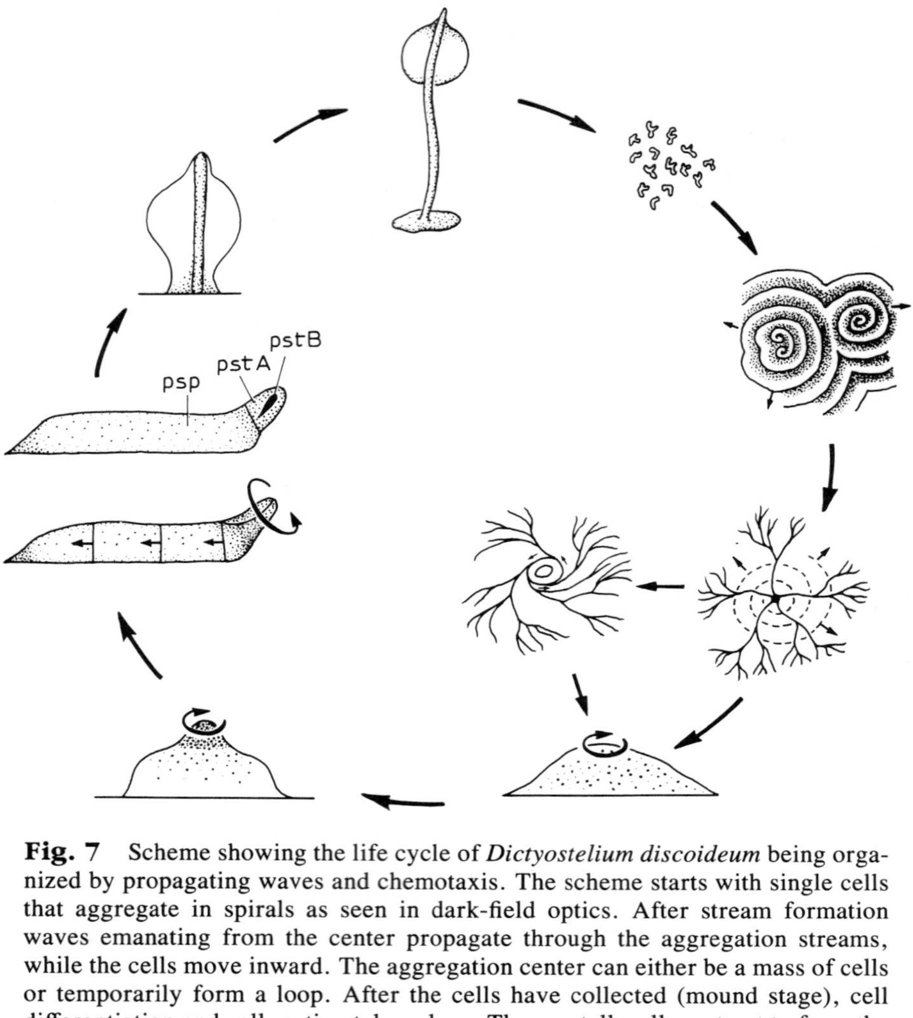

Fig. 7 Scheme showing the life cycle of *Dictyostelium discoideum* being organized by propagating waves and chemotaxis. The scheme starts with single cells that aggregate in spirals as seen in dark-field optics. After stream formation waves emanating from the center propagate through the aggregation streams, while the cells move inward. The aggregation center can either be a mass of cells or temporarily form a loop. After the cells have collected (mound stage), cell differentiation and cell sorting takes place. The prestalk cells sort out to form the tip. The tip is a twisted scroll wave. The wave rotates along its long axis as indicated by the black arrow. The mound now extends up in the air and forms a slug, which falls over and migrates away. The prestalk region stays a twisted scroll wave which decomposes on arrival in the prespore zone in planar waves, due to the lower excitability of the prespore cells. The core of the slug is a region of low cAMP, and this favors PST B expression, while the periphery of the tip with high cAMP favors PST A expression. The slug then converts by a series of not yet understood morphogenetic changes into a fruiting body, during which process the differentiation in the final cell types spore and stalk cells takes place.

would explain tip inhibition and formation of new slugs in grafting experiments. Furthermore, cells in the prespore zone can break down the external cAMP less efficiently than can cells in the prestalk zone, since they have less phosphodiesterase (Otte et al., 1986). This would leave them always being partly adapted. As a consequence, their autonomous oscillation frequency will be lower and the amplitude of the released signal will be smaller than that from the prestalk cells. A spiral wave collects all the cells into a mound during aggregation. There all the fast-responding cells (prestalk cells) sort out to the top of the mound, which leads to the formation of a three-dimensional twisted spiral organizer. The prestalk cells move toward the central core of the spiral. Since the spiral is twisted, this results in the formation and elongation of the tip. The circumference of the inner filament is determined by the minimal refractory period of the cells, with the diameter of the tip determined by the average refractory period of the cells. The twisted scroll wave then decomposes into planar waves in the less excitable prespore region, analogous to that observed in the Belousov–Zhabotinski reaction.

F. Models of Slug Movement and Behavior

One central question in slug morphogenesis is the mechanism of slug movement (i.e., how the motile effort of all cells in the slug is coordinated to result in migration). How is the fine tuning of cell movement achieved that results in the characteristic movement and shape changes? Several models of slug migration have been proposed. An intellectually attractive model is the ''inverse fountain flow model'' (Odell and Bonner, 1986). In this model cells in the periphery of the slug move forward until they reach the tip. From there they migrate again backward through the central core of the slug. Rotation of all cells in a slug is ruled out by the existence of a stable prespore/prestalk pattern, which would require a continuous redifferentiation of the cell types, which is not observed in the time scale of slug migration. If one assumes that this type of movement is restricted to either the tip or prespore zone, the prestalk zone has to be excluded since it is up in the air most of the time and one can not imagine how it can contribute to slug migration. If cell circulation takes place in the prespore zone, one would expect the cells in the core to be stationary in respect to the substratum and the cells in the periphery to move at twice the speed of the slug. This is not compatible with our observations since in our measurement all cells move with approximately slug speed.

In another model, the ''squeeze-pull'' model, it is proposed that cells gain traction from the slime sheath and the slug is embedded in an epithelial layer of specialized cells that can locally contract in a radial direction that squeezes the anterior forward and can draw up the posterior part of the slug (Williams et al., 1986). This model would not explain the differences in movement behavior in the prestalk and prespore zone, and periodic cell movement would not necessarily follow. Furthermore, we found that slugs migrating under a coverslip in mineral oil can move without changing their diameter.

We propose that cells move chemotactically to periodic signals coming from the tip. The mechanism of coherent cell movement must be very similar to that in multicellular aggregation streams. The prestalk cells in the tip rotate in a twisted spiral and probably do not contribute motive force to slug migration. They are lifted above the substratum most of the time and serve as pacemaker. Motive force is produced mainly by the prespore cells, possibly through traction on an extracellular matrix between the cells which forms a continuum with the slime sheath surrounding the slug and which is left behind as a trail. Prestalk cells can produce a considerable motive force when placed in an agar tunnel (Inouye and Takeuchi, 1980). Their motive force per unit volume is even greater than that of a whole slug, which means that they move more vigorously than prespore cells. This is the prerequisite for the maintenance of a stable prestalk/prespore sorting pattern during development.

The active turning behavior in photo- and thermotaxis (Fischer et al., 1984) can be explained by differential cell movement due to slight modulation of the chemotactic signal or alternatively by a modulation of the intensity of the chemotactic response. Local changes in the relay properties of the cells will lead to distortion of the wavefronts and will be followed by changes in the direction of chemotactic movement. For example, the periodic up and down lifting of the slug tip (Williams et al., 1986) can be explained by differences in the rate of cell movement. Cells at the bottom move faster than upper cells due to loss of cAMP by diffusion in the substratum. As a result, the tip would be lifted up in the air until it falls back to the substratum and the process repeats itself again.

One factor that has been shown to influence morphogenesis drastically, most likely by modulating cell movement, is ammonia, a gas produced by starving cells. Ammonia orients the direction of fruiting body formation. Fruiting bodies will bend away from ammonia sources and move toward ammonia sinks. It probably acts by locally increasing the rate of cell movement (Bonner et al., 1986). Ammonia has been reported to stimulate both the rate of movement of vegetative amoebae and cells in aggregation streams at low concentrations and inhibit cell movement at high concentrations (Bonner et al., 1989). We have found that ammonia can increase the amplitude of optical density waves in early aggregation, which we interpreted as a more intense response of cells to the chemotactic signal (Siegert and Weijer, 1989).

G. Role of Periodic Signals in Differentiation

It is well known that cAMP is not only involved in cellular communication but also in gene expression of both aggregation and slug stage-specific genes. The aggregation-specific genes, many of which are involved in the oscillatory cAMP pathway (i.e., cAMP receptors, cAMP phosphodiesterase, adenylate cyclase, and G proteins) are efficiently induced by periodic cAMP signals, while most of them are repressed by continuous flux of cAMP (Gerisch, 1987). Prespore specific genes are efficiently induced by both periodic and continuous signals (Schaap et al., 1986). Prestalk cell gene expression requires exposure to cAMP during the aggregation stage of development, followed by exposure to the morphogen DIF (Williams, 1988). Recently, it has been found that the prestalk zone of the slug consist of at least two classes of prestalk cells, PST A and PST B cells (Williams et al., 1989; Jermyn et al., 1989). Expression of the PST A cell markers is restricted to the front 10% of the slug; PST B cell markers are expressed in a central funnel that occupies 10 to 20% of the length of the slug (Fig. 7). Recently, it was reported that a specific combination of the morphogens DIF-1 and cAMP is necessary for the expression of the different cell markers (Berks and Kay, 1990): PST A differentiation is stimulated by cAMP and DIF-1, PST B differentiation is stimulated by DIF-1 and inhibited by cAMP, and prespore differentiation is stimulated by cAMP and inhibited by DIF-1.

With our concept of signal propagation in slugs we can offer an explanation for this peculiar pattern of cell differentiation. Cells in the center of the twisted scroll wave, which travels through the slug's tip, are adapted most of the time and therefore produce very little cAMP. Cells surrounding the central filament should give a response to every cAMP wavefront. This would lead to a low cAMP concentration in the central funnel and a higher cAMP concentration in the periphery. The low cAMP concentration in the central funnel will favor the expression of PST B, while the higher cAMP in the periphery favors expression of PST A.

VI. CONCLUSIONS

It is clear that there has to be a supercellular control mechanism to coordinate the behavior of a mass of single cells in the slime mold. *Dictyostelium* is first a two-dimensional and then later in development a three-dimensional excitable medium. On the basis of the data described above, we believe that all of *Dictyostelium* development is controlled by periodic signals, wave propagation, and chemotaxis. Due to the developmental regulated increase in oscillation frequency and the corresponding decrease in wave propagation velocity, the system is able to organize both large aggregation fields and the much smaller slugs. The precise course of morphogenesis must be based on slight local changes in signal propagation and chemotactic responses.

Do Periodic Signals Controls Morphogenesis in Other Systems?

There are many examples of morphogenesis in the nonliving world that take place in excitable media. Well known are target patterns and spiral wave formation in the Belousov–Zhabotinski reaction, and the formation of hurricanes and spiral galaxies (Schulman and Seiden, 1986). One would, however, assume that morphogenesis in the living world involves excitable systems (Gierer and Meinhardt, 1972). Although it is clear that many biological patterns are composed of spatially periodic structures, it is less clear whether these are organized by temporarily periodic signals. Many biological processes show a temporal periodicity at all levels of organization. At the level of populations there are cycles of infectious diseases (Murray et al., 1986), reproduction, and extinction. Circadian rhythms occur in both population behavior and in individual's behavior and physiology. Individuals can show reproductive cycles, which are most often based on the reproduction of cells, which in itself again involves—among others—periodic DNA synthesis. Fewer examples are known of periodic information transmission as hormonal regulation and information processing in the nervous system. Only very limited data are available about the role of periodic signals in morphogenesis. The best documented example is certainly the cellular slime mold *Dictyostelium discoideum*. However, there is increasing evidence that the formation of periodic structures such as somites in vertebrates is controlled by a wave-like signal. This signal synchronizes the cell cycles of all the cells that will become allocated to a particular somite (Primmett et al., 1989). To what extent periodic cell movement and chemotaxis play a role in the morphogenesis of other organisms is still open to discussion. In some vertebrates, including the zebrafish and the mouse, there is extensive cell mixing before and after gastrulation. The significance and mechanism of cell movement in these animals is not yet well studied. It will therefore remain open to future experimentation to test the role of periodic signals and chemotactic cell movement in morphogenesis.

ACKNOWLEDGMENT

We thank the DFG (WE 1127) for support.

REFERENCES

Alcantara, F., and Monk, M. (1974). *J. Gen. Microbiol.*, *85*: 321.
Berks, M., and Kay, R. R. (1990). *Development*, *110*: 977.
Bonner, J. T., Suthers, H. B., and Odell, G. M. (1986). *Nature*, *323*: 630.
Bonner, J. T., Har, D., and Suthers, H. B. (1989). *Proc. Natl. Acad. USA*, *86*: 2733.
Brenner, M., and Thoms, S. (1984). *Dev. Biol.*, *101*: 136.
Clark, R. L., and Steck, T. L. (1979). *Science*, *204*: 1163.
Devreotes, P. N. (1989). *Science*, *245*: 1054.
Dockery, J. D., Keener, J. P., and Tyson, J. J. (1988). *Physica D*, *30*: 177.
Durston, A. J. (1974a). *Dev. Biol.*, *38*: 308.
Durston, A. J. (1974b). *Dev. Biol.*, *37*: 225.
Durston, A. J. (1976). *Nature*, *263*: 126.
Durston, A. J., and Vork, F. (1979). *J. Cell Sci.*, *36*: 261.
Fisher, P. R., Dohrmann, U., and Williams, K. L. (1984). *Mod. Cell Biol.*, *3*: 197.
Foerster, P., Müller, S. C., and Hess, B. (1988). *Science*, *241*: 685.
Foerster, P., Müller, S. C., and Hess, B. (1990). *Development*, *109*: 11.
Gerisch, G. (1987). *Annu. Rev. Biochem.*, *56*: 853.
Gerisch, G., Hülser, D., Malchow, D., and Wick, U. (1975). *Philos. Trans. R. Soc. Lond. B*, *272*: 181.
Gierer, A., and Meinhardt, H. (1972). *Kybernetik*, *12*: 30.
Inouye, K., and Takeuchi, I. (1980). *J. Cell Sci.*, *41*: 53.
Jermyn, K. A., Duffy, K. T. I., and Williams, J. G. (1989). *Nature*, *340*: 144.
Keener, J. P., and Tyson, J. J. (1986). *Physica D*, *21*: 307.
MacWilliams, H. K. (1982). In *Developmental Order: Its Origin and Regulation* (S. Subtelny and P. B. Green, eds.), Alan R. Liss, New York, pp. 463–483.
Matsukuma, S., and Durston, A. J. (1979). *J. Embryol. Exp. Morphol.*, *50*: 243.
Murray, J. D., Stanley, E. A., and Brown, D. L. (1986). *Proc. R. Soc. Lond. B*, *229*: 111.
Odell, G. M., and Bonner, J. T. (1986). *Philos. Trans. R. Soc. Lond. B*, *312*: 487.
Otte, A. P., Plomp, M. J. E., Arents, J. C., Janssens, P. M. W., and Van Driel, R. (1986). *Differentiation*, *32*: 185.
Pagola, A., Vidal, C., and Ross, J. (1988). *J. Phys. Chem.*, *92*: 163.
Primmett, D. R. N., Norris, W. E., Carlson, G. J., Keynes, R. J., and Stern, C. D. (1989). *Development*, *105*: 119.
Raper, K. B. (1940). *J. Elisha Mitchell Sci. Soc.*, *56*: 241.
Rubin, J., and Robertson, A. (1975). *J. Embryol. Exp. Morphol.*, *33*(1): 227.

Schaap, P., Van Lookeren Campagne, M. M., Van Driel, R. Spek, P., Van Haastert, P. J. M., and Pinas, J. (1986). *Dev. Biol.*, *118*: 52.
Schulman, L. S., and Seiden, P. E. (1986). *Science*, *233*: 425.
Siegert, F., and Weijer, C. J. (1989). *J. Cell Sci.*, *93*: 325.
Siegert, F., and Weijer, C. J. (1991). *Physica D*, *49*: 224.
Spemann, H. (1938). *Embryonic Development and Induction*, Yale University Press, New Haven, Conn. (reprinted in 1967 by Harper, New York).
Sternfeld, J., and David, C. N. (1981). *Differentiation*, *20*: 10.
Takeuchi, I., Kakutani, T., and Tasaka, M. (1988). *Dev. Genet.*, *9*: 607.
Tomchik, K., and Devreotes, P. (1981). *Science*, *212*: 443.
Varnum-Finney, B., Edwards, K. B., Voss, E., and Soll, D. R. (1987). *J. Cell Motil. Cytoskeleton*, *8*: 7.
Weijer, C. J., McDonald, S. A., and Durston, A. J. (1984). *Differentiation*, *28*: 9.
Williams, J. G. (1988). *Development*, *103*: 1.
Williams, K. L., Vardy, P. H., and Segel, L. A. (1986). *BioEssays*, *5*: 148.
Williams, J. G., Jermyn, K. A., and Duffy, K. T. (1989). *Development Suppl.*, 91.
Yamaguchi, T., and Müller, S. C. (1991). *Physica D*, *49*: 40.

8
Cells into Slugs: Oscillatory Temporal Signals, Taxis, and Morphogenesis in *Dictyostelium discoideum*

Michael G. Vicker *University of Bremen, Bremen, Germany*

I. POSITIONAL INFORMATION IN MORPHOGENESIS

The concept that patterning is guided by positional information obtained by each cell is fundamental to developmental biology, partly because Wolpert (1969) envisaged positional information together with a specific mechanism for its delivery: namely, a spatial concentration gradient of an effector molecule, a "morphogen," spanning a small embryonic "field." By reading the ambient morphogen concentration, a cell's developmental fate would be diverted to one type at one concentration, another type at a lower one, and so on along the gradient's path.

However, signal gradients seem unnecessary in three other views of patterning:

1. In a modified "clock and wavefront" model (Cooke and Zeeman, 1976), cell growth and synchronous oscillations in an embryonic field are confined to a progress zone (e.g., the tip of a developing limb) (Wolpert, 1984). This growth region leaves a trail of cells, but as they fall outside its influence they become phase-fixed, lose their mitotic potential, and differentiate.
2. Goodwin and Cohen (1969) suggested that an intrinsic graded property along one axis of the embryo (e.g., the supply of nutrients from the yolk) induces a corresponding phase difference in some cellular oscillatory activity.

3. Positional information in the ''clockface'' model (French et al., 1976) is related to a cell's angular location or ''phase'' around the circumference of a tissue (e.g., a limb). Wounding experiments show that the missing positional information arises as new tissue grows between two cut surfaces (i.e., intercalary regeneration).

How might morphogenic information be transmitted and translated into cell behavior and tissue pattern, what role might gradients play, and can cells read them? The notion of morphogen gradients is indebted to late-nineteenth-century efforts to explain accumulation by bacteria and phagocytes in terms of taxis and ''chemoattractant'' gradients: ideas that still survive as the biological paradigm ''chemotactic gradient'' (see Vicker, 1989). Therefore, this investigation will begin by examining gradient development, taxis, and kinesis in amoebae of *Dictyostelium discoideum*, a classical model of taxis and morphogenesis. Starving amoebae begin to synthesize and secrete cyclic AMP (cAMP) and an enzyme that hydrolyzes it. Within 6 hours oscillator, or ''pacemaker,'' cells express rhythmic cAMP pulses about every 10 min, which entrain the secretions from other cells (Durston, 1974). Oscillation models, applying to the population as a whole, depend on receptor desensitization and frequency coupling (Li and Goldbeter, 1990). Each signal is relayed distally across the ''field,'' inducing an episode of taxis as up to 10^5 cells advance toward a pacemaker in concentric or spiral patterns (e.g., Siegert and Weijer, 1990). The cells cohere into streams and the streams into an aggregate, while the population is differentiating into, roughly, prespore and prestalk cell types. These sort out later into the posterior and anterior regions of the slug, the cylindrical aggregate about a millimeter long, which may begin to locomote by an unknown mechanism with an anterior pacemaker region. Aggregation is determined by taxis, which is supposed to be induced by a cAMP spatial gradient in aggregation fields (Gerisch et al., 1975) and in slugs (MacWilliams, 1984).

II. GRADIENT DEVELOPMENT, SIGNALS, AND CELL RESPONSES

The criteria for a signal's identity are worth consideration (e.g., as a spatial gradient). Since it is the signal's form rather than the molecular morphogen or taxen per se that is of interest here, it is not trivial that molecules diffusing from some point will tend to form a concentration gradient once turbulence is suppressed. However, this should not be construed as evidence that the gradient itself can be ''read'' by cells or supply positional or directional information. Yet taxen gradients are thought to be perceived even when no steeper than 1% across a cell (Mato et al., 1975), a theoretically supported figure (Berg and Purcell, 1977) provided that the cells are moving (Lauffenburger, 1982). Some authors have suggested that cells perceive the gradient by comparing concentrations either between two receptors (Zigmond, 1978) or with one receptor as it transverses the field (Gerisch et al., 1975). The dynamics of emerging tissue patterns require a morphogen gradient development of appropriate kinetics and a limited field size (Crick, 1970). Therefore it would be useful to know the kinetics and steepness profiles of developing gradients.

To illustrate these dynamics, a gradient developing by molecular diffusion was numerically simulated from a constant source of cAMP across a field of 150 μm, the size used experimentally below. The cAMP concentration, [cAMP], increases exponentially everywhere until the gradient nears linearity in a few seconds (Fig. 1a), resembling actual events (Vicker, 1981). Several possible signals for cells may be delivered by it [e.g., temporal $\partial C/\partial t$, spatial concentration $\partial C/\partial x$, or relative spatial $(\partial C/\partial x)/C$ gradients]. Steepness, as $\partial C/\partial x$, initially increases everywhere then decreases upfield until equilibrium when it becomes a constant throughout (Fig. 1b). Thus spatiotemporal properties differ qualitatively between the up- and downfield regions. Taxis should be limited to or stronger upfield if cells can only read the especially steep gradient that appears only during development, but this signal may be too brief to induce a reaction. If cells can respond to the persistent but weaker mature signal (E), the degree of taxis should be similar throughout the field (Vicker, 1989, 1990b).

However, a relative spatial gradient appears to be a more realistic signal; because cellular desensitization to the stimulus evolves quite soon (Dinauer et al., 1980) (e.g., ligand-driven receptor aggregation on the cell surface alone may greatly reduce sensitivity) (Berg and Purcell, 1977). The concentration profile of the relative gradient is hyperbolic: steeper for the first few seconds upfield, then downfield, where it persists (Fig. 1c). Since adaptive processes will operate essentially only in the mature gradient, tactic responses should then be at least greatest downfield.

The delivery of various signals by developing gradients has serious consequences for investigations of taxis, because, with a few exceptions, cells have usually been confronted experimentally with developing and/or turbulent signals in low-viscosity media. Although molecular diffusion has always been assumed to govern gradient development, in low-viscosity media the stimulant's spatiotemporal kinetics is essentially unpredictable, being driven by convection, turbulance, and stimulant flow, with

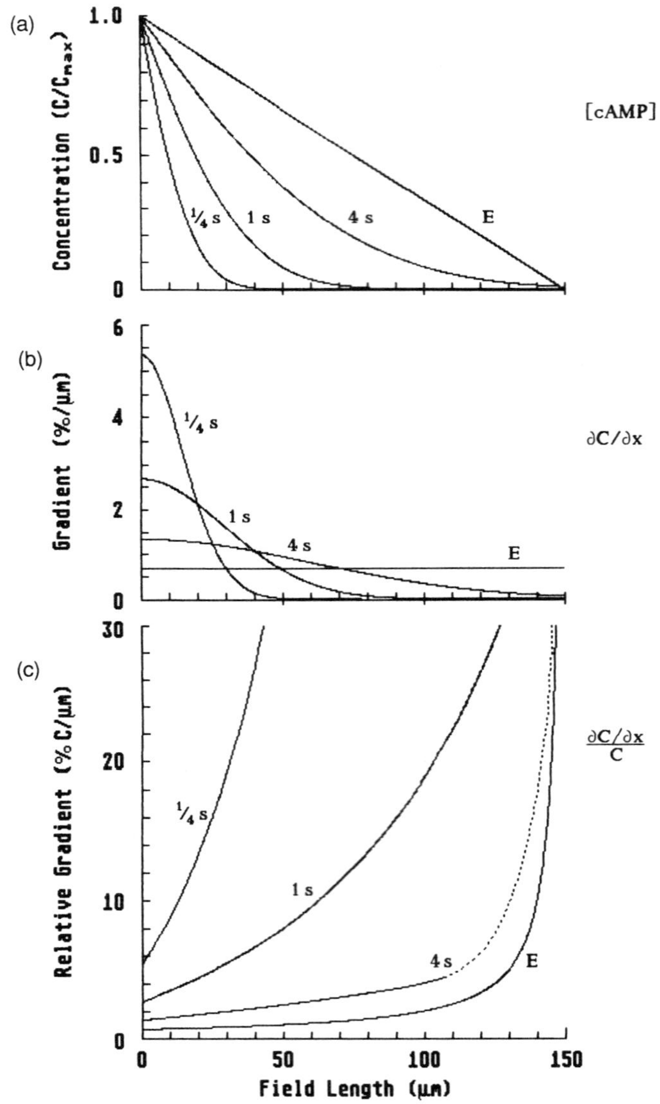

Fig. 1 Cyclic AMP distributions during gradient development were generated numerically using the expressions of Brdička (1963) at $\frac{1}{4}$ s, 1 s, 4 s, and E (equilibrium), after Vicker (1989). (a) The [cAMP] gradient: C is the [cAMP], x is the field position, $C_{\min} = 0$, and $C_{\max}$ is a constant. The ''sink'' affects the development even before 4 s. The steepness of (b) the spatial gradient and (c) the relative spatial gradient are presented. The dashed tail of the 4-s curve in (c) is extrapolated and the sign of the slopes has been rectified throughout.

molecular diffusion playing a subsidiary role (Vicker, 1981). This reveals critical weaknesses in theories of gradient perception, because developing, chaotic, and unstable gradients all deliver temporal concentration changes.

III. TAXIS AND KINESIS IN *DICTYOSTELIUM DISCOIDEUM*

The cell reactions predicted above may be compared with the behavior of *D. discoideum* populations confronted with temporal and spatial signals of cAMP. The point was to examine cell migration within micropore filters, which allow cell movement and support turbulence-free cAMP signals. The cells were distributed evenly across the thickness of the filter before they were exposed to cAMP. Taxis as defined here requires the directed turning of a motile cell in response to a directed signal (Vicker, 1990a).

After exposure to a single signal pulse as a developing gradient, the entire cell population shifted upfield during a brief period of locomotion (Fig. 2), indicating that cell responses are similar at all field positions. The shift is greater if cells are allowed to respond longer (Vicker, 1990b). The effect of a spatial gradient signal was assessed using filters initially infused on both sides with cAMP for 30 min to allow time for cellular adaptive reactions to cease. Medium without cAMP was then infused into the downfield compartment to generate a spatial gradient as the [cAMP] decreased everywhere across the field. Thus the signal was a spatial gradient lacking a cAMP pulse. Taxis was not induced despite the development of a steep gradient. Rather, the population tended to shift downfield within 45 min, and this degree of accumulation remained steady for at least 2 h, as long as the gradient was maintained (Vicker, 1990b).

Cell accumulation downfield in a stable cAMP gradient demonstrates that, adapted or not, cells cannot read the gradient. However, the steepness ($\partial C/\partial x$) of this gradient is relatively shallow compared to upfield steepness in the positively developing gradient. Nonetheless, the upfield accumulation of cells induced in the developing gradient cannot be due to their perception of only its steepest regions, because cell accumulation movements should then be restricted to the upfield region. Time-lapse films of cells exposed to a cAMP impulse show that cells immediately produce a pseudopod toward the signal source, turn, and move in that direction, apparently without testing the gradient (Swanson and Taylor, 1982; Vicker, unpublished). Although cell behavior in filters are beyond direct observation, the wholesale upfield shift in the population distribu-

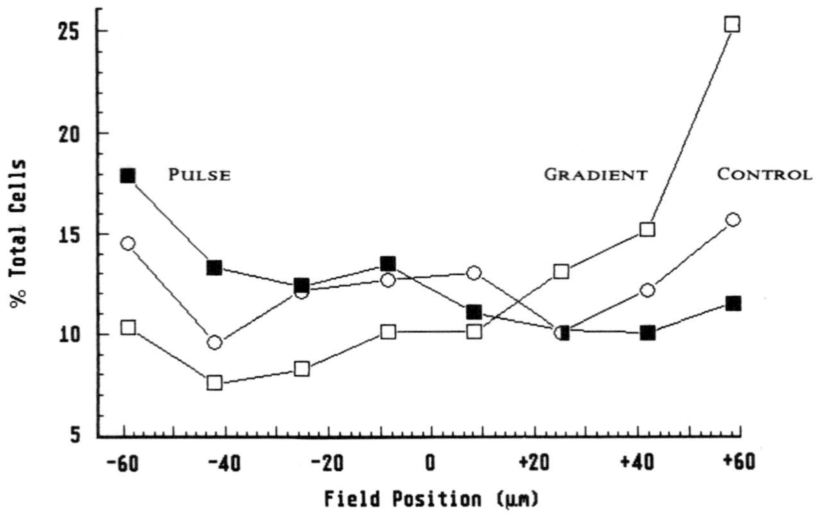

Fig. 2 Effect of a cAMP pulse and a cAMP spatial gradient on population distributions of *D. discoideum*. Cells in the early aggregative stage were initially evenly distributed in microprore filters 150 μm thick, with a pore size of 5 μm, and incubated in 30 n*M* cAMP for 30 min. To generate a *spatial gradient* (□) from this isotropic cAMP distribution, the media at the downfield side of the filter (at +75 μm) was replaced with buffered medium without cAMP and the gradient was maintained by a constant 0.5 mL/min inflow and outflow of media with or without 30 n*M* cAMP, respectively (Vicker, 1990b). After 45 min of locomotion in the gradient, the cells were fixed, stained, and mounted. To induce a signal *pulse* (■), the cells were first incubated in cAMP-free media 20 min before cAMP was rapidly mixed in the compartment at −75 μm to give a [cAMP] of 1 n*M*. After 40 s the filters were washed three times in buffered medium and incubated in medium for 7 min before fixation, staining, and mounting. The *control* population (○) received neither pulse nor gradient. The nominal even distribution is 12.5%. Cell positions were determined in optical sections through the filters (Vicker et al., 1984). Between 350 and 620 cells were counted from five different points in each filter.

tion following a directed pulse of cAMP is, reasonably, due to taxis (Vicker et al., 1984). The signal for this upfield accumulation is a temporal one.

The behavior of individual cells in gradients lacking an increasing temporal signal component demonstrates that gradients in themselves are not signals. Cells respond independent of gradient steepness. The cell distribution in a stable cAMP gradient (Fig. 2) is generated by differences in [cAMP]-dependent speed (*orthokinesis*) and locomotory persistence (*klinokinesis*), but individual cell locomotion remains biased-random (Vicker et al., 1984, 1986; Vicker, 1989). The elevated [cAMP] upfield is optimal for cell speed and persistence (Vicker, 1990b), and cells evacuate that region and accumulate where they move slowest. The cell distribution depends alone on the momentary affect of the ambient [cAMP] on random motility at each position in the gradient. Results obtained using computer simulations of population behavior and simple rules of ambient concentration reading support these conclusions (Vicker et al., 1984).

Apparently, temporal stimuli induce reactions, which are adapted or absent after some time in isotropic or spatial gradient fields (i.e., tactic reactions adapt and kinetic ones do not). Similar differences in physiological reactions to rapidly increasing versus constant [cAMP] have been identified in *Dictyostelium*, and two cAMP perception systems have been hypothesized (Wurster and Butz, 1983). As a rapid [cAMP] increase envelopes one side of a cell first, presumably, it induces a pseudopod to develop from there, accounting for reasonably accurate cellular directionality. However, the same signal will pass on across the entire 5 to 15 μm of the cell. Some process must necessarily inhibit perception of anything but the reactions at the initial contact site. One possible way depends on how pseudopods often appear at fairly predictable time intervals and positions, and how their duration seems related to the directional persistence time of cell locomotion (Vicker, unpublished). Thus a cAMP pulse might effect a resetting of the spatiotemporal course of a hypothetical pseudopod oscillator. One candidate for such an oscillator is the dynamic network of cAMP signal-sensitive actin elements in the *Dictyostelium* cytoskeleton (Condeelis et al., 1988). Actin's self-organizing feedback character may drive its rhythmical assembly and disassembly, which has been observed in vitro and modeled (Alt, 1985).

IV. OSCILLATOR-DEPENDENT BEHAVIORAL–MORPHOGENIC ELEMENTS

The qualitatively different reactions of cells to spatial gradients and isotropic concentrations of cAMP, on the one hand, and pulse or impulse signals, on the other, suggest how morphogenesis may depend on oscillator-regulated intercellular signals and cellular behavior (Vicker and Rensing, 1987). The tissue pattern and locomotion of the *D. discoideum* slug offers a convenient example. A single cAMP pulse initiates cellular cAMP signal relay and actin polymerization (Condeelis et al., 1988), which generates a consecutive series of brief behavioral reactions in a significant order: cell contraction (Futrelle et al., 1982), increased cell cohesion (Wurster and Kurzenburger, 1989), and taxis (directed pseudopod production, cell orientation, enhanced locomotion), each of which adapt (relax) in turn (Fig. 3a). As a cAMP signal transits a cell mass, this series of "morphogenic elements" (Table 1) will produce two behavioral consequences. First, variations in taxis (differential taxis) will arise between cells able to respond to every pulse or every other pulse signal initiated by the pacemaker and will induce these cells to sort out (Matsukuma and Durston, 1979). These "advanced" and "retarded" cell classes correspond to prestalk and prespore cell types, respectively (McDonald, 1984). Sorting-out may account for the anterior–posterior patterning of these tissues within the slug and for the phase difference observed between the two regions (Durston, 1974; Weijer et al., 1984). Biphasic harmonics has also been observed in cell streams during *Dictyostelium* aggregation (Siegert and Weijer, 1990).

The second consequence is slug migration, a notoriously difficult task for a limp, structureless tissue. Consider that each morphogenic element arises, then transforms into the next one in a spatiotemporal sequence along the slug, somewhat like the red spiral bands, which seem to shift distally along a rotating barber pole. The mass of cells might organize its own movement, because each "band" of contracted cells would provide a relatively stiff section across the slug's diameter. This "platform" will translate the coordinated stretching and pseudopodial activities of the cell band immediately anterior to it (i.e., the cells of the previous platform band as they begin taxis) into a forward projection of the entire anterior cell mass and thus into slug elongation and locomotion. This would occur essentially in the slug's anterior "nose," which is seen alternatively to lift from and sink onto the substrate, and in which cAMP oscillations may have a relatively low amplitude (Wang et al., 1988). Cycles of lifting, projecting, and falling (under its own weight) will allow the slug tip to progress and to tow the posterior mass (Fig. 3b).

The scheme depicted here is essentially identical to cell behavior during aggregation and streaming, with the exception that cell contraction and sorting-out play roles in the slug. It requires no additional mechanisms, as in other models of slug motility (e.g., Williams et al., 1986), and is only put forward as a minimal model. The complexities of cell locomotion in slugs make it difficult to test these ideas at present. For example, the role of cell streams due to cell cohesion must be accounted for and cell movement in the anterior region seems to proceed in a forward spiral motion [perhaps around a central core cell region (Jermyn et al., 1989)], while posterior tissue moves in straighter trajectories (see Clark and Steck, 1979, and Chapter 7, this volume).

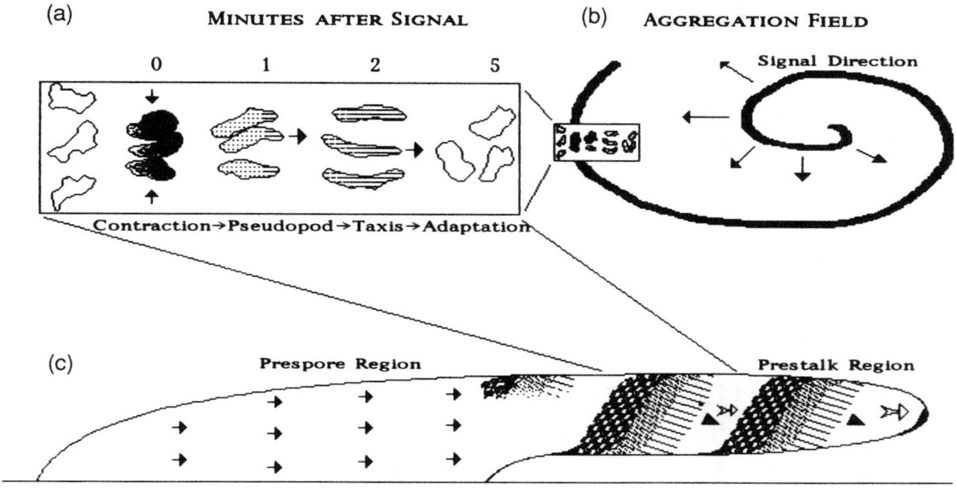

Fig. 3 Scheme in which cell behavioral reactions to cAMP pulses generate cell aggregation, patterning, and slug locomotion. (a) The sequence of cell reactions to one cAMP pulse is contraction and increased cohesion, directed pseudopod extension (stretching), differential taxis and sorting-out, relaxation, and deadaptation. In any one cell, each reaction succeeds its predecessor as the cell adapts simultaneously to the signal. As visualized here, the reactions form a sequence from left to right, because time after the cAMP pulse also proceeds from left to right. (b) Pulsatile signals are relayed outward by the population as bands of cells move in the opposite direction. The behavioral elements lead to aggregation around the core cell group, which, in the case of spiral signals, cannot be a pacemaker. (c) The pattern of cell behavior generated in the aggregation field may similarly regulate cell behavior in the slug. Just after reception of the cAMP signal, which spirals along the slug, the dark band of contracted and firmly cohering cells provides a semirigid "platform" upon which the taxis and stretching by cells (direction given by dark arrow heads) immediately ahead of it will induce cells to (1) sort out according to their intrinsic frequency and (2) press anterior tissue forward (open arrows). The slug's "nose," which alternatively lifts and falls onto the substrate, will continuously advance while lifted and thus fall somewhat farther along each time. This operation will pull and guide the posterior, prespore tissue, where the cAMP pulse-amplitude is also reduced. Correspondingly, cell movement there may be expected to be increasingly linear (small arrows), but not entirely passive, because "it takes all the running . . . to keep in the same place." (Lewis Carroll, *Through the Looking Glass*)

Table 1 Morphogenic (Behavioral) Elements in Slugs[a]

Cellular contraction
Increased cell cohesion
Directed pseudopod production
Taxis
Sorting-out
Relaxation
Deadaptation

[a]Listed according to their order of appearance after induction by tactic signals. Deadaptation to the signal follows behavioral relaxation. Cyclic AMP signal relay is not shown.

D. discoideum morphogenesis requires cAMP pulse relay from an anterior pacemaker and intrinsic phase differences in signal perception and relay frequency between its two main cell groups. The rhythm of taxis and derivative cell behavior (including cell contraction, cohesion, and sorting-out) is crucial for the positioning and polarity of cells during morphogenic maneuvers and has implications for other systems, because this constellation of elements depends specifically on temporal signaling and adaptive signal perception and cannot be ascribed to a signal gradient. These morphogenic elements have been observed in a wide variety of developmental systems. The polar quality that characterizes the clockface model (French et al., 1976) might arise, along a cylindrical organ, from the proximal–distal signal transmission as a spiral, which would produce centripetal phase-related differences in positional information (Glass, 1977; and personal communication). The rotations reported in insect limb (French, 1976) and epidermis (Nübler-Jung, 1974) transplants, perhaps due to differential cell adhesion, might also be evidence of directed, temporal signaling. A directed "vortex" of coordinated, rhythmic cell migration characterizes *Drosophila* gastrulation (Minden et al., 1989). Waves of pulsatile activity accompany locomotion, aggregation, and sorting-out of mesenchymal cells during sea urchin gastrulation (Kinnander and Gustafson, 1960; Solursh, 1986). A cAMP pulse induces signal relay, greater cohesiveness, and attraction in embryonic chick cells (Robertson et al., 1978; Gingle, 1977; Robertson and Gingle, 1977). Visible waves may transverse chick and fish embryos (Robertson, 1979; Stern and Goodwin, 1977). During amphibian gastrulation, the directional migration of mesenchymal cells is often pulsatile (Kubota and Durston, 1978). During chick and frog somitogenesis, discrete blocks of mesodermal cells undergo rearrangement from an initially lateral to a parallel orientation in an anteroposterior (A-P) wave. The movement is accompanied by cell contraction, increased cell adhesion, and polarization (Sanders, 1986). Each somite appears with a species-specific frequency (e.g., 140 min in *Rana*), apparently controlled from the outgrowing tail tip (Cooke and Elsdale, 1980; Elsdale and Davidson, 1983). Before neural crest cells invade the somite apex, they round (contract?), extend processes, then appear along the neural axis in an A-P wave, becoming increasingly adhesive and oriented to it. Finally, they migrate in unison, entering cell streams (Erickson, 1986). These phenomena offer surprisingly analogous parallels to behavior in *Dictyostelium*. Developing somites and limbs exemplify progress zones and seem analogous to the *Dictyostelium* slug, the difference being that after continuous growth in the progress zone tip a record of periodic cell activity is laid down as a banded ''track'' of differentiated tissue, while in slugs each oscillation induces periodic locomotion. These examples seem to possess a number of elements of cell behavior in common.

V. GRADIENTS AND MORPHOGENESIS

Many genes in *D. discoideum* are specifically regulated either by pulses or by constant concentrations of cAMP. Thus cAMP serves in taxis, morphogenesis, and differentiation. Currently, the evidence linking a cAMP gradient with any of these processes is equivocal at best (e.g., if one exists within the slug, its orientation appears to be posterior–anterior) (Wang et al., 1988). The putative ability of a morphogen to affect developmental fate in a concentration-dependent, steplike fashion, as attributed to the mesoderm-inducing protein XTC-MIF of *Xenopus* (Green and Smith, 1990), is typical of cell surface receptor–mediated stimulation (Robertson and Grutsch, 1987); for example, the neutrophil leukocyte taxen formyl-Met-Leu-Phe stimulates locomotion at n*M* doses and granule release and motility inhibition at μ*M* doses, presumably as a consequence of the parallel pathways and/or crosstalk involving various second messenger systems (Lackie and Lawrence, 1987).

But indications that gradients of putative morphogens invariably elicit more than one developmental fate are often weak. For example, gradients of retinoic acid derivatives (RA) have been associated with several morphogenic events (e.g., in limb and digit patterns). However, the saturation of the nuclear RA receptors at the lowest [RA] in these tissues essentially extinguishes the signal potential of the gradient (Summerbell and Maden, 1990). It is an intriguing possibility that the function of RA in pattern regulation might be related to its growth-inducing properties. The role of growth has been recognized in progress zones (Wolpert, 1984) and in intercalary pattern regeneration (French et al., 1976). A role for a morphogen gradient in digit patterning (Tickle et al., 1975) has become increasingly unclear, since the induction of mirror-symmetrical pattern duplications and cell growth in both juxtaposed tissues (Muneoka and Murad, 1988) plainly contravene the gradient signaling hypothesis. Alternatively, these experimental results might be explicable in terms of intercalary regeneration. It should also be noted that various models proposed by Meinhardt (1988) may often be satisfied, under certain circumstances, by substituting temporal signal exchanges for the gradient signal exchanges, which he also explores. In contradistinction to the preceding examples, spatial gradients of DNA-binding proteins appear exquisitely efficient in transmitting positional information in early fly development (Driever and Nüsslein-Vollhard, 1988). This fact may be related to the syncytial character of the embryological field, whereas the plasmalemma of cellular systems might inhibit gradient signals by prescribing different signal requirements (e.g., that they be endocrine).

Variation in cell responses to temporal signals will either lead to phase separation or be suppressed by frequency domination. Although positional information is often very accurate, reading error arising from responses to ambient morphogen concentrations in gradient fields would cloud spatial patterning. Such "stochastical variation" is usually treated statistically by biologists, who perhaps regard it as a universal nuisance, assumed to obscure "real" trends, and every effort is made to minimize its influence. Yet it might indicate uncoupled, unsynchronized oscillators. Variations in signal reading, of the magnitude usually encountered in dose-response trials, would require that reading accuracy be assured by special mechanisms, such as the ability of cells to cluster or communicate to integrate, or average out, signal or individual cell response "noise" (Vicker et al., 1984; Meinhardt, 1988).

VI. CONCLUSIONS

Rhythmic, directed temporal signals of cAMP are specifically required to organize a series of adaptive cell behavioral reactions necessary for *Dictyostelium* morphogenesis and, possibly, slug motility. The rhythm of cohesion and differential taxis plays a key role in sorting-out and the generation of tissue patterning, but cannot be induced by a cAMP spatial gradient. These reactions manifest general principles of oscillator-dependent morphogenesis common to *Dictyostelium* and many other developing systems in which similar morphogenic elements of cell behavior need to be regulated.

REFERENCES

Alt, W. (1985). In *Temporal Order* (L. Rensing and N. I. Jaeger, eds.), Springer-Verlag, Berlin, p. 163.

Berg, H. C., and Purcell, E. M. (1977). *Biophys. J.*, *20*: 193.

Brdička, R. (1963). *Grundlagen der Physikalischen Chemie*, Deutscher Verlag der Wissenschaft, Berlin, p. 294.

Clark, R. L., and Steck, T. L. (1979). *Science*, *204*: 1163.

Condeelis, J., Hall, A., Bresnick, A., Warren, V., Hock, R., Bennett, H., and Ogihara, S. (1988). *Cell Motil. Cytoskeleton*, *10*: 77.

Cooke, J., and Elsdale, T. (1980). *J. Embryol. Exp. Morphol.*, *58*: 107.

Cooke, J., and Zeeman, E. C. (1976). *J. Theor. Biol.*, *58*: 455.

Crick, F. (1970). *Nature* (*Lond.*), *225*: 420.

Dinauer, M. C., Steck, T. L., and Devreotes, P. N. (1980). *J. Cell Biol.*, *86*: 554.

Driever, W., and Nüsslein-Vollhard, C. (1988). *Cell*, *54*: 83.

Durston, A. J. (1974). *Dev. Biol.*, *37*: 225.

Elsdale, T., and Davidson, D. (1983). *J. Embryol. Exp. Morphol.*, *76*: 157.

Erickson, C. A. (1986). In *Developmental Biology*: *A Comprehensive Synthesis*, Vol. 2 (L. W. Bowder, ed.), Plenum Press, New York, p. 481.

French, V. (1976). *Roux's Arch. Dev. Biol.*, *179*: 57.

French, V., Bryant, P. J., and Bryant, S. V. (1976). *Science* (*Washington, D.C.*), *193*: 969.

Futrelle, R. P., Traut, J., and McKee, W. G. (1982). *J. Cell Biol.*, *92*: 807.

Gerisch, G., Hülser, D., Malchow, D., and Wick, U. (1975). *Philos. Trans. R. Soc. Lond. B*, *272*: 181.

Gingle, A. R. (1977). *Dev. Biol.*, *58*: 394.

Glass, L. (1977). *Science* (*Washington, D.C.*), *198*: 321.

Goodwin, B. C., and Cohen, M. H. (1969). *J. Theor. Biol.*, *25*: 49.

Green, J. B. A., and Smith, J. C. (1990). *Nature* (*Lond.*), *347*: 391.

Jermyn, K. A., Duffy, K. T. I., and Williams, J. G. (1989). *Nature* (*Lond.*), *340*: 144.

Klinnander, H., and Gustafson, T. (1960). *Exp. Cell Res.*, *19*: 278.

Kubota, H. Y., and Durston, A. J. (1978). *J. Embryol. Exp. Morphol.*, *44*: 71.

Lackie, J. M., and Lawrence, A. J. (1987). *Biochem. Biopharm.*, *36*: 1941.

Lauffenburger, D. A. (1982). *Cell Biophys.*, *4*: 177.

Li, Y. X., and Goldbeter, A. (1990). *J. Theor. Biol.*, *146*: 355.

McDonald, S. A. (1984). *Roux's Arch. Dev. Biol.*, *194*: 50.

MacWilliams, H. K. (1984). In *Pattern Formation*: *A Primer in Developmental Biology* (G. M. Malacinski and S. V. Bryant, eds.), Macmillan, New York, p. 127.

Mato, J. M., Losada, A., Nanjundiah, V., and Konijn, T. M. (1975). *Proc. Natl. Acad. Sci. USA*, *72*: 4991.
Matsukuma, S., and Durston, A. J. (1979). *J. Embryol. Exp. Morphol.*, *50*: 243.
Meinhardt, H. (1988). *Development Suppl.*, *104*: 95.
Minden, J. S., Agard, D. A., Sedat, J. W., and Alberts, B. M. (1989). *J. Cell Biol.*, *109*: 505.
Muneoka, K., and Murad, E. H. B. (1988). *Development*, *99*: 521.
Nübler-Jung, K. (1974). *Nature (Lond.)*, *248*: 610.
Robertson, A. (1979). *J. Embryol. Exp. Morphol.*, *50*: 155.
Robertson, A., and Gingle, A. R. (1977). *Science (Washington, D.C.)*, *197*: 1078.
Robertson, A., and Grutsch, J. F. (1987). *J. Theor. Biol.*, *125*: 41.
Robertson, A., Grutsch, J. F., and Gingle, A. R. (1978). *Science (Washington, D.C.)*, *199*: 990.
Sanders, E. J. (1986). In *Developmental Biology: A Comprehensive Synthesis*, Vol. 2 (L. W. Bowder, ed.), Plenum Press, New York, p. 449.
Siegert, F., and Weijer, C. J. (1990). *J. Cell Sci.*, *93*: 925.
Solursh, M. (1986). In *Developmental Biology: A Comprehensive Synthesis*, Vol. 2 (L. W. Bowder, ed.), Plenum Press, New York, p. 396.
Stern, C. D., and Goodwin, B. C. (1977). *J. Embryol. Exp. Morphol.*, *41*: 15.
Summerbell, D., and Maden, M. (1990). *Trends Neurosci.*, *13*: 142.
Swanson, J. A., and Taylor, D. L. (1982). *Cell*, *28*: 225.
Tickle, C. A., Summerbell, D., and Wolpert, L. (1975). *Nature (Lond.)*, *254*: 199.
Vicker, M. G. (1981). *Exp. Cell Res.*, *136*: 91.
Vicker, M. G. (1989). *J. Cell Sci.*, *92*: 1.
Vicker, M. G. (1990). In *Biological Motion* (W. Alt and G. Hoffmann, eds.), *Lecture Notes in Biomathematics*, Vol. 89, Springer-Verlag, Berlin, p. 518 (a) and p. 487 (b).
Vicker, M. G., and Rensing, L. (1987). In *Temporal Disorder in Human Oscillatory Systems* (L. Rensing, U. an der Heiden, and M. C. Mackey, eds.), Springer-Verlag, Berlin, p. 24.
Vicker, M. G., Schill, W., and Drescher, K. (1984). *J. Cell Biol.*, *94*: 2204.
Vicker, M. G., Lackie, J. M., and Schill, W. (1986). *J. Cell Sci.*, *84*: 263.
Wang, M., Van Driel, R., and Schaap, P. (1988). *Development*, *103*: 611.
Weijer, C. J., Duschl, G., and David, C. N. (1984). *J. Cell Sci.*, *70*: 111.
Williams, K. L., Vardy, P. H., and Segel, L. A. (1986). *BioEssays*, *5*: 149.
Wolpert, L. (1969). *J. Theor. Biol.*, *25*: 1.
Wolpert, L. (1984). In *Pattern Formation: A Primer in Developmental Biology* (G. M. Malacinski and S. V. Bryant, eds.), Macmillan, New York, p. 3.
Wurster, B., and Butz, U. (1983). *J. Cell Biol.*, *96*: 1566.
Wurster, B., and Kurzenburger, W. (1989). *Differentiation*, *41*: 1.
Zigmond, S. H. (1978). *J. Cell Biol.*, *77*: 269.

9
Intracellular Oscillations and Pattern Formation in the Cell Behavior of *Physarum*

Tetsuo Ueda *Nagoya University, Nagoya, Japan*

I. INTRODUCTION

A. Aim

An amoeboid cell is thought to be one of the most primitive forms of life, probably because there are no definite organs for sensation or for motility in this organism. So its protoplasm may be regarded as a solution of proteins and organelles. Yet this protoplasmic solution is able to sense external environments, judge the information sensed, and behave accordingly as an integrated unity, just as higher animals do (Ueda and Kobatake, 1982a).

What are the differences between a nonliving solution and a living protoplasm? Self-organization, which was once thought to be characteristic only of a living organism, should be a key notion. But recently, phenomena of self-organization have also become popular in nonliving systems: The Belousov–Zhabotinskii (BZ) reaction, the glycolysis, and so on, and studies of these phenomena have developed new notions concerning rich dynamic behaviors of systems far away from equilibrium: sustained oscillations, pattern formation, wave propagation, entrainments, chaos, and so on. We believe that these notions should be applicable to living systems. Here we show that the cell behavior of an amoeboid cell is based on the spatiotemporal organization of metabolism.

B. Use of a Giant Amoeboid Cell, the True Slime Mold

To make clear the mechanism of coordinated cell behavior in terms of physical chemistry, chemical organization in both space and time of a living cell should be clarified. This seems very difficult in a normal-sized cell. The plasmodium of *Physarum polycephalum* is a giant multinuclear organism and migrates with amoeboid motility (Fig. 1). Morphologically, the protoplasm of the plasmodium extends like a sheet in the front, and intricate networks of veins develop toward the rear, in which protoplasmic sol flows to and fro with a period of a few minutes. Each small part of the plasmodium is potentially able to behave just like the entire organism, because when separated from the mother organism, each part develops small organisms similar to the larger one. With this large organism we may study the temporal variation (Section II) and also the spatial distribution of various intracellular chemical components (Section III). Dynamic behavior of the thickness oscillation throughout the organism is also analyzed in relation to information processing and coherence (Section IV).

II. OSCILLATIONS

A. Chemical Oscillations Accompanying the Rhythmic Contractions

Motility of the plasmodium is based on the actomyosin system, and the rhythmic contraction of the plasmodium seems to be brought about by the rhythmic variation of the modifiers of actomyosin, such as Ca^{2+} and ATP. The question we raise here is whether or not the chemical oscillation is limited to modifications of motility itself. So changes in the concentration with time for a wide range of chemicals were examined (Mori et al., 1987; Ueda et al., 1986, 1987, 1991).

Time courses of the intracellular concentrations of cAMP and NADH are shown in Fig. 2, where the small frontal part in a directionally migrating plasmodium is analyzed during a time interval. Both cAMP and NADH oscillate accompanying the rhythmic contraction. The phase of the cAMP oscillation is advanced about one-third of the period, while that of the NADH oscillation is delayed about one-third of the period with respect to the motility oscillation. Since cAMP and NADH do not seem to act directly on one-third actomyosin, and since these chemicals are important regulators in many metabolic pathways, we may conclude that the protoplasm as a whole is oscillating.

This conclusion is confirmed by the oscillation of phospholipid contents. The time course of phospholipid contents in the cell homogenate is shown in Fig. 3. Phosphatidic acid (PA) and phosphatidylserine (PS) oscillate with a period of a few minutes, while phosphatidylinositol (PI) and cardiolipin (CL) remain relatively constant. Phosphatidylcholine

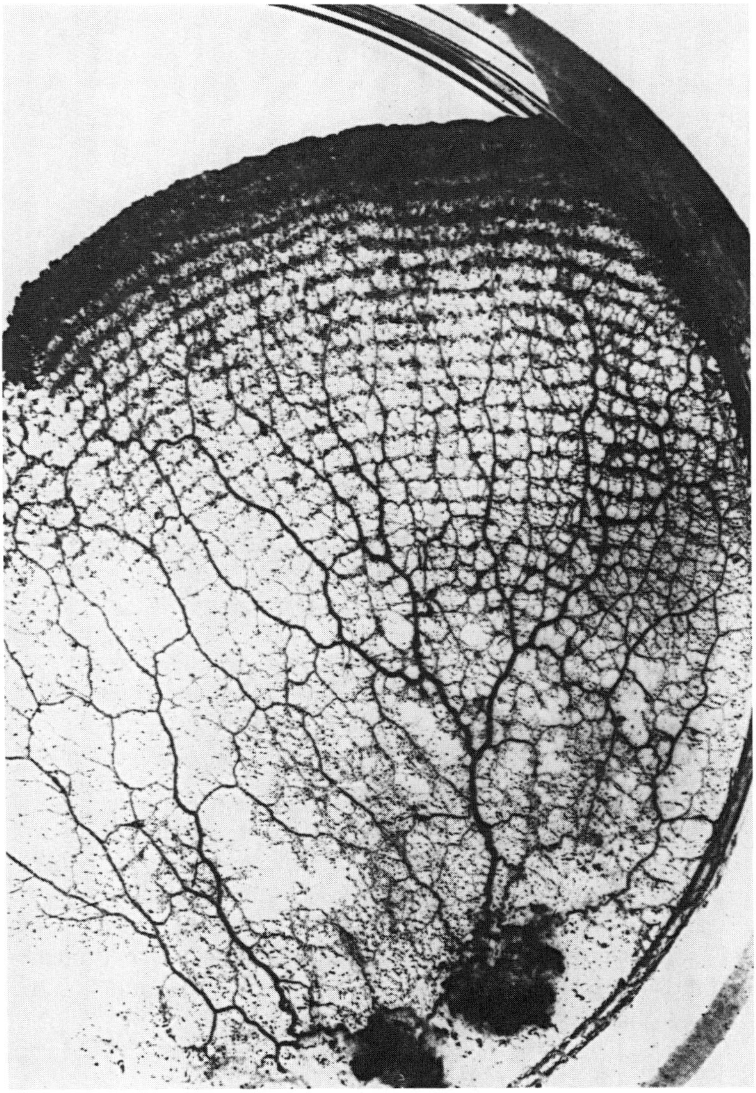

Fig. 1 Plasmodium migrating (upward) in a petri dish (ϕ = 15 cm). A mass of protoplasm was initially placed at the lower part of the figure.

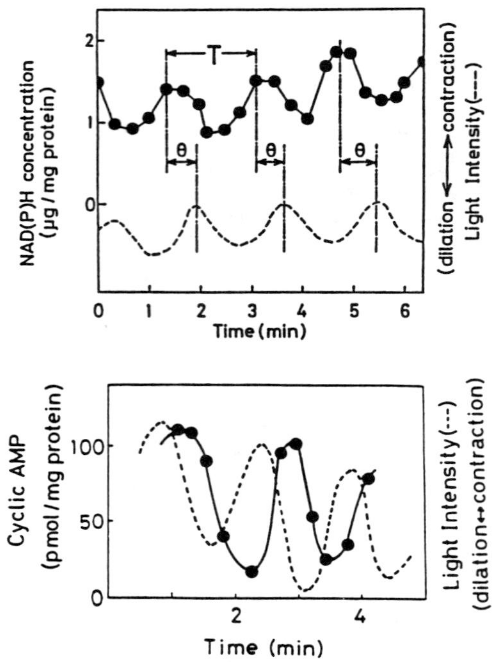

Fig. 2 Oscillations in NADH and cAMP concentrations accompanying the rhythmic contraction of the plasmodium. (After Mori et al., 1987; Ueda et al., 1986.)

(PC) and phosphatidylethanolamine (PE), major phospholipids in this organism, are gradually decreasing. Phospholipids are major components of membranes, and the amoeboid motility involves the deformation of the cell surface. So there is good reason to believe that the phospholipid metabolism is actively participating in forming the polarity in cell behavior. This hypothesis is tested in Section III. Involvement of mitochondria in keeping the oscillation is suggested (Satoh et al., 1982, 1984).

B. Factors Affecting the Period of Oscillation

The period of the oscillation is affected by many environmental factors (Ueda et al., 1987). Anaerobic conditions (N_2 atmosphere) do not stop the oscillation but make the period of ATP oscillation as long as 20 to 30 min (Fig. 4a). Ethanol brings about an interesting phenomenon: A slower oscillation appears, superimposing the faster rhythm (Fig. 4b). Thus the organism shows a double periodicity. The organism also shows a much

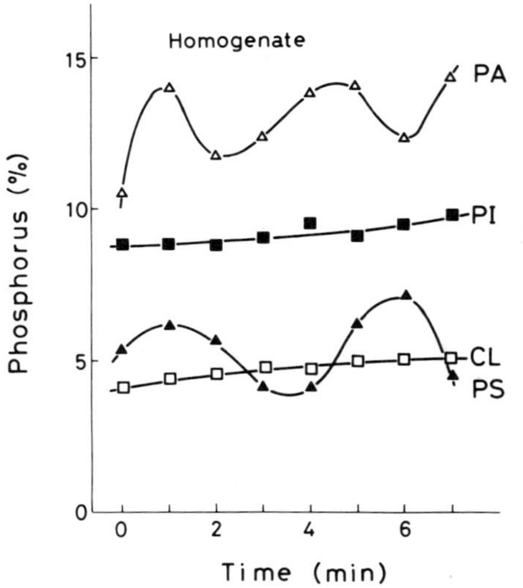

Fig. 3 Oscillations in the contents of some phospholipids in the cell homogenate. (After Ueda et al., 1991.)

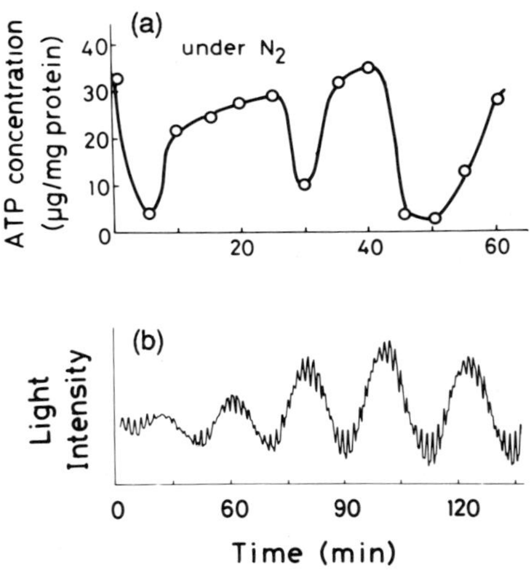

Fig. 4 Long-period oscillation of ATP under N_2 (a) and multiperiodicity of the rhythmic contraction in the presence of ethanol (b). (After Ueda et al., 1987.)

slower rhythm of about 5 h (Akitaya et al., 1985), and a cell cycle rhythm of about 10 h. Hence the protoplasm is in the state of multiperiodicity.

There is a temperature-independent component in the period of the oscillation (Ueda and Kobatake, 1980, 1982a,b). The period of the isometric tension T depends on the amplitude F of the oscillation as follows:

$$T = \begin{cases} aF + T_0 & (\text{for } F < F_c) \\ T_s & (\text{for } F > F_c) \end{cases}$$

Here a, T_0, and T_s are constants independent of F but dependent on environmental factors. T_0 is temperature compensated (independent of temperature). Temperature independence is regarded as an important property of circadian rhythms. We see here that the rhythmic contraction is much faster than the circadian and shows temperature independence. Effects of chemicals are classified into two groups. Attractants such as glucose and amino acids make the period slightly faster, while the repellents make it slightly slower.

The notion that the metabolism in amoeboid cells is oscillating may be extended to other amoeboid cells. Cell shape changes periodically, accompanying cell locomotion: This includes leukocytes, macrophages, *Amoeba proteus*, and the cellular slime mold *Dictyostelium discoideum* (Ueda and Kobatake, 1983; Satoh et al., 1985). Therefore, oscillation phenomena are ubiquitous among amoeboid cells.

Why oscillations? In the following we answer this question from two points of view. One is considering the source (motive force) of pattern formation (Section III), and the other the mechanism for information processing (Section IV).

III. PATTERN FORMATION

A. Chemical Patterns in Relation to the Cell Behavior

The spatial distribution of phospholipid contents is shown in Fig. 5, where the organism is migrating directionally leftward in the figure (Ueda et al., 1992). PA is low at the frontal part and becomes higher toward the rear. PE is high at the front, and toward the rear decreases as if to compensate the PA increase. Thus the polarity in cell behavior is related to the polar distribution of phospholipid contents. Taking into account the fact that the phospholipid content oscillates (Fig. 4), it is likely that phospholipid metabolism takes part actively in self-organizing the spatial pattern.

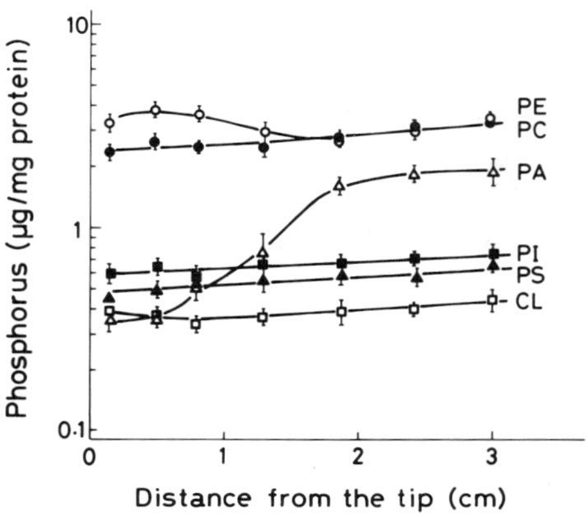

Fig. 5 Spatial distribution of phospholipid contents in a plasmodium migrating unidirectionally (leftward in the figure). (After Ueda et al., 1992.)

Patterns in the distribution of ATP concentration vary in accord with cell behavior (Fig. 6) (Ueda et al., 1987). When the organism is migrating unidirectionally, the ATP concentration is high at the frontal part and decreases toward the rear (a). When the organism is extending concentrically, the ATP distribution looks like a round trough, being high at the peripheral region (b). When the organism is extending like a fan, the distribution looks like the shell of an oyster, still on the average being high at the advancing front (c).

The ATP distribution changes from one pattern to another in response to external stimuli. When the frontal part of the unidirectionally migrating plasmodium is locally irradiated with repulsive blue light, the distribution pattern of ATP turns from polar to wavy, resulting in an inhibition of the forward direction and then an avoidance reaction (Mori et al., 1986).

Thus we may conclude that various chemical patterns correspond to various types of cell behavior. How is the chemical pattern related to cell behavior? To answer this question, the dynamic behavior in the distribution of the cytoskeleton should be studied.

B. Organization of Chemical Patterns, Formation of Cytoskeleton, and Cell Behavior

The cytoskeleton in this organism appears as birefringent fibrils under a polarizing microscope. Effects of UV irradiation on the formation of

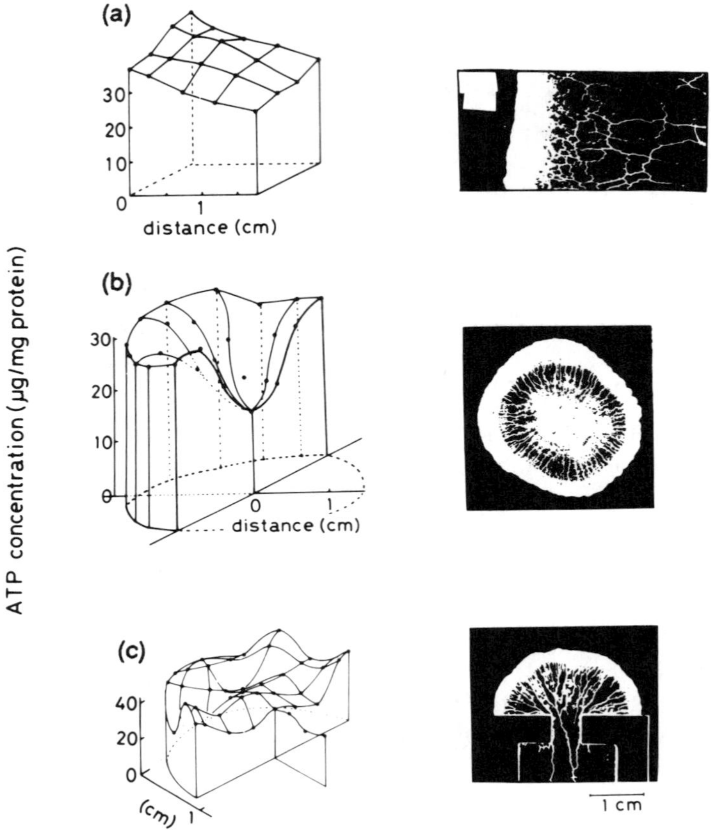

Fig. 6 Various ATP patterns accompanying various cell behavior. (After Ueda et al., 1987.)

fibrils are shown in Fig. 7 (Ueda et al., 1988a; Ueda, 1990b). The birefringent fibrils appear dark when vertical and bright when horizontal. The fibrils become abundant and thick after UV irradiation, and contract soon. This explains why the organism moves away from UV irradiation: The protoplasmic sol is squeezed away from the stimulated region when locally stimulated.

The distribution of birefringent fibrils is studied quantitatively by applying microcomputer image processing: the fibrous structure is extracted and the spatial distribution in the amount of fibrils is determined (Ueda et al., 1990a). The distributions of the fibrils and ATP concentra-

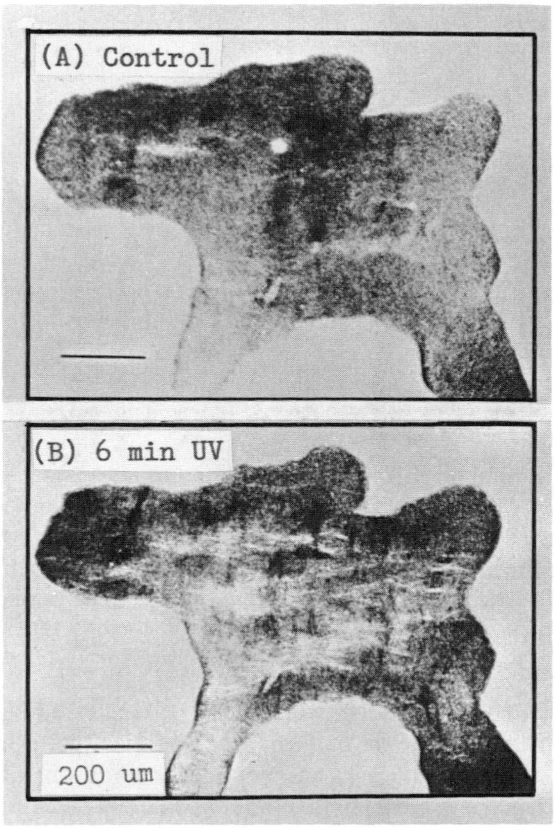

Fig. 7 Formation of the birefringent fibrils in the plasmodium after UV irradiation: (A) control; (B) 6-min UV. (After Ueda et al., 1988.)

tion are compared in Fig. 8, where the organism extends uniformly and the polarity is induced by applying a dc electric field. Soon after stimulation, ATP concentration decreases just inside the anodal side and becomes higher toward the cathodal side, with undulations in between. Birefringent fibrils become abundant just inside the anodal side and decrease toward the cathode, with undulations in between. So, roughly speaking, birefringent fibrils become abundant where ATP concentration decreases. In other words, self-organization of chemical patterns may cause reorganization of the cytoskeleton, and hence cell behavior.

This mechanism of cell behavior gives a clear answer why the cell is able to behave coordinately and also as an integrated unity, a question that

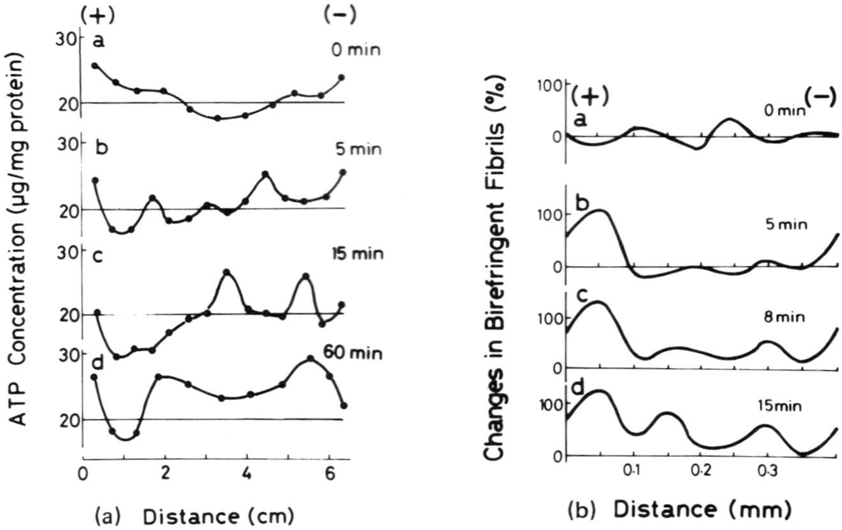

Fig. 8 Comparison of the distributions of ATP (a) and birefringent fibrils (b). The organism is subjected to a dc electric field at time zero. (After Ueda et al., 1990a.)

the previous theories of amoeboid cell motility could not solve. Based on the theory of self-organization of chemical patterns, the chemical pattern itself is generated by both kinetic parameters and boundary conditions of the system (Nicolis and Prigogine, 1977). In other words, the spatial pattern itself is nothing but a reflection of the system as a whole.

IV. CELL BEHAVIOR WITH COUPLED OSCILLATORS

A. Phase Wave in Relation to Information Processing

Due partly to active contraction and partly to passive flow of the protoplasm, the thickness of the plasmodium oscillates everywhere in the organism. Thus the organism is regarded as a system of coupled oscillators, and its dynamics are analyzed in relation to cell behavior.

A typical example of the response is shown in Fig. 9. When the organism is stimulated with repellent blue light, the thickness oscillation is modified (Mori et al., 1986). First, the phase of the oscillation at the stimulated point (P_1) becomes opposite to that at the unstimulated point

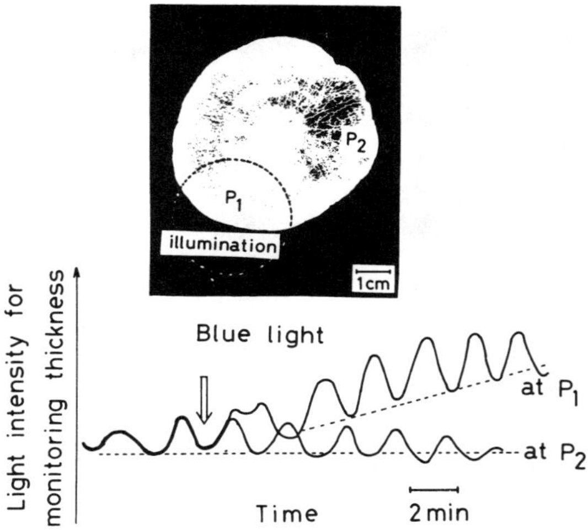

Fig. 9 Responses of the thickness oscillation to repellent blue light in a concentrically extending plasmodium. (After Mori et al., 1986.)

(P_2). This occurs within one period of the oscillation. Then the thickness at P_1 becomes thinner and thinner, indicating that the protoplasm moves away. A similar phase shift occurs when stimulated with attractants, so there is no way to discriminate attraction and repulsion by phase shift alone.

A typical example of the spatial organization of oscillations is shown in Fig. 10, where the organism is stimulated locally with an attractive warm temperature. Phases become opposite in the stimulated region, and at the same time waves begin to propagate away from the stimulated region. A propagating wave in this system is expressed as follows, with the approximation that the waveform does not change during transmission:

$$I(r,t) = I(kr - wt)$$

The direction of transmission, or phase vector k, is calculated from

$$\frac{k}{w} = -\frac{\text{grad}(I)}{dI/dt}$$

Values of k are averaged over 5 min, and k vectors are displayed by a mark (Matsumoto et al., 1986; Kobatake et al., 1989; Ueda et al., 1990a). Typical arrangements of k vectors are shown in Fig. 11. The

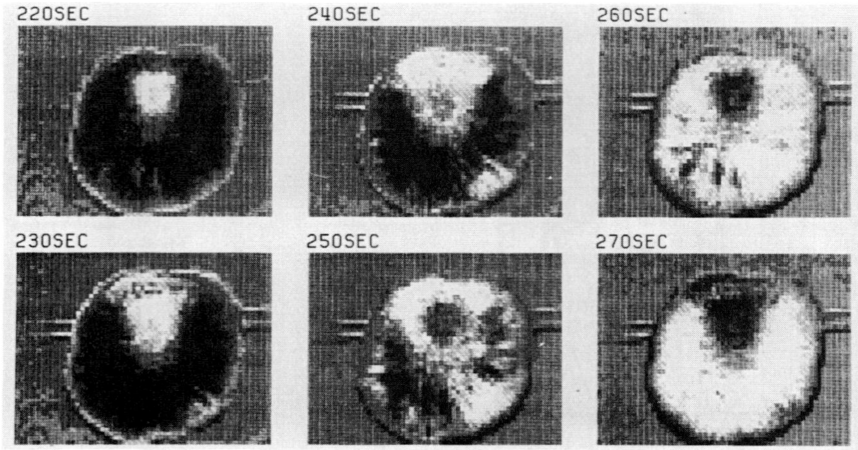

Fig. 10 Variation of the oscillation patterns after the organism is stimulated locally with attractive warm temperature. Gray levels correspond to the amplitude of the oscillation. (After Matsumoto et al., 1986.)

phase vectors are oriented outward from the stimulated region in the case of attractant stimulation. Contrary to this, a repulsive stimulation causes an orientation of k in the direction of the stimulated region. Thus the attraction and repulsion correspond to the orientation of phase gradient vectors, or direction of wave propagation.

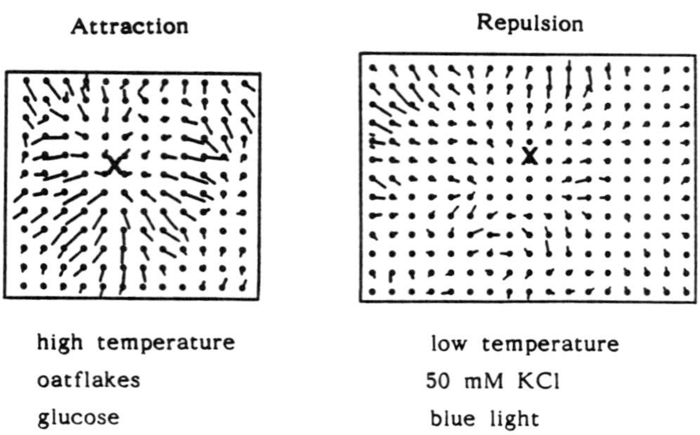

Fig. 11 Orientation patterns of phase gradient vectors in response to various attractants and repellents. (After Matsumoto et al., 1986.)

What happens when the orientation of phase gradient vectors is modified? This is achieved by a periodic external stimulation (e.g., low temperature), slightly faster than the intrinsic oscillation of the organism (Matsumoto et al., 1988). The intrinsic oscillation is entrained by the external one, and waves begin to propagate away from the stimulated region. Then the protoplasm moves toward the lower temperature. Thus the behavior is modified by artificial arrangement of *k* through entrainment.

Our results show that the information sensed is integrated at the level of the interacting oscillators. Coupled oscillator models are widely explored for explaining synchronization of heart beats, olfactory recognition, pattern recognition in brain, and so on. Thus our notion described here may have a wide significance in many fields where dynamic cooperativity plays a role in a multioscillator system.

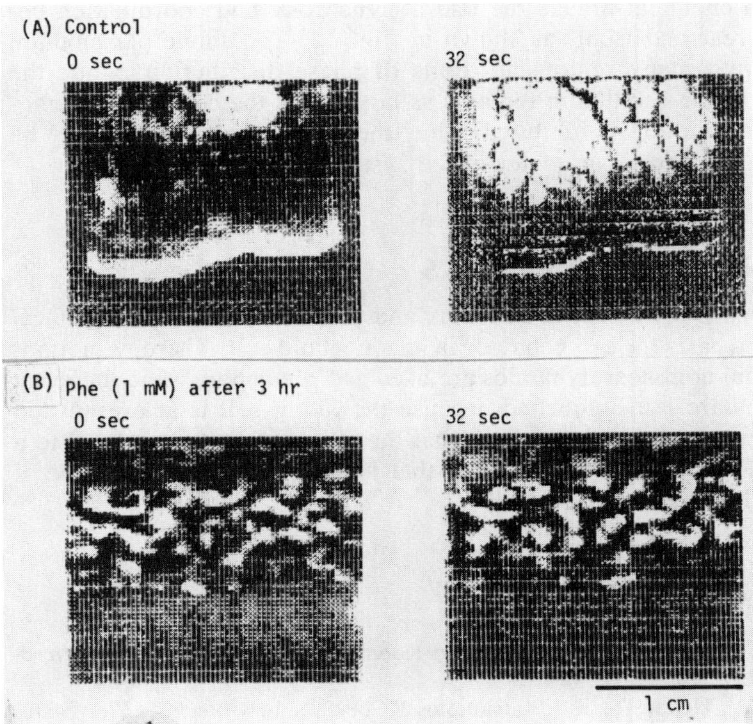

Fig. 12 Changes in the oscillation patterns in the presence of amino acids: (A) control; (B) Phe (1 m*M*) after 3 h. (After Ueda et al., 1988b.)

B. Cell Locomotion and Coherence

Locomotion of the organism is inhibited with amino acids, and the inhibition depends on the concentration and fits Hill-type plots:

$$v = \frac{v_0}{1 + (KC)^n}$$

Here v and v_0 are locomotion velocities in the presence (at the concentration C) and absence of chemicals, respectively. K and n are parameters, expressing sensitivity and cooperativity, respectively (Ueda et al., 1988b).

The effects of chemicals appear in two distinct ways in terms of oscillation patterns: amplitude and pattern modulations. In the former case, the amplitude of the oscillation decreased as the concentration of a chemical decreased. It is trivial that when the component element loses its activity, there should be no activity in the system. Interesting is the case in which a chemical breaks the lateral synchrony and coordination between the rear and front, as shown in Fig. 12. The single plasmodium separates into many regions in terms of phase distribution, while the amplitude of the oscillation remains as large as in the control. Nonetheless, the locomotion is inhibited. Thus the locomotion velocity may be regarded as an order parameter of the system.

V. CONCLUDING REMARKS

We have just shown that oscillations and morphogenesis in a chemical system are a basis for cell behavior in an amoeboid cell. Thereby, notions derived from nonlinear dynamics are used as "metaphor," and the exact mechanisms are left untouched because the living cell is at present too complex a system. Yet we believe that the "living" state is related to a higher system complexity, and hope that further study along this line of thought will proceed.

REFERENCES

Akitaya, T., Ohsaka, S., Ueda, T., and Kobatake, Y. (1985). *J. Gen. Microbiol.*, *131*: 195.

Kobatake, Y., Ueda, T., and Matsumoto, K. (1989). In *Molecular Electronics* (F. T. Hong, ed.), Plenum Press, New York, p. 223.

Matsumoto, K., Ueda, T., and Kobatake, Y. (1986). *J. Theor. Biol.*, *122*: 339.

Matsumoto, K., Ueda, T., and Kobatake, Y. (1988). *J. Theor. Biol.*, *131*: 175.

Mori, Y., Ueda, T., and Kobatake, Y. (1986). *Protoplasma*, *135*: 31.
Mori, Y., Ueda, T., and Kobatake, Y. (1987). *Protoplasma*, *139*: 141.
Nicolis, G., and Prigogine, I. (1977). *Self-Organization in Nonequilibrium Systems*, Wiley, New York.
Satoh, H., Ueda, T., and Kobatake, Y. (1982). *Cell Struct. Funct.*, *7*: 275.
Satoh, H., Ueda, T., and Kobatake, Y. (1984). *Cell Struct. Funct.*, *9*: 37.
Satoh, H., Ueda, T., and Kobatake, Y. (1985). *Exp. Cell Res.*, *156*: 79.
Ueda, T. (1990a). In *Molecular and Biological Physics of Living Systems* (R. K. Mishra, ed.), Kluwer Academic, Norwell, Mass., p. 133.
Ueda, T. (1990b). *Trends Photochem. Photobiol.*, *1*: 25.
Ueda, T., and Kobatake, Y. (1980). *Eur. J. Cell Biol.*, *23*: 37.
Ueda, T., and Kobatake, Y. (1982a). In *Cell Biology of Physarum and Didymium* (H. C. Aldrich and J. W. Daniel, eds.), Academic Press, New York, p. 111.
Ueda, T., and Kobatake, Y. (1982b). *J. Theor. Biol.*, *97*: 87.
Ueda, T., and Kobatake, Y. (1983). *Exp. Cell Res.*, *147*: 466.
Ueda, T., Matsumoto, K., Akitaya, T., and Kobatake, Y. (1986). *Exp. Cell Res.*, *162*: 486.
Ueda, T., Mori, Y., and Kobatake, Y. (1987). *Exp. Cell Res.*, *169*: 191.
Ueda, T., Mori, Y., Nakagaki, T., and Kobatake, Y. (1988a). *Photochem. Photobiol.*, *47*: 271.
Ueda, T., Nakagaki, T., and Kobatake, Y. (1988b). *Protoplasma*, *Suppl. 1*: 51.
Ueda, T., Nakagaki, T., and Yamada, T. (1990). *J. Cell Biol.*, *110*: 1097.
Ueda, T., Inagaki, H., and Yoshida, M. (1992). (To be published).

10
Growth Pulsations in Hydroid Polyps: Kinematics, Biological Role, and Cytophysiology

L. V. Beloussov and N. I. Kazakova *Moscow State University, Moscow, Russia*

Ju. A. Labas *Institute of Evolutionary Morphology and Ecology of Animals, Russian Academy of Sciences, Moscow, Russia*

I. INTRODUCTION

A pulsatory mode of growth and shaping in the vegetative generation of hydroid polypes was first observed by Saint-Hilaire (1930). Later his work was forgotten and the phenomenon has been rediscovered in the 1960s (Beloussov, 1961; Wyttenbach et al., 1965). It consists of a series of elongations–contractions of growing tips with a periodicity of several minutes (Wyttenbach, 1968, 1974; Wyttenbach et al., 1973; Beloussov et al., 1972; Labas et al., 1981). These may be designated as growth pulsations (GPs). At least two opposite phases can be distinguished in each GP: the extension phase, associated with the increase in the rudiments surface roughness, and the retraction phase, during which the rudiments surface is smoothed. GPs should be distinguished from the related but in no way identical phenomenon of periodic contractions of the proximal parts of hydroid coenosarc causing hydroplasmic streaming (Hale, 1964). The latter are much less regular than GPs and their periods are roughly twice as long as GP periods (Beloussov et al., 1972). However, if the extension phases of both coincide the tip extension is much greater than usual (Wyttenbach, 1973).

GP is a fascinating example of a periodic process involved directly in the development of form, suggesting nonlinear morphogenetic mechanisms. In this paper we review the data related mainly to the cellular kinematics, morphogenetical role, and cytophysiological mechanisms of

GPs, as observed in thecate hydroids of *Obelia* genus (*O. longissima* Pall., *O. loveni* Allm., *O. flexuosa* Hincks) and in *Dynamena pumila* L. collected by the authors in the tidal zone of the White Sea. Our intention is to show that GPs are representative of a large and important class of oscillatory phenomena involved in morphogenesis, rather than being an isolated phenomenon (see Table 1).

II. CELLULAR KINEMATICS OF GP

According to time-lapse filming and light-optical histology (Beloussov et al., 1972; Beloussov, 1973; Beloussov et al., 1989), GP are associated with periodic reorientations of both ecto- and endodermal cells in the tip regions of outgrowths (Fig. 1). In the above-mentioned species these regions are extended to 300 to 400 μm below growing tips and are characterized by close adhesion of the apical surfaces of ectodermal cells to the outer skeleton, perisarc. The latter is permanently secreted at the tip and hardens gradually in the proximal direction. In the region where cell reorientations are taking place, the perisarc is still deformable.

In each extension GP phase the cells are turning themselves from oblique toward the perpendicular direction, in relation to the stem axis (Fig. 1A–C). In the ectoderm their apical ends remain attached to the inner perisarc surface, whereas their basal ends are shifted distally, pulling the anchored basal membrane and pushing out the rudiment tip (Fig. 1A, vertical arrows). In most (although not in all) of the samples studied the cell turnings are spread as a proximodistal wave, with an average rate of about 50 to 90 μm/min. That means that in each GF, first, the most proximal cells of the tip region are turned toward the perpendicular orientation, then more distal ones, and so on (Fig. 1F). During the shorter retraction GP phase (in *O. longissima* not more than 20 to 30 s of the total 6- to 8-min GP period) the basal cell ends are moving proximally (Fig. 1D), whereas the apical ends are rapidly (with a rate of about 1 μm/s) shifted distalward and change their attaching points on the inner perisarc surface (Fig. 1D,E). Although a growing tip is retracted at this phase, it retains a part of its linear advance achieved during the previous GP extension phase. This is obviously due to the stiffness of the upper perisarc cusp.

III. THE RELATIONS BETWEEN GP PATTERNS AND RUDIMENT SHAPES

The "shapes" of GP records, that is, their impulse/periods relations (IPR) are both stage- and species-specific (Wyttenbach et al., 1973; Be-

Table 1 Oscillatory and Wavelike Events in Developing and Growing Objects

Object	Oscillatory component		Wave component		Reference
	Description	Period length (min)	Description	Rate (μm/min)	
Acrasiacea (*Dictyostelium*)	Oscillations of [cAMP]	5	Chemotactic relay	40–50	Gerisch, 1968
Red algae (*Polyides*)	Oscillations of linear dimensions	18	Unknown	Unknown	Labas et al., 1981
Brown algae (*Fucus*)	Oscillations of linear dimensions	6–8	Unknown	Unknown	Labas et al., 1981
Sponges (*Tethya*)	Oscillations of linear dimensions	6	Unknown	Unknown	Labas et al., 1981
Hydroid polyps	Oscillations of linear dimensions	5–15	Proximodistal wave of cell reorientation	40–90	Beloussov et al., 1989
Freshwater hydra	Oscillations of linear dimensions	5–7	Unknown	10–100	Campbell, 1980
Blastoderm of chick embryo	Cell contraction and tracking	2–3[a]	Waves of cell contraction	200[a]	Stern and Goodwin, 1977
Eggs of Crustacea (*Pollycipes*)	Cortical peristaltic	2	Waves of cortical contraction	14–20	Vacquier, 1981
Eggs of amphibians	Cortical contractions	30	Waves of cortical contraction	60	Yoneda et al., 1982
Eggs of ezox (*Teleostei*)	Ooplasmic ingressions	1–22[a]	Rotatory waves of ooplasmic ingressions	3800– –300[b]	Reznitchenko, 1982

[a]Measured at 37°C.
[b]Measured in the interval 5–35°C. The extreme values are given. Other objects measured at 15–22°C.

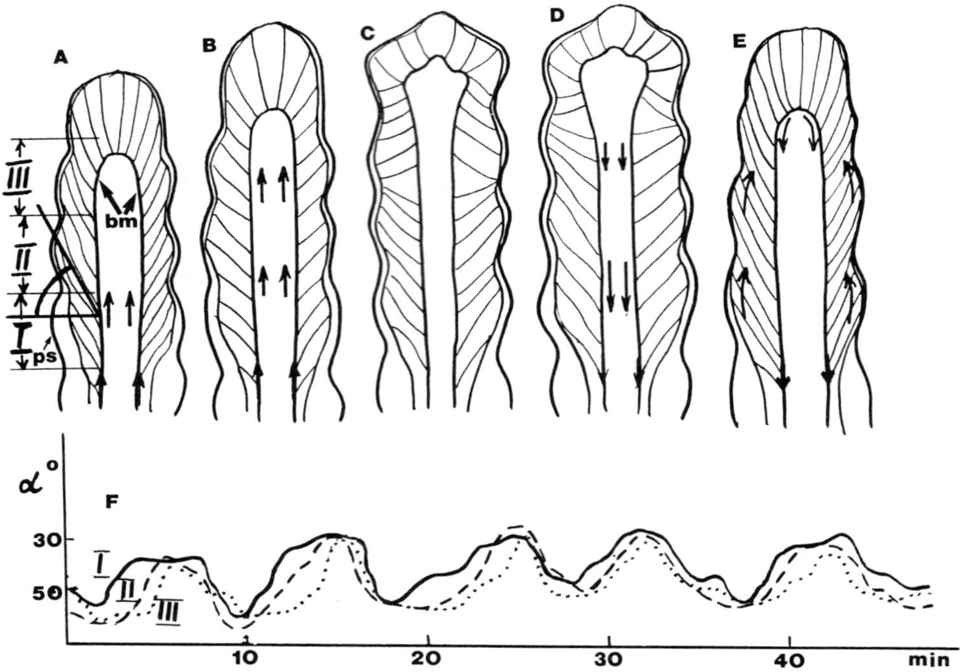

Fig. 1 Cell reorientations within a GP (a summary of light-optical and time-lapse studies). A–E, successive phases of a GP (only ectoderm is shown). Arrows denote the shifts of the basal membrane and the apical cell parts. bm, basal membrane; ps, perisarc. F is a diagram of the proximodistal waves of cell rotations within several successive GPs. Abscissa: time, min; Ordinate: angle between cell axes and the transversal stem axis in three successive proximodistal zones I to III (shown in A). (From Beloussov et al., 1980, 1989, modified.)

loussov et al., 1980, 1984). A simple rule can be suggested relating IPR to the rudiment shapes: The greater the relative IPR values (the relations of the shadowed squares in Fig. 2A–E to the period–amplitude products of the same GP, as shown by rectangles), the greater are also the rudiment width–length relations (Fig. 2). As argued by Beloussov and Lakirev (1988, 1991) IPR values are roughly proportional to the work of a plastic deformation of the upper perisarc by the reorienting cells within each GP. The width–length relation of a growing rudiment is proportional just to this work.

We have tested this hypothesis by computer simulation of hydroid morphogenesis under different pulsatory regimes employing either "nar-

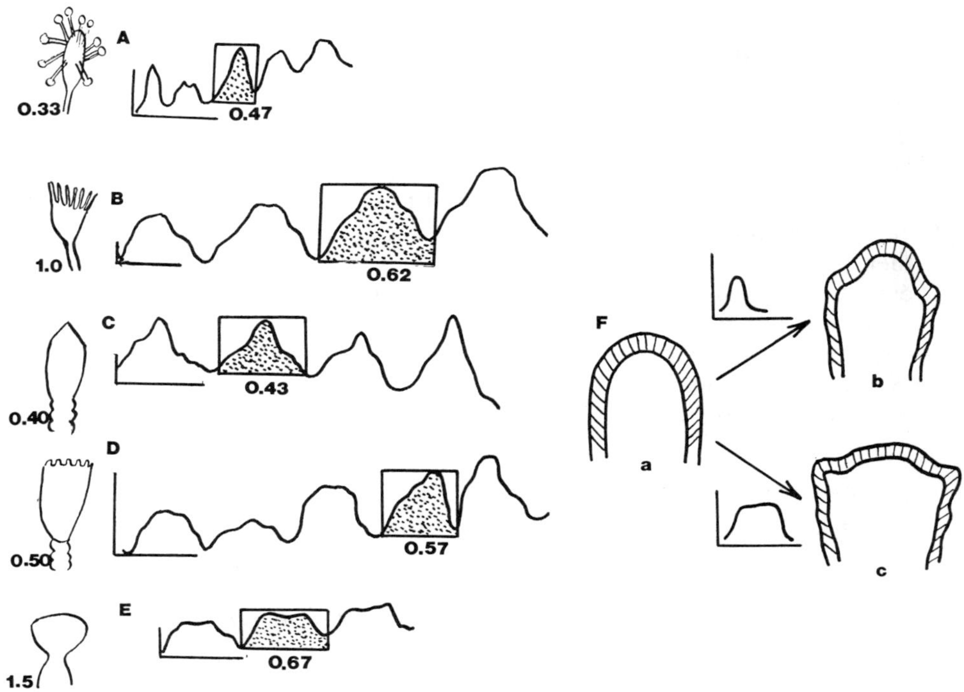

Fig. 2 Correlations between species-specific hydranth shapes and GP patterns. A, *Coryne loveni*; B, *Eudendrium rameum*; C, *Campanulina lacerata*; D, *Obelia loveni*; E, *Dynamena pumila*. In all the species the width/length hydranth relations (expressed by numbers at the left row of figures) correlate with IPR values (figures under GP records). F, Results of computer modeling of the morphogenesis of an idealized rudiment (a) by imitating either "narrow" (b) or "wide" (c) pulsations. Scale bars in GP records are 10 min/10 μm. (From Beloussov et al., 1984; Beloussov and Lakirev, 1988.)

row" or "wide" pulsations (Beloussov and Lakirev, 1988, 1991) and could reproduce the same IPR/shape correlation (Fig. 2F). This supports the idea that the mechanics of hydroid morphogenesis may be compared with a glass-blow technology (deformations of a soft distalmost perisarc by precisely scheduled periodical pressure impulses exerted by reorienting cells). The correlations between GP "shapes" and those of the rudiments were revealed in the experiments described below (see Fig. 4 and the corresponding comments).

IV. CYTOPHYSIOLOGICAL MECHANISMS OF GP

Electron microscopic studies showed the differences in the structure of the intracellular vacuolar system at opposite GP phases. Whereas at the height of the extension phase a lot of small rounded vacuoles could be seen, at the retraction phase they looked to be elongated and fused with each other into prolonged channels opened to intercellular spaces and then under the perisarc (Beloussov et al., 1989). This suggests that in each GP extension phase cell vacuoles swell by an osmotically driven water uptake, whereas within each retraction phase the water is leaking out from the fused vacuoles to the intercellular spaces. The idea of the involvement of ionic and osmotical mechanisms in GP is supported by the following observations (Beloussov et al., 1984, 1989 and unpublished results):

1. The reduction in external tonicity by seawater dilution from 755 mOsm (normal for White Sea water) to 530 mOsm led to the rapid cessation of GP and to smooth rudiment extension up to a stationary state (Fig. 3A). A lot of small spherical vacuoles appeared inside ectodermal cells. Within an hour the samples adapted to decreased tonicity and renewed GP. When transferred to the normal tonicity (which they perceived now as a hypertonicity) they retracted enormously (Fig. 3B). Cells shrank and vacuoles fused with each other into vast prolonged channels. Hence the decreased tonicity perfectly imitated the normal extension phase, and the increased tonicity did the same with the retraction phase of GP.
2. When diluting seawater 2.5 times by isotonic NaCl solution (thus increasing sodium concentration in relation to other ions without changing the normal medium tonicity), the result was the same as after tonicity decrease; that is, the rudiment was gradually extending (Fig. 3C, arrow 1).
3. The addition of 2×10^{-4} ouabain, which is expected to increase $[Na^+]_{out}$, led to approximately the same results as sodium increase (Fig. 3D).
4. On the other hand, the blockers of Na^+/Cl^- cotransport, 2×20^{-4} furosemid and ethacrynic acid inhibited the progressive growth and arrested the samples in the state, which corresponded, both by GP records and histologically, to the retraction GP phase.

The same was observed after exchanging the external Na^+ by Li^+ (Fig. 3E) or Cl^- by the poorly penetrating sulfate ion (Fig. 3C, 2). When seawater was diluted 2.5 times by mannosite, keeping the normal tonicity

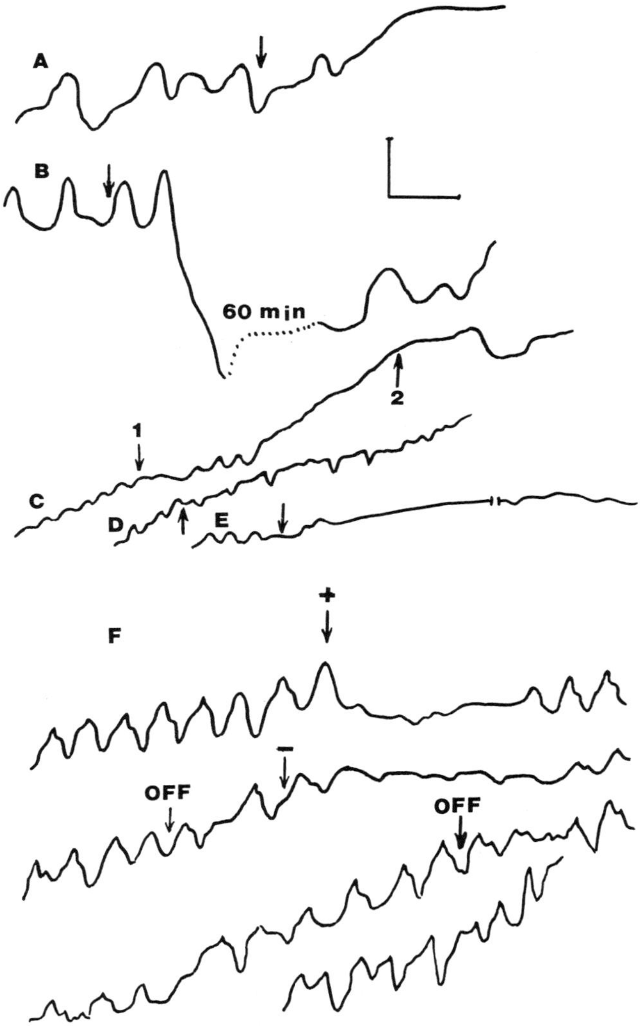

Fig. 3 Effects of osmotical and ionic changes (A–E) and electrical fields (F) upon GP in *O. longissima* samples. Arrows indicate the beginning of an agent's action. A, Shift toward hypotonicity arrests GP at the extension phase; B, shift to normal tonicity, after adaptation to hypotonicity results in pronounced retraction; C, effects of NaCl increase (1) and NaCl to Na_2SO_4 exchange (2); D, effects of ouabain addition; E, effects of LiCl addition; F, effects of the descending (anode from the top side) and ascending (cathode from the top side) constant current electrical fields upon GP. Scale bars as in Fig. 2. (From Beloussov et al., 1989; Labas et al., 1987.)

but reducing ionic strength, the rudimentary and abnormally frequent pulsations appeared unassociated with the progressive growth.

It is worth mentioning that GP and progressive stem growth proceeded normally even after a fivefold increase in magnesium concentration. In the isotonic (in relation to the seawater) $MgCl_2$ solution (255 m*M*) free of Na^+, K^+, and Ca^{2+}, GP proceeded with an increased rate and amplitude for about 40 min, although progressive growth ceased immediately.

Taken together, the data above argue for a leading role for the sodium chloride–mediated osmotical pressure as a driving force in pulsatory extension of hydrid rudiments. The transverse cell turnings at the extension phase appear as direct results of their shifts from a parallelogram-like shape toward a shape approximately rectangular, due to osmotically driven swelling (Beloussov et al., 1989).

What might be the factors determining periodical shifts from extension to contraction phases and back again? Employing Stern's (1984) idea about the "+, −" feedback relations between sodium and calcium concentrations in the chick embryo blastodermal cells, we have studied the effects of agents modulating intracellular calcium upon GP. We found that in 10^{-5} solutions of the Ca^{2+} ionophore, A23187, GP gradually ceased at the phase of the moderate retraction; on the other hand, in the presence of 10 M^{-5} L-*cis*-diltiazemchloride (blocker of Ca^{2+} channels) the pulsations were arrested at the extension phase. That makes it probable that the retraction phase is associated with an increase and the extension phase with a decrease in $[Ca^{2+}]_{in}$. This was supported by direct measurement of this value at opposite GP phases by using a fluorescent indicator, Qwin 2 AM (Labas et al., 1989).

In some of these experiments, the following unexpected observation was useful: It was found that the external electrical field of a constant current exerted a distinct and sharply polarized influence upon GP (Labas et al., 1987). The field of the constant ascending currents (cathode at the stem tip) of 1 to 3 V/s·m voltage and 20 to 150 μA/mm current density within a few minutes arrested the tips at the extension GP phase, whereas the same current with descending field (anode from the tip side) arrested GP at the retraction phase (Fig. 3F). In both cases not only external kinematics, but also the cell's ultrastruture, were very similar to those in normal or osmotically induced GP phases. Intracellular calcium concentrations showed the same patterns: They were high in samples exposed to descending fields and low in those exposed to ascending fields. The polar effects of external electrical fields seem to be associated with a distoproximal (as well as apicobasal) electrophysiological polarity of the growing stem cells (Labas et al., 1987).

Taking these data together, we suggest the following provisional scheme of GP regulation. As a basic continuous process, the osmotically driven water flow inside vacuoles should be considered. The associated increase in intracellular sodium (by Stern's suggestion, cited above), and even more probably, the stretching of cell membranes caused by osmotic swelling leads, due to activation of the mechanosensitive calcium channels (Christensen, 1987), causes an increase in $[Ca^{2+}]_{in}$. This increase should stimulate the fusion of vacuoles with each other and the cell membrane, as well as contractile reactions. All of that will stimulate water flow out of cells and hence a transition to the retraction phase.

GP can also be affected by such regulatory factors as cyclic nucleotides and neurotransmitters (Beloussov et al., 1984). The effects of two neuromimetics may be of interest as further arguments on the dependence between GP pattern and rudiment shapes. In *Dynamena pumila*, 10^{-6} to 10^{-3} *M m*-cholinomimetic, F-2268, together with 10^{-4} *M* eserine or proserine (cholinesterase inhibitors), led to reduction in the GP retraction phases; correspondingly, the growing tips became wider than normal and the lateral (hydranths) rudiments were selectively stimulated (Fig. 4A, C–E). In contrast, β-adrenomimetic isoprotherenol in the same concentration range induced ''narrow'' GP (extension phase abbreviated) and stimulated the exclusively longitudinal growth of the central stem tips, with the lateral buds of the hydranths being reduced (Fig. 4, F–H). That supports the foregoing conclusions on the existence of causal relations between a GP pattern and the shapes of growing tips and indicates the existence of different regulatory patterns for GP in the rudiments of the hydranths and stems.

V. OUTLOOK

GP in hydroid polyps are certainly not an isolated phenomenon in nature. This is evident from the uncomplete list of oscillatory processes with similar periodicites and wave propagation rates taking place in developing systems. The osmotical nature of the oscillations is highly probable in algae, sponges, and chick blastoderm as well as in hydroids. By means of several personal communications, we have learned that mammalian blastocysts also pulsate pumping water inside and outside the cavity, probably by means of hydroid-like mechanisms, and periodic water inflows and active outflows have been discovered in plant roots (Lazareva et al., 1986). These developmentally and physiologically important and evolutionarily ancient events undoubtedly deserve greater interest from biologists.

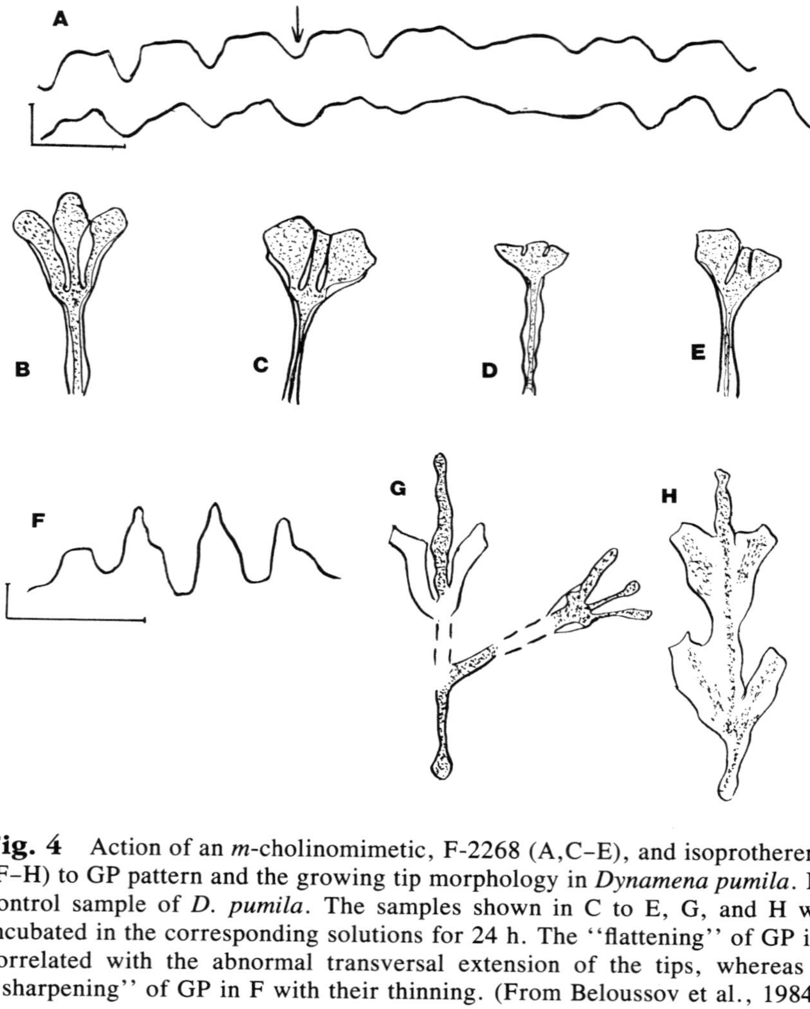

Fig. 4 Action of an *m*-cholinomimetic, F-2268 (A,C–E), and isoprotherenole (F–H) to GP pattern and the growing tip morphology in *Dynamena pumila*. B: a control sample of *D. pumila*. The samples shown in C to E, G, and H were incubated in the corresponding solutions for 24 h. The "flattening" of GP in A correlated with the abnormal transversal extension of the tips, whereas the "sharpening" of GP in F with their thinning. (From Beloussov et al., 1984.)

REFERENCES

Beloussov, L. V. (1961). *Dokl. Akad. Nauk SSSR, 136*: 1490 (in Russian).

Beloussov, L. V. (1973). *Publ. Seto Mar. Biol. Lab., 20*: 315.

Beloussov, L. V., and Lakirev, A. V. (1988). In *Thermodynamics and Pattern Formation in Biology* (I. Lamprecht and A. I. Zotin, eds.), Walter de Gruyter, New York, p. 321.

Beloussov, L. V., and Lakirev, A. V. (1991). *J. Theor. Biol., 152*: 455.

Beloussov, L. V., Badenko, L. A., Katchurin, A. L., and Kurilo, L. F. (1972). *J. Embryol. Exp. Morphol.*, *27*: 317.
Beloussov, L. V., Badenko, L. A., and Labas, Ja. A. (1980). In *Developmental and Cellular Biology of Coelenterates* (P. Tardent and R. Tardent, eds.), Elsevier/North-Holland Biomedical Press, Amsterdam, p. 175.
Beloussov, L. V., Labas, Ju. A., and Badenko, L. A. (1984). *Zh. Obshch. Biol.*, *45*: 796 (in Russian).
Beloussov, L. V., Labas, Ju. A., Kazakova, N. I., and Zaraisky, A. G. (1989). *J. Exp. Zool.*, *249*: 258.
Campbell, R. D. (1980). In *Developmental and Cellular Biology of Coelenterates* (P. Tardent and R. Tardent, eds.), Elsevier/North-Holland Biomedical Press, Amsterdam, p. 421.
Christensen, O. (1987). *Nature*, *330*: 66.
Gerisch, G. (1968). *Curr. Top. Dev. Biol.*, *3*: 157.
Hale, L. J. (1964). *J. Embryol. Exp. Morphol.*, *12*: 517.
Labas, Ju. A., Beloussov, L. V., Badenko, L. A., and Letunow, V. N. (1981). *Dokl. Akad. Nauk SSSR*, *257*: 1247 (in Russian).
Labas, Ju. A., Beloussov, L. V., Kazakova, N. I., and Badenko, L. A. (1987). *Ontogenez (Sov. J. Dev. Biol.)*, *18*: 154.
Labas, Ju. A., Zinchenko, W. P., Beloussov, L. V., and Kazakova, N. I. (1989). *Zh. Obshch. Biol.*, *50*: 504.
Lazareva, N. P., Borisova, T. A., and Zcholkewitsch, V. N. (1986). *Dokl. Akad. Nauk SSSR*, *289*: 761 (in Russian).
Reznitchenko, P. N. (1982). *Funkcionalnie Transformatsii v Ontogenese Nischich Pozvonotchnich*, Nauka, Moscow, pp. 1–216 (in Russian).
Saint-Hilaire, K. (1930). *Zool. Jahrb. Abt. Allg. Zool. Physiol. Tiere*, *47*: 512.
Stern, C. D. (1984). *J. Theor. Biol.*, *107*: 229.
Stern, C. D., and Goodwin, B. C. (1977). *J. Embryol. Exp. Morphol.*, *41*: 15.
Vacquier, V. D. (1981). *Dev. Biol.*, *84*: 1.
Wyttenbach, C. R. (1968). *J. Exp. Zool.*, *167*: 333.
Wyttenbach, C. R. (1973). *J. Exp. Zool.*, *186*: 79.
Wyttenbach, C. R. (1974). *Am. Zool.*, *14*: 699.
Wyttenbach, C. R., Crowell, S., and Suddith, R. L. (1965). *Biol. Bull.*, *129*: 429.
Wyttenbach, C. R., Crowell, S., and Suddith, R. L. (1973). *J. Morphol.*, *139*: 363.
Yoneda, M., Kobayakawa, Y., Kubota, H. Y., and Sakai, M. (1982). *J. Cell Sci.*, *54*: 35.

11
Wavelike Propagation of cAMP and Ca^{2+} Signals: Link with Excitability and Oscillations

Albert Goldbeter and Geneviève Dupont *Université Libre de Bruxelles, Brussels, Belgium*

I. INTRODUCTION

Wherever oscillations occur in spatially homogeneous conditions, wavelike phenomena may arise as soon as diffusion processes begin to play a prominent role. Thus the Belousov–Zhabotinsky reaction, which provides the prototype of chemical oscillations, readily gives rise to propagating waves in an unstirred medium (Winfree, 1972). Oscillations and waves in chemical systems can be viewed, respectively, as temporal and spatiotemporal dissipative structures representing different modes of nonequilibrium self-organization in open systems (Nicolis and Prigogine, 1977). In biology, the close link that exists between the two kinds of phenomena is well exemplified by neural and cardiac dynamics (Winfree, 1987; Murray, 1989).

Such a link also exists for rhythmic processes of nonelectrical nature involved in intercellular communication. Thus cAMP oscillations in *Dictyostelium* cells (Gerisch and Wick, 1975) underlie the wavelike aggregation of these amoebae after starvation. The periodic generation of cAMP signals by aggregation centers transforms into a propagating wave of cAMP (Tomchik and Devereotes, 1981) which then triggers a wave of cellular chemotactic movement (Gerisch, 1971; Alcantara and Monk, 1974; Devreotes, 1982). The propagation of the cAMP wave depends on the relay of the signal by chemotactically responding cells (Shaffer, 1962).

While wave propagation of the cAMP signal occurs in the extracellular medium in *Dictyostelium*, a corresponding phenomenon occurs for

Ca^{2+} signals within the cytosol of a variety of cells, either spontaneously or upon external stimulation. A number of recent observations indicate that in a variety of cells, Ca^{2+} signaling occurs in a space-dependent manner, in the form of propagating Ca^{2+} waves (Gilkey et al., 1978; Busa and Nuccitelli, 1985; Kort et al., 1985; Takamatsu and Wier, 1990; Jacob, 1990a; Rooney et al., 1990; Cornell-Bell et al., 1990). The phenomenon is closely associated with the oscillations of cytosolic Ca^{2+} that are triggered in these cells by a rise in extracellular Ca^{2+} or by stimuli such as hormones or neurotransmitters (Berridge and Galione, 1988; Berridge et al., 1988; Cuthbertson, 1989; Jacob, 1990b; Petersen and Wakui, 1990; Berridge, 1990).

The purpose of this chapter is to examine briefly the link that relates excitability and oscillations to wavelike propagation of signals in intercellular communication. We shall deal, successively, with extracellular cAMP waves in aggregating *Dictyostelium* cells, and with intracellular Ca^{2+} waves. The two phenomena will be related to the occurrence of cAMP and Ca^{2+} oscillations. In both cases, models that account for the molecular mechanism of the intracellular oscillations can be extended, by incorporation of diffusion, to throw light on the origin of the wavelike phenomena.

Waves of cAMP and Ca^{2+} provide cell signaling devices which, although much slower than electrical ones, are faster than simple diffusion (see also Meyer, 1991). Such second messenger waves are likely to play important roles in development. While such a role is well established for the case of intercellular signaling during aggregation of *Dictyostelium* amoebae, the function of Ca^{2+} waves remains to be determined, but they clearly have the potential of allowing for the rapid transfer of information between distant regions in a given cell (see, e.g., Kasai and Augustine, 1990) or even between adjacent cells, as exemplified by astrocytes (Cornell-Bell et al., 1990) and glial cells (Charles et al., 1991).

II. OSCILLATIONS AND WAVES OF cAMP IN *DICTYOSTELIUM* CELLS

A. Wavelike Aggregation of Cells and cAMP Waves

The cellular slime mold *Dictyostelium discoideum* represents a particularly useful model in developmental biology (Bonner, 1967; Loomis, 1975). This is also the case for our purpose here, which is to discuss the molecular bases of wavelike phenomena in intercellular communication. After starvation, these amoebae indeed aggregate by a chemotactic response to pulses of cyclic AMP (cAMP) emitted by cells that behave as aggregation centers; as a result, the amoebae placed on agar move toward the centers in concentric or spiral waves whose periodicity is on the order of 5 to 10 min (Shaffer, 1962; Gerisch, 1971, 1987; Alcantara and Monk, 1974; Durston, 1974; Devreotes, 1982).

B. Model for cAMP Relay and Oscillations

A model proposed by Martiel and Goldbeter (1987) on the basis of cAMP-induced cAMP synthesis and receptor desensitization accounts for the autonomous oscillations of cAMP (Gerisch and Wick, 1975) and for the relay of suprathreshold signals of cAMP (Roos et al., 1975) observed both in cellular suspensions and during aggregation on agar. The model (Fig. 1) takes into account the binding of extracellular cAMP to a membrane receptor, the ensuing activation of adenylate cyclase, which catalyzes the synthesis of cAMP from intracellular ATP, the transport of cAMP into the extracellular medium, and its hydrolysis by phosphodiesterase. Of primary importance in this model are self-amplification in cAMP synthesis and the cAMP-induced receptor desensitization through reversible phosphorylation (Vaughan and Devreotes, 1988).

The three variables of the model are intracellular and extracellular cAMP and the fraction of active, nondesensitized cAMP receptor. Provided that some rough approximations are made, the number of variables can be reduced from three to only two: the concentration of extracellular cAMP and the fraction of active receptor. This two-variable system can

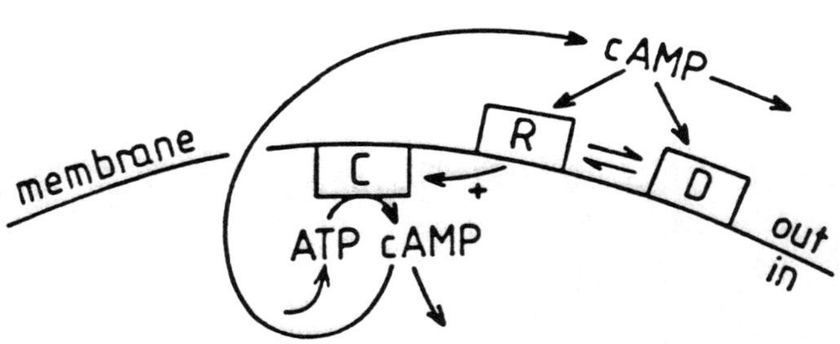

Fig. 1 Model for cAMP signaling in *Dictyostelium discoideum* based on receptor desensitization. Extracellular cAMP binds to the active (R) and desensitized (D) states of the receptor; the former process triggers the activation of adenylate cyclase (C). Arrows relate to the transport of cAMP into the extracellular medium and to its hydrolysis by phosphodiesterase. (From Martiel and Goldbeter, 1987.)

be viewed as the true core of the mechanism for relay and oscillations of cAMP in *Dictyostelium* (Martiel and Goldbeter, 1987; Goldbeter, 1990).

The model shows how sustained oscillations (Fig. 2a) around an unstable steady state arise from conjunction of the self-amplified synthesis of cAMP and the counteracting effect of receptor desensitization. In conditions where the steady state is stable but close to becoming unstable for slightly different values of the parameters, a phenomenon of excitability occurs; the system is then capable of amplifying suprathreshold perturbations in a pulsatory manner before returning to the stable steady state (Fig. 2b). Such behavior, which accounts for the relay of suprathreshold pulses of cAMP in the course of aggregation (Shaffer, 1962; Roos et al., 1975), reflects the excitability of the signaling system and is characterized by an absolute as well as a relative refractory period. Signal transduction between the receptor and the cyclase involves the action of G proteins (Snaar-Jagalska and Van Haastert, 1990); their incorporation into the model yields similar results as to relay and oscillatory behavior (J. Halloy, B. Bodholdt, and A. Goldbeter, manuscript in preparation).

The necessary link between excitable and oscillatory behavior is best demonstrated by means of phase-plane analysis (Martiel and Goldbeter, 1987; Goldbeter, 1990). Such an analysis is analogous to that applied by Fitzhugh (1961) to nerve cells, which can either be excitable or function as pacemakers: in the former case, a single action potential is generated following a depolarizing stimulus of sufficient amplitude, while in the latter a train of action potentials are produced spontaneously. Thus in neurons as in *Dictyostelium*, the two closely linked modes of dynamic behavior originate from a common molecular mechanism and occur for slightly different values of the parameters.

C. Link Between Relay, Oscillations, and cAMP Waves

The wavelike aggregation of *Dictyostelium discoideum* cells has been investigated theoretically by Tyson and Murray (1989), who resorted to the model for cAMP oscillations based on receptor desensitization (Martiel and Goldbeter, 1987), to which they added diffusion of extracellular cAMP; as a first approximation, the amoebae were assumed to remain immobile while cAMP diffuses away into the surrounding medium.

An appropriate choice of initial conditions and of parameter values then leads to the emergence of spatiotemporal patterns of cAMP (Fig. 3). These waves take the form of concentric, ''target patterns'' as long as initial conditions and biochemical parameters remain isotropic (J. Halloy and A. Goldbeter, unpublished results); spiral waves can occur when the

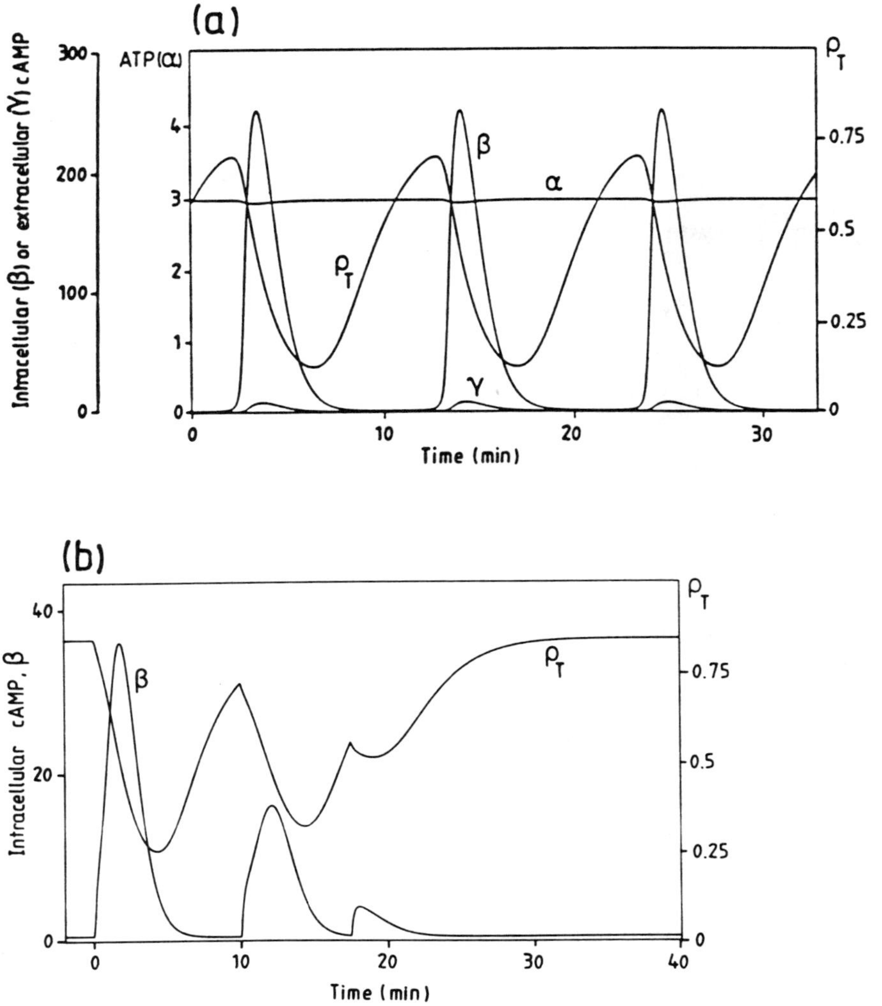

Fig. 2 Oscillations (a) and relay (b) of cAMP in the model for cAMP signaling in *D. discoideum* schematized in Fig. 1. Shown are the evolution of intracellular (β) and extracellular (γ) cAMP, ATP (α), as well as the fration of active, nondesensitized cAMP receptor (ρ_T). In (b), the response to three successive cAMP stimuli of similar magnitude illustrates the existence of a refractory period resulting from receptor desensitization. (From Martiel and Goldbeter, 1987.)

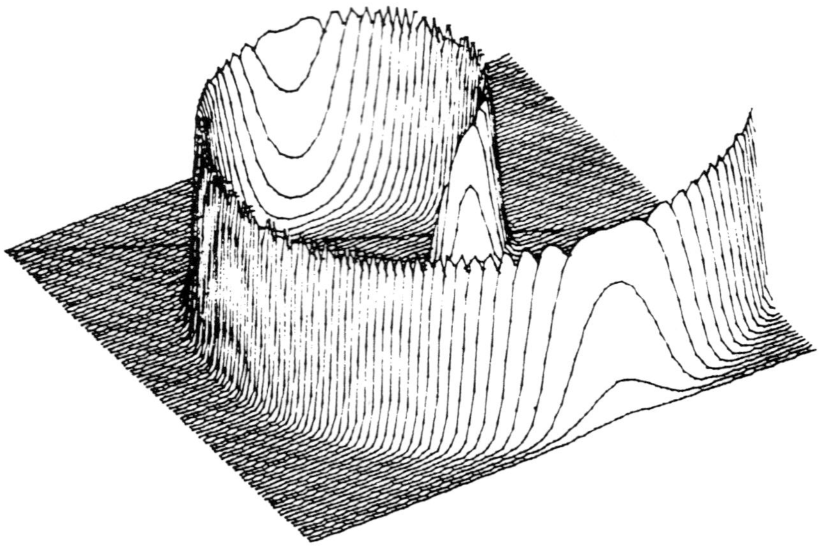

Fig. 3 Wavelike propagation of cAMP in a two-dimensional medium. The data have been obtained by means of the two-variable version of the model for cAMP signaling shown in Fig. 1, supplemented with cAMP diffusion into the extracellular medium. (From Tyson, 1989; Tyson and Murray, 1989; Tyson et al., 1989.)

symmetric patterns are broken by an obstacle or by the presence of refractory cells (Tyson, 1989; Tyson and Murray, 1989; Tyson et al., 1989). Concentric or spiral waves of cAMP have also been obtained in a model based on an alternative mechanism for cAMP relay and oscillations (Monk and Othmer, 1990). Similar patterns can also be generated by means of cellular automata (Markus and Hess, 1990; Gerhardt et al., 1990; A. Pang, personal communication), which simulate in a discrete manner the periodic stimulation of excitable cells that enter a refractory phase after generating a pulse of cAMP.

III. OSCILLATIONS AND WAVES OF CYTOSOLIC Ca^{2+}

A. Cytosolic Ca^{2+} Oscillations

Intracellular oscillations of Ca^{2+} ions have been the focus of much research in recent years. These oscillations occur in the form of a train of Ca^{2+} spikes, either spontaneously or after stimulation by an external signal such as a hormone or a neurotransmitter. The oscillations occur in a variety of cells, with periods ranging from less than 1 s up to 30 min (for recent reviews, see Berridge and Galione, 1988; Berridge et al., 1988; Cuthbertson, 1989; Jacob, 1990b; Petersen and Wakui, 1990; Berridge, 1990, as well as the special issue devoted to the phenomenon by the journal *Cell Calcium*, volume 12, number 2/3, 1991).

B. Model for Intracellular Ca^{2+} Oscillations

The mechanism of Ca^{2+} oscillations appears to involve the synthesis of inositol 1,4,5-trisphosphate (IP_3) that follows binding of the stimulatory ligand to the membrane receptor. According to Berridge (1990), the rise in IP_3 in turn triggers the release of a certain amount of Ca^{2+} from an IP_3-sensitive intracellular store. The analysis of a two-variable model (see the scheme in Fig. 4) taking into account the Ca^{2+} input from the external medium, the extrusion of Ca^{2+} from the cell, and the exchange of Ca^{2+} between the cytosol and an IP_3-insensitive store shows that sustained oscillations may arise (Fig. 5) through a mechanism of Ca^{2+}-induced Ca^{2+} release (Endo et al., 1970; Fabiato and Fabiato, 1975) from the IP_3-insensitive store once the Ca^{2+} secretion triggered by IP_3 has reached a

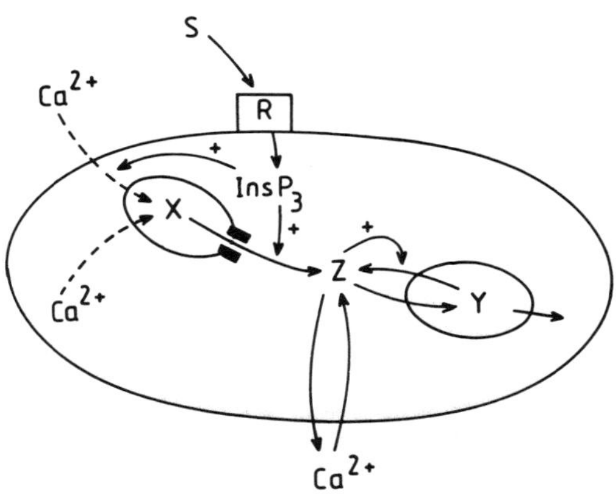

Fig. 4 Minimal model for Ca^{2+} oscillations based on Ca^{2+}-induced Ca^{2+} release. A signal (S) triggers the synthesis of IP_3, which enhances the release of Ca^{2+} from an IP_3-sensitive store (X); cytosolic Ca^{2+} (Z) is pumped into an IP_3-insensitive compartment (Y) from which it is released in a process activated by cytosolic Ca^{2+}. Other arrows relate to transport of Ca^{2+} into and out of the cell. (From Dupont and Goldbeter, 1989; Goldbeter et al., 1990; Dupont et al., 1991.)

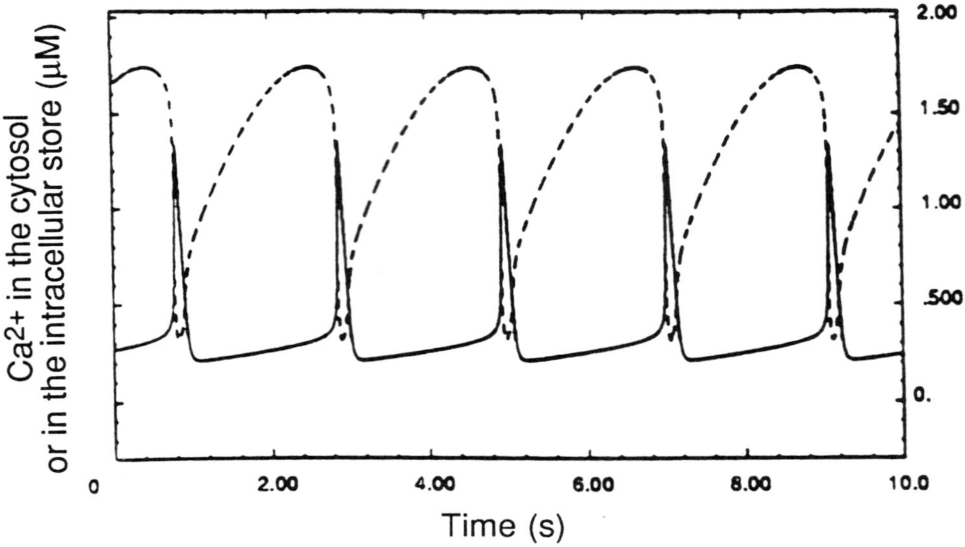

Fig. 5 Oscillations of Ca^{2+} generated by the model of Fig. 4. Shown are the concentrations (defined with respect to total cell volume) of cytosolic Ca^{2+} (solid line) and of Ca^{2+} in the Ca^{2+}-sensitive stores (dashed line).

sufficient level (Dupont and Goldbeter, 1989; Goldbeter et al., 1990; Dupont et al., 1991).

The model accounts for the observation that the frequency of the Ca^{2+} spikes increases with the extent of stimulation; the latter is measured by the saturation function of the IP_3 receptor, given that the level of IP_3 rises with the external signal. In contrast with an alternative model based on a feedback of Ca^{2+} on IP_3 production (Meyer and Stryer, 1988), a specific prediction of the analysis is that sustained oscillations in cytosolic Ca^{2+} may occur in the absence of a concomitant, periodic variation in IP_3. Such a prediction agrees with a number of experimental observations (Capiod et al., 1987; Wakui et al., 1989, 1990). On the other hand, the model is related to that proposed by Kuba and Takeshita (1981) for membrane potential oscillations in sympathetic neurons treated with caffeine; there also, oscillations were due to Ca^{2+} spikes resulting from the mechanism of Ca^{2+}-induced Ca^{2+} release, but the role of IP_3, discovered since, was not considered.

C. Ca^{2+} Wave Propagation

Propagating Ca^{2+} waves have been observed in the cortical layer of fertilized eggs (Gilkey et al., 1978; Busa and Nuccitelli, 1985) as well as in the cytoplasm of cardiac cells (Kort et al., 1985; Takamatsu and Wier, 1990), endothelial cells (Jacob, 1990a), hepatocytes (Rooney et al., 1990; Thomas et al., 1991), and astrocytes (Cornell-Bell et al., 1990; Cornell-Bell and Finkbeiner, 1991). The idea that such a phenomenon arises from the propagation of a diffusing Ca^{2+} front generated by Ca^{2+}-induced Ca^{2+} release (Busa et al., 1985; Kort et al., 1985; Berridge, 1990) has been tested in theoretical models for wave propagation in oocytes (Cheer et al., 1987) and cardiac cells (Backx et al., 1989), where Ca^{2+}-induced Ca^{2+} release was represented in an empirical manner. The latter process has also been used in modeling the mechanochemistry of actomyosin gels to account for a variety of morphogenetic phenomena (Oster and Odell, 1984).

The minimal model for Ca^{2+} oscillations allows us to investigate in further detail the link between IP_3 signaling, Ca^{2+}-induced Ca^{2+} release, and wave propagation. When assuming, for simplicity, that Ca^{2+}-sensitive Ca^{2+} pools are distributed along a single space dimension, the incorporation of cytosolic Ca^{2+} diffusion into the equations of the two-variable model for Ca^{2+} oscillations shows that the wave propagation of a Ca^{2+} spike requires a finely tuned balance between reaction and diffusion processes (Dupont et al., 1991).

A typical result shown in Fig. 6 was obtained by considering the longitudinal section of a cell containing five Ca^{2+}-sensitive Ca^{2+} pools distributed at regular intervals; between these pools cytosolic Ca^{2+} propagates by diffusion. When the first pool, located on the left extremity, is stimulated by increasing the influx of extracellular Ca^{2+} or by raising the level of IP_3 to reflect the effect of an external signal, the resulting spike generated by Ca^{2+}-induced Ca^{2+} release is seen to propagate along the cell toward the right extremity. The sequential firing of the five Ca^{2+}-sensitive pools in the course of time is reflected by the alignment of the peaks on the diagonal in Fig. 6.

The velocity of Ca^{2+} waves varies in cells of different sizes and shapes. Thus the wave propagates at a rate of the order of 10 $\mu m/s$ on the surface of *Medaka* oocytes, whose diameter is on the order of 1.1 mm (Gilkey et al., 1978) and at a rate close to 70 to 100 $\mu m/s$ in the cytoplasm of cardiac cells whose length is on the order of 100 μm (Kort et al., 1985; Takamatsu and Wier, 1990). In Fig. 6, the wave propagation rate is close to 80 $\mu m/s$ when taking a cell length of 107 μm and a Ca^{2+} diffusion coefficient of 400 $\mu m^2/s$ (Kushmerick and Podolsky, 1969). The model

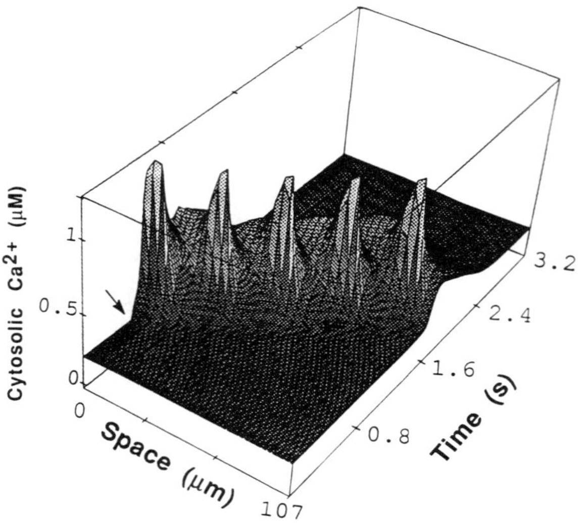

Fig. 6 Propagation of a Ca^{2+} spike along a one-dimensional medium containing five equidistant Ca^{2+}-sensitive Ca^{2+} stores. The system is stimulated on the left extremity by transiently raising the input of Ca^{2+} into the cytosol (see the text for details). (From Dupont et al., 1991.)

therefore predicts the correct magnitude of the wave propagation rate observed in cardiac cells when the size of the system and the period of oscillations—on the order of 1 s—match the characteristics of these cells.

Waves of Ca^{2+} have also been obtained by Meyer and Stryer (1991) in their model based on a cross-activation of Ca^{2+} and IP_3. These authors attribute the primary role in wave propagation to IP_3, mainly because they consider a much lower value (6 $\mu m^2/s$) for the Ca^{2+} diffusion coefficient, while, in line with other studies (Cheer et al., 1987; Backx et al., 1989), we took a higher value, on the order of that observed in water (Kushmerick and Podolsky, 1969), given that pumping of Ca^{2+} into the intracellular stores, which decreases the effective value of the diffusion coefficient, is already taken into account in our model for Ca^{2+} signaling. The value considered for the Ca^{2+} diffusion coefficient also matches the value determined by Lechleiter et al. (1991), whose analysis of spiral waves of Ca^{2+} by means of cellular automata supports Ca^{2+}-induced Ca^{2+} release as the mechanism underlying wave propagation in *Xenopus* oocytes.

Because the period and half-width of the Ca^{2+} spike are short in cardiac cells, the wave in Fig. 6 propagates in the form of a sharp, narrow

Ca^{2+} peak. In other cells, such as hepatocytes (Rooney et al., 1990), endothelial cells (Jacob, 1990a), or oocytes (Gilkey et al., 1978; Busa and Nuccitelli, 1985), the wave has a different appearance (Meyer, 1991) as the level of Ca^{2+} is seen to increase progressively, starting from one cell extremity, until spreading is such that the whole cell reaches a high level of Ca^{2+}, after which the Ca^{2+} concentration decreases overall and a new front is formed. Numerical simulations indicate (Thomas et al., 1991; Dupont and Goldbeter, 1992) that the model based on Ca^{2+}-induced Ca^{2+} release can generate such a wavelike pattern when the period of oscillations and the half-width of Ca^{2+} spikes are larger, of the order of those observed in hepatocytes and endothelial cells.

The data of Fig. 6 were obtained in conditions where the initial influx of Ca^{2+} is only transient, so that a single wave will propagate from left to right. If the stimulus is maintained at an appropriate value, the spatiotemporal phenomenon is sustained; successive waves then propagate along the cell from the stimulation site as described by Rooney et al. (1990). These results are currently being extended to the propagation of Ca^{2+} signals in two dimensions (Dupont and Goldbeter, 1992) and over a spherical surface. Other questions of interest to be investigated by means of this model are the dependence of Ca^{2+} waves on the spatial distribution of Ca^{2+}-sensitive Ca^{2+} pools as well as the relation between the propagation rate, the size of the cell, and the period of Ca^{2+} oscillations.

IV. CONCLUDING REMARKS

The temporal and spatial aspects of nonequilibrium self-organization are closely intertwined. On the one hand, excitability and oscillations are necessarily linked, as these two modes of dynamic behavior originate from a common mechanism and occur in closely related conditions (i.e., for slightly different parameter values). On the other hand, once a mechanism for oscillations and excitability exists, diffusion processes readily allow for the appearance of spatiotemporal organization; steady inhomogeneous patterns may also arise in such systems (Nicolis and Prigogine, 1977; Murray, 1989).

The link between excitability, oscillations, and propagating waves was illustrated here by two examples pertaining to intercellular communication: the occurrence of wavelike cAMP signaling in aggregation fields of *Dictyostelium* amoebae, and the intracellular propagation of Ca^{2+} signals triggered in a number of cell types by external stimuli. These two examples share some properties but differ in several aspects. While in both cases the spatial pattern is closely associated with the occurrence of excitable or oscillatory behavior in spatially homogeneous conditions—

oscillations of cAMP and of Ca^{2+} indeed represent two of the best-known examples of rhythmic behavior at the cellular level—a first difference between the two cases is that cAMP waves occur in the extracellular medium in *Dictyostelium* amoebae whereas for Ca^{2+} the phenomenon takes place within the cell. The reason for such a difference lies in the physiological function of the waves: cAMP signaling controls the chemotactic movement of surrounding amoebae, while the targets of the Ca^{2+} response to external stimulation reside in the cytosol. The difference extends to the type of regulation involved in generating the oscillations as the destabilizing feedback is based on extracellular cAMP and on intracellular Ca^{2+}, respectively.

A peculiarity of the Ca^{2+} signaling system is that the Ca^{2+}-sensitive Ca^{2+} pools within the cytosol are distributed at discrete intervals. Since these pools play a prominent role in the mechanism of oscillations, which relies on Ca^{2+}-induced Ca^{2+} release, wave propagation will crucially depend on their spatial distribution. Thus if the pools lie too far apart, the amount of Ca^{2+} discharged from one of them may be too low to trigger the release of Ca^{2+} from the next pool when it reaches the latter by diffusion: the occurrence of Ca^{2+} waves therefore depends on a fine balance between reaction and diffusion terms. A sufficiently dense distribution of Ca^{2+} pools would ensure that Ca^{2+} waves are readily triggered by external stimulation. In *Dictyostelium* amoebae, similarly, the density of cells has to be sufficiently high for wave patterns to occur.

The distinction between the triggering of a single response and that of a train of spikes in the respective cases of excitability and oscillations finds its counterpart in spatial organization. Thus conditions can be found in which either a single wave or a train of waves are propagated. Sustained waves of cAMP and Ca^{2+} will occur as long as an appropriate stimulation is present, while a single wave will follow a transient perturbation or a stimulus allowing for a unique response but too weak to support repetitive waves.

Given the close link that exists between wave propagation and sustained oscillations, the comprehension of the spatial phenomenon clearly requires a detailed knowledge of the molecular mechanism of the cellular rhythm. Prior understanding of the latter is a natural first goal since the temporal dynamics is always simpler than the spatiotemporal one, from both a mathematical and a physical point of view.

Of particular importance with regard to the mechanism of wave propagation are the two factors whose interplay gives rise to excitability and oscillations: self-amplification through regulatory feedback and its limitation associated with refractoriness. The latter property is brought about by reversible desensitization in *Dictyostelium* cells and by the transient depletion of the Ca^{2+} intracellular stores in the case of Ca^{2+} signaling. These processes govern the interspike dynamics of cAMP and Ca^{2+} in the course of oscillations as well as the time interval between successive waves.

ACKNOWLEDGMENTS

This work was supported by the Belgian National Incentive Program for Fundamental Research in the Life Sciences (Convention BIO/08), launched by the Science Policy Programming Services of the Prime Minister's Office (SPPS).

REFERENCES

Alcantara, F., and Monk, M. (1974). *J. Gen. Microbiol.*, *85*: 321.

Backx, P. H., de Tombe, P. P., van Deen, J. H. K. Mulder, B. J., and ter Keurs, H. E. D. J. (1989). *J. Gen. Physiol.*, *93*: 963.

Berridge, M. J. (1990). *J. Biol. Chem.*, *265*: 9583.

Berridge, M. J., and Galione, A. (1988). *FASEB J.*, *2*: 3074.

Berridge, M. J., Cobbold, P. H., and Cuthbertson, K. S. R. (1988). *Philos. Trans. R. Soc. Lond. B*, *320*: 325.

Bonner, J. T. (1967). *The Cellular Slime Molds*, Princeton University Press, Princeton, N.J.

Busa, W. B., and Nuccitelli, R. (1985). *J. Cell Biol.*, *100*: 1325.

Busa, W. B., Ferguson, J. E., Joseph, S. K., Williamson, J. R., and Nuccitelli, R. (1985). *J. Cell Biol.*, *101*: 677.

Capiod, T., Field, A. C., Ogden, D. C., and Sanford, C. A. (1987). *FEBS Lett.*, *217*: 247.

Cheer, A., Vincent, J. P., Nuccitelli, R., and Oster, G. (1987). *J. Theor. Biol.*, *124*: 377.

Cornell-Bell, A. H., and Finkbeiner, S. M. (1991). *Cell Calcium*, *12*: 185.

Cornell-Bell, A. H., Finkbeiner, S. M., Cooper, M. S., and Smith, S. J. (1990). *Science*, *247*: 470.

Cuthbertson, K. S. R. (1989). In *Cell to Cell Signalling: From Experiments to Theoretical Models* (A. Goldbeter, ed.), Academic Press, London, pp. 435–447.

Devreotes, P. N. (1982). In *The Development of Dictyostelium discoideum* (W. F. Loomis, ed.), Academic Press, New York, pp. 117–168.

Dupont, G., and Goldbeter, A. (1989). In *Cell to Cell Signalling: From Experiments to Theoretical Models* (A. Goldbeter, ed.), Academic Press, London, pp. 461–474.

Dupont, G., and Goldbeter, A. (1992). *BioEssays*, *14:* 1.

Dupont, G., Berridge, M. J., and Goldbeter, A. (1991). *Cell Calcium*, *12*: 73.

Durston, A. J. (1974). *Dev. Biol.*, *37*: 225.

Endo, M., Tanaka, M., and Ogawa, Y. (1970). *Nature*, *228*: 34.
Fabiato, A., and Fabiato, F. (1975). *J. Physiol.* (*Lond.*), *249*: 469.
Fitzhugh, R. (1961). *Biophys. J.*, *1*: 445.
Gerhardt, M., Schuster, H., and Tyson, J. J. (1990). *Science*, *247*: 1563.
Gerisch, G. (1971). *Naturwissenschaften*, *58*: 430.
Gerisch, G. (1987). *Annu. Rev. Biochem.*, *56*: 853.
Gerisch, G., and Wick, U. (1975). *Biochem. Biophys. Res. Commun.*, *65*: 364.
Gilkey, J. C., Jaffe, L. F., Ridgway, E. B., and Reynolds, G. T. (1978). *J. Cell Biol.*, *76*: 448.
Goldbeter, A. (1990). *Rythmes et Chaos dans les Systèmes Biochimiques et Cellulaires*, Masson, Paris.
Goldbeter, A., Dupont, G., and Berridge, M. J. (1990). *Proc. Natl. Acad. Sci. USA*, *87*: 1461.
Jacob, R. (1990a). *Cell Calcium*, *11*: 241.
Jacob, R. (1990b). *Biochim. Biophys. Acta*, *1052*: 427.
Kasai, H., and Augustine, G. A. (1990). *Nature*, *348*: 735.
Kort, A. A., Capogrossi, M. C., and Lakatta, E. G. (1985). *Circ. Res.*, *57*: 844.
Kuba, K., and Takeshita, S. (1981). *J. Theor. Biol.*, *93*: 1009.
Kushmerick, M. J., and Podolsky, R. J. (1969). *Science*, *166*: 1297.
Lechleiter, J., Girard, S., Peralta, E., and Clapham, D. (1991). *Science*, *252*: 123.
Loomis, W. F. (1975). *Dictyostelium discoideum*: *A Developmental System*, Academic Press, New York.
Markus, M., and Hess, B. (1990). *Nature*, *347*: 56.
Martiel, J. L., and Goldbeter, A. (1987). *Biophys. J.*, *52*: 807.
Meyer, T. (1991). *Cell*, *64*: 675.
Meyer, T., and Stryer, L. (1988). *Proc. Natl. Acad. Sci. USA*, *85*: 5051.
Meyer, T., and Stryer, L. (1991). *Annu. Rev. Biophys. Biophys. Chem.*, *20*: 153.
Monk, P. B., and Othmer, H. G. (1990). *Proc. R. Soc. Lond. B*, *240*: 555.
Murray, J. D. (1989). *Mathematical Biology*, Springer-Verlag, Berlin.
Nicolis, G., and Prigogine, I. (1977). *Self-Organization in Nonequilibrium Systems*, Wiley, New York.
Oster, G. F, and Odell, G. M. (1984). *Physica*, *12D*: 333.
Petersen, O. H., and Wakui, M. (1990). *J. Membr. Biol.*, *118*: 93.
Rooney, T. A., Sass, E. J., and Thomas, A. P. (1990). *J. Biol. Chem.*, *265*: 10792.
Roos, W., Nanjundiah, V., Malchow, D., and Gerisch, G. (1975). *FEBS Lett.*, *53*: 139.
Shaffer, B. M. (1962). *Adv. Morphog.*, *2*: 109.
Snaar-Jagalska, B. E., and Van Haastert, P. J. M. (1990). *Mol. Cell. Biochem.*, *92*: 177.
Takamatsu, T., and Wier, W. G. (1990). *FASEB J.*, *4*: 1519.
Thomas, A. P., Renard, D. C., and Rooney, T. A. (1991). *Cell Calcium*, *12*: 111.

Tomchik, K. J., and Devreotes, P. N. (1981). *Science*, *212*: 443.
Tyson, J. J. (1989). *Cell to Cell Signalling: From Experiments to Theoretical Models* (A. Goldbeter, ed.), Academic Press, London, pp. 521–537.
Tyson, J. J., and Murray, J. D. (1989). *Development*, *106*: 421.
Tyson, J. J., Alexander, K. A., Manoranjan, V. S., and Murray, J. D. (1989). *Physica D*, *34*: 193.
Vaughan, R., and Devreotes, P. N. (1988). *J. Biol. Chem.*, *263*: 14538.
Wakui, M., Potter, B. V. L., and Petersen, O. H. (1989). *Nature*, *339*: 317.
Wakui, M., Osipchuk, Y. V., and Petersen, O. H. (1990). *Cell*, *63*: 1025.
Winfree, A. T. (1972). *Science*, *175*: 634.
Winfree, A. T. (1987). *When Time Breaks Down.* Princeton University Press, Princeton, N.J.

IIIA
Periodic Timing of Morphogenetic Processes: Single or Repeated Transient Processes

12 Morphogenesis of *Acetabularia*: Coordinate Development, Spatial, and Circadian Control

Thérèse Vanden Driessche *Université Libre de Bruxelles, Rhode Saint-Genèse, Belgium*

I. INTRODUCTION

The aim of this chapter is to show how growth and morphogenesis of *Acetabularia* are controlled by both spatial and temporal distribution of regulatory molecules as well as by development. Many data will be presented in a table and figures and shall refer to general recent reviews rather than to the original articles.

II. THE DEVELOPMENTAL CYCLE

Acetabularia (Fig. 1) is a giant unicellular and uninucleate green alga. From the zygote, usually in the spring, a tiny alga emerges, its apex growing geotropically negative and phototropically positive while small rhizoidal outgrowths develop from the basal part surrounding the nucleus in the opposite direction (they will secure the alga to its support). The alga grows very slowly during the first 3 months or so, then enters the rapid growth phase, which lasts for about 2 months, then slowly again just before primordium initiation [its length varies from 3 to 5 cm, depending on the medium and culture conditions (Vanden Driessche et al., 1988a)]. In turn, the primordium develops into a species-specific gametangium 1 cm in diameter, the cap, which displays an elaborate morphology. The successive hair whorls (which are always formed prior to the cap) are iterative expressions of morphogenesis, their number depending on environmental conditions. When the cap has reached its full size, the nucleus undergoes meiosis followed by an astonishing wave of mitoses, which

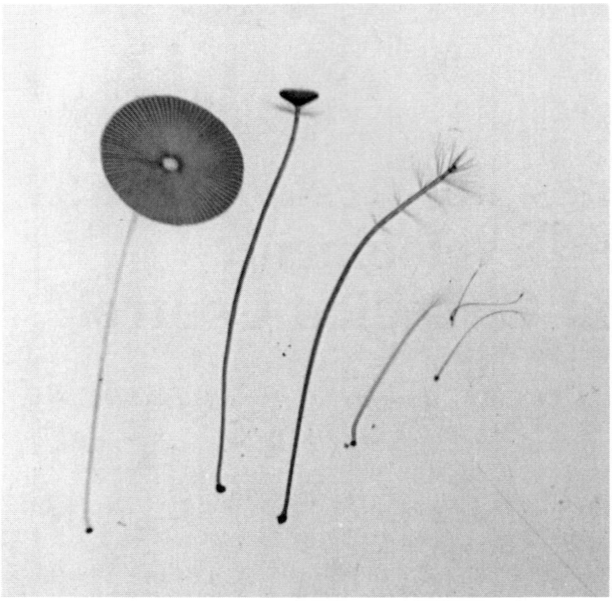

Fig. 1 *Acetabularia* at various developmental stages.

results in 10^8 secondary nuclei. They migrate from the rhizoid up to the cap, along the several-centimeter-long stalk, bringing along, in the intense cytoplasmic movement, all the other cytoplasmic organelles. The nuclei get into the cap rays and become compartmentalized, forming winter-resistant cysts (they have a cellulosic cell wall, whereas the other cell walls are formed of mannans). The nucleus of each cyst divides several times and finally contains numerous haploid nuclei, each differentiating into a biflagellate gamete. In spring, the cysts are freed from the disrupted cap, open up, and release the gametes. The major developmental signal is blue light, on which both growth and cap formation depend (Clauss, 1977), but the speed of development also depends on the intensity of the light, temperature, and water quality. It is known that the cytoplasm and the nucleus exchange signals: If the stalk is torn away shortly before meiosis, the nucleus retracts and does not divide until a new cap has been formed. One of the most remarkable features of *Acetabularia* is survival of the anucleate fragment, which develops a (sterile) species-specific cap. Brachet (1960, 1965) showed that this is due to long-lived mRNAs.

All around the cell cycle, the number of enzymes and regulatory molecules are synthetized in a stage-dependent way (much of this information has been obtained by Schweiger and colleagues; see Berger et al., 1987). The lipid composition also varies (Dubacq, 1973). In addition, ethylene, auxin, and abscissic acid levels change (Vanden Driessche et al., 1988b; Kof and Kefeli, 1979).

III. POLARITY AND DEVELOPMENT

Besides the obvious polarity in morphology, ultrastructural features (Van Gansen and Boloukhère, 1965, Werz, 1965) and physiological activities are distributed according to a positive or negative apicobasal gradient or accumulate in either the apex or the rhizoid. First, there is a polar distribution of genetic information: Around the nucleus appear the newly synthetized mRNAs, but they migrate along the stalk in a light-dependent way and accumulate at the apex (Kloppstech and Schweiger, 1975a). Newly synthesized 80S ribosomes also migrate in the apex, but independently and much more slowly (Kloppstech and Schweiger, 1975b). In addition, the chloroplasts contribute to the nucleic acid gradient: The DNA content is higher in the apical chloroplasts than in lower ones. The apical chloroplasts are endowed with a greater capacity to replicate their DNA than are the rhizoidal ones; this is also true of their transcriptional activity (Lüttke and Bonotto, 1982). Indeed, the chloroplasts differ in ultrastructure and in almost every physiological activity (see Vanden Driessche and Cotton, 1986). The polar distribution of organelles and their various physiological activities are given in Table 1. The two studies cited in the table are interesting in that they show not only the gradient in concentration in either peroxidases or ascorbic acid content but also how the gradient is altered at certain developmental stages.

Small algae are strongly and entirely stained intensely blue with a benzidine reagent (a reagent for peroxidases); of a single hair whorl, the basal article only becomes densely blue. As the alga grows, the rhizoid with its rhizoidal outgrowths remains deeply stained, with the distal part less and less so. The hair whorls repeat the gradient: Much peroxidase activity is detected in the basal articles and a decreasing activity in the successive ones. The apex is almost unstained. When the primordium is formed, it appears deep blue. The mature cells possess rhizoidal outgrowths, of which only the extreme ends are stained, and a fully developed cap, slightly stained at the corona inferior (the part under the cap) and the edges of the cap rays. The exact role(s) of peroxidases in the development of *Acetabularia* is not fully understood, but it has been shown that ascorbate peroxidase is the main isozyme. These experiments

Table 1 Polar Distribution of Organelles and Molecules[a]

	Apex versus base or type
mRNAs	+
Ribosomes	+
Some proteins	+
Transcriptional activity (U* incorporation)	
Translational activity (80S and 70S ribosomes)	+
Dictyosomes	+
Chloroplasts	
Morphology	Young
Multiplication rate	+
Chlorophyll *a/b*	−
Polysaccharide granules	−
Stroma: density	+
Photosynthetic rate	+
Sensitivity to light (growth response)	+
Presumed IP_3 receptors (preliminary experiments)	+
Calcium concentration	+
Calmodulin concentration	+
Phosphorylation rate	−
Peroxidase concentration	−
Ascorbic acid concentration	−

[a]For references, see Lüttke and Bonotto (1982) and Vanden Driessche and Cotton (1986).

suggest that their polar distribution is somehow implicated in development (Vanden Driessche et al., 1984).

Moreover, similar changes in ascorbic acid content have been observed with Tauber's reagent and the same stage-specific pattern as that of peroxidase activity: Both enzyme and substrate decrease with a more advanced developmental stage, both being essentially periplasmic in their location. However, very young algae display an ascorbic acid gradient at the same time that they show an equal peroxidase activity all along the alga. In contrast, cap-bearing algae still display a peroxidase gradient when there is no longer an ascorbic acid gradient. Finally, the anucleate portion of a newly fragmented alga stains intensely with Tauber's reagent on its basal part (Castillo et al., 1986). Again, these changes suggest that polarity is somehow important in development: a contention strongly supported by a totally different approach, based on bioelectrical measurements. The transcellular current of *Acetabularia* implies that the quality and properties of the plasmalemma varies according to a gradient (the transmembrane potential difference is higher at the apex than it is at the base of the stalk). The interesting point is that clamping the potential difference between apex and base to zero totally abolishes regeneration and growth no longer occurs (Novak and Bentrup, 1972). Recently, M. Chaignon, in our laboratory, detected by autoradiography a gradient in labeling (apically positive) after incubation with a tritiated phosphoinositol precursor. This suggests a gradient in phosphoinositols receptors (which could also provide direct evidence for changes in the plasmalemma).

Thus *Acetabularia* displays a number of polarly distributed activities and molecules or ions (the gradient being apicobasal negative, as for nucleic acids or apicobasal positive as for those of peroxidase and ascorbic acid or that of phosphorylation activity (Goodwin and Peteromichelakis, 1979). At least some of them display dramatic changes at crucial steps in development. The maintenance of polarity depends partially on phytohormones. This has been shown in *Acetabularia* by experiments with morphacstin, a competitive inhibitor of auxin (Vanden Driessche, 1974).

IV. TEMPORAL MORPHOLOGY AND DEVELOPMENT

Acetabularia displays an elaborate temporal morphology (i.e., most of its physiological functions are modulated by circadian rhythmicity). The numerous rhythms displayed by the chloroplasts have been examined by Vanden Driessche and Cotton (1986); since they might not be directly relevant to the topic of this chapter, we only mention them here. Presumably more important are rhythms entirely or partially involving the cytoplasm and the plasmalemma (very little is known about possible temporal changes of the tonoplast). This series of circadian rhythms (CRs) is presented in Fig. 2.

The transcellular potential is modulated by a circadian rhythm, which may be monophasic or biphasic in both nucleate and anucleate fragments. The direction of the current is always determined by the regenerating zone. In addition, spontaneous pulses of 1 to 3 min (involving K^+ efflux) are superimposed on the steady transcellular current; they occur more

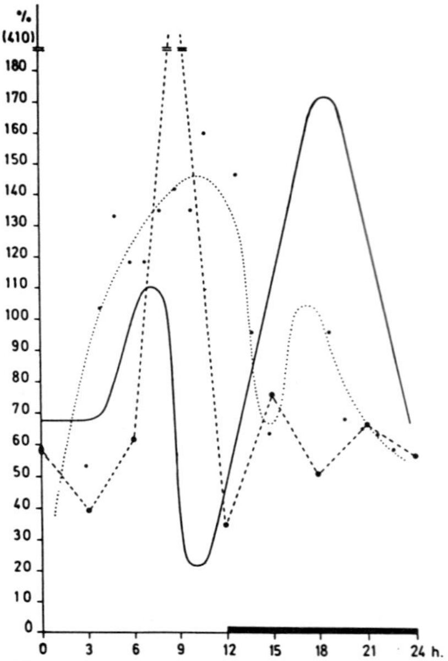

a

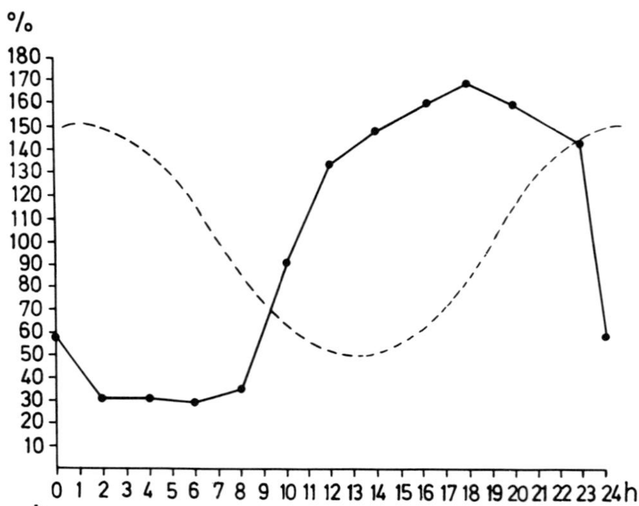

b

Fig. 2 Temporal morphology of *Acetabularia* (chloroplast circadian activities and ultrastructure features being excluded). (a) Cyclic AMP content: solid line (individual points not shown). Transcellular bioelectric current: dotted line

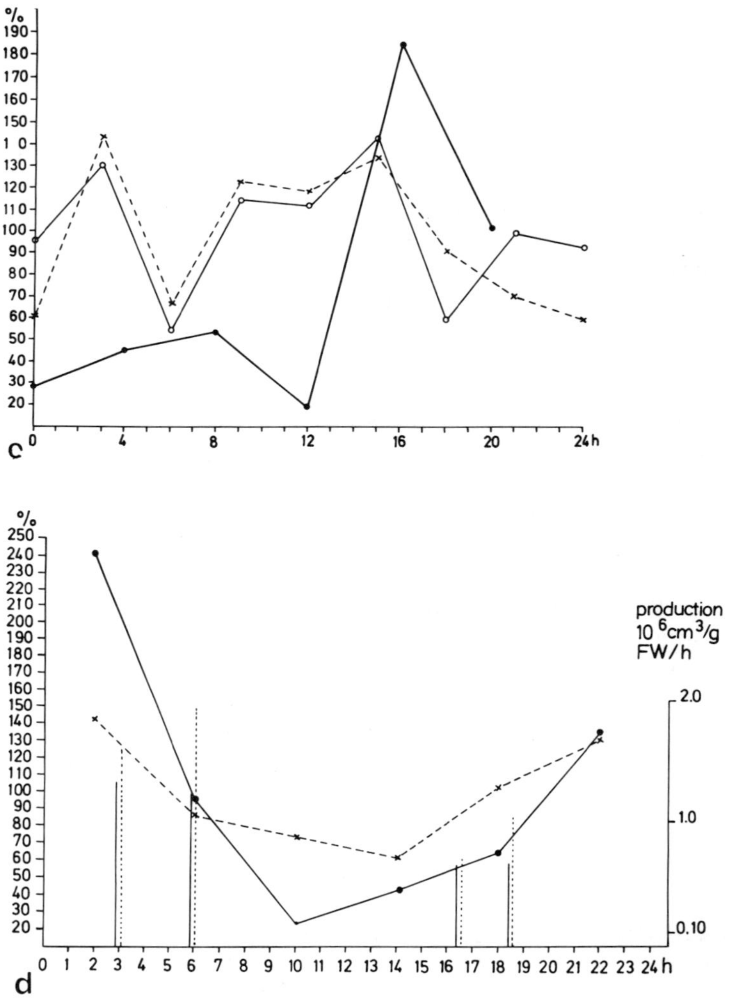

with small black dots. Oleic content: dashed line with large solid dots. (b) Cytoplasmic movement: solid line with black dots. Synthesis of Protein 230 (as explained in the text): dashed line (mathematically shown to be a sinusoid). (c) NAD^+ content: dashed line with x. $NADP^+$ content: solid line with open circles. β-Galactosidase activity: solid line with solid dots. (d) Capacity of *Acetabularia* to convert 1-aminocyclopropane-1-carboxylic acid (ACC) to ethylene: solid line with solid dots. Ethylene retention: dashed line with x. Vertical bars represent the production of ethylene at four different time points (dotted bars) and ethane (solid bars). (From Vanden Driessche and Cotton, 1986.)

frequently during the light period than during the dark period (Novak and Bentrup, 1972; Novak and Sironval, 1975, 1976; Broda and Schweiger, 1981). The oleic content of the alga, which is concerned primarily with the plasma membrane, always displays a biphasic pattern, a major peak occurring around 9 h [circadian time (CT)] and a minor one at 15 or 16 h (Jerebzoff and Vanden Driessche, 1983). The β-galactosidase activity rhythm is biphasic, the two peaks occurring at 16 and 24 h CT (T. Vanden Driessche et al., in press). The content of the two oxidized forms of the pyridine nucleotides NAD^+ and $NADP^+$ is modulated by circadian rhythms of high amplitude, which are biphasic (A. Van Langendonckt et al., 1991).

Much emphasis has been put on the periodic appearance of a thylakoid polypeptide, P230 (because its molecular weight is 230 kDa) (Hartwig et al., 1985): this protein is synthetized on 80S ribosomes (the protein synthesis on 70S ribosome is known to be irrelevant for chloroplast CR) in the correct phase with regard to the peak of photosynthesis and inhibited by cycloheximide precisely at the time when the photosynthesis rhythm is sensitive to this inhibitor (Karakashian and Schweiger, 1976).

Cytoplasmic movement differs dramatically between the light and dark periods and is controlled by an endogenous CR (Koop et al., 1978). At least one of the important phytohormones, ethylene, displays circadian rhythmicity: in ethylene emission, in its capacity to be formed from ACC, and in ethylene retention (Vanden Driessche et al. 1988). Cyclic AMP displays a high-amplitude circadian rhythm in content which is interesting in more than one issue (see the next section). However, much remains to be learned on the origin and maintenance of polarity (for a discussion, see Schnepf, 1986).

V. DEVELOPMENTAL REGULATION AND CIRCADIAN RHYTHMS

In the particular case of cAMP (a regulatory molecule whose exact function in plants remains unknown), the biphasic rhythm is of high amplitude (Figs. 2 and 3) in *Acetabularia* during their rapid growth phase, but at this stage only. Algae developing a cap are no longer rhythmic in cAMP content (Vanden Driessche et al., 1976).

Interestingly, algae in their rapid growth phase are sensitive to theophyllin, papaverin, and isoproterenol, all substances that influence the endogenous cAMP content (Vanden Driessche, 1976). If one considers the amount of endogenous cAMP, it appears that very young algae and cap-bearing algae have fairly high levels of cAMP, whereas rapidly growing algae have a much smaller content, considerably less than at the other stages (Minder and Vanden Driessche, 1978). Thus it is the algae at a particular developmental stage, which contain a small quantity of cAMP, which are theophylline-sensitive and rhythmic. This suggests that the rhythm must have a special significance in *Acetabularia* physiology. We have observed large changes that affect ethylene metabolism during both the developmental and circadian cycles.

Finally, a strong argument in support of the importance of proper temporal relationships between *Acetabularia* and its environmental light–dark (LD) schedule (setting the phases of its various CRs) concerns the dramatic effects on growth and morphogenesis of LD cycles of 2 : 2, 4 : 4, and especially, 6 : 6 (Vanden Driessche and Guisset, 1988).

VI. DISCUSSION AND CONCLUSIONS

We have compiled data showing that developmentally regulated processes appear to be dependent in some way on spatial and temporal parameters. Cap morphogenesis occurs at the apex where the major ingredients are assembled: mRNAs, ribosomes, a relatively high concentration of photoreceptors inducing growth and morphogenesis (blue light triggers tropism and differentiation in *Acetabularia*), and the IP_3–Ca^{2+}–calmodulin transducing chain (leading eventually to the regulation of RNA expression; Rodland et al., 1990). Indeed, Ca^{2+} has been shown to be a morphogen, as it is in other eukaryotic systems (Goodwin and Pateromichelakis, 1979; Rasmussen, 1981). Ca^{2+} evokes typically pleiotropic effects: It acts as a second messenger (as we shall see below), and it regulates microtubule assembly (as we shall also see below), microfilament activity (and thus, notably, cytoplasmic streaming), and ion transport activity (of Ca^{2+} but also of other ions). Two very important pieces of information have to be added to this straightforward scheme. First, in 1985, Goodwin and colleagues deduced from their study on hair whorl spacing in media of different Ca^{2+} concentrations that a standing wave of free Ca^{2+} around the tip must occur. Indeed, Ca^{2+} may induce sustained effects by virtue of an increase in cycling (Alkon and Rasmussen, 1988). The second and more recent conclusion is that high-frequency Ca^{2+} oscillations generate phosphorylation signals that are frequency encoded (Goldbeter et al., 1990).

It has often been observed that the same crosspoints in metabolism or molecules with regulatory functions can be modulated by both circadian and high-frequency oscillations. In the case of *Acetabularia*, high-frequency oscillations in Ca^{2+} have not been examined, but we know that in the chloroplast, the ion content varies in a circadian way and that both La^{3+}, which substitutes for Ca^{2+}, and lindane, which interferes with inositol metabolism, affect ulterior cap formation when used to treat cells

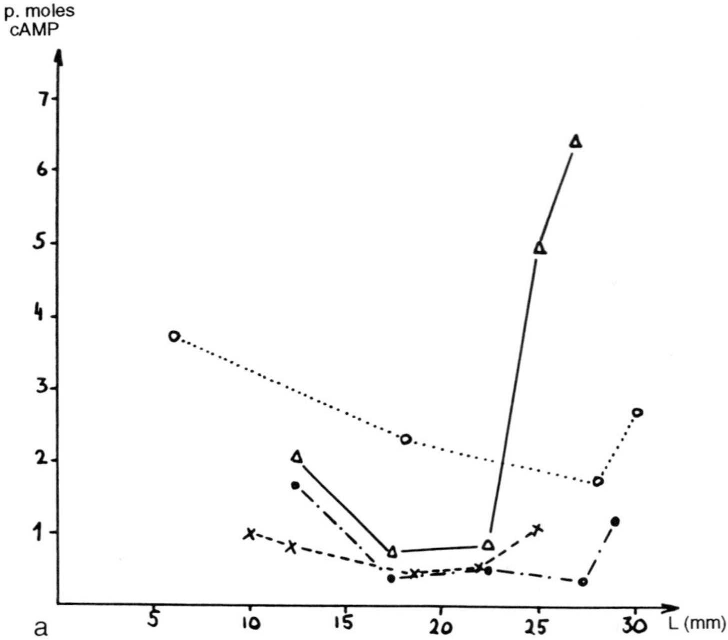

a

Fig. 3 (a) Developmental changes in cAMP content in *Acetabularia.* A series of individual experiments show that slowly growing algae (their length at this stage is generally smaller than 10 mm) have a higher cAMP content than algae in their rapid growth phase (i.e., between 15 and 20–28 mm in length, depending on the experiment) and than in algae developing a cap in which there is a dramatic increase in cAMP content. All measurements were performed between 0 and 3 h. (b) CR in cAMP content. (c) Effect of theophyllin at various concentrations on *Acetabularia* growth. Control, central solid curve; 2×10^{-3} *M*, lower dash-dotted curve; 10^{-3} *M*, dashed line; 10^{-4} *M*, upper solid curve. (From Gaspar, Kevers, and Vanden Driessche, unpublished data.)

for short periods of time, and interestingly, they do so in a circadian time-dependent way (Fig. 4).

Cap morphogenesis in *Acetabularia* also in some way involves microtubules (MTs); colchicine, which has no effect on growth, totally inhibits cap formation. A comparison of nucleate and anucleate fragments in normal and in colchicine-supplemented medium was made by Werz (1969), who also studied the morphogenetic capability for restoration by the nucleus. Reconsidering these interesting experiments, it can now be suggested that either a nucleation factor or MT-associated proteins (MAPs) are short-lived and synthetized on a short-lived mRNA, whereas

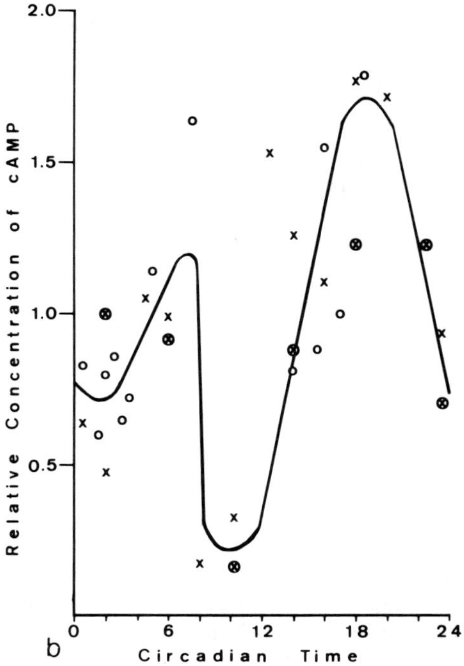
2.0
1.5
1.0
0.5
Relative Concentration of cAMP
0
6
12
18
24
Circadian Time

b

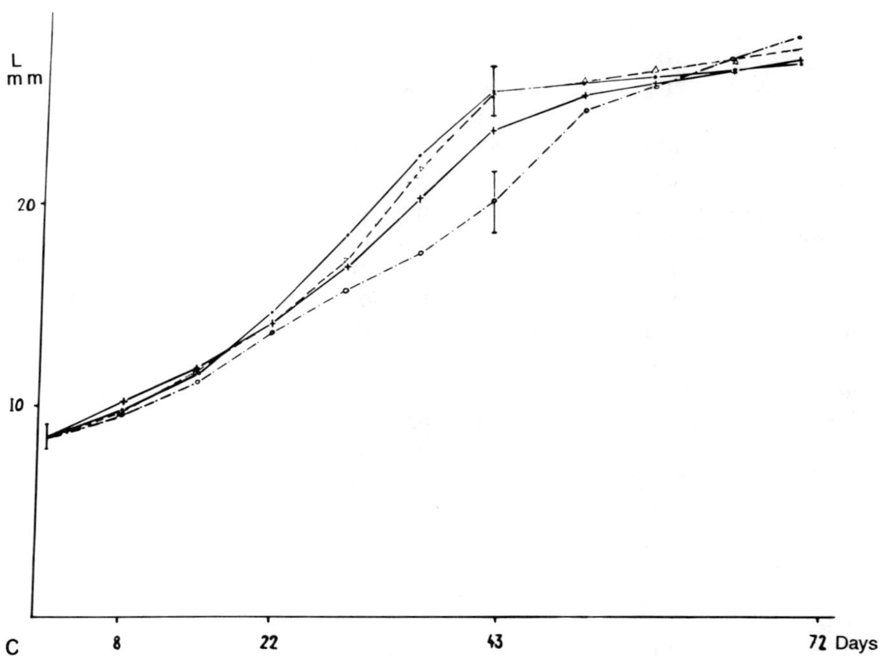
L
mm
20
10
8
22
43
72 Days

C

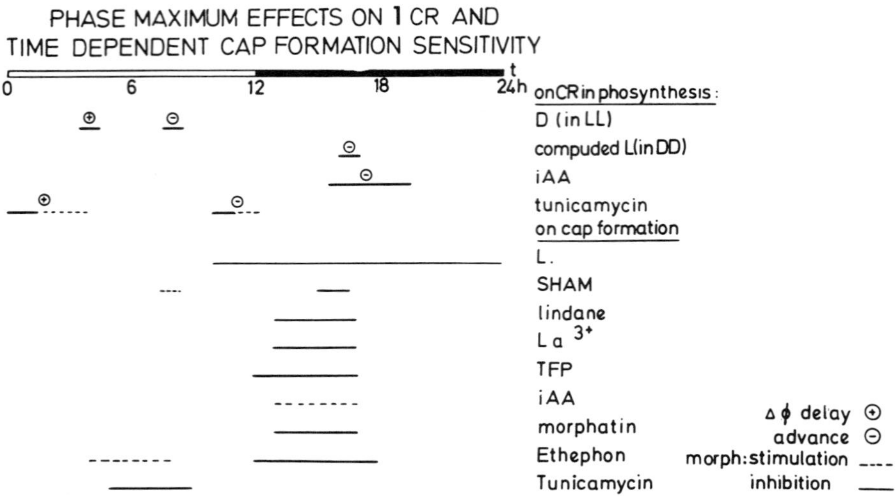

Fig. 4 Sensitivity map of *Acetabularia*. The effects on one particular CR (photosynthesis) are shown in the upper part: the phase of the CR may be advanced (−) or delayed (+). The effects on cap formation (occurring several days or weeks after the algae were returned to a normal medium) are given in the lower part: it may be accelerated (dashed line) or delayed (solid line).

the precursors (or rather, the monomeric form of tubulin) are very stable in the cell. Later in development, MTs anchor the nucleus in the center of a small portion of cytoplasm (the cyst-to-be) and somehow prevent chloroplasts and mitochondria from invading the central zone around the nucleus (Woodcock, 1971; Menzel and Elsner-Menzel, 1990).

It is known that Ca^{2+}, calmodulin, and phosphorylation regulate MT assembly directly or indirectly (see Olmsted, 1986), as well as microfilaments (see Traas, 1990). Thus Ca^{2+} might also act at this level. In addition, Mandelkow and colleagues (1989) observed spatial patterns of oscillation in MTs by their assembly and disassembly in the presence of GTP as cofactor (the experiments were carried out in vitro). Thus *Acetabularia*, in which MTs are obviously labile, could serve as a model for investigating possible relationships between MTs and morphogenesis.

In another unicellular organism, *Euglena*, experimental evidence suggests that the photosynthetic reactions could be coupled to the CR control by calmodulin (Lonergan, 1986).

VII. FINAL COMMENTS

The intricacy of the morphogenesis regulation is far from elucidated. However, dynamic parameters such as oscillations presumably play an important role. Feedback loops may provide control and cross-control on the physiological state of the cell before allowing morphogenesis (such as cap formation), to take place. Many of the critically involved pathways are interrelated, which has important implications for stability as well as for appropriate activity levels. Rhythms affect spatially distributed molecules. Many molecules regulating morphogenesis and rhythm parameters are dependent on the developmental stage in *Acetabularia*.

ACKNOWLEDGMENTS

We are grateful to H. Alexandre for his critical reading of the manuscript and thank D. Francks for preparing the figures. The unpublished experimental material was supported by a Crédit aux Chercheurs du Fonds national de la Recherche scientifique to TVD and by an Action de Recherche Concertée to the Laboratory.

REFERENCES

Alkon, D. L., and Rasmussen, H. (1988). *Science*, *239*: 998.

Berger, S., de Groot, E. J., Neuhaus, G., and Schweiger, M. (1987). *Eur. J. Cell Biol.*, *44*: 349.

Brachet, J. (1960). *The Biological Role of Ribonucleic Acids*, Elsevier, Amsterdam, p. 144.

Brachet, J. (1985). *Molecular Cytology*, Vol. 1, *The Cell Cycle*, Academic Press, New York, p. 424.

Broda, H., and Schweiger, H. G. (1981). *Protoplasma*, *105*: 352.

Castillo, F. J., Cotton, G., Kevers, C., Greppin, H., Gaspar, T., and Vanden Driessche, T. (1986). *Differentiation*, *33*: 17.

Clauss, H. (1977). *Ann. Eur. Symp. Photomorphog.* (*Bet Dagan*): 16.

Dubacq, J. P. (1973). *Protoplasma*, *76*: 373.

Goldbeter, A., Dupont, G., and Berridge, M. J. (1990). *Proc. Natl. Acad. Sci. USA*, *87*: 1461.

Goodwin, B. C., and Pateromichelakis, S. (1979). *Planta*, *145*: 427.

Goodwin, B. C., Murray, J. M., and Baldwin, D. (1985). *Acetabularia 1984* (S. Bonotto, F. Cinelli, and R. Billiau, eds.), Belgian Nuclear Center, Mol, Belgium, p. 101.

Hartwig, R., Schweiger, M., Schweiger R., and Schweiger, H. G. (1985). *Proc. Natl. Acad. Sci.* USA. *82*: 6899.

Jerebzoff, S., and Vanden Driessche, T. (1983). *C.R. Acad. Sci.* (*Paris*), *296*: 319.

Karakashian, M., and Schweiger, H. G. (1976). *Exp. Cell Res.*, *98*: 303.

Kevers, C., Vanden Driessche, T., and Gaspar, T. (1989). *Saussurea Soc. Bot. Geneve*, *19*: 121.

Kloppstech, K., and Schweiger, H. G. (1975a). *Differentiation*, *4*: 115.

Kloppstech, K., and Schweiger, H. G. (1975b). *Protoplasma*, *83*: 27.

Kof, E., and Kefeli, V. (1979). *Developmental Biology of Acetabularia* (S. Bonotto, V. Kefeli, and S. Puiseux-Dao, eds.), Vol. 45, Elsevier/North-Holland Biomedical Press, Amsterdam.

Koop, H. U., Schmid, R., Heunert, H. H., and Milthaler, B. (1978). *Protoplasma*, *97*: 301.

Lonergan, T. A. (1986). *Plant Physiol.*, *82*: 226.

Lüttke, A., and Bonotto, S. (1982). *Int. Rev. Cytol.*, *77*: 205.

Mandelkow, E., Mandelkow, E. M., Hotani, H., Hess, B., and Müller, S. C. (1989). *Science*, *246*: 1291.

Menzel, D., and Elsner-Menzel C. (1990). *Protoplasma*, *157*: 52.

Minder, C., and Vanden Driessche, T. (1978). *Differentiation*, *10*: 165.

Novak, B., and Bentrup, F. W. (1972). *Planta*, *108*: 227.

Novak, B., and Sironval, C. (1975). *Plant Sci. Lett.*, *5*: 183.

Novak, B., and Sironval, C. (1976). *Plant Sci. Lett.*, *6*: 273.

Olmsted, J. B. (1986). *Annu. Rev. Cell Biol.*, *2*: 421.

Rasmussen, H. (1981). In *Calcium and the Cell Function*, Vol. 4 (W. Y. Cheung, ed.), Academic Press, New York, p. 1.

Rodland, K. D., Muldoon, L. L., Lenormand, P., and Magun, B. E. (1990). *J. Biol. Chem.*, *265*: 11000.

Schnepf, E. (1986). *Annu. Rev. Plant Physiol.*, *37*: 23.

Traas, J. A. (1990). In *The Plant Plasma Membrane* (C. Larsson and I. M. Moller, eds.), Springer-Verlag, Berlin, p. 269.

Vanden Driessche, T. (1974). *Protoplasma*, *81*: 323.

Vanden Driessche, T. (1976). *Differentiation*, *5*: 119.

Vanden Driessche, T., and Cotton, G. (1986). *Endocytobiosis Cell Res.*, *3*: 275.

Vanden Driessche, T., and Guisset, J. L. (1988). *Arch. Int. Physiol. Biochim.*, *96*: 43.

Vanden Driessche, T., Doege, K. J., Minder, C., and Cairns, W. L. (1976). In *Chronopharmacology* (A. Reinberg and F. Halberg, eds.), Pergamon Press, Elmsford, N.Y., p. 291.

Vanden Driessche, T., Lause, P., Guisset, J.-L., Glory, M., and Petiau-de Vries, G. M. (in press). *J. Interdiscip. Cycle Res.*

Vanden Driessche, T., Lateur, L., Rzmieniwski, P., and Guisset, J. L. (1988a). *Chronobiol. Int.*, *5*: 1.

Vanden Driessche, T., Kevers, C., Collet, M., and Gaspar, T. (1988b). *J. Plant Physiol.*, *133*: 635.

Vanden Driessche, T., Kevers, C., Gaspar, T., and Caubergs, R. (1984). *Differentiation*, *27*: 175.

Van Gansen, P., and Bolouk hère, M. (1965). *J. Microsc.* (*Paris*), *4*: 347.

Van Langendonckt, A., de Solan, C., Jerebzoff, S., and Vanden Driessche, T. (1991). *J. Interdiscip. Cycle Res.*, *122*: 223.

Werz, G. (1965). *Brookhaven Symposia in Biology*, *18*: 185.

Werz, G. (1969). In *Inhibitors, Tools in Cell Research* (T. Bücher and H. Sies, eds.), Springer-Verlag, Berlin, p. 167.

Werz, G. (1974). *Int. Rev. Cytol.*, *38*: 319.

Woodcock, C. L. F. (1971). *J. Cell Sci.*, *8*: 611.

13
Regulation of Cell Division Cycles by Circadian Oscillators: Role of the cAMP Pathway

Leland N. Edmunds, Jr. *State University of New York, Stony Brook, New York*

I. INTRODUCTION

The cell division cycle (CDC) of eukaryotes is a complex cascade of developmental and morphogenetic events that culminate in cell duplication (Cross et al., 1989). Rapid progress has been made within the past several years toward dissecting the regulatory network that provides precise coordination of these processes. A unifying view has emerged: The CDC consists of transitions from one regulatory state to another (Lewin, 1990; Murray and Kirschner, 1989). These transitions initiate the modification of substrates that determine the physical state of the cell and are themselves feedback controlled. The basic mechanisms, as elucidated by a combination of genetic and biochemical analyses, appear to have been conserved in organisms ranging from unicells such as yeast (Andrews and Herskowitz, 1990; Draetta, 1990), to *Xenopus*, clams, and starfish eggs, and to cultured human cells (D'Urso et al., 1990).

Two crucial transition control points exist in the G_1 phase and at the G_2/M boundary. The M phase is characterized by the activation of a kinase [MPF, or maturation (M phase)-promoting factor, so-named for its ability to induce cell division when injected into *Xenopus* oocytes], which consists of two component subunits. One is the $p34^{cdc2}$ protein (the product of the *Schizosaccharomyces pombe* gene *cdc2*), which is able to phosphorylate casein and histone H1 in vitro and has maximal kinase activity at mitosis. The other is cyclin, whose cellular concentration fluctuates during the cell cycle: It accumulates gradually during interphase, forms a complex with $p34^{cdc2}$, and activates the protein kinase; it is then degraded during mitosis, turning off kinase and MPF activity. The system thus behaves like an oscillator, or "clock," which is reset to its interphase state during mitosis. The first cell divisions in sea urchin eggs (Pines and Hunt, 1987), in frog oocytes (Swenson et al., 1986), and probably also in *Drosophila* embryos (O'Farell et al., 1989) are triggered by the accumulation of cyclins above a critical concentration. Less is known about the initiation of DNA synthesis, but a complex comprising p34*cdc2*, and possibly other cyclin-like molecules, may play a role (Cross et al., 1989; Murray and Kirschner, 1989).

There is also strong evidence that an autonomous oscillator(s) (independent of either p34*cdc2*, cyclins, or both) is part of, or is coupled to, the cell division cycle. Thus when the *S. pombe* cell cycle is arrested by a temperature-sensitive *cdc2* mutation, the activity of the enzyme nucleotide diphosphate kinase (NDPK) and the rate of CO_2 production still undergo periodic oscillations, with a period that approximates the length of the CDC (Creanor and Mitchison, 1986; Novak and Mitchison, 1986). Klevecz (1976) has shown that mammalian cell cycle lengths do not exhibit a smooth distribution, but are clustered (or quantized) at intervals of 3 to 4 h. He explained these data by the existence of a quantal subcycle, driven by an autonomous clock whose period would be approximately 4 h. The duration of the quantal cycle (G_q) does not vary with temperature over the range 6 to 7°C, whereas cell-cycle duration varies greatly (Klevecz and King, 1982). Such temperature compensation of G_q (Q_{10} of 1.25 to 1.15 in Chinese hamster V79 lung fibroblasts) is incompatible with the idea that cell-cycle duration would depend only on the accumulation of cellular components such as cyclins.

The notion of timers and clocks as a possible means of integrating the disparate concepts of cell-cycle transit, exit, and arrest is not new (Mitchison, 1974; Klevecz, 1976; Wille, 1979; Edmunds and Adams, 1981; Gilbert, 1981; Edmunds, 1984a; Skehan, 1988). Inasmuch as mitosis is a periodic event of short duration relative to the total length of the CDC, it is not surprising that various types of *autonomous* biochemical and macromolecular oscillators have been proposed to underlie this and other "landmarks" comprising the CDC (see Edmunds and Adams, 1981; Edmunds, 1984a). These range from those of the relaxation type, in which a single continuous variable accumulates or declines in the cell, triggering some event in the CDC when it reaches a critical threshold and then resetting to a baseline so that the cycle starts again (e.g., Sachsenmaier, 1976), to those exhibiting limit-cycle dynamics (Kauffman and Wille, 1975; Klevecz, 1976; Gilbert, 1981; Shymko et al., 1984). In the latter, a central clock, characterized by a self-sustained oscillation of at least two continuously varying biochemical species, would maintain stable periodic behavior and coordinate the timing of events comprising the CDC despite transient perturbations in phase; it would continue to function even if mitosis and division were blocked.

Although the majority of these putative "clocks" have periods in the neighborhood of minutes to hours, they well may be relevant to circadian oscillators (having longer period lengths of about 20 to 28 h). Circadian clocks represent an intensively investigated class of autonomous, self-sustained cellular oscillator (Edmunds, 1988). Circadian rhythms have been documented in organisms ranging from unicellular eukaryotes to higher plants and animals and concern activities as varied as leaf movements in plants, locomotor activity in insects and mammals, conidiation in fungi, photosynthesis in unicellular algae, and many enzyme activities. Of particular relevance to this discussion is the fact that the CDC of many algae, fungi, and protozoans also exhibit circadian rhythms of cell division (Edmunds, 1984a,b): Cell division occurs only during restricted intervals of the circadian cycle—often at the times corresponding to the dark intervals in a diurnal light–dark (LD) cycle such as LD:12,12 (the repetition of 12 h of light followed by 12 h of dark). This "gating" phenomenon is thought to reflect an interaction between the circadian clock and the cell cycle oscillator. Circadian rhythms of cell division indeed exhibit the formal properties of other circadian rhythms (Edmunds and Laval-Martin, 1984): entrainability by diurnal LD cycles; persistence in constant light (LL) or in constant darkness (DD), with a period that only approximates 24 h; phase shiftability by light signals; and temperature compensation of the free-running period.

The mechanism whereby the circadian oscillator and higher-frequency oscillators, such as G_q, interact with the p34^{cdc2}/cyclin pathway for the control of cell division is an important question that must be addressed. Cyclic AMP (3′5′-cyclic adenosine monophosphate), which exhibits bimodal circadian rhythmicity in the algal flagellate *Euglena gracilis* and which is known to play a pivotal role in cellular regulation (Whitfield et al., 1987) and to have the capacity to control certain rate-limiting steps in the progression of the CDC in many cell types (Whitfield, 1990), may effect such a coupling between oscillator and CDC, participating in the gating of CDC events to specific phases of the circadian cycle.

Indeed, a number of earlier studies have suggested the involvement of cAMP in circadian clock function (Engelmann and Schrempf, 1980). Circadian variations in cellular cAMP levels have been observed in *Neurospora* (Hasunuma et al., 1987), *Tetrahymena* (Dobra and Ehret, 1977), and *Acetabularia* (Vanden Driessche et al., 1979). Increasing the levels of cAMP in *Neurospora* by treatments with several methylxanthines resulted in a lengthening of the period of the conidiation rhythm (Feldman, 1975). In contrast, decreasing the level of endogenous cAMP by inhibiting adenylate cyclase with quinidine failed to change the period (Feldman, 1975; Feldman and Dunlap, 1983), a finding corroborated by the observation that mutants (*crisp-1*) having significantly reduced levels of adenylate cyclase and cAMP entrained normally to a diurnal light–dark cycle (LD) and free-ran with an unaltered periodicity (Feldman, 1979). Therefore, neither adenylate cyclase, cAMP phosphodiesterase, nor cAMP could be classified as "gears" of the underlying circadian oscillator in this system (Goto et al., 1985). More recent work, however, implicates cAMP in circadian timing in both *Neurospora* (Hasunuma, 1984a,b) and *Aplysia* (Eskin et al., 1982; Eskin and Takahashi, 1983; Lotshaw and Jacklet, 1983, 1987), and we have found that this species may play an important role in *Euglena* (Carré et al., 1989b), although we suspect that it lies downstream from the clock, providing a coupling link to the CDC that it regulates.

We have chosen *Euglena gracilis* (strain Z), a facultative photoautotroph, to test our hypothesis so as not to have to confront the complexities of higher organisms and levels of organization. This eukaryotic, unicellular system is particularly advantageous in that it is well characterized physiologically and biochemically (Buetow, 1968a,b, 1982, 1989) and has been shown to display many other circadian periodicities (Edmunds and Halberg, 1981; Edmunds, 1982), including the extensively documented rhythm of cell division (Edmunds and Laval-Martin, 1984), our primary overt reference cycle. We are thus in an advantageous position to study the nature and interaction of CDC oscillators and circadian clocks at both the formal and metabolic level in a well-defined cellular system.

II. CELL DIVISION RHYTHMICITY IN WILD-TYPE AND MUTANT STRAINS OF *EUGLENA*

The CDCs of many algae, fungi, and protozoans exhibit persisting circadian rhythms of cell division, or "hatching" (Edmunds, 1978; Edmunds and Adams, 1981). Division occurs only at a certain phase of the circadian cycle—often the times ("subjective" nights) in DD or LL corresponding to the dark intervals in an environmentally synchronizing LD cycle. This "gating" phenomenon, reflecting an interaction between a circadian oscillator and the CDC, has been perhaps investigated most intensively in the eukaryote *Euglena*. The *formal* properties of circadian clocks—in particular, entrainability by diurnal and non-24-h light (Cook and James, 1960; Edmunds, 1965, 1978; Edmunds and Funch, 1969a,b;

Ledoigt and Calvayrac, 1979) and temperature (Terry and Edmunds, 1970a,b), persistence (Edmunds, 1966), initiation, phase shiftability (Edmunds et al., 1982), phase singularity (Malinowski et al., 1984), and temperature compensation (Anderson et al., 1985)—have been found to characterize the circadian rhythm of cell division in photoautotrophically batch- and continuously cultured wild-type *Euglena gracilis* Klebs (strain Z) (reviewed by Edmunds and Laval-Martin, 1984).

In earlier work (Laval-Martin et al., 1979), we used a higher-frequency LD cycle (LD:3,3) to elicit circadian (τ = 28 to 30 h) rhythms of cell division in the wild-type strain. To obtain phase-response curves (PRCs) for 3-h light pulses, we routinely deleted one of the 3-h dark intervals. In fact, this created a 9-h light interval—perhaps too long to constitute a true "pulse" (and possibly generating unwanted side effects). Therefore, we have utilized shorter, LD:1,1 cycles (T = 2 h), which also elicit circadian periods (τ = 25 to 26 h), and have found that 5-h light pulses, obtained by deleting one (or two) of the now shorter dark intervals, also yield similar PRCs for the division rhythm (Edmunds and Laval-Martin, 1984). These high-frequency regimens, having symmetric photo- and scotophases and even properly constructed random regimes, have proved particularly useful in that they (1) provide no information to the cells with regard to 24-h time [although we do not exclude a possible modulation of the period to some extent (Adams et al., 1984)], yet elicit free-running circadian rhythms ($T \ll \tau \simeq$ 24 h) not only of cell division, but also of motility (Schnabel, 1968) and of photosynthetic capacity and chlorophyll content (Laval-Martin et al., 1979; Edmunds and Laval-Martin, 1981); and (2) afford an amount of light during a 24-h time span identical to that received in a full-photoperiod LD:12,12 entraining reference cycle ($T = \tau$ = 24 h). Thus the total duration of light afforded for growth is held constant, reflected in the comparable step sizes of the respective growth curves, while the signaling information inherent in the light pulses can be varied to manipulate the light-sensitive circadian oscillator.

To eliminate a cellular compartment in which the clock might lie and to obviate possible signaling influences of the higher-frequency LD cycles used previously to elicit rhythmicity and provide energy for growth, we have chosen the achlorophyllous ZC mutant of *Euglena*—an obligate organotroph—for a series of studies parallel to those being carried out in the wild-type (Carré et al., 1989a). Division rhythmicity could be entrained by LD:12,12 (Fig. 1) or even by a one-pulse skeleton photoperiod (LD:1,23), and the rhythm free-ran in ensuing DD for at least 8 days (τ = 25.5 h), provided that the overall growth rate (g) was more than 24 h (see Fig. 2). Similar, though less extensive results had been obtained

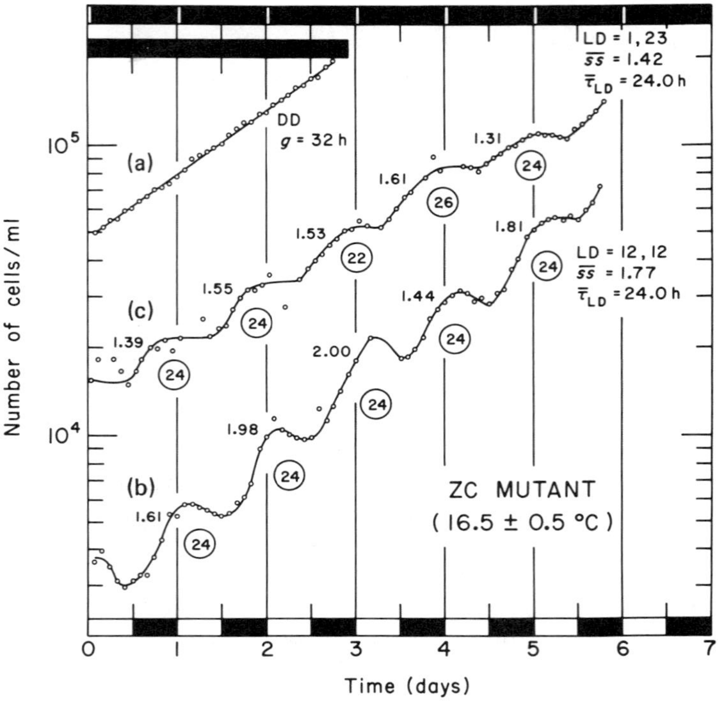

Fig. 1 Entrainment of the cell division rhythm by diurnal LD cycles in populations of the achlorophyllous ZC mutant of *Euglena gracilis* (strain Z), batch-cultured at 16.5°C on a mineral medium supplemented with ethanol (0.1% v/v): (a) exponential growth of cells maintained in DD; (b) gating of cell division to the dark intervals in LD:12,12; (c) entrainment of the cell division rhythm by a one-pulse skeleton photoperiod (LD:1,23). Step sizes (*ss*) are given on the left of the corresponding step, and the estimated period (τ) of each oscillation (interval between successive onsets of cell division) is indicated in a circle. Note for curves (b) and (c) that the average period ($\overline{\tau}$) of the rhythm matched exactly that (24 h) of the LD cycle, although the average period of the individual cell cycle (the generation time of the population, *g*) was longer than 30 h. Values of *ss* were less than 2, indicating that not every cell divided during each division burst. (From Carré et al., 1989a.)

earlier with other photosynthesis mutants of *Euglena* (Edmunds, 1975, 1978; Edmunds et al., 1976; Jarrett and Edmunds, 1970).

A "strong" (type 0) PRC for light signals in the ZC mutant strain (Fig. 3) was generated from 15 phase-shifting experiments (Fig. 5) (Carré et al., 1989a). Phase delays ($-\Delta\phi$) were obtained between circa-

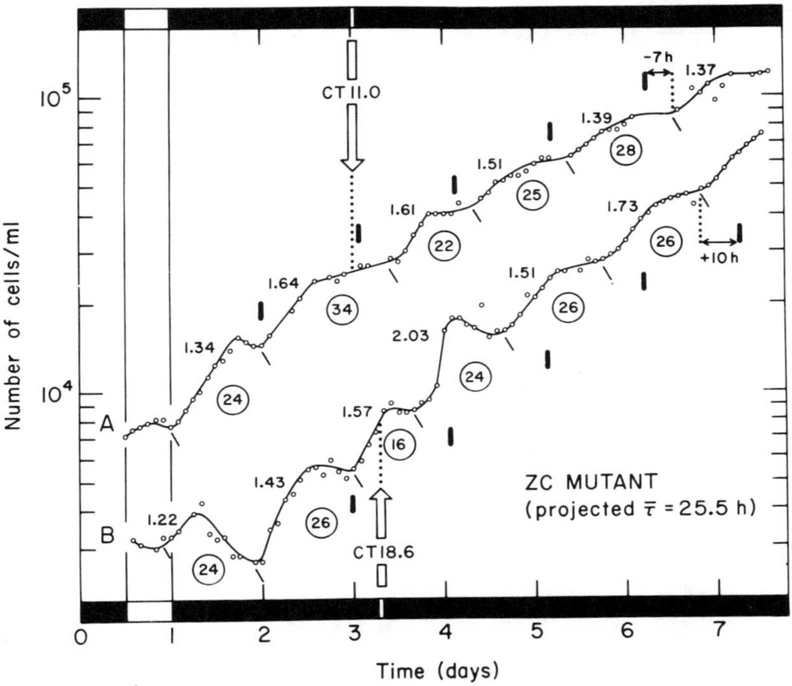

Fig. 2 Examples of phase-shifting experiments and methodology used in the calculation of phase advances ($+\Delta\phi$) and delays ($-\Delta\phi$) obtained after perturbation of the rhythm of cell division of the ZC mutant (free-running in DD) by 1-h light (3000 lux) signals given at different circadian times. Two cultures (A, B) synchronously dividing in DD, were perturbed by light pulses, the middle of which occurred at either hour 24.5 or hour 7.0 (corresponding to circadian times CT 11.0, or 18.6, respectively) following the preceding onset of cell division (taken to fall at CT 12). After transients had subsided, $\bar{\tau}$ of the oscillations was approximately 25.5 h. The difference in phase ($\Delta\phi$, here given in real time) between the shifted rhythm (vertical dotted lines) and that projected (solid markers) for the rhythm before it was perturbed then was measured and considered to represent the steady-state phase shift engendered by the light signal. The value of τ (25.5 h) used to calculate the phase of the unperturbed rhythm was the pooled average obtained for the whole family of experiments. Other notations are as for Fig. 1. (From Carré et al., 1989a.)

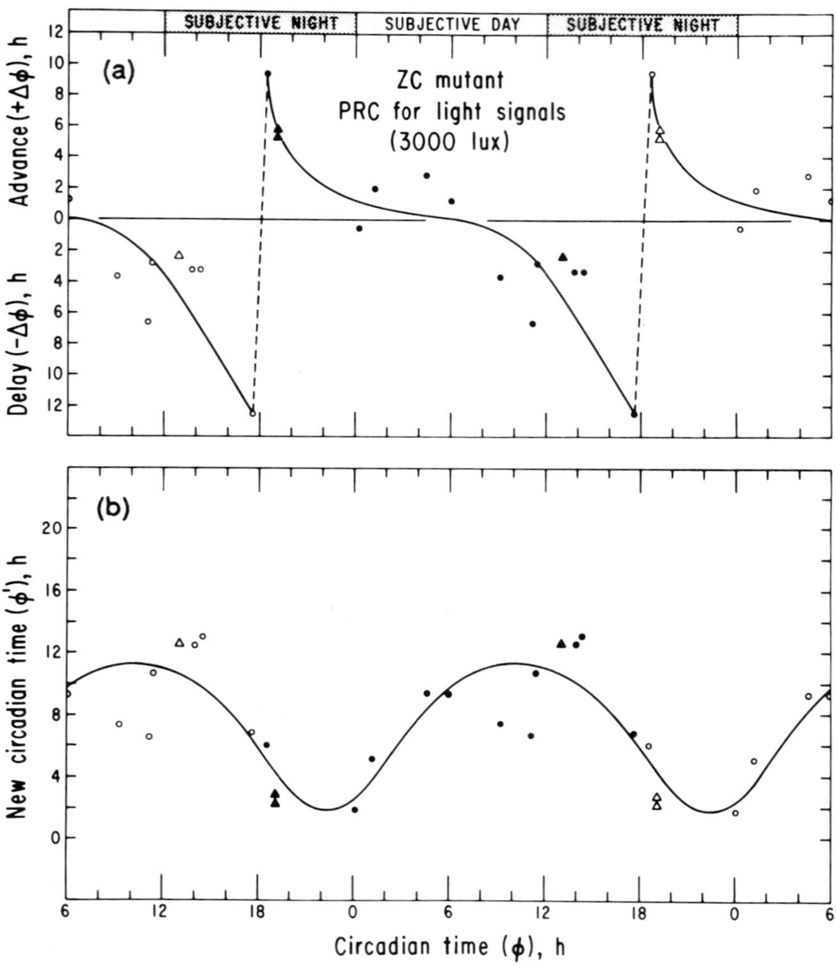

Fig. 3 Phase-response curve (a) for the resetting action of light (3000 lux) signals on the free-running rhythm of cell division of the achlorophyllous ZC mutant of *E. gracilis* maintained in DD. The steady-state phase shifts ($\pm\Delta\phi$) engendered by 1-h (o) and 3-h (Δ) pulses were determined as described in Fig. 2 and were normalized to 24 h. Results were double-plotted (open symbols) as a function of the circadian time (CT, modulo 24) at which the midpoint of the signal was given. The discontinuity (breakpoint, or phase jump, indicated by a dotted line) is inherent to this type of plot. In curve (b), the resultant new CT ($\phi' = \phi \pm \Delta\phi$) following the perturbation is plotted as a function of the CT (ϕ) at which the signal was given, resulting in a smooth sinusoidal curve. (From Carré et al., 1989a.)

dian times (CT) 08 and 18, and phase advances ($+\Delta\phi$) between CT 18 and 0; the discontinuity ("breakpoint") inherent to this type of plot (curve a, dashed line) occurred at about CT 00. The PRC for the mutant resembled that obtained previously with the wild-type strain (Edmunds et al., 1982). These studies therefore provide even more conclusive evidence for the role of a circadian oscillator in the control of the CDC and have effectively circumvented the problem of the dual use of imposed LD cycles: As an energy source, or substrate, for growth, and as a timing cue (*Zeitgeber*) for the underlying clock. They demonstrate further that the presence of a functional chloroplast compartment is unnecessary for circadian timekeeping in *Euglena* and suggest that phase shifting the clock by light occurs via similar pathways in photosynthetic nonphotosynthetic cell types.

This short review has demonstrated the key role of a circadian oscillator(s) in the control of the CDC in *Euglena*, taken to be representative of other eukaryotic microorganisms (and, perhaps, multicellular systems). Mitosis would not be an essential part of the oscillator but would lie downstream from it: Blockage of cell division should not stop the system from oscillating (Edmunds and Adams, 1981; Edmunds and Laval-Martin, 1984). Thus cell division would be a "hand" of the underlying clock, which we have exploited in our research. We have recently tested this hypothesis in two ways: (1) If the division rhythm (free-running in LD:3,3) was stopped due to low initial levels of vitamin B_{12}, and if this inhibition subsequently was released by readdition of B_{12}, the rhythm started up again in phase with an unperturbed control; and (2) if a pulse of lactate was given to a free-running culture, temporarily accelerating the CDC and overriding circadian oscillator controls (Jarrett and Edmunds, 1970), the phase of the rhythm when it was finally restored after the substrate had been depleted was in phase with that of an unperturbed control. These results are consistent with those found previously for the in-phase restoration of rhythmicity in the P_4ZUL mutant free-running in LL by the addition of sulfur-containing compounds to the medium (Edmunds, et al., 1976).

III. BIOCHEMICAL ANALYSIS OF THE CIRCADIAN OSCILLATOR

Several years ago we put forward a working hypothesis for a closed control loop that would constitute a light-entrainable, self-sustaining circadian oscillator in which NAD^+, the mitochondrial (or other) Ca^{2+} transport system, Ca^{2+}-calmodulin, NADP phosphatase, and NAD kinase, and perhaps cAMP together would constitute "gears" of the clock underlying the overt rhythm of cell division (and of other variables, or "hands") in eukaryotic microorganisms (Goto et al., 1985). This model was based in part on results obtained previously for the duckweed *Lemna gibba* and on more recent findings in the algal flagellate *Euglena gracilis* (Edmunds et al., 1987).

In brief, we have shown that the following:

1. The in vivo levels of NAD^+ and $NADP^+$ (or NADPH) in both synchronously dividing and stationary-phase, free-running cultures of *Euglena* oscillated with a 27-h circadian period; the NAD^+ rhythm (as well as the division rhythm) could be reset in a similar manner by a 3-h light pulse or by pulses of various chemicals that would be expected to increase or decrease the level of NAD^+ (Goto et al., 1985).

2. Circadian rhythmicity in the activities of both NAD kinase and NADP phosphatase were found in dividing and nondividing cultures of the achlorophyllous ZC mutant of *Euglena* free-running in DD (Laval-Martin et al., 1990a); these rhythms were ascribable to changes in the affinities of the two enzymes for their substrates, NAD^+ and $NADP^+$ (Laval-Martin et al., 1990b).

3. Pulses of various agents that would be expected to cause transitory increases or decreases in $[Ca^{2+}]$ produced phase shifts of the free-running rhythm of cell division, whose sign and magnitude were dependent on the circadian time (CT) that the drug pulse was applied (Goto et al., 1985). Further, although high or low $[Ca^{2+}]$ in the external medium did not affect the period of the division rhythm, high or low *pulses* of Ca^{2+} phase-shifted the oscillation, and the phase-response curves (PRCs) thus obtained were mirror images of each other (Fig. 4), with that for LoCa pulses' resembling the PRC for light signals (Tamponnet and Edmunds, 1990).

This regulatory scheme—not for the clock but for one oscillator in what is probably a cellular clock shop, or network—includes the following three steps:

1. NAD^+ would enhance the rate of net Ca^{2+} efflux from the mitochondria (or other compartment) into the cytoplasm, resulting 6 h (90°) later in a maximal concentration of cytosolic Ca^{2+}. Alternatively (step 1′), a photoreceptor (phytochrome or a blue-light photoreceptor in plants, or perhaps a blue-light photoreceptor in *Euglena* and animal cells) stimulated by (red or blue) light would enhance either net Ca^{2+} efflux from a cell compartment or net Ca^{2+} influx across the plasmalemma into the cytoplasm.

2. Ca^{2+} would immediately form an activated Ca^+–calmodulin complex in the cytoplasm (maximal level at 90°).

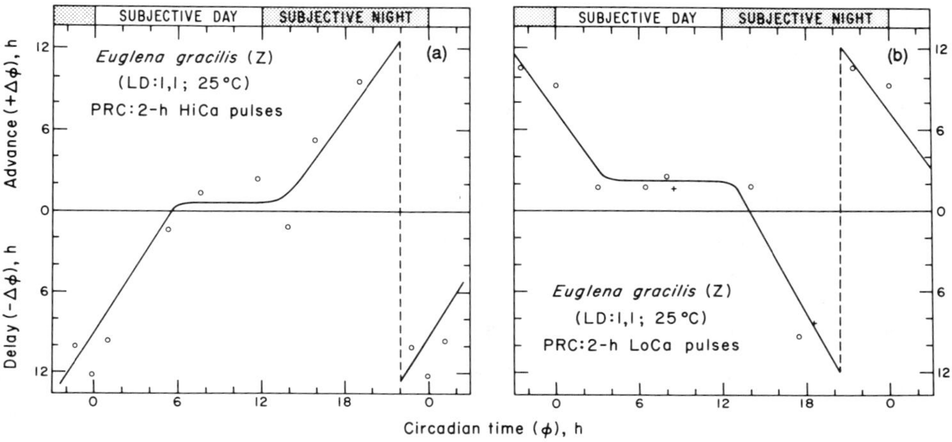

Fig. 4 Phase-response curves (PRCs) for the action of 2-h HiCa (10 m*M*) pulses (a) and 2-h LoCa (2 μ*M*) perturbations (b) on the free-running rhythm of cell division in photoautotrophic cultures of *E. gracilis* (Z) maintained in LD:1,1. Steady-state phase shifts ($\pm\Delta\phi$), determined as described in the legend of Fig. 2, are plotted as a function of CT (normalized to 24 h) at which the midpoint of the signal was given. The breakpoint, indicated by dashed line between CT 20 and CT 22, would disappear if the curve were drawn in monotonic form with all phase shifts being treated as delays. Crosses (curve b) are points from a LD:3,3 cycle (Goto and Edmunds, unpublished results). (From Tamponnet and Edmunds, 1990.)

3. This active form of Ca^{2+}–calmodulin would decrease the rate of net production of NAD^+ by both activating NAD^+ kinase and inhibiting $NADP^+$ phosphatase in the cytoplasm so that the rate would become maximal 12 h later (at 270°) when Ca^+–calmodulin reaches its minimum level. After 6 h more, when the in vivo level of NAD^+ then becomes lowest (at 0°), the regulatory sequence would be closed and the cycle would repeat. Should additional experimental data demand it, other elements can be inserted into the proposed loop and the oscillator simply expanded (as, indeed, may be necessary in *Euglena* for the case of cAMP (Carré et al., 1989b).

For the moment our model does not purport to explain either the long period length or its steady-state temperature compensation characteristic of circadian rhythmicity, primarily because of a lack of hard data. Limit cycles can display a circa-24-h τs given a particular set of parameter values, some sort of deposition effect, or a network of interacting intracellular oscillators; alternatively, other time-consuming processes involving transcription, translation, or membrane-based ion transport may play a role (Engelmann and Schrempf, 1980; Edmunds, 1988). Similar explanations could be invoked as compensatory mechanisms for the temperature-dependent steps (such as the NAD^+ kinase- and $NADP^+$ phosphatase-catalyzed reactions) in our model. This model shares several aspects in common with the calcium-cycle model of Kippert (1987), in which the Ca^{2+} concentration gradient between cytoplasm and mitochondrial matrix alters cyclically (attributed simply to their counteracting Ca^{2+} transport pathways) and to that of Lakin-Thomas (1985) based on oscillations in intracellular Ca^{2+} compartmentation. Kippert (1987) speculates that endogenous rhythms and the calcium system of intracellular signaling coevolved as a consequence of mutual interactions among partners in a cellular symbiotic consortium developing toward the eukaryotic cell, with its defined nucleoplasm and mitochondria (Levandowsky, 1981). Thus the primitive endogenous clock would have provided a device for internal timekeeping and temporal coordination between still largely autonomous host and endosymbiont compartments. A homeostatic, self-regulated, oscillatory regimen would have helped preserve this developing association from destabilizing, chaotic fluctuations that otherwise might have developed in the complex network of feedback mechanisms (Decroly and Goldbeter, 1982).

IV. COUPLING BETWEEN OSCILLATOR AND CELL CYCLE: THE ROLE OF cAMP

In recent studies (Carré et al., 1989b), we have shown that cultures of the achlorophyllous ZC mutant of *Euglena*, displaying free-running, circadian rhythms of cell division in DD (see Figs. 1 and 2), also exhibit bimodal circadian variations in their level of cAMP (Fig. 5). These oscillations in cAMP concentration can be phase-shifted by light signals in a manner predicted by the PRC derived previously for the phase-resetting of the cell division by light pulses (see Fig. 3). These results strongly suggest that cAMP levels are controlled by the same endogenous clock as that regulating the CDC. Maximum cAMP levels occur at the beginning of the light period (CT 0 to 2, when cells are in G_1) and at the onset of darkness (CT 12 to 14, corresponding to the onset of mitosis). These variations, however, appear to be under the control of the circadian oscillator rather than to be cell cycle–dependent, since they persist, independently of the CDC, after the cultures have reached the stationary phase of growth (Fig. 5b,c). We believe that such a periodic, bimodal cAMP signal (initiated by the circadian oscillator) may participate in the gating of CDC events to specific phases of the circadian cycle.

Indeed, cAMP seems to have the capacity to control some of the rate-limiting steps of cell cycle progression in many cell types (Dumont et al., 1989; Whitfield, 1990). It has been found to stimulate the proliferation of some cells but to have the opposite effect, or no effect at all, on others. Such contradictory findings might be explained by a variation in different systems of the cAMP concentration required for optimal stimulation. Genetic experiments in the yeast *Saccharomyces* (Matsumoto et al., 1983) and pharmacological studies in mammalian cells (for reviews, see Boynton and Whitfield, 1983; Whitfield et al., 1987) have shown that a transient rise and the ensuing fall of the cAMP level are necessary for the initiation of DNA synthesis, and that a second cAMP surge is correlated with the onset of mitosis. Similar changes of cAMP concentration at different phases of the CDC have been shown to occur in *Euglena* (Carell and Dearfield, 1982).

Thus cAMP may play a central role as a coupling link for control of the CDC by the circadian clock. If this hypothesis holds true, conditions that would override the control of cAMP level by the oscillator should also cause perturbations of the cell division rhythm. One would expect transient increases of cAMP concentration (effected by exogenous addition of cAMP, or of the cAMP analogs isobutylmethylxanthine and forskolin) to cause either an advance or a delay of the following burst of division; the division rhythm should then return to its original phase. Indeed, we have obtained no evidence thus far that cAMP pulses can cause significant steady-state $\Delta\phi$ values of the division rhythm in the ZC mutant of *Euglena* free-running in DD. Rather, we have demonstrated transient, phase-dependent effects of cAMP on division rhythmicity: cAMP seems to cause either a shortening or a lengthening of the CDC, depending on the phase of the CDC when the drug is given (Carré and Edmunds, 1990; Edmunds and Carré, 1990). Delays of division rhythm were obtained when cAMP was given from CT 03 to 09, and advances were obtained when cAMP was given between CT 16 and 22. Maximum effects were obtained at CT 06 to 08 and at CT 18 to 20, corresponding to times when endogenous cAMP levels were minimal (Carré et al., 1989b; see Fig. 5). These effects on the cell division rhythm were only transitory, and after 48 h, cell division occurred at the same phase as in unperturbed controls, a finding indicating that the circadian oscillator itself was unaffected. Dose-response curves (DRCs) were also derived at CTs when maximum $+\Delta\phi$ or $-\Delta\phi$ values of the CDC had been obtained. We found that as little as 1 to 5 n*M* cAMP was enough to perturb cell cycle progression. These experiments strongly suggest that control of the CDC in *Euglena* is at least partly mediated by the circadian variations of the cellular cAMP level demonstrated previously (Fig. 5).

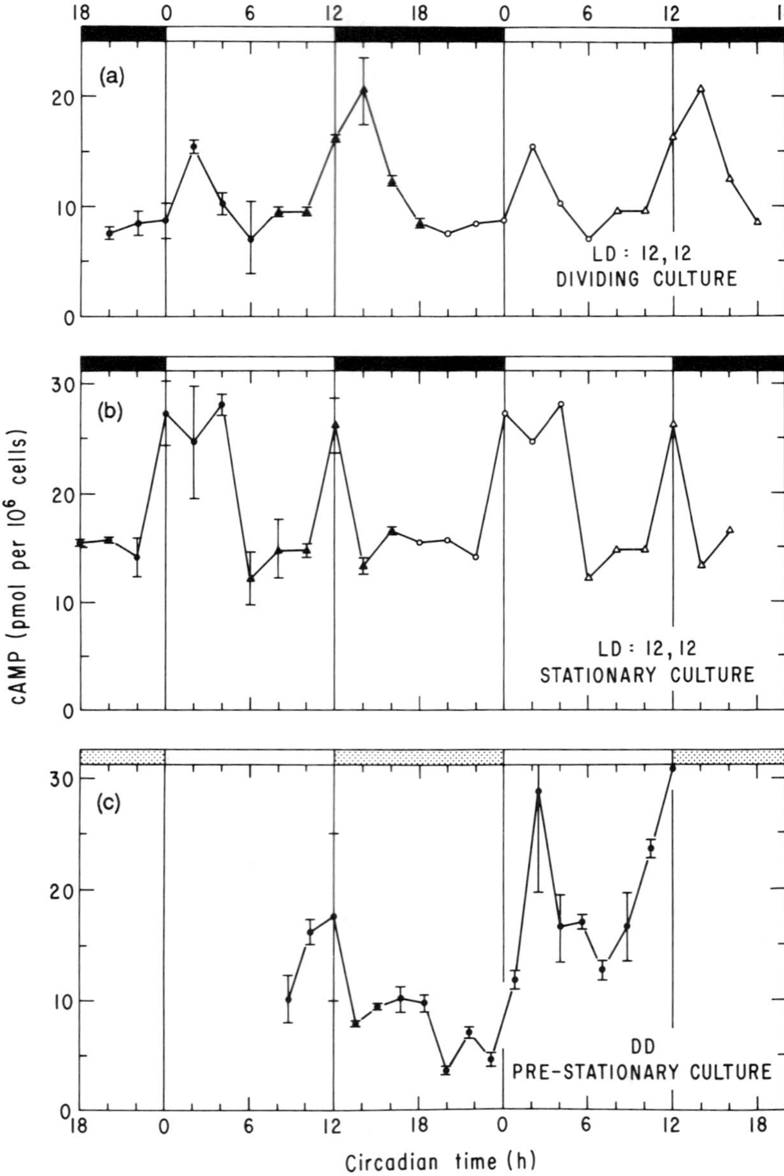
(a)
LD : 12, 12
DIVIDING CULTURE
(b)
LD : 12, 12
STATIONARY CULTURE
(c)
DD
PRE-STATIONARY CULTURE
cAMP (pmol per 10^6 cells)
Circadian time (h)

We have begun to characterize the cellular receptor for cAMP from the ZC mutant of *Euglena*, and it appears to be a cAMP-dependent kinase (Carré and Edmunds, 1992). Saturation studies indicate the presence of two types of binding sites (K_d of 10 and 30 nM). The cAMP-binding activity eluted from a DEAE-cellulose column as a single peak, a finding which suggests that these two cAMP binding sites are likely to be part of the same protein. A cAMP-stimulated kinase activity copurified with the cAMP-binding activity on a DEAE-cellulose column. To our knowledge, this is the first demonstration of a cAMP-dependent kinase in a photosynthetic organism. This kinase phosphorylated the Kemptide peptide (a substrate specifically phosphorylated by cAMP-dependent kinases from mammalian cells), and was inhibited by the protein kinase inhibitor-protein from bovine heart. We plan now to measure the activation of the kinase as a function of CDC phase and then to attempt to demonstrate a correlation between the stimulation of the kinase and the effects of cAMP or of cAMP analogs on cell cycle progression.

In *Euglena*, however, cAMP oscillates with the same periodicity in nonproliferating cells as in actively growing, synchronously dividing cells (see Fig. 4), but at a slightly higher level. Thus changes in cAMP levels do not always serve as signals for cell cycle progression. Rather, cAMP must modulate other biochemical pathways that are essential for cell cycle progression, such as the activation of the *cdc2* protein kinase (Draetta and Beach, 1988; Gautier et al., 1989). Indeed, Mendenhall et

---

Fig. 5 Entrained and free-running circadian oscillations of the cAMP level in synchronously dividing and in nondividing cultures of the achlorophyllous ZC mutant of *Euglena gracilis* (strain Z), grown at 16.5°C on a mineral medium supplemented with ethanol (0.1% v/v): (a) diurnal variations of the cAMP level in cultures exhibiting rhythmic cell division, entrained to a 24-h period by a LD:12,12 cycle; (b) persistence of the bimodal 24-h oscillation of the cAMP level after the cells had ceased dividing; (c) variations of the cAMP level in a culture displaying a free-running rhythm of cell division in DD and just entering the stationary growth phase. For Experiments (a) and (b), cell extracts were prepared every 2 h over a 12-h time span from two cultures entrained by out-of-phase LD cycles, thus scanning the entire circadian period. Results were double-plotted (open symbols) as a function of the circadian time. In Experiment (c), cell extracts were prepared every 2 h over a 34-h time span. The onset of the last burst of cell division (also corresponding to the first peak on the graph) was used as a reference for CT 12, and the interval between the first and the third peak gave the estimated value (27 h) for τ. To facilitate the comparison with curves A and B, the results were then normalized to a 24-h period. (From Carré et al., 1989b.)

al. (1987) have found that the *S. cerevisiae cdc28* gene product (homolog of the p34^{cdc2} protein of fission yeast) loses its protein kinase activity when adenylate cyclase mutants are shifted to the restrictive temperature. This result suggests that cAMP plays a role in the pathway leading to MPF activation in budding yeast and may play a similar role in *Euglena*. Homologs of p34 (the *cdc2* gene product) have recently been found in several algae and higher plants (John et al., 1989). If *Euglena* also has a homolog of p34, we plan to assay for its histone H1 kinase activity as a function of the phase of the CDC, in unperturbed cultures as well as in cultures that have received a cAMP signal.

But what upstream controls regulate the level of cAMP, downstream from the clock? Preliminary results from a series of experiments undertaken to elucidate this question for the ZC mutant of *Euglena* can be summarized as follows (Tong et al., 1991).

A. Adenyl Cyclase and Phosphodiesterase Rhythms

A circadian rhythm in the level of adenyl cyclase (AC) has been found in both dividing and stationary cultures in *Euglena*, which has approximately the same pattern as that of cAMP. In dividing cultures, the two peaks were found at CT 02 and CT 12 to 14 and the troughs at CT 08 and CT 20. In stationary cultures, the situation was similar except that the mean value of the oscillation was slightly higher. A rhythm in phosphodiesterase (PDE) activity was also present, which had an inverse phase relation to that of AC: the two peaks of PDE activity appeared at CT 06 to 08 and CT 18 to 20 in both dividing and nondividing cultures, while the two minima occurred at CT 02 to 04 and CT 12 to 14. Thus the rhythms of both AC and PDE appear to be responsible for generating the circadian oscillations in cAMP content.

B. Effect of cAMP and cGMP on the Level of PDE Activity

It has been proposed that cGMP may be a regulatory factor for the level of intracellular cAMP level by stimulating the hydrolysis of cAMP (Goldberg et al., 1983; Hartzell and Fischmeister, 1986). Our results indicate that cGMP can stimulate in vitro PDE activity somewhat, which in turn hydrolyzes cAMP. With the addition of 0.01 to 10 μM of cGMP, PDE activity increased by about 60 to 70%. Addition of cAMP also activated PDE, but to a lesser degree (50%). When both cAMP and cGMP were added at the same concentration, PDE activity was increased by an amount (80%) greater than that caused by either cAMP or cGMP added alone.

C. Effect of Ca^{2+} and Calmodulin on the Level of PDE Activity

It is known that there is a so-called Ca^{2+}–calmodulin-sensitive PDE that has a high affinity for cAMP (Beavo et al., 1982; Brothers et al., 1982). Addition of 0.1 to 10 m*M* of Ca^{2+} to *Euglena* cultures stimulated PDE activity; maximum effect (twofold increase over basal levels) was found at 10 μ*M* in cell extracts. EGTA, a chelator of Ca^{2+}, had the opposite effect (30% decrease) within the range 0.01 to 10 μ*M*. Similar results have been obtained with calmodulin and its antagonist W7 over the linear ranges 0.6 to 600 U/mL and 0.01 to 10 μ*M*, respectively. Interestingly, when Ca^{2+} and W7 were added together, PDE activity remained unchanged, a finding that implies that the effect of Ca^{2+} on PDE was mediated via calmodulin.

D. Effects of Ca^{2+}, cGMP, and Forskolin on the Level of AC Activity

Ca^{2+} (10 μ*M* to 10 m*M*) caused a twofold decrease in the in vitro AC activity of *Euglena* extracts. This finding may reflect another pathway by which Ca^{2+} regulates cAMP level in vivo, in addition to stimulating cAMP PDE. Surprisingly, cGMP at concentrations of 1 μ*M* to 1 m*M* inhibited AC activity by about 30% also; the mechanism of this action is unknown. On the other hand, forskolin (10^{-4} to 10^{-5} *M*) was found to increase AC activity by a factor of 2.

In sum, these recent results suggest that the CDC of *Euglena* is modulated by cAMP, whose rhythmic activity, in turn, is generated by periodic fluctuations in the activities of AC and PDE. We hypothesize that these two enzymes are regulated by an upstream circadian oscillator, perhaps by signals of Ca^{2+}, cGMP, or both.

V. FUTURE DIRECTIONS

To pursue our results suggesting a coupling role for cAMP between the circadian oscillator and the CDC, we have begun a two-pronged approach. On the one hand, we are investigating how the clock confers rhythmicity upon the adenyl cyclase (AC)–cAMP–phosphodiesterase (PDE) system; on the other, we are attempting to determine how the AC–cAMP–PDE pathway interacts with and modulates the CDC. First, the regulation of the level and rhythmicity of cAMP by AC, PDE, and other second messengers (such as cGMP and Ca^{2+}) will be studied in synchronously dividing and nondividing cultures of the ZC mutant of *Euglena*, maintained in DD, by examining the ability of pulses of various drugs and chemicals that are known to activate or inhibit these biochemical species to phase-shift the rhythms of cAMP level, AC and PDE activities, and cell division. In the downstream analysis, we characterize the cellular receptor for cAMP, which preliminary results indicate is a cAMP-dependent kinase, and attempt to correlate kinase stimulation and the effects of cAMP or cAMP analogs on CDC progression. Finally, we search for a homolog of p34 (the *cdc 2* protein kinase gene product of fission yeast) in *Euglena* and assay for its histone H1 kinase activity at different CDC phases in synchronous cultures perturbed by a cAMP pulse.

ABBREVIATIONS

LL, continuous illumination; DD, continuous darkness; LD, light–dark cycle; LD:x,y, a repetitive light–dark cycle comprising x hours of light and y hours of dark; T, period of an LD cycle or other periodic *Zeitgeber* (environmental cue); CR, circadian rhythms; CO, circadian oscillator; τ, average period of a free-running rhythm in constant conditions; ϕ, phase of a rhythm; ϕ_r, phase reference point, or phase marker; $+\Delta\phi$, $-\Delta\phi$, change (advance, delay) in phase (phase shift); ZT, environmental (*zeitgeber*) time (where ZT 0 corresponds to the onset of light); CT, circadian time (CT 0 indicates the phase point of a free-running rhythm, which has been normalized to 24 h, that corresponds to the phase point occurring at the onset of light in a LD:12,12 reference cycle); PRC, phase-response curve [plot of the phase shift of a free-running circadian rhythm engendered by a perturbing light (or other) signal as a function of the CT at which it was applied]; g, average generation (doubling) time of a population of cells; ss, average step size, or factorial increase in cell concentration (plateau to plateau) after a phased or synchronized division step.

ACKNOWLEDGMENTS

This work was supported in part by grants to L. Edmunds from the National Science Foundation (DCB-9105752 and DCB-8901944) and the National Institutes of Health (RR05736). Particular credit is due to Dr. Jian Tong and Isabelle Carré for the use of their recent published and unpublished results.

REFERENCES

Adams, K. J., Weiler, C. S., and Edmunds, L. N., Jr. (1984). In *Cell Cycle Clocks* (L. N. Edmunds, Jr., ed.), Marcel Dekker, New York, pp. 395–429.

Anderson, R. W., Laval-Martin, D. L., and Edmunds, L. N., Jr. (1985). *Exp. Cell Res.*, *157*: 144.

Andrews, B. J., and Herskowitz, I. (1990). *J. Biol. Chem.*, *265*: 14057.

Beavo, J. A., Hansen, R. S., Harrison, S. A., Hurwitz, R. L., Martins, T. J., and Mumby, M. C. (1982). *Mol. Cell Endocrinol.*, *28*: 387.

Boynton, A. L., and Whitfield, J. F. (1983). *Adv. Cyclic Nucleotide Res.*, *15*: 193.

Brothers, V. M., Walker, N., and Bourne, H. R. (1982). *J. Biol. Chem.*, *257*: 9349.

Buetow, D. E. (ed.) (1968a,b, 1982, 1989). *The Biology of Euglena*, Vols. I, II, III, IV, Academic Press, New York.

Carell, E. F., and Dearfield, K. L. (1982). *Life Sci.*, *31*: 249.

Carré, I. A., and Edmunds, L. N., Jr. (1990). *ASCB/EMBO Conf. Cell Cycle*, March 24–28. Bethesda, Md. (Abstract 1230).

Carré, I. A., and Edmunds, L. N., Jr. (1992). *J. Biol. Chem.*, *267*: 2135.

Carré, I. A., Oster, A. S., Laval-Martin, D. L., and Edmunds, L. N., Jr. (1989a). *Curr. Microbiol.*, *19*: 223.

Carré, I. A., Laval-Martin, D. L., and Edmunds, L. N., Jr. (1989b). *J. Cell. Sci.*, *94*: 267.

Cook, J. R., and James, T. W. (1960). *Exp. Cell Res.*, *21*: 583.

Creanor, J., and Mitchison, J. M. (1986). *J. Cell Sci.*, *86*: 207.

Cross, F., Roberts, J., and Weintraub, H. (1989). *Annu. Rev. Cell Biol.*, *5*: 341.

Decroly, O., Goldbeter, A. (1982). *Proc. Natl. Acad. Sci. USA*, *79*: 6917.

Dobra, K., and Ehret, C. F. (1977). In *Proc. 12th Int. Conf. Int. Soc. Chronobiol.*, Washington, D.C., 1975, Publishing House I1 Ponte, Milan, Italy, pp. 589–594.

Draetta, G. (1990). *Trends Biochem.*, *15*: 378.

Draetta, G., and Beach, D. (1988). *Cell*, *54*: 17.

Dumont, J. E., Jauniaux, J. C., and Roger, P. P. (1989). *Trends Biochem. Sci.*, *14*: 67.

D'Urso, G., Marraccino, R. L., Marshak, D. R., and Roberts, J. M. (1990). *Science*, *250*: 786.

Edmunds, L. N., Jr. (1965). *J. Cell. Comp. Physiol.*, *66*: 147.

Edmunds, L. N., Jr. (1966). *J. Cell. Physiol.*, *67*: 35.

Edmunds, L. N., Jr. (1975). In *Les Cycles cellulaires et leur blocage chez plusieurs protistes* (M. Lefort-Tran and R. Valencia, eds.), Colloques Int. C.N.R.S. 240, Centre National de la Recherche Scientifique, Paris, pp. 53–67.

Edmunds, L. N., Jr. (1978). In *Aging and Biological Rhythms* (H. V. Samis, Jr., and S. Capobianco, eds.), Plenum Press, New York, pp. 125–184.

Edmunds, L. N., Jr. (1982). In *The Biology of Euglena*, Vol. 3 (D.E. Buetow, ed.), Academic Press, New York, pp. 53–142.

Edmunds, L. N., Jr. (ed.) (1984a). *Cell Cycle Clocks*, Marcel Dekker, New York.

Edmunds, L. N., Jr. (1984b). In *Advances in Microbial Physiology*, Vol. 25 (A. H. Rose and D. W. Tempest, eds.), Academic Press, New York, pp. 61–148.

Edmunds, L. N., Jr. (1988). *Cellular and Molecular Bases of Biological Clocks*, Springer-Verlag, New York.

Edmunds, L. N., Jr., and Adams, K. J. (1981). *Science*, *211*: 1002.

Edmunds, L. N., Jr., and Carré, I. A. (1990). *2nd Meet. Soc. Res. Biol. Rhythms*, May 9–13, Amelia Island, Jacksonville, Fl. (Abstract 31, p. 37).

Edmunds, L. N., Jr., and Funch, R. R. (1969a). *Science*, *165*: 500.

Edmunds, L. N., Jr., and Funch, R. R. (1969b). *Planta*, *87*: 134.

Edmunds, L. N., Jr., and Halberg, F. (1981). In *Neoplasms: Comparative Pathology of Growth in Animals, Plants, and Man* (H. E. Kaiser, ed.), Williams & Wilkins, Baltimore, pp. 105–134.

Edmunds, L. N., Jr., and Laval-Martin, D. L. (1981). In *Photosynthesis*, Vol. 6 (G. Akoyunoglou, ed.) (*Proc. 5th Int. Congr. Photosynthesis*, Halkidiki, Greece, 1980), Balaban International Science Services, Philadelphia, pp. 313–322.

Edmunds, L. N., Jr., and Laval-Martin, D. L. (1984). In *Cell Cycle Clocks* (L. N. Edmunds, Jr., ed.) Marcel Dekker, New York, pp. 295–324.

Edmunds, L. N., Jr., Jay, M. E., Kohlmann, A., Liu, S. C., Merriam, V. H., and Sternberg, H. (1976). *Arch. Microbiol.*, *108*: 1.

Edmunds, L. N., Jr., Tay, D. E., and Laval-Martin, D. L. (1982). *Plant Physiol.*, *70*: 297.

Edmunds, L. N., Jr., Laval-Martin, D. L., and Goto, K. (1987). In *Endocytobiology III* (J. L. Lee and J. F. Fredrick, eds.), *Ann. N.Y. Acad. Sci.*, *503*: 459.

Engelmann, W., and Schrempf, M. (1980). *Photochem. Photobiol. Rev.*, *5*: 49.

Eskin, A., and Takahashi, J. S. (1983). *Science*, *220*: 82.

Eskin, A., Corrent, G., Lin, C.-Y., and McAdoo, D. J. (1982). *Proc. Natl. Acad. Sci. USA*, *79*: 660.

Feldman, J. F. (1975). *Science*, *190*: 789.

Feldman, J. F. (1979). In *Biological Rhythms and Their Central Mechanism* (M. Suda, O. Hayaishi, and H. Nakagawa, eds.), Elsevier/North-Holland Biomedical Press, Amsterdam, pp. 57–66.

Feldman, J. F., and Dunlap, J. C. (1983). *Photochem. Photobiol. Rev.*, *7*: 319.

Gautier, J., Matsukawa, T., Nurse, P., and Maller, J. (1989). *Nature*, *339*: 626.

Gilbert, D. A. (1981). *S. Afr. J. Sci.*, *77*: 541.

Goldberg, N. D., Ames, A., III, Gander, J. E., and Walseth, T. F. (1983). *J. Biol. Chem.*, *258*: 9213.

Goto, K., Laval-Martin, D. L., and Edmunds, L. N., Jr. (1985). *Science*, *228*: 1284.

Hartzell, H. C., and Fischmeister, R. (1986). *Nature*, *323*: 273.

Hasunuma, K. (1984a). *Proc. Jpn. Acad. Sci.* (*Ser. B*), *60*: 260.

Hasunuma, K. (1984b). *Proc. Jpn. Acad. Sci. (Ser. B)*, *60*: 377.
Hasunuma, K., Funadera, K., Shinohara, Y., and Wanatabe, M. (1987). *Curr. Genet.*, *12*: 127.
Jarrett, R. M., and Edmunds, L. N., Jr. (1970). *Science*, *167*: 1730.
John, P. L. C., Sek, F. J., and Lee, M. G. (1989). *Plant Cell*, *1*: 1185.
Kauffman, S. A., and Wille, J. J. (1975). *J. Theor. Biol.*, *55*: 47.
Kippert, F. (1987). *Ann. N.Y. Acad. Sci.*, *503*: 476.
Klevecz, R. R. (1976). *Proc. Natl. Acad. Sci. USA*, *73*: 4012.
Klevecz, R. R., and King, G. A. (1982). *Exp. Cell Res.*, *140*: 307.
Lakin-Thomas, P. L. (1985). Ph.D. thesis, University of California, San Diego.
Laval-Martin, D. L., Shuch, D. J., and Edmunds, L. N., Jr. (1979). *Plant Physiol.*, *63*: 495.
Laval-Martin, D. L., Carré, I., and Edmunds, L. N., Jr. (1990a). *Arch. Biochem. Biophys.*, *276*: 443.
Laval-Martin, D. L., Carré, I., and Edmunds, L. N., Jr. (1990b). *Chronobiol. Int.*, *7*: 99.
Ledoigt, G., and Calvayrac, R. (1979). *J. Protozool.*, *26*: 632.
Levandowsky, M. (1981). *Ann. N.Y. Acad. Sci.*, *361*: 369.
Lewin, B. (1990). *Cell*, *61*: 743.
Lotshaw, D., and Jacklet, J. W. (1983). *Fed. Proc.*, *42*: 575 (Abstr. 229.11).
Lotshaw, D., and Jacklet, J. W. (1987). *Comp. Biochem. Physiol.*, *86C*: 27.
Malinowski, J. R., Laval-Martin, D. L., and Edmunds, L. N., Jr. (1984). *J. Comp. Physiol. B*, *155*: 257.
Matsumoto, K., Uno, I., and Ishikawa, T. (1983). *Exp. Cell Res.*, *146*: 151.
Mendelhall, M. D., Jones, C. A., and Reed, S. I. (1987). *Cell*, *50*: 927.
Mitchison, M. (1974). In *Cell Cycle Controls* (G. M. Padilla, I. L. Cameron, and A. Zimmerman, eds.), Academic Press, New York, pp. 125–142.
Murray, A. W., and Kirschner, M. W. (1989). *Science*, *246*: 614.
Novak, B., and Mitchison, J. M. (1986). *J. Cell Sci.*, *86*: 191.
O'Farell, P. H., Edgar, B. A., Lakich, D., and Lehner, C. F. (1989). *Science*, *246*: 635.
Pines, J., and Hunt, T. (1987). *EMBO J.*, *6*: 2987.
Sachsenmaier, W. (1976). In *The Molecular Basis of Circadian Rhythms* (J. W. Hastings and H.-G. Schweiger, eds.), Dahlem Konferenzen, Berlin, pp. 409–420.
Schnabel, G. (1968). *Planta*, *81*: 49.
Shymko, R. M., Klevecz, R. R., and Kauffman, S. A. (1984). In *Cell Cycle Clocks* (L. N. Edmunds, Jr., ed.) Marcel Dekker, New York, pp. 273–293.
Skehan, P. (1988). *Biochem. Cell Biol.*, *66*: 467.
Swenson, K. I., Farell, K. M., and Ruderman, J. V. (1986). *Cell*, *47*: 861.
Tamponnet, C., and Edmunds, L. N., Jr. (1990). *Plant Physiol.*, *93*: 425.
Terry, O. W., and Edmunds, L. N., Jr. (1970a). *Planta*, *93*: 106.
Terry, O. W., and Edmunds, L. N., Jr. (1970b). *Planta*, *93*: 128.
Tong, J., Carré, I. A., and Edmunds, L. N., Jr. (1991). *J. Cell Sci.*, *100*: 365.

Vanden Driessche, T., Doege, K. J., Minder, C., and Cairns, W. L. (1979). In *Chronopharmacology: Advances in the Biosciences*, Vol. 19 (A. Reinberg and F. Halberg, eds.), Pergamon Press, Oxford, pp. 291–299.

Whitfield, J. F. (1990). *Calcium, Cell Cycles, and Cancer*, CRC Press, Boca Raton, Fl.

Whitfield, J. F., Durkin, J. P., Franks, D. J., Kleine, L. P., Raptis, L., Rixon, R. H., Sikorska, M., and Walker, R. (1987). *Cancer Metastasis Rev.*, *5*: 205.

Wille, J. J., Jr. (1979). In *Biochemistry and Physiology of Protozoa*, Vol. 2, 2nd ed. (M. Levandowsky and S. H. Hutner, eds.), Academic Press, New York, pp. 67–149.

14
Temporal and Spatial Gene Expression in Relation to Cell Division in the Pea Embryo

Angela J. Hauxwell, Fiona M. K. Corke, and Trevor L. Wang *John Innes Institute, Norwich, England*

I. INTRODUCTION

The angiosperm seed is the culmination of complex morphogenetic changes that provide an encapsulated miniature organism with its own supply of nutrients for the next generation. In many species, all the structures necessary to produce a fully formed plant are present in the embryo. We have chosen to investigate seed development in peas (*Pisum sativum* L.) in relation to storage product accumulation, as pea has many advantages; not only is it a relatively large seed but its embryology is well documented and it is well characterized genetically. In some varieties, by the time the seed is mature, all the vegetative nodes will be present in the plumule and the only new structures that will be formed are the nodes bearing reproductive organs. Over the last 10 to 20 years most work on embryo development has concentrated on gene expression during the later (especially maturation) phases. However, it is during the early stages of development that the major morphogenetic changes occur, and it is these changes that provide the blueprint for the final plant. We have attempted to answer some questions concerning cellular development during early cotyledon formation where the laying down of storage products appears to be concomitant with an increase in nuclear DNA content and a truncation of the cell division cycle.

II. EMBRYOLOGY OF *PISUM SATIVUM* L.

Prior to fertilization the egg sac contains three antipodal cells at the chalazal end, two synergids and one egg cell at the micropylar end (through which the pollen tube bearing male gametes enters), and one central cell containing two polar nuclei. The embryology is of crucifer type (Johri, 1984), the fertilized egg cell dividing transversely into two approximately equal cells—an apical and a basal cell—the basal one giving rise to a pair of elongate suspensor cells. The apical cell further divides, the lower one eventually producing two large, multinucleate, highly vacuolate bulbous cells, which will form the neck of the suspensor, while the upper one gives rise to the embryo proper (Cooper, 1938a,b). The elongated suspensor cells become multinucleate by synchronous division and are considered to provide nutrients to the very young embryo. The occurrence of transfer cell-like convolutions at the base of the suspensor supports this conclusion. At later stages the abaxial surface of the embryo's cotyledons and the inner surface of the seed coat, the testa, become transfer cells. There is no cytoplasmic continuity between the embryo and the mother tissues, so all nutrient flow has to be via such transfer processes. There are plasmadesmatal connections, however, between suspensor and embryo.

The cells of the embryo continue to divide to form a globular structure, with a lightly staining central ground meristem that gives rise to the procambium and a darkly staining outer zone or protoderm. The overall structure becomes heart shaped, delineating the position of the two structures that will form the main storage organs of the embryo: the cotyledons. At the same time, the vertical axis is defined as this contains, at its base, the root meristem and, toward the cleft of the heart, what will become the shoot meristem. The cotyledons continue to develop by cell division, but the number of dividing cells gradually decreases so that growth becomes mainly by expansion. There is much genetic variation for the timing of this cessation of division (Hedley and Smith, 1985). In legumes, in particular, the cotyledon cells enlarge considerably and the division cycle becomes truncated so that the DNA content of cotyledon cells rises dramatically to, in the pea, as much as 128C. There is no polyploidy and the increase in DNA is known as endoreduplication.

The fusion of the other male gamete nucleus with the two polar nuclei of the central cell produces a triploid endosperm. The nuclei continue to divide to form the multinucleate endosperm, partially cytoplasmic, partially liquid, but there is evidence for some cellularization via the formation of thin strands of cellulose between testa and embryo (Marinos, 1970). As the embryo expands, the endosperm disappears (hence the pea is described as exalbuminous), so that the embryo eventually abuts the testa and their respective layers of transfer cells lie against one another. By the time the embryo is mature and starting to dry, the cotyledons have rounded and accumulated considerable reserves of starch (ca. 50% of seed dry weight) and protein (ca. 25%). They bear an abaxial surface to the seed coat, with their adaxial surfaces adjacent and enclosing the embryonic shoot apex (plumule). The embryonic root (radicle) lies against the testa in the pocket originally occupied by the globular embryo and elongated suspensor cells. The cotyledons themselves will have differentiated into epidermal cells, parenchymatous storage cells, and relatively (compared with other species) underdeveloped vascular tissues. The germination of pea is hypogeal, with the cotyledons remaining underground. They do not, therefore, become photosynthetic as in some other legumes: for example, *Glycine max* and *Phaseolus vulgaris*.

III. MITOSIS AND GENE EXPRESSION DURING DEVELOPMENT

As mentioned previously, the majority of the growth of the embryo is by cell expansion. Concomitant with this cell expansion is the deposition of storage products. Although the storage products are usually considered to be protein, starch, and lipid, DNA is also accumulated in the cotyledon cells as their nuclei contain greatly increased levels of DNA (Johri, 1984). The possibility that the accumulation of DNA also has a storage function is supported by the finding that the transcriptional activity of the storage protein genes is not related to the increase in availability of DNA template (Raghavan, 1986). In addition, the increase in nuclear DNA level is due to endoreduplication of the entire genome, rather than amplification of selected genes (Cullis and Davies, 1975).

The increase in DNA level above the normal diploid level is not a prerequisite for storage protein synthesis as shown when DNA synthesis is inhibited (Corke et al., 1990b), although mitotic cells are never seen to contain storage protein. It must be emphasized, however, that while it is known that the developing embryo is not a homogeneous tissue, the localization of mitotic cells in the pea embryo has not been described in the same detail as in animal systems such as *Drosophila*. In the developing *Drosophila* embryo, mitosis occurs synchronously within the embryo until the fourteenth mitotic cycle. Synchrony is normally observed in the plant embryo only during suspensor development, as mentioned earlier. After this point the embryo consists of discrete ''mitotic domains'' each of which show a wavelike spread of mitosis across, but within the boundary of, the domain. In addition, between different domains there is variation in the duration of the mitotic cycle (Foe, 1989). Although such clear mitotic domains cannot be seen within the cotyledon, there must be some signal that determines the individual developmental pathway of cells that make up the heterogeneous but obviously ordered population.

In our study of the relationship of cellular DNA level and the deposition of the storage protein vicilin during embryo development, it became clear that (1) cells in mitosis do not deposit vicilin, (2) the DNA content of cells that do contain the protein is higher than that of the usual diploid level, but (3) the deposition of storage protein begins early in the amitotic phase of cell development (Corke et al., 1987). When the spatial distribution of cells transcribing vicilin genes was determined, a distinct pattern was seen. The mRNA for vicilin was first localized by in situ hybridization in the upper adaxial region of early-stage developing pea embryos (Fig. 1). With increasing fresh weight, the mRNA could be detected in a teardrop-like wave pattern, from the adaxial region to the abaxial region of the cotyledon (Fig. 2a). Vicilin mRNA was not localized in all cell types and was restricted to the parenchyma cells of the cotyledons and axis. The mRNA was never localized in the vascular cells of the axis, the

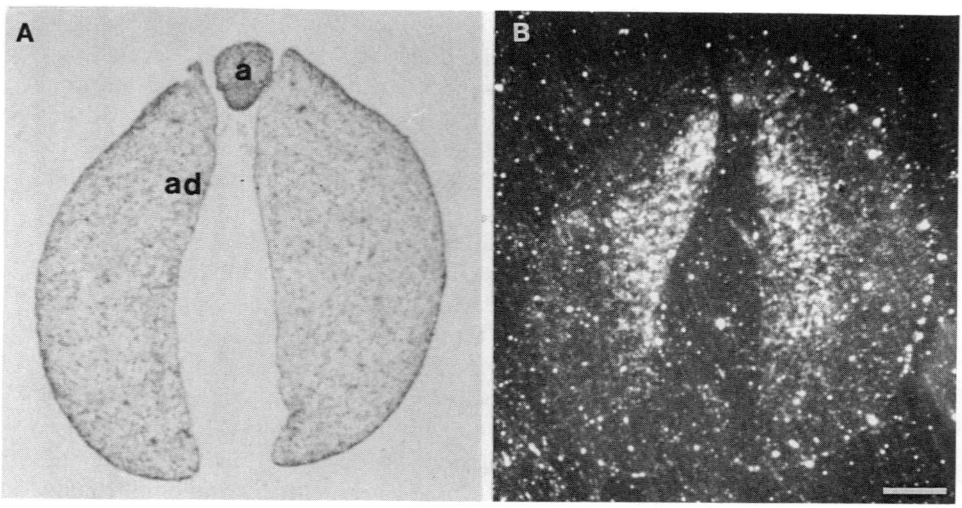

Fig. 1 Longitudinal section of a 10-mg JI 181 pea embryo, in which vicilin mRNA has been localized by in situ hybridization, using a tritium-labeled cRNA probe. (a) bright-field illumination shows the nuclear staining and cell outlines; (b) dark-field image shows the hybridization signal (white grains). Vicilin mRNA is present in the cells of the adaxial region of the cotyledon [ad] but is absent from the embryonic axis [a]. Bar = 400 μm

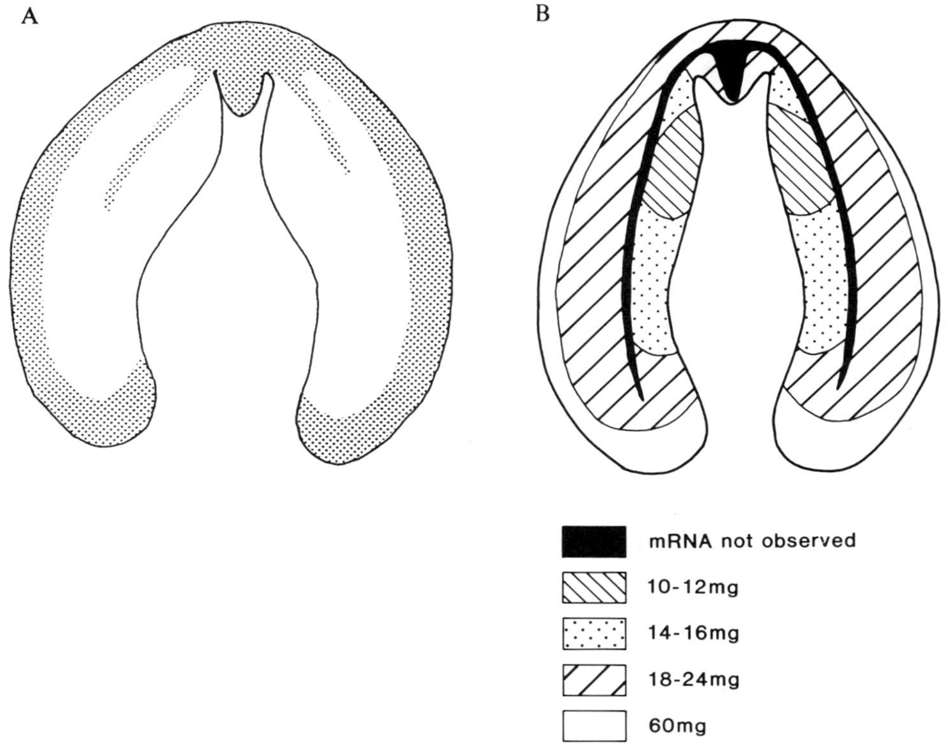

Fig. 2 Localization of mitotic activity and vicilin mRNA in longitudinal sections: (a) diagrammatic representation of the overall pattern of mitotic activity (shaded areas) in a 10-mg pea embryo; (b) line diagram to show the regions in which vicilin mRNA was localized by in situ hybridization, at different stages during development. Vicilin mRNA was localized to regions lacking mitotically active cells.

epidermis, or the provascular cells, even during the midmaturation phase of development. It was found that during the wave, the regions actively transcribing vicilin genes lacked dividing cells (Hauxwell et al., 1990), mitotic cells being confined increasingly to certain regions of the cotyledons: near the abaxial surface, the provascular strands, and the embryonic axis (Fig. 2b).

A spatial distribution pattern has also been reported for storage product mRNAs in soybean embryos (Perez-Grau and Goldberg, 1989). The mRNA for β-conglycinin (and other proteins) was first localized in midmaturation soybean embryos, in the parenchyma cells of the abaxial region of the cotyledons. A ''wavelike'' pattern of mRNA localization was seen to spread with increasing maturation, from this region to the adaxial region, in the direction opposite to that reported in pea. The cell type specificity, however, was the same as that found in pea.

There are few other reports of spatial patterns of mRNA localization akin to those reported in pea and soybean, but gradients of mRNA and of the respective protein have been found in *Drosophila*. The *bicoid* protein and mRNA are localized in the egg in a gradient, being most abundant at the anterior pole with the concentration dropping toward the posterior pole. This gradient has been shown to determine the pattern of development at the anterior part of the embryo (St. Johnston et al., 1989), and in *bicoid* mutants the anterior structures are absent.

Although there are no established morphological mutants for pea embryos, one way to alter the pattern of development is via tissue culture. When isolated embryos are placed in a defined medium at a very early stage in development, the levels of DNA endoreduplication, vicilin, and mRNA for vicilin are increased (Corke et al., 1990a; Yang et al., 1990). The spatial distribution pattern of this protein and mRNA are also altered (Fig. 3a) compared with that found in vivo (Fig. 3b) (Hauxwell, 1990). The alteration of mRNA and protein localization reflects a disruption of the normal processes of cellular development in culture, and the changes observed can be explained by an aging of cells and an increase in recruitment of those producing storage protein (Yang et al., 1990). The aging is also indicated by the premature and rapid cessation of cell division in embryos placed in culture (Fig. 4) and a change in the cellular population (Ambrose et al., 1987), the latter showing a shift in its distribution away from meristematic-sized cells to those with an increased area and nuclear DNA content. The manipulation of storage protein production in this way supports the conclusion that its initiation is controlled embryonically and is independent of a maternal signal (Raghavan, 1986).

IV. CELLULAR PERIODICITY AS A GENERATOR OF EMBRYO DEVELOPMENT

From previous sections it is evident that mitotic activity is important in embryo development, not only in establishing a basal level of cells, but also in molding changes in morphology and cell biochemistry. Throughout development there is both temporal and spatial periodicity in cell division activity. Thus the proembryo (considered to be that stage prior to the development of cotyledon primordia) retains mitotic activity in its cells but gives rise, via its ground meristem and procambium, to other meristems, such as those of the root and shoot (Raghavan, 1986). The

A

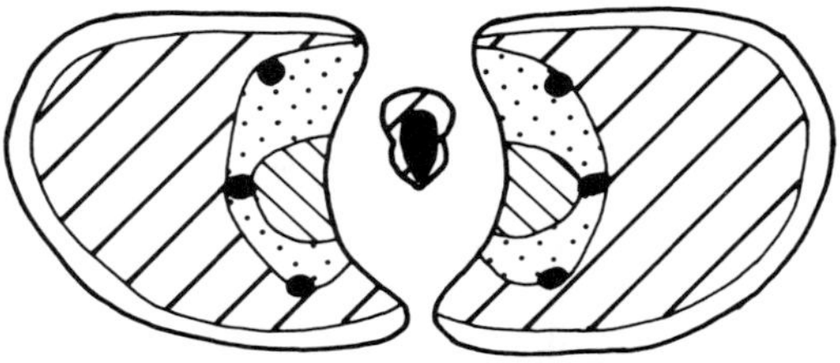

B

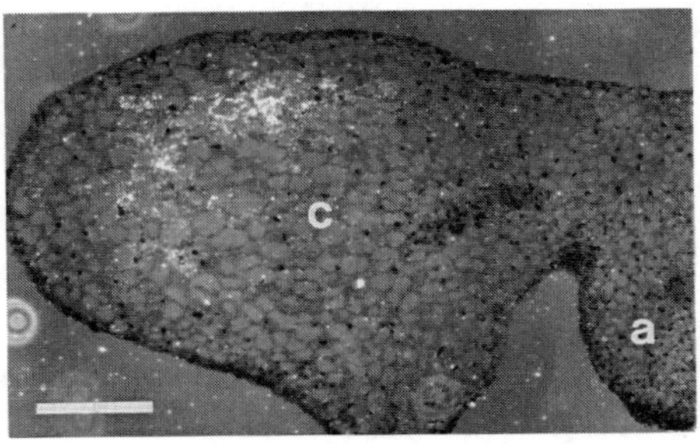

Fig. 3 Comparison of the localization of vicilin mRNA in transverse sections of in vivo- and in vitro-grown embryos: (A) line diagram to show the regions, in which vicilin mRNA has been localized, at different stages during development; (B) transverse section from the upper cotyledon region of a 7.3-mg (final weight) embryo cultured for 4 days, showing one cotyledon (C) and part of the embryonic axis (A). Bar = 220 μm. In vitro-grown embryos showed differences from in vivo-grown ones in (i) the position of cells within the embryo containing vicilin mRNA in (ii) the minimum embryo age when vicilin mRNA could first be detected. (From Hauxwell, 1990.)

cotyledons are considered to be generated from the shoot meristem, which retains mitotic activity throughout the life of the embryo until the structure becomes quiescent due to desiccation. The cotyledons themselves, however, show temporal and spatial variation for mitotic activity, as demonstrated in Section III. Clearly, in the early stages of cotyledon development all cells must be mitotically active, but it is likely, if one extrapolates the data for pea to the point of initiation of the cotyledon, that some parenchyma cells cease division relatively early and enter the truncated cell cycle (endoreduplication) characteristic of the family. Nevertheless, to generate the morphology of the final seed and keep pace with the expanding cells of the parenchyma, mitotic activity must continue for

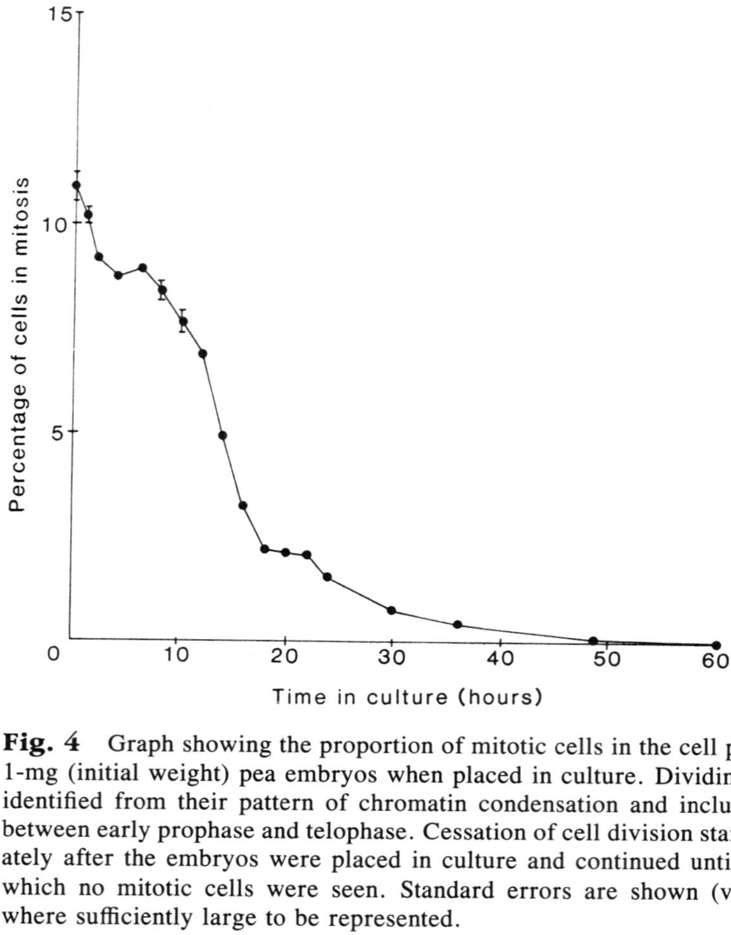

Fig. 4 Graph showing the proportion of mitotic cells in the cell population of 1-mg (initial weight) pea embryos when placed in culture. Dividing cells were identified from their pattern of chromatin condensation and included all cells between early prophase and telophase. Cessation of cell division started immediately after the embryos were placed in culture and continued until 60 h, after which no mitotic cells were seen. Standard errors are shown (vertical bars) where sufficiently large to be represented.

some time in at least some cells of the outer regions, especially the epidermis. When overall changes in the cell number of the embryos of pea (Hedley and Smith, 1985) and other species (cited in Raghavan, 1986) are considered, it is also clear that these show periodic changes since there are definite lag phases. This must indicate oscillations in overall mitotic activity.

Within the cotyledons, operation of the truncated mitotic cycle is not understood. There is evidence for oscillations between DNA replication and a resting state, but it is unclear whether the cycles are contiguous, at a fixed rate, or whether they vary. It is also unclear when in the cycle storage protein genes are expressed. It is clear, however, that DNA endoreduplication and storage protein deposition occur at the same time, but there is no evidence that they are causally related. The cessation of cell division in the cotyledon parenchyma cells, however, may be a cue for the initiation of storage protein synthesis.

In work on another legume seed, soybean, Perez-Grau and Goldberg (1989) noted that events leading to the specification of the cotyledons and the activation of storage protein gene transcription are temporally separated, the latter not occurring until at least 100,000 cotyledon cells are formed. The transcription of the genes in soybean, like pea, occurs in a wave and thus shows temporal and spatial periodicity. In contrast, however, the wave is in the opposite direction to that in pea and hence it is unlikely, as suggested by these authors, that the direction of the wave is an intrinsic feature of cotyledon development. What, then, initiates the wave of activation? Perez-Grau and Goldberg suggested that the activation of transcription may be induced by either the transmission of signals cell to cell, a gradient of regulatory substances in the direction of the wave, or an intrinsic clock mechanism. From the evidence in pea, a transmitted activating gradient from cell to cell is not necessary, since a gradient in age from the cessation of cell division (and hence a ''clock'') already exists. The mechanism of the clock could simply be the counting of division/endoreduplication cycles. Thus the oldest cells in pea, as indicated by their DNA contents and cell size (which would be related to the number of cycles), would occur toward the adaxial surface, which is where storage protein mRNA is first localized.

There is no evidence to support such a hypothesis in other species simply because the information required on mitotic activity and DNA levels in the cotyledons is absent. The information that is available for soybean (Meinke et al., 1981) is of insufficient detail to argue against such a mechanism. At present, the data strongly indicate that the pattern of development in the pea embryo is a function of temporal and spatial periodicities in mitotic activity and that the accumulation of storage protein is an intrinsic function of the cells (rather than the organ) themselves.

REFERENCES

Ambrose, M. J., Wang, T. L., Cook, S. K., and Hedley, C. L. (1987). *J. Exp. Bot.*, *38*: 1909.

Cooper, G. O. (1938a). *Bot. Gaz.*, *99*: 584.

Cooper, D. C. (1938b). *Bot. Gaz.*, *100*: 123.

Corke, F. M. K., Hedley, C. L., Shaw, P. J., and Wang, T. L. (1987). *Protoplasma*, *140*: 164.

Corke, F. M. K., Hedley, C. L., and Wang, T. L. (1990a). *Protoplasma, 155*: 127.

Corke, F. M. K., Hedley, C. L., and Wang, T. L. (1990b). *Protoplasma, 155*: 136.

Cullis, C. A., and Davies, D. R. (1975). *Genetics, 81*: 485.

Foe, V. E. (1989). *Development, 107*: 1.

Hauxwell, A. J. (1990). Ph.D. thesis, University of East Anglia.

Hauxwell, A. J., Corke, F. M. K., Hedley, C. L., and Wang, T. L. (1990). *Development, 110*: 283.

Hedley, C. L., and Smith, C. M. (1985). In *The Pea Corp* (P. D. Hebblethwaite, M. C. Heath, and T. C. K. Dawkins, eds.), Butterworth, London, pp. 329–338.

Johri, B. M. (ed.) (1984). *Embryology of Angiosperms*, Springer-Verlag, Berlin.

Marinos, N. G. (1970). *Protoplasma, 70*: 261.

Meinke, D. W., Chen, J., and Beachy, R. N. (1981). *Planta, 153*: 130.

Perez-Grau, L., and Goldberg, R. B. (1989). *Plant Cell, 1*: 1095.

Raghavan, V. (1986). *Embryogenesis in Angiosperms*, Cambridge University Press, Cambridge.

St. Johnston, D., Driever, W., Berleth, T., Richstein, S., and Nüsslein-Volhard, C. (1989). *Development, 107(Suppl.)*: 13.

Yang, L. J., Barratt, D. H. P., Domoney, C., Hedley, C. L., and Wang, T. L. (1990). *J. Exp. Bot., 41*: 283.

15
Recurrent Activation Cycles, Not Localized Factors, Determine Cellular Bifurcations in Early Mosaic Development of Nematodes

Karl B. Moritz *Zoological Institute, University of Munich, Munich, Germany*

Helmut W. Sauer *Texas A&M University, College Station, Texas*

I. INTRODUCTION

Development has been described as the emergence of complex forms that arise from a single-cell embryo, the fertilized egg, by a regular sequence of events known as embryogenesis. These events include cell divisions, pattern formation (morphogenesis), specification of different cell fates, and cell differentiation. Two alternative philosophical views have dominated the history of developmental biology—preformation and epigenesis—which can be associated with two kinds of embryos: mosaic and regulative, respectively. However, it became apparent that the opposing views can be explained by the phenomenon of cell commitment, which happens early in mosaic and late in regulative embryos. This conclusion was reached by experimental embryologists during the last hundred years (Sauer, 1980).

More recently, molecular and developmental geneticists have focused on the role of genes during embryogenesis. The theory of differential gene expression provides a solid concept for the mechanism of cell differentiation. Moreover, the identification of regulatory genes and the deduction of complex networks of control genes allow insight into the mechanism of continuous regulation of differentiation. The success of developmental genetics has led some investigators to believe in the existence of developmental programs that are inscribed, stored, and propagated in the genome. This reductionistic view is a modern version of preformation. On the other hand, it has long been postulated that development progresses by complex interactions between the cytoplasm and the nucleus, and among cells in the whole embryo. This holistic view is closer to epigenesis since new programs are generated during morphogenesis which are not stored in the information of the genome. To fully comprehend development, both concepts must be explored with a specific focus on cell-fate specification and cell commitment.

The central dogma of prelocalization neatly explains how cells become different and how cell lineages are established in the earliest cleavage divisions of an egg cell. According to this favored notion, informational macromolecules are stored in the egg, and they become segregated into different cells during development. Consequently, preformed local factors are inherited by different cell lineages, where they can act as intrinsic signals in various pathways of autonomous cell-fate specification. Rigid lineage determination by presumed local factors during an invariant cleavage pattern in the nematode *Caenorhabditis elegans* is a prime example of the dogma of prelocalized determinants.

It is the purpose of this brief review to introduce a different concept of cell-fate specification in the early development of nematodes which is centered on repeated cell polarizations, followed by segregative (i.e., asymmetric) cell divisions. Consequently, just a recurring quantitative change leads to bifurcations that generate the qualitatively different early cell lineages in the absence of local informational molecules. This concept was proposed almost a century ago by Theodor Boveri (1910b) in a little-remembered but brilliant analysis of the *Ascaris* embryo, where he demonstrated that a general change in the cytoplasmic milieu, rather than local specific factors, can explain the epigenetic process of cell-fate specification. Specifically, a small difference in newly born daughter cells would lead to a dramatically different developmental fate of their descendants.

This view of bifurcative steps, which is defined within the framework of the theory of deterministic chaos is also consistent with the result of many recent experiments on *C. elegans*, the other nematode embryo system. A testable model is presented which assigns to the cell cycle a crucial role in cellular decision making which leads beyond the current emphasis on cell proliferation control and cancer.

II. NEMATODES AS A DEVELOPMENTAL SYSTEM

Caenorhabditis elegans is a small free-living soil nematode. Ease of cultivation, its rapid generation time of about 3 days, its simple organization, and cell constancy at adulthood make the worm a most useful model organism for the study of many problems in developmental biology. The description of the entire cell-lineage tree during the ontogeny of the transparent animal on a cell-by-cell basis has revealed that the pattern of cell divisions is essentially invariant between individuals (Fig. 1). The small genome of 8×10^7 bp is contained in six chromosomes and comprises 3000 to 4000 genes. The recent developmental and genetic analyses have made this small worm a key model among the "big four" in developmental biology, in addition to fruit fly, frog, and mouse. (For an excellent review, readers are referred to Strome, 1989.) *Ascaris*, the classic experimental model nematode, has left us a legacy of crucial information regarding the early establishment of the soma and the germline, and the important phenomenon of chromatin diminution in somatic cells (Boveri, 1899; Moritz, 1967a). The current challenge is to elucidate how embryos work at the molecular and cellular level, and to integrate the directive role of genes with T. Boveri's system theory of epigenetic cell-fate specification (Boveri, 1910b).

A. Cell Lineages of the Early Embryo

During the first cleavage cycle of *C. elegans* a number of cytoplasmic rearrangements occur in the zygote, or the P_0 cell (Fig. 2a), which manifest its polar organization (Strome and Wood, 1983). These include a series of membrane contractions at the anterior cell hemisphere, which culminate in a transient constriction, termed pseudocleavage (Fig. 2a), and the segregation of cytoplasmic germline-specific granules, the P granules, to the posterior egg hemisphere. The bipolar spindle rotates into the anteroposterior (animal–vegetal) axis, which becomes the primary axis of the embryo. Then the spindle moves toward the posterior end of the embryo immediately before furrowing. Consequently, the zygote undergoes an unequal cleavage division (Fig. 2a), generating two blastomeres that differ in size, cell cycle period, further cleavage pattern, and developmental fate (Fig. 1). The larger anterior blastomere is the first-order somatic founder cell AB, the smaller posterior cell is the germline blastomere P_1. This is the first of a series of segregative cleavage divisions of the embryonic germline represented by P_0, P_1, P_2, and P_3 that generate four somatic founder cells: AB, EMS, C, and D. With the

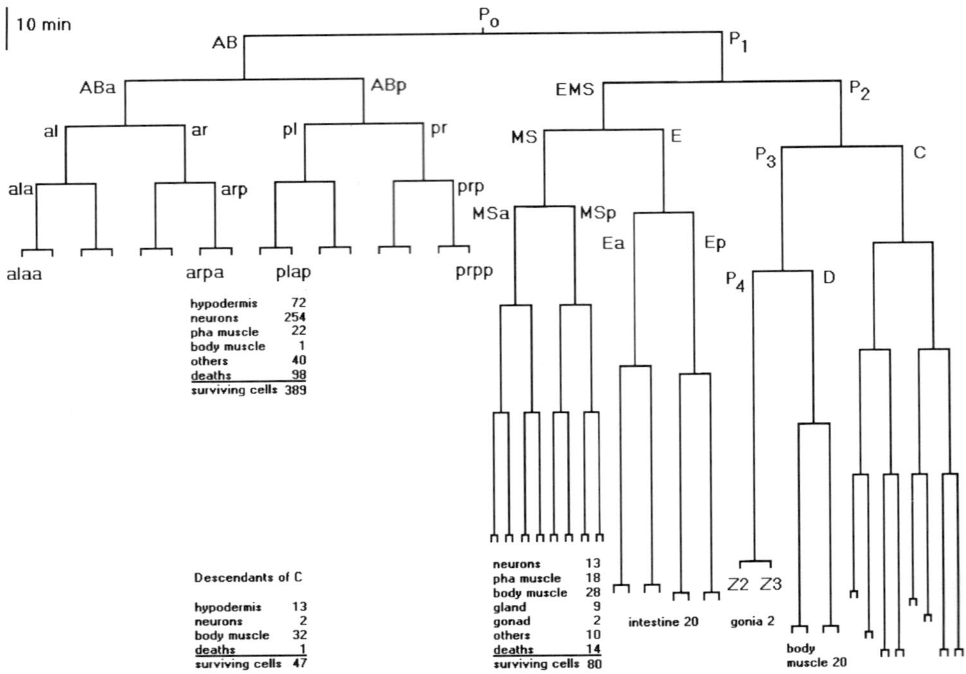

Fig. 1 Early lineage tree of the *C. elegans* embryo. Each horizontal line represents a cell division, each vertical line represents the cells' lifetime. Blastomeres of the germline lineage (P-lineage) have longer cell-cycle periods than their presomatic sisters. a/p, Anterior–posterior positioning; l/r, left–right positioning of sisters. The lineage is drawn with anterior to the left and posterior to the right. Note that the somatic founder blastomeres C and D stem from the posterior hemispheres of the respective germline blastomeres, P_2 and P_3. Below each founder blastomere, type and number of cells are indicated that derive from it at hatching. The (hermaphrodite) first larvae consists of 558 cells.

exception of EMS, which divides at the third cleavage step by a polar segregative division into the larger anterior cell, MS, and a smaller posterior sister, E, all somatic founder cells cleave by symmetrical divisions into apparently equipotent daughter cells. P_4 divides equally into the progenitors of the germ cells, Z_2 and Z_3. The P granules are segregated exclusively into the germline daughters. The unequal polar cell divisions, which give rise to smaller germline daughters and larger somatic founder cells, result from asymmetrical positioning of the spindle. The blasto-

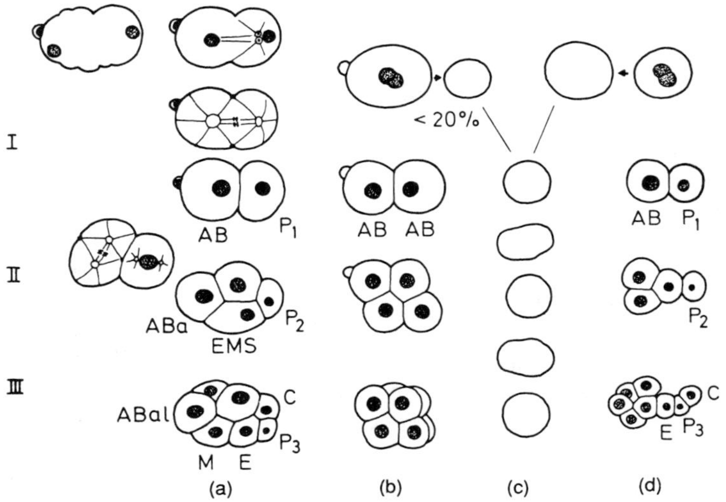

Fig. 2 Cleavage pattern of a normal embryo and egg-cell fragments. Generation and position of daughter cells, presented during the first cleavage steps (I, II, III) leading to the eight-cell stage embryo. Anterior pole is marked by the polar body. (a) Normal egg: the series of invariant segregative cell divisions (topogenesis) is shown. From top to bottom: membrane activities at the anterior pole indicate cell polarization; the oocyte pronucleus moves to the posterior pole (late pseudocleavage stage); asymmetric position of the spindle at metaphase; unequal cleavage gives rise to the larger somatic founder AB and the smaller germline cell P_1. AB cleaves by a nonpolar symmetric division into equipotent daughter cells, ABa and ABp. P_1 cleaves by a polar unequal division into the larger EMS and the smaller P_2. After the third cleavage the asymmetric 8-cell stage is formed by characteristic cell-movements of the ABa and ABp daughters. EMS divides unequally into the anterior MS and the posterior E. P_2 cleaves off the somatic founder C in posterior direction. Pattern and timing of cleavage are used to define the qualities of living cells in experimental embryos. (b) Anterior fragment of the polarized zygote: equal, synchronous cleavage with short cell cycle periods characteristic of the somatic AB lineage. (c) Anucleate fragments (cytoplasts) oscillate in step with cleavage cycles of whole cells. (d) Posterior fragment of the polarized zygote: normal cleavage pattern with respect to cleavage planes in soma and germline.

meres are arranged into a typical form of the early embryo by a process called topogenesis.

B. Cell-Cycle Periods Correlate with Cell Fates

The prospective fates of each of the founder cells (Fig. 1) are known from elaboration of the complete embryonic lineage tree (Sulston et al., 1983). Only the intestine and germ cells are clonally derived, from E and P_4, respectively. Mesodermal and ectodermal tissues are polyclonal in origin.

Following its birth, each of the five somatic founder cells exhibits a characteristic cell-cycle period (Deppe et al., 1978). It is always shorter than that of the respective germline sister cell (see Fig. 1). The period correlates with the order of generation of the founder cells and their position. For example, the gradually elongating cell cycles in the germline are also apparent in the germline-derived somatic founders, and central positioning (E descendants) correlates with an elongated rhythm of cell divisions. Cells that descend from the same somatic founder divide initially at the same rate and slow down coordinately. Later, in embryogenesis, asynchronies occur within the same lineage in an invariant manner. Those programmed changes in cycle timing correlate with diversification of developmental fates. The Caa daughters, for example, exhibit different cycle timing (Fig. 1). The Caaa descendants divide by rapid synchronous divisions and differentiate into hypodermal cells. Caap, which has a longer period than its sister Caaa, gives rise to different descendants: The progeny cell, with a much longer cycle (Caapa), undergoes an extremely asymmetric division. Its smaller, karyoplast-like daughter is committed to die, while the surviving cell differentiates into a nerve cell. Programmed cell death after an asymmetric cell division is a common feature in several lineages. In conclusion, the correlation between asynchrony and subsequent alternate specifications of sister cells may indicate short-ranging cell interactions by which one sister cell becomes exclusively instructed to execute a certain developmental program.

C. Early Cell Fates Are Autonomously Specified

Analysis of the cleavage pattern and developmental potential of isolated blastomeres and early partial embryos is a common assay to test whether the specification of cells depends on intrinsic or extrinsic factors. These experiments revealed a complete lack of regulative potential of partial embryos (Laufer et al., 1980). Instead, the cleavage pattern of early blastomeres, cultured in isolation, suggests that these cells become determined by internal cues rather than by position or exogenous signals. For example, isolated P_1 cells (Fig. 3a) undergo the same series of germline-typic unequal divisions as P_1 of the two-cell embryo does, thus generating the presomatic cell EMS and P_2. Then EMS divides by a polar asymmetrical cleavage into MS and E, and P_2 into P_3 and C. Partial embryos derived from P_1 show muscle cell functioning, and intestinal granules are formed in the progeny of E cells. Therefore, not only early cleavage pattern and cell-cycle periods but cell differentiation can occur independently of the normal neighboring cells. Hence founder-cell fates become internally specified. This cell-autonomous specification in the normal spatiotemporal order has been interpreted repeatedly as evidence for the existence of macromolecular determinants, localized in the zygote, that confer developmental qualities to those cells to which they are partitioned (Davidson, 1990). Alternatively, a quantitative difference of a factor released at the end of the cell cycle may function as the initial signal by which daughter cells could be specified to different developmental pathways (Moritz, 1990). The results of the following experiments are more consistent with the second concept of intracellular signaling.

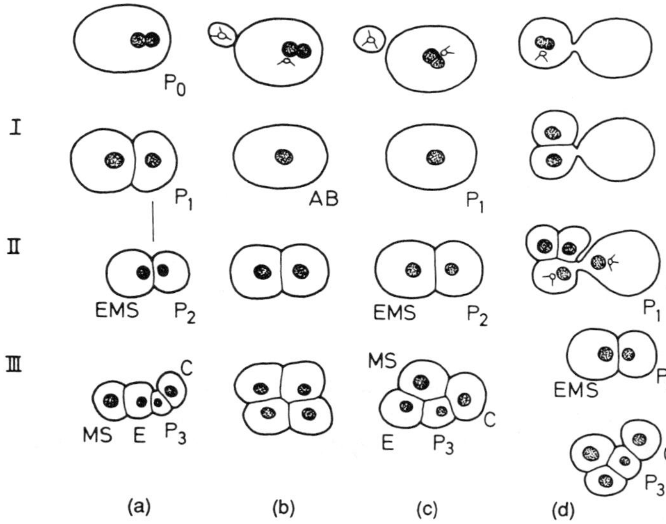

Fig. 3 Modes of cell-fate specification: (a) cleavage of a P_1 cell, after isolation, reveals autonomous cell-fate specification; (b,c) suppression of cytokinesis of first cleavage step (achieved by removal of one centrosome) results in alternative cell fates [i.e., either AB (b) or P_1 (c) quality]; (d) cell-cycle-dependent cell-fate specification in the absence of nucleus and centriole.

D. Cell Fragments Have Regulatory Properties

Extrusion of up to 20% cytoplasm (i.e., the presumptive total P_2 cell mass) from either the anterior or posterior region of the fully polarized zygote does not interfere with normal cleavage and cytodifferentiation in later embryogenesis (Laufer and von Ehrenstein, 1981). Even later, when all the presumptive P_4 cytoplasm was extruded, the embryos developed into mature and fertile adults. From these observations it was concluded that cytoplasmic determinants, if they exist, are not prelocalized in an *unbound* form in the zygote. Because large membrane fragments were extruded concommitantly with the exovates, exclusive membrane prepatterning also seems highly improbable. However, according to the alternative concept of sequential cell-fate specification by transient quantitative signals, normal development—as observed—should not be affected. Extrusion of much larger portions of anterior cytoplasm (up to 50%) from the zygote at the time of pronuclear conjugation does not affect the potential of posterior egg fragments (containing the pronuclei and the centrosomes) to undergo the canonical unequal cleavage divisions of the entire zygote (Fig. 2d). If, however, a critical mass of *posterior* cytoplasm (at least 25% of the total egg cytoplasm) is removed, the normal cleavage potential is always lost (Schierenberg, 1988). Those egg fragments always pass through a series of symmetric and synchronous cleavage divisions like AB descendants do in the whole embryo (Fig. 2b). Germline-typic, unequal cleavage potential is never reestablished. This difference in regulative capacity of anterior and posterior fragments indicates asymmetry of the polarized zygote.

Significantly, a small posterior fragment (about 20% of total cytoplasm), isolated from the zygote prior to final polarization of the zygote, is able to execute the early unequal cleavage program (Schierenberg, 1988). Hence a zygotic fragment can self-organize regulatively, as seen from the normal cell-proportioning of cell descendents and their normal cell-cycle timing throughout early cleavage. The results of cytoplasmic extrusion experiments reveal an extensive cell-intrinsic regulative potential, and thus are inconsistent with the concept of prelocalization of blastomere-determining macromolecules. The polarization process, however, drastically restricts the potency to regulate.

E. Posterior Egg Fragments Show Alternative Fates

Posterior egg fragments display two completely different cleavage patterns (Schierenberg, 1987): They either "regulate" and divide by an unequal cleavage into AB and P_1 (Fig. 2d); or alternatively, they cleave into equal-sized cell daughters. Their further cleavage pattern clearly indicates that both blastomeres have gained P_1 quality (this is a case of lineage duplication).

Lineage duplication was also observed with posterior P_1 fragments of a two-cell stage. If those fragments cleave equally, both daughters execute the cleavage program of P_2 blastomeres. The duplication of lineages (after equal divisions of posterior fragments of P_0 and P_1) clearly indicates that (1) blastomere specification depends on the number of cell cycles, and (2) recurrent (i.e., proliferative) cell division cycles do not occur during the stage of segregative cleavages. This means that a P_0 cell can never divide and maintain P quality.

F. Suppression of One Cleavage Step Results in Extinction of Mother and Emergence of Daughter-Cell Quality

In the sea urchin embryo it had been shown that the developmental clock can run independently of cleavage divisions (see Hörstadius, 1973). For example, when two or one cleavage divisions are skipped, the micromeres at the vegetal pole form at the four- or eight-cell stage rather than at the normal 16-cell stage. This clearly demonstrates that progressive changes take place in the cytoplasmic system, which controls the timing of micromere formation, apparently independent of cell division. Since in the sea urchin embryo each of the four blastomeres is still omnipotent after two cleavage divisions, one could argue that this result is possible only because of nonsegregative (i.e., proliferative) early cleavage divisions in regulative eggs. It would reflect merely a clock mechanism, coupled to the cell-cycle oscillator, while other developmental instructions might be programmed separately. Therefore, the question arises what quality (developmental potential) displays the zygote of a mosaic egg known to undergo asymmetric divisions, subsequent to inhibition of cytokinesis of the first cleavage step. Would P_0 quality persist undisturbed or become destroyed as a consequence of interactions of somatic (AB) and germline (P_1) domains exposed to each other? Or has the uncleaved egg progressed to either one of the two daughter-cell fates (i.e., AB or P_1)?

This question has been addressed in several experimentally derived egg types of *Ascaris*. Its early cleavage pattern is the same as that of *C. elegans* (Figs. 1 and 2). However, commitment of blastomeres to somatic fate can be detected easily due to a drastic genome reduction in the somatic cells, referred to as chromatin diminution (Moritz, 1967a; Teschke et al., 1991). Consequently, cell qualities in the early embryo can be clearly distinguished. It was found that subsequent to cleavage inhibition, the one-cell embryo displays either AB or P_1 quality (Seidl et al., 1988). In conclusion, this result demonstrates that the first cell-cycle step, in the absence of cell division, generates one of the two different cell qualities. Also in *Ascaris*, lineage duplication, at the cost of segregative division at first cleavage, can be induced easily by a simple centrifugation experiment (Fig. 4d). In case of equal cleavage (caused by centrifugation), both blastomeres always gain P_1 quality (Boveri, 1910a; Moritz, 1967b). If the

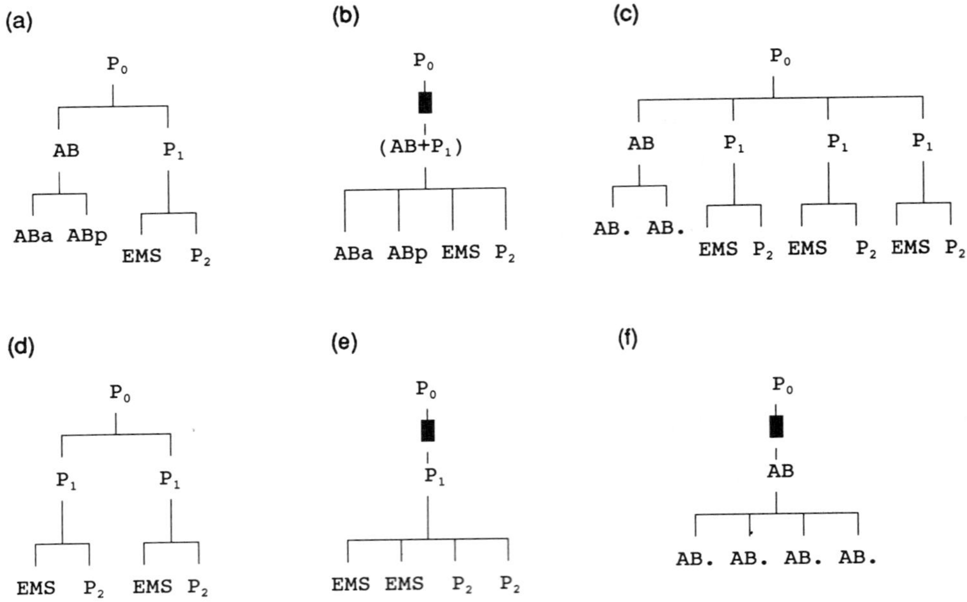

Fig. 4 Interpretative lineage trees of normal and manipulated eggs. (a) Normal and (b) Predicted lineages after suppression of first cleavage (black bar) and subsequent second cleavage into four blastomeres (simultanvierer) under the presumption of prelocalized determinants. (c) Simultanvierer produced by the first cleavage of a dispermic *Ascaris* egg. Three types have been observed: 1 AB + 3 P_1's (shown); 2 AB's + 2 P_1's, 3 AB's + 1 P_1. Hence egg segments undergo normal cellular bifurcation. (d) Lineage duplication resulting from nonpolar cleavage (parallel to the polarity axis) of the egg cell (*Ascaris*) achieved by centrifugal force. (e) After suppression of cytokinesis of the first cleavage egg cells (*C. elegans*) gain daughter-cell quality, either P_1 (e) or AB (f). Because of undisturbed duplication of centrioles the one-cellular embryo cleaves into four cells at the next (second) cleavage step. As a result, lineage duplication is always observed.

cleavage plane deviates from the egg axis, the resulting two-cell stage undergoes normal cleavage and chromatin diminution in the somatic founder blastomeres. This embryo, despite stratification of the visible constituents caused by centrifugation, develops into a normal, hatching larvae (Moritz, 1967b). It seems that cellular bifurcative decision making requires an anisotropy of the egg that is independent of stratifiable cytoplasmic constituents.

In *C. elegans*, acquisition of daughter cell quality of the whole egg cell was induced by blocking the first cleavage step (Schierenberg, 1988). One centrosome was removed from the zygote by extrusion through a small hole burned with a laser microbeam into the anterior region of the eggshell (Fig. 3b,c). A monopolar spindle was formed in those manipulated eggs and no furrowing occurred. Prior to the second cleavage step the centrosome duplicates and a bipolar spindle is formed. As the further cleavage pattern clearly indicates, the cell no longer has P_0 quality. It either passes through exclusively symmetrical and synchronous divisions which are characteristic for the more proliferative cleavage behavior of AB descendents in the undisturbed embryo (Fig. 3a), or the cell cleaves like a P_1 blastomere of the normal two-cell stage, generating the somatic founder cells EMS, C, and D in the correct sequence and positioning by asymmetrical cleavage divisions (Fig. 3c). In this case the zygote undergoes the correct bifurcative decision without cellular separation of anterior and posterior cytoplasmic domains.

The question arises whether nuclei and/or centrioles are necessary for correct cellular decision making. This is apparently not the case, as shown by the following experiment (Schierenberg, 1988). The cytoplasm of the zygote can easily be separated into two domains (Fig. 3d). A smaller anterior portion of cytoplasm containing the pronuclei and the centrosomes was separated from a larger posterior cytoplast. After the anterior fragment had passed through the second cleavage cycle, the anucleate cytoplast was provided with one nucleus and one centrosome. The reestablished blastomere, consisting exclusively of posterior cytoplasm cleaves unequally, indicating germline quality, as expected (Fig. 3d). From its further cleavage pattern one has to conclude that the anucleate posterior cytoplast had passed through one cell cycle, thereby changing its P_0 quality into that of a P_1 blastomere.

Prolonged cold treatment (2 to 4°C) of zygotes at the time of pronular conjugation uncouples the cytoplasmic cell cycle from cytokinesis, thus preventing furrowing of the egg cell. After release from cold, karyokinesis occurs, as seen in the formation of two cleavage nuclei. During the following cleavage cycle, both centrosomes duplicate and a tetrapolar spindle forms. Cytokinesis segments the one-cell embryo into four simultaneously arising blastomeres (simultanvierer). We can ask whether the qualities of the blastomeres in simultanvierers resemble those of the normal four-cell stage. From the prelocalization hypothesis of blastomere specification (according to which nondiffusable macromolecules determine the fates of blastomeres to which they are partitioned) and the observed regulative potential of the egg cell and early blastomeres (revealed by cytoplasm extrusion experiments) one would expect that all cell qualities of the normal four-cell stage are generated (Fig. 4b). This, however, is not the case. Two alternatives can be observed: (1) Only one AB-like lineage becomes established (Schierenberg, 1991), as indicated by short cell-cycle periods and synchronous and equal cleavage divisions of all four sisters of the simultanvierer (Fig. 4f); and (2) more than one lineage is established (Moritz, in preparation). Depending on the positioning of the four asters with respect to the anteroposterior egg axis, either one or two blastomeres incorporates posterior cytoplasm (Fig. 4e). Those blastomeres apparently gain a normal P_2 fate. In the case of two P_2 cells one would predict that the other two cells have EMS (and not AB) quality. This has been demonstrated by a lineage-specific cytomarker, characteristic of E-cell progeny, the intestine. These findings argue strongly against the prelocalization concept, because the AB lineage has become extinct in case 2 (Fig. 4b). Instead, the P_0 cell is able to establish the fate of either daughter cell of the normal two-cell stage after cleavage suppression.

Simultanvierers have been observed long ago by Boveri in *Ascaris* eggs that were fertilized by two sperms. These dispermic eggs divide into four cells at the *first* cleavage step. Either one, two, or three of the four blastomeres have P_1 quality, while the remaining cells have AB quality. One type is shown in Fig. 4c. This was one argument that led Boveri (1910b) to abandon specific organ-forming substances for a quantitative change of the milieu as a necessary prerequisite for cellular bifurcation.

G. Appearance of Cell-Autonomous Markers Confirms Cell-Fate Specification in Cleavage-Arrested Embryos

Qualities of blastomeres are defined by ancestry and progeny. As a consequence of invariant cleavage pattern, early blastomeres can be identified by their topology, which results from spindle positioning, cleavage timing, and short-range movements. The question arises whether blastomeres in abnormal position of experimental embryos (as defined by their cleavage potential) are developmentally equivalent to those in the normal embryo. Several cell-autonomous cytomarkers are available that are characteristic of the differentiation potential in certain lineages.

For example, in cleavage-arrested embryos (achieved by continuous treatment with cytochalasin) only the ancestors of the intestinal cells are able to synthesize gut granules. They are the P_1 cell of the arrested two-cell stage, the EMS of the four-cell stage, and the E of the eight-cell stage. While P_1, which is ancestral to several other lineages (Fig. 1), also expressed hypodermal and muscle markers, the E cell produced only gut-specific granules (Cowan and McIntosh, 1985). In addition, it was shown that cell differentiation was an exclusive process: Each single (cleavage-arrested) blastomere expressed only one particular cytospecific marker, although it may have the potency to express several.

These findings are consistent with the concept of autonomous cell specification. Moreover, they have repeatedly provided an argument for the concept of prelocalized cytoplasmic gene regulatory domains which are segregated into the appropriate blastomeres by means of an invariant cleavage geometry (Davidson, 1990). However, several results of cleavage-arrest experiments are inconsistent with a simple prelocalization concept. For example, the "omnipotent" zygote, P_0, expresses only the hypodermal marker after continuous cleavage arrest. The predicted segregation pathway for muscle determinants is $P_0 \rightarrow P_1 \rightarrow$ EMS + $P_2 \rightarrow$ MS, C + P_3. However, an arrested EMS is unable to express muscle differentiation potential, but in its MS daughter this potential reappears (Cowan and McIntosh, 1985). Hence this differentiation potential is not segregated in a simple linear fashion. A polar cleavage division of EMS appears to be necessary to reactivate the muscle-specific potential in the larger cell daughter MS and establish gut-specific potential in the smaller daughter E. Transverse, nonpolar spindle orientation in EMS (and hence equal division) results in loss of E quality. It remains to be shown that after induced symmetrical cleavage of EMS, both daughters display MS fate (i.e., lineage duplication).

As discussed earlier, each lineage displays a unique cell division rhythm. In cleavage-arrested P_0 cells, or P_0 cytoplasts, the cell-cycle oscillations are like those of the AB lineage in the undisturbed embryo (Schierenberg and Wood, 1985). Since alterations in cell cycle periods correlate with certain determinative events, it becomes clear why a cleavage-arrested P_0 can no longer follow several differentiation pathways. Instead, it executes one of the autonomously specified ectodermal programs (e.g., hypodermal cytodifferentiation). In P_1 cells of cleavage-blocked two-cell stages, the cell-cycle periods are initially shorter than those of the germline lineage and closer to those of EMS, E, and MS.

This cycle timing corresponds with the respective different somatic pathways which have been observed in cleavage-arrested P_1 cells.

One of the myogenic regulatory genes coding for MyoD, which belongs to the group of HLH (helix–loop–helix) proteins and binds to defined regulatory regions of muscle-specific genes has been cloned from *C. elegans* (Krause et al., 1990). The nematode gene product, Ce MyoD, is a nuclear protein that is exclusively expressed in body wall muscle cells. A Ce MyoD-β-galactosidase fusion reporter gene was expressed in all lineages that establish body wall muscle cells. Gene activity was also observed as early as in all four MS descendents of the 28-cell embryo, although none of these blastomeres is exclusively ancestral to body muscle cells. However, thereafter the MyoD gene activity dropped to undetectable levels in the MS progeny and reappeared in those MS descendants which are finally committed to body muscle. Therefore, this histospecific regulatory gene, which belongs to an auto- and cross-regulatory set of genes, can be prematurely expressed in progenitors, suppressed in progeny cells, and finally reappear in the terminally differentiating cells. Hence conditions in lineages may arise that destroy a network of activated regulatory genes. How MyoD expression is triggered by further upstream regulators remains to be shown.

After application of aphidicolin (a DNA polymerase α inhibitor) P_1 and EMS (of arrested two- and four-cell stages, respectively) never synthesize gut granules (Edgar and McGhee, 1988). At the eight-cell stage expression of gut granules by the intestinal founder E becomes aphidicolin resistant. However, as seen from the low frequency of gut differentiation, E cells shortly after their birth are highly aphidicolin sensitive, while E cells late in the cell cycle are mostly aphidicolin resistant with respect to their cytodifferentiation potential. These findings indicate that the embryo must pass through four cell cycles to establish E quality and that this differentiation potential matures early in the E cell cycle.

H. Cytoplasm of a Somatic Cell, Transferred to a Germline Cell, Destroys Germline Quality

Cold treatment of polarized zygotes for a period that still allows the first cleavage to occur, frequently results in altered spindle positioning and exovate formation prior to furrowing (K. B. Moritz, in preparation). Those exovates arise from the anterior hemisphere, and they are constantly resorbed by the blastomeres a few minutes after cleavage. Thereby cytoplasm transfer has been achieved. In one type of transfer a trefoil configuration arises with nuclei only in the two posterior blastomeres. The anterior anucleated cytoplast fuses with one of the posterior blastomeres. The fusion product behaves like a normal AB cell, and those embryos develop normally. In a second type of transfer, anterior cytoplasm is later resorbed by the smaller P_1 cell. It produces cell sisters of equal size in which germline quality is always lost, as revealed by equal, synchronous divisions characteristic of the AB lineage.

Laser-induced cell fusion experiments (Schierenberg, 1984; Schierenberg and Wood, 1985) done at several stages gave essentially the same results: Somatic cytoplasm, transferred to a germline cell, causes shorter cycle periods of the recipient and, finally, loss of germline quality. Cell fragments and cytoplasts display the same period as an undisturbed cell of the same quality (Fig. 2c). Therefore, cell-cycle periods are independent of cell size. Consequently, the cell-cycle oscillatory system appears to be controlled by the concentration of a factor that is released in the cytoplasm at the end of a cell cycle. Different concentrations of this factor may cause different cell-cycle periods and cell fates. Specifically, during the cell polarization process, which results in segregative cleavage, conditions should arise which establish different concentrations of that factor in the daughter cells. One can speculate that this factor causes the crisis required for a bifurcation event in the segregative cell cycle.

III. POLARIZATION GENERATES THE CONDITIONS FOR CELLULAR BIFURCATIONS

Experiments on early blastomere determination have raised three fundamental questions: How does the fertilized egg become asymmetrically organized (i.e., polarized)? How does the spindle become oriented parallel to the polarity axis of the cell? What primary signal triggers commitment of the cell daughters to different fates? None of these questions has been answered satisfactorily at the molecular level.

Asymmetrical positioning of the nucleus at the anterior pole in the mature oocyte indicates a weak polarity. Sperm entry at the posterior pole induces activation of the oocyte. Paternal-effect mutants suggest that sperm functions are essential for egg activation, polarization, and spindle positioning (Hill et al., 1989). Drug experiments with one-cell embryos indicate that the microfilament-motility system participates in the generation of zygote asymmetry. Transient perturbation of the microfilament array during pronuclear migration prevents membrane activities, P granule posterior location, and asymmetrical spindle positioning. Those eggs produce blastomeres of altered identity by an equal first division (Hill and Strome, 1990).

Positioning of the contractile ring, which determines the plane of cleavage, is dictated by spindle microtubules, as demonstrated by orientation of the cleavage furrows in a tetrapolar mitosis (rather than the normal bipolar) of simultanvierers. The graded defects of maternal-effect mutant phenotypes (which perturb the generation of asymmetry) affecting multiple lineages (Kemphues et al., 1988; Schnabel and Schnabel, 1990) are more compatible with the concept of cell-fate specification by quantitative differences in the level of a signal than by segregating "determinants," which should affect certain lineages only. In summary, the polar organization of the zygote and germline blastomeres is sufficient to induce unequal cleavage as a consequence of polar spindle orientation and, concommitantly, cellular segregation of developmental potential to the arising cell daughters.

IV. A MODEL OF BIFURCATIVE CELL-FATE SPECIFICATION

Most blastomere manipulation experiments suggest that cell-fate specification is cell-cycle dependent and occurs as a bifurcative process following an intracellular signaling cascade. It is well known that activated phospholipase C (PLC) hydrolyzes inositol phospholipids to generate inositol triphosphates (IP_3), which serve as an intracellular messenger. IP_3 activates the release of calcium ions (Ca^{2+}) from internal stores. Information is encoded in the magnitude, frequency, and spatial organization of changes in the concentration of cytoplasmic free calcium. There is evidence that PLC can be stimulated by Ca^{2+}. Therefore an initial weak Ca^{2+} activation would amplify itself. The following presents a brief model [supported by molecular data of Jansen and Moritz (1986), Seidl and Moritz (1990), and Moritz (1990)] which involves bifurcative cellular decision making by differential activation in seven steps:

1. We propose that PLC activation is stronger at the anterior pole where the meiotic divisions take place.
2. The localized IP_3-dependent Ca^{2+} release stimulates Ca^{2+}-mediated aggregation processes in the cortex. These aggregations lead to the polar organization of the zygote and a polar transport in anterior–posterior direction. The transport finally causes inactivation of PLC through Ca^{2+}-dependent inhibitors at the posterior pole, whereby the inhibitor coats the phospholipids, thus blocking IP_3 release.
3. The polar Ca^{2+} release results in a higher concentration of bound Ca^{2+} in aggregates in the anterior than in the posterior hemisphere.

4. The migration of the centrosome–nuclear complex to the cell center in the polarized egg is mediated by a higher degree of microtuble disassembly on the cortex near the anterior pole. The competition between the equivalent asters for a preferred microtuble attachment site could be biased in one direction, thus generating a torque that mediates spindle rotation.
5. Different activities regulating microtuble assembly–disassembly on opposite cell poles shift the spindle toward the posterior end of the cell. The asymmetric spindle positioning causes unequal cleavage by contractile ring formation at a subequatorial position.
6. At anaphase a Ca^{2+} burst results transiently in different concentrations of cytoplasmic free Ca^{2+} in the arising daughter cells.
7. The quantitative difference determines different ranges of a Ca^{2+} signal-dependent cascade which leads to the alternative patterns of enzyme activities in the daughter cells.

In summary, the model describes cell-fate specification as a bifurcative process without seeking refuge to localized "determinants."

REFERENCES

Boveri, T. (1899). *Festschrift C. von Kuppfer*, Gustav Fischer, Jena, Germany, pp. 383–340.

Boveri, T. (1910a). *Roux's Arch. Dev. Biol.*, *30*: 101.

Boveri, T. (1910b). *Festschrift R. Hertwig*, Vol. III, Gustav Fischer, Jena, Germany, pp. 133–214.

Cowan, A. E., and McIntosh, J. R. (1985). *Cell*, *41*: 923.

Davidson, E. H. (1990). *Development*, *108*: 365.

Deppe, U., Schierenberg, E., Cole, T., Krieg, C., Schmitt, D., Yoder, B., and von Ehrenstein, G. (1978). *Proc. Natl. Acad. Sci. USA*, *75*: 376.

Edgar, L. G., and McGhee, J. G. (1988). *Cell*, *53*: 589.

Hill, D. P., and Strome, S. (1990). *Development*, *108*: 159.

Hill, D. P., Shakes, D. C., Ward, S., and Strome, S. (1989). *Dev. Biol.*, *136*: 154.

Hörstardius, S. (1973). *Experimental Embryology of Echinoderms*, Clarendon Press, Oxford, pp. 35–40.

Jansen, P., and Moritz, K. B. (1986). *Naturwissenschaften*, *73*: 739.

Kemphues, K. J., Priess, J. R., Morton, D. G., and Cheng, N. (1988). *Cell*, *52*: 311.

Krause, M., Fire, A., Harrison, S. W., Priess, J. R., and Weintraub, H. (1990). *Cell*, *63*: 907.

Laufer, J. S., and von Ehrenstein, G. (1981). *Science*, *211*: 402.

Laufer, J. S., Bazzicalupo, P., and Wood, W. B. (1980). *Cell*, *19*: 569.

Moritz, K. B. (1967a). *Roux's Arch. Dev. Biol.*, *159*: 31.

Moritz, K. B. (1967b). *Roux's Arch. Dev. Biol.*, *159*: 203.
Moritz, K. B. (1990). *Verh. Dtsch. Zool. Ges.*, *83*: 467.
Sauer, H. W. (1980). *Entwicklungsbiologie*, Springer-Verlag, New York.
Schierenberg, E. (1984). *Dev. Biol.*, *101*: 240.
Schierenberg, E. (1987). *Dev. Biol.*, *122*: 452.
Schierenberg, E. (1988). *Roux's Arch. Dev. Biol.*, *197*: 282.
Schierenberg, E. (1991). *Roux's Arch. Dev. Biol.*, *199*: 437.
Schierenberg, E., and Wood, W. B. (1985). *Dev. Biol.*, *107*: 337.
Schnabel, R., and Schnabel, H. (1990). *Development*, *108*: 107.
Seidl, C., and Moritz, K.B. (1990). *Naturwissenschaften*, *77*: 482.
Seidl, C., Bauer, M., and Moritz, K. B. (1988). *Roux's Arch. Dev. Biol.*, *197*: 307.
Strome, S. (1989). *Int. Rev. Cytol.*, *114*: 81.
Strome, S., and Wood, W. B. (1983). *Cell*, *35*: 15.
Sulston, J. E., Schierenberg, E., White, J. G., and Thomson, J. N. (1983). *Dev. Biol.*, *100*: 64.
Teschke, C., Solleder, G., and Moritz, K. B. (1991). *Nucleic Acids Res.*, *19*: 2677.

16
Oscillations, Photoperiodism, and Insect Development

David S. Saunders *University of Edinburgh, Edinburgh, Scotland*

I. INTRODUCTION

Insect postembryonic development is orchestrated by an interacting hormonal system involving precisely timed releases of neuropeptides and feedback loops between ecdysteroids, juvenile hormones, the endocrine glands producing them, and their target tissues (Gilbert et al., 1980; Bollenbacher, 1988). Part of this complex regulatory system is governed by endogenous oscillations interacting with the environment to provide a meaningful temporal organization for development. For example, circadian oscillators within the brain and/or endocrine system are entrained by the light–dark cycle to govern daily periodicities in egg hatch, larval molt, pupation, or pupal eclosion (adult emergence), each controlled by a cerebral clock, nerve activity, and hormone release (Truman and Riddiford, 1970; Saunders, 1982a).

In contrast to overt circadian rhythmicity, photoperiodism—in its strictest sense—involves a response to changes in day length (or more frequently, night length) to produce seasonally appropriate changes in development. In insects, the most frequent responses include the appearance of overwintering or diapause stages, and seasonal morphs, as in aphids. Also controlled by photoperiod, however, may be the number of larval instars, the rate of development, and fecundity (Saunders, 1982a). Like overt rhythms these phenomena are controlled by neural and endocrine events, diapause occurring as an alteration or "block" to molting, metamorphosis, or ovarian maturation.

The photoperiodic time measurement (PPTM) inherent in such phenomena is not obviously of an oscillatory nature. Diapause, for example, may occur only once in an insect's life cycle, when autumnal day lengths fall below a well-defined critical value. Nevertheless, the relationship between PPTM and circadian rhythmicity has been recognized for over 50 years (Bünning, 1936) and is commonly referred to as "Bünning's hypothesis." In this chapter we review current ideas supporting Bünning's hypothesis in insect photoperiodism and outline some of the ways in which circadian and other endogenous oscillations may be involved in diapause induction.

II. THE PHOTOPERIODIC RESPONSE

Photoperiodic regulation of diapause has been recorded in several hundred insect species from about 15 orders (Saunders, 1982a). Within this array there is an enormous variety of responses (Danks, 1987), but for the sake of simplicity, the phenomenon may be reduced to a simple switch from uninterrupted growth and development during the favorable long days (or short nights) of summer, to the onset of diapause as the days shorten (or the nights lengthen) in the fall (Fig. 1).

Although different insect species usually overwinter at a characteristic stage of development, diapause may occur in the egg (= embryo), larva, or pupa, where it is most often expressed as a block to further

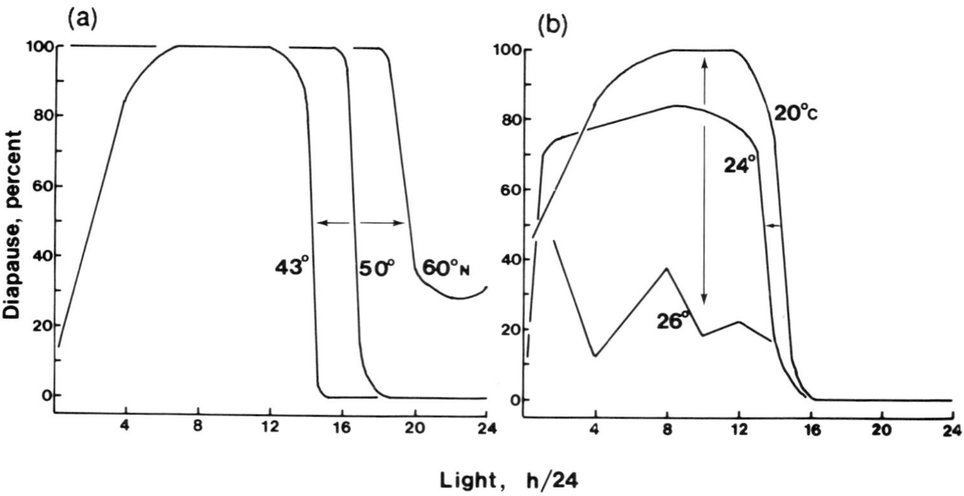

Fig. 1 Photoperiodic response curves for (a) the moth *Acronycta rumicis* from three latitudes, showing an increase in the critical day length in more northerly populations, and (b) the blowfly *Calliphora vicina*, showing a reduction of diapause incidence and a shortening of the critical day length at higher temperature. [(a) After Danilevskii, 1965; (b) after Kenny and Saunders, 1991.]

molting; or it may occur in the adult as a block to gonadal maturation. With some exceptions, larval and pupal diapauses occur in response to falling titers of the ecdysteroid molting hormones, and ovarian diapause to a fall in the juvenile hormones that facilitate uptake of yolk proteins by the developing oocyte (Denlinger, 1985). In both cases, upstream regulation is usually by the brain failing to release the appropriate peptide neurohormone that is required to activate either the prothoracic glands (ecdysteroids) or corpora allata (juvenile hormones). Available evidence suggests that all of the important elements of the photoperiodic mechanism, including photoreceptors, the clock distinguishing short from long nights, the "counter" mechanism accumulating inductive cycles, and the downstream effectors, are contained within the brain and retrocerebral glands (Bowen et al., 1984; Hasegawa and Shimizu, 1987).

The photoperiodic response curve (PPRC) describes the responses of a population of insects to the full range of photoperiods, including continuous darkness (DD) and continuous light (LL) (Fig. 1). The most important aspect of the response is the critical day length (or night length) which separates the diapause and nondiapause pathways. This value varies according to both latitude and temperature. More northerly populations of a particular species have a longer critical day length (Fig. 1a) appropriate because summer day lengths in northerly latitudes are longer, yet autumnal frosts arrive earlier. High temperature tends to lower the overall incidence of diapause and may move the critical day length to shorter values (Fig. 1b).

Apart from the ecologically important critical day length, PPRCs frequently show a fall in diapause in very short photoperiods and in DD. The entire shape of the PPRC thus needs to be explained in physiological terms and by any theory for time measurement.

III. EXPERIMENTAL EVIDENCE FOR THE OSCILLATORY NATURE OF PHOTOPERIODIC TIME MEASUREMENT

The diapause responses of an insect exposed to experimental (or naturally changing) photoperiods in 24-h cycles reveal a simple threshold response (Fig. 1), with no evidence for an underlying oscillator. Such oscillatory responses may be demonstrated, however, by raising insects in abnormal light–dark cycles whose periods may differ from 24 h. The most commonly used of these are (1) cycles in which the light (L) and the dark (D) components are varied independently of each other; and (2) cycles comprising two light pulses defining a "skeleton" photoperiod. These experimental designs have been applied to a wide range of animals and plants with broadly similar results, and these results have been reviewed on numerous occasions (Saunders, 1982a; Follett and Follett, 1982). Here we confine our discussions to some of the most informative experiments: the so-called Nanda–Hamner or resonance protocols (Nanda and Hamner, 1958; Saunders and Lewis, 1987b), and those using two pulses of light per cycle to define skeleton photoperiods, particularly those in the zone of bistability (Pittendrigh, 1966; Saunders, 1975).

A. Nanda–Hamner, or Resonance Experiments

In this type of experiment, different groups of insects are exposed throughout their sensitive periods to a range of LD cycles in which L is held at a constant value per cycle (e.g., at 12 h), but the duration of D is varied widely (say, 6 to 72 h). At the end of the experiment, diapause incidence or its equivalent is assessed in each subset of the experiment, and the results expressed as a function of cycle length (T).

Examples of Nanda–Hamner results are shown in Fig. 2, some negative and some positive. An example of a negative response is that of the aphid *Megoura viciae* (Fig. 2a), in which the long-day (or summer) response, leading to the production of the viviparous and parthenogenetic forms (virginoparae), occurs when D is less than about 9 h, whereas a short-day (or autumnal) response leading to the production of sexual, egg-laying forms (oviparae) occurs when D exceeds that value. In this species ovipara production is more or less constant once D exceeds a critical night length. There is no evidence of an oscillatory response in longer T values, and the night-length measuring system is regarded as a simple, nonoscillatory, "hourglass" (Lees, 1973) (Table 1, model I).

In the flesh fly *Sarcophaga argyrostoma*, which has a pupal diapause, a similar hourglass-like response is seen in Nanda–Hamner experiments performed at low ($<16°C$) temperatures (Fig. 2b). At higher temperature, however, a positive or oscillating response appears with peaks of high diapause incidence occurring at T values close to 24, 48, and 72 h (multiples of the presumed period of the circadian system τ), but troughs of low diapause occurring in T cycles close to 36, 60, and 84 h (Saunders, 1973, 1982b). A similar periodic response is observed with the blow fly *Calliphora vicina*, which has a larval diapause (Kenny and Saunders, 1991).

Figure 2c shows the result of exposing *S. argyrostoma* larvae to regimes containing only 1 h of light per cycle. In this case, a peak of high diapause occurs at T 24 h (close to the period of the presumed circadian oscillator, τ) but also at T 12 h, which is about half the circadian period and where two pulses per 24 h define a skeleton photoperiod of 12 h (see below).

The results of Nanda–Hamner experiments can therefore be variable and complicated. At this juncture, however, suffice it to say that, when positive, the high diapause response is observed when the period of the light cycle (T) is close to multiples or submultiples of the circadian period (τ), but diapause incidence is low when T is far from τ. Superficially, this lends support to Bünning's general hypothesis that photoperiodic time measurement is a function of the circadian system. The way in which circadian oscillations are involved, however, is far from clear. Are they providing the "clockwork" as Bünning originally suggested (Table 1, models IV to XI), or is circadian organization merely modulating the output from some other, perhaps nonoscillatory timer (Table 1, model II or III)?

B. Skeleton Photoperiods: Two Pulses per Cycle

There are several different types of skeleton regimes, but here we will consider two: those comprising two pulses of short and equal duration per cycle, defining a symmetrical skeleton, and those comprising one longer pulse (say, 10 or 12 h), and a shorter pulse (say, 1 h), which scans the night to provide an asymmetrical skeleton or night interruption regime (Saunders, 1982a).

Night interruption experiments have been performed with a number of insect species, and their interpretation has given rise to one of the simplest and most useful models for photoperiodic time measurement, that of external coincidence (Table 1, model IV). In the flesh fly *S. argyrostoma*, as in many other species, the night-scanning shorter pulse picks out two positions in the otherwise diapause-inducing long night, where illumination produces nondiapause or long-day effects (Saunders, 1975) (Fig. 3). The interpretation of these two points (often called points A and B) follows Pittendrigh and Minis (1964), who first observed that such responses were consistent with the entrainment of a circadian oscillation to such two-pulse regimes.

Pittendrigh and Minis (1964) observed that the pupal eclosion rhythm of *Drosophila pseudoobscura*, when entrained to a cycle such as LD 12:12, reached a steady state in which the beginning of the "subjective night" (defined as circadian time, Ct 12) coincided with the end of the 12-h light phase. Supplementary light pulses falling early in the subjective night then caused phase delays ($-\Delta\phi$), whereas those falling in the late subjective night caused phase advances ($+\Delta\phi$), as shown in a typical

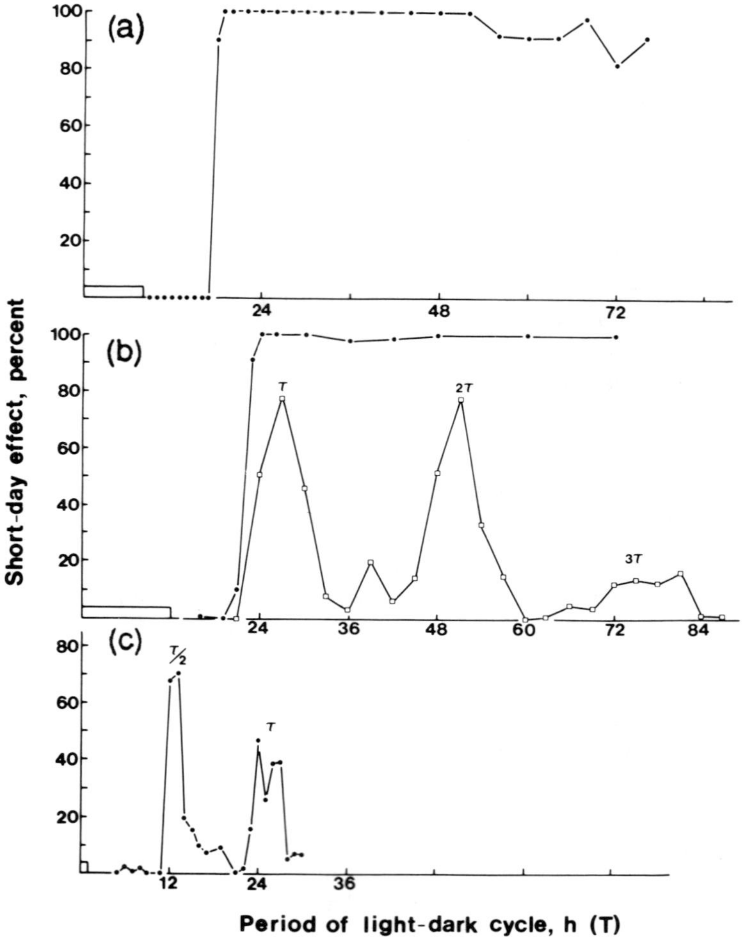
(a)
(b)
(c)
Short-day effect, percent
100
80
60
40
20
24
36
48
60
72
84
12
T
2T
3T
T/2
Period of light-dark cycle, h (T)

phase-response curve (Fig. 4a). The interpretation of night interruption experiments was as follows: A particular light sensitive (photo-inducible) phase (ϕ_i) is postulated to occur at the end of the critical night length—about 9.5 h after the end of the "main" photophase (i.e., at Ct 12 + 9.5 = Ct 21.5, late in the subjective night). When the scanning light pulses fall in the early subjective night (Fig. 4b) they cause increasingly large phase delays until ϕ_i crosses the "dawn" threshold of the following main photophase, giving rise to nondiapause development at point A. Pulses falling in the late subjective night (Fig. 4c), however, cause phase advances until, at point B, the 1 h scanning pulse coincides directly with ϕ_i. Since all photophases longer than about 12 h reset the oscillation to a phase equivalent to Ct 12, measurement of night length in natural light cycles occurs because the dawn threshold tracks back to illuminate ϕ_i (point B) as day length increases. In this model, therefore, point B marks a phase where photochemical events occur that ultimately lead to an endocrine cascade regulating nondiapause development. For this reason the model is called external coincidence (Table 1, model IV) because it involves coincidence between a particular phase of a circadian oscillation and external light.

Results of symmetrical skeletons may also be interpreted in terms of external coincidence and, in doing so, provide some of the strongest evidence we have that diapause responses are regulated by endogenous oscillators.

Working with the overt rhythm controlling pupal eclosion in populations of *D. pseudoobscura*, Pittendrigh (1966) showed that two 15-min

---

Fig. 2 Results of Nanda–Hamner experiments in which groups of insects were exposed to a series of different light–dark cycle lengths (period *T* hours), each containing a constant photophase (L), but a variable duration of darkness (D). (a) Incidence of "autumnal" ovipara producers in the aphid *Megoura viciae* with a constant photophase of 8 h (15°C). A switch to the production of oviparae occurred when the duration of darkness per cycle exceeded 9.5 h; thereafter it remained high. (b) Incidence of pupal diapause in the flesh fly *Sarcophaga argyrostoma* with a constant photophase of 12 h. ●——● at a low temperature, □——□ at a higher temperature. At the higher temperature, peaks of high diapause incidence were observed at circadian intervals (*T* 24, 48, and 72 h). (c) Incidence of diapause in *S. argyrostoma* in cycles with a constant photophase of 1 h. Peaks of high diapause incidence occurred at *T* 24 and *T* 12. The oscillatory profiles in (b) and (c) are regarded as positive Nanda–Hamner responses, indicating a role for circadian oscillations in photoperiodic time measurement (see the text). [(a) After Lees, 1965; (b) data from Saunders, 1986.].

Table 1 Some Models for the Insect Photoperiodic Clock

1. Hourglass clocks (model I)	Lees, 1973
2. Circadian involvement	
A. Nonclock role for the circadian system	
(i) Resonance (model II)	Pittendrigh, 1972
(ii) Hourglass-clock oscillator-counter model (model III)	Vaz Nunes and Veerman, 1982
B. Clock role for the circadian system	Bünning, 1936
(i) External coincidence	
(a) ϕ_i is a pacemaker phase (model IV)	Pittendrigh and Minis, 1964
(b) ϕ_i is a slave phase (model V)	Vaz Nunes et al., 1991
(c) Multioscillator version (model IV)	Saunders, 1982a
(d) Damped oscillator version (model VII)	Lewis and Saunders, 1987
(ii) Internal coincidence	Pittendrigh, 1960
(a) Pacemaker–pacemaker interactions in a complex pacemaker (model VIII)	Pittendrigh and Daan, 1976
(b) Pacemaker–pacemaker interactions in a complex pacemaker uncoupled (model IX)	Tyshchenko, 1966
(c) Pacemaker–slave interactions (model X)	Beck, 1974
(d) Slave–slave interactions (model XI)	Pittendrigh et al., 1984

light pulses less than about 11 h apart (in each cycle) were accepted as a complete photoperiod of the same duration. For example, a light regime of LD $\frac{1}{4}:7\frac{1}{2}:\frac{1}{4}:16$ was was equivalent to LD 8:16, the flies emerging from their puparia in a well-defined peak close to the first 15-min light pulse, or close to the start of a simulated 8-h photophase, as if this signaled dawn (Fig. 5a). Flies exposed to skeletons longer than about 14 h, on the other hand, accepted the *shorter* interval as "day." That is, for an initial LD $\frac{1}{4}:15\frac{1}{2}:\frac{1}{4}:8$ the population underwent a phase jump and then emerged close to the *second* pulse, as if in an 8-h photophase (Fig. 5d).

Skeleton photoperiods between 11 and 14 h, however, were found to be open to two different interpretations, the steady state finally adopted depending on two initial variables: (1) the *phase* (circadian time) which the first pulse in the train illuminated, and (2) the value of the first *interval*

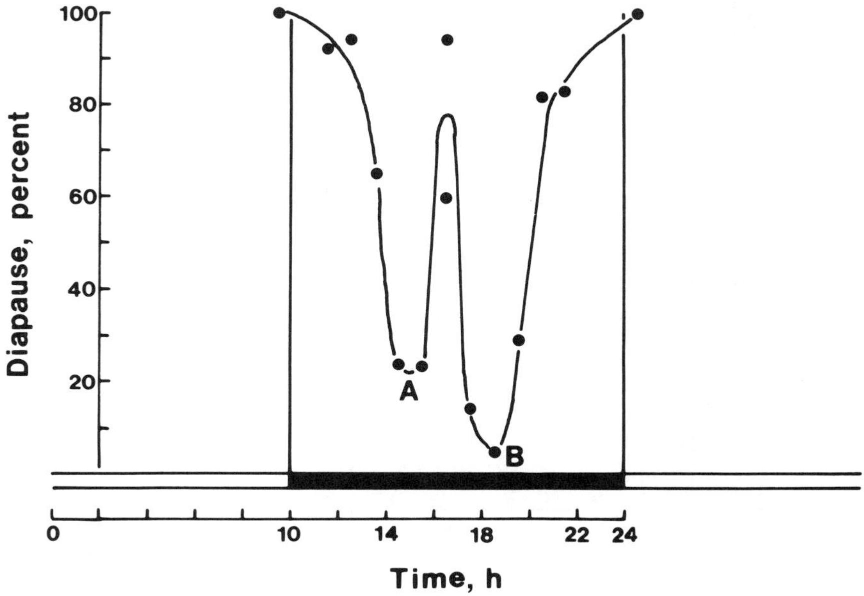

Fig. 3 Results of a "night interruption" experiment with the flesh fly *S. argyrostoma*. Larvae were raised in a short-day (= long night) cycle of LD 10:14, with the 14-h night systematically interrupted in each cycle by a scanning 1-h pulse of light. Plotted points indicate the incidence of pupal diapause obtained with the scanning pulse at different phases in the night. A and B: the early and late points at which the scanning pulse induced a low incidence of diapause resembling a long-day (or short-night) response. (After Saunders, 1975.)

presented. For example, using a skeleton that could be read as either 11 or 13 h (i.e., LD $\frac{1}{4}:10\frac{1}{2}:\frac{1}{4}:13$), the shorter interpretation was adopted as day when the 11-h interval was presented first and the first pulse fell in the late subjective night (between Ct 22 and 03), but the longer interval was accepted when the first pulse fell in the *early* subjective night (between Ct 11 and 16). Similarly, when the pulse train began with the 13-h interval, the longer interpretation was adopted as day when the first pulse fell in the late subjective night (between Ct 22 and 03), but the shorter interpretation when the first pulse fell in the early subjective night (between Ct 11 and 16) (Fig. 5b,c). The region between 11 and 14 h was called the zone of bistability, and the performance of the system in response to the two light pulses was fully predictable in terms of entrainment of a circadian oscillation to the skeleton light regimes.

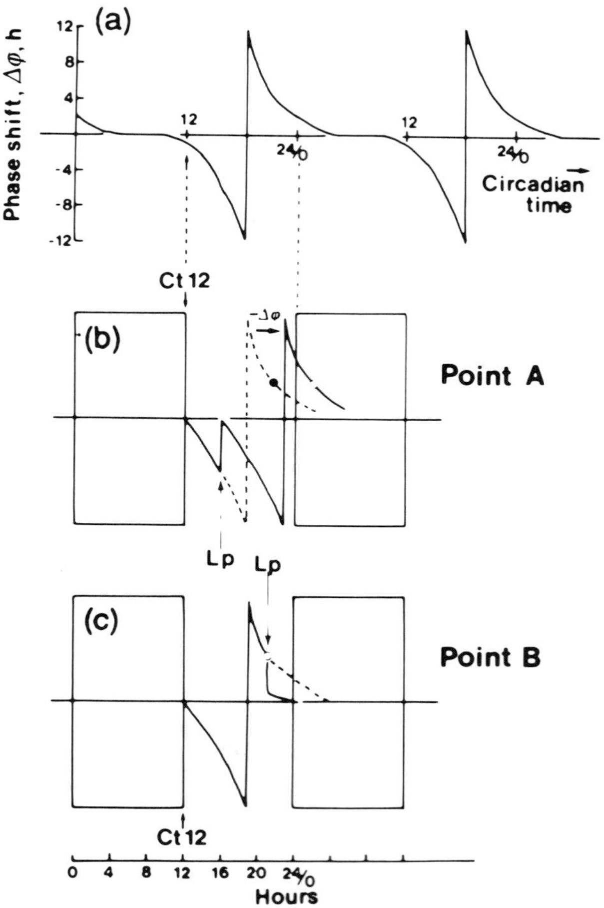

Fig. 4 Explanation of night interruption experiments (as in Fig. 3) in terms of entrainment of a circadian oscillation to light pulses. (a) Representative phase response curve (for the rhythm of pupal eclosion in *Drosophila pseudoobscura*). A single 15-min light pulse causes phase delays ($-\Delta\phi$) when it falls in the early subjective night (Circadian time 12 to 18), but a phase advance ($+\Delta\phi$) when it falls in the late subjective night (Ct 18 to 24). Ct 12 = beginning of subjective night. (b) Derivation of point A in the night interruption experiment. A light pulse (Lp) falling in the early subjective night causes $-\Delta\phi$, delaying a photo-inducible phase (ϕ_i) into the light of the next photophase. (c) Derivation of point B in a night interruption experiment. A light pulse (Lp) falling in the late subjective night eventually coincides directly with the photo-inducible phase (ϕ_i). ●, ϕ_i in the dark; 0, ϕ_i illuminated. Illumination of ϕ_i leads to nondiapause development. (After Pittendrigh and Minis, 1964.)

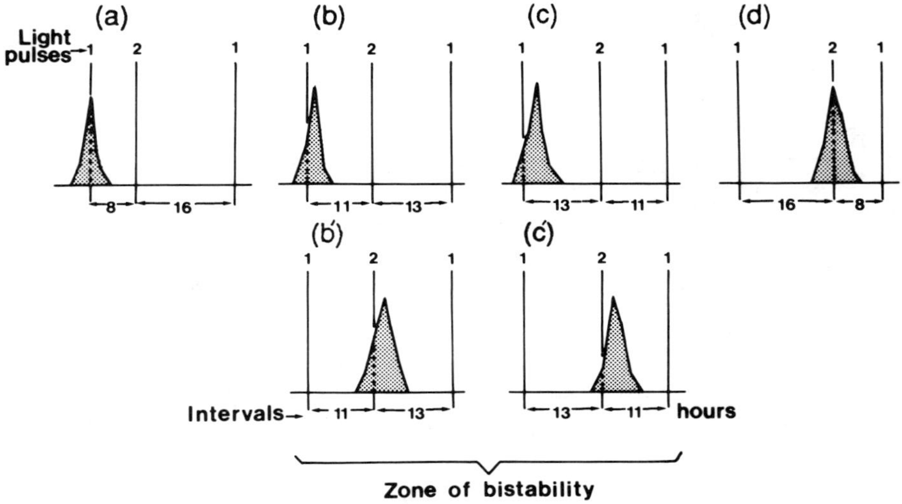

Fig. 5 Zone of bistability, as it appears in an overt circadian rhythm (pupal eclosion of *Drosophila pseudoobscura*) entrained by a symmetrical skeleton photoperiod comprising two short pulses of light per 24 h. (a) Two pulses initially 8 h apart. A unique phase relationship develops irrespective of the circadian phase illuminated by the first pulse, and the flies emerge close to P_1 as if that signalled 'dawn'. (d) Two pulses initially 16 h apart. A similar unique phase relationship develops after a series of transient cycles until the shorter interval (8 h) is accepted as 'day' and the flies emerge close to P_2. (b,c) Two pulses initially 11 and 13 h apart (within the zone of bistability). When the first entraining pulse falls in the late subjective night the flies emerged close to P_1, but when the first entraining pulse fell in the early subjective night (b′ and c′) they emerged close to P_2. (Data after Pittendrigh, 1966.)

Predictions arising from entrainment within the zone of bistability have been extended to photoperiodism using pupal diapause induction in the flesh fly *Sarcophaga argyrostoma* (Saunders, 1975, 1981) and larval diapause in the blowfly *Calliphora vicina* (Vaz Nunes et al., 1990). In *S. argyrostoma*, for example, cultures of larvae were exposed to the skeleton regimes, LD 1:9:1:13 or LD 1:13:1:9, with the first pulse in each case starting at a series of different circadian times (Fig. 6). These light cycles were continued throughout larval development and the resulting pupae examined for diapause (arrested pupal development) or nondiapause (pharate adult development). The rationale behind this experiment was that if the shorter of the two interpretations (i.e., 11 h) was accepted as day, this would be diapause inductive, whereas if the longer interval (i.e.,

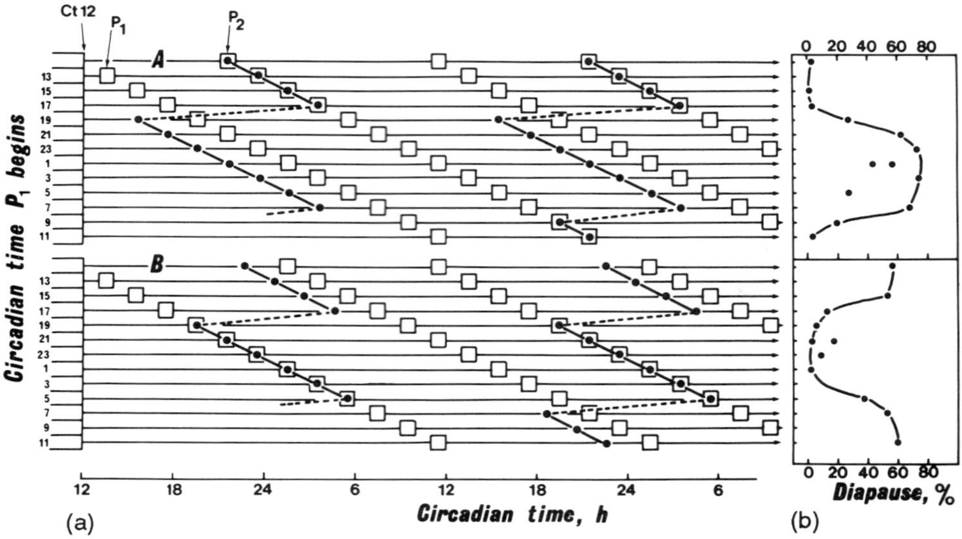

Fig. 6 (a) Zone of bistability phenomenon as applied to photoperiodic induction of pupal diapause in *S. argyrostoma*. Reading from left to right, groups of larvae were transferred from continuous light (LL) to continuous darkness (DD) (= Ct 12), and then exposed to a symmetrical skeleton photoperiod composed of two 1-h light pulses per 24 h, with the first pulse (P_1) in the pulse train starting at different circadian times after the LL/DD transfer. A, cultures exposed to an initial 11-h skeleton (LD 1:9:1:13); B, cultures exposed to an initial 15-h skeleton (LD 1:13:1:9). (b) incidence of pupal diapause observed at the end of the experiment. Note that diapause incidence is low when the photo-inducible phase, ϕ_i (● in the main panels) is illuminated, but diapause incidence is high when ϕ_i falls in the dark. (From Saunders, 1981.)

15 h) was accepted, this would be read as a long-day, or nondiapause, photoperiod (see Fig. 1).

The results showed precisely this. With the shorter interval provided first, long-day responses were obtained when the first pulse in the train fell between Ct 9 and 17, but a short-day response (high diapause) when the first pulse fell between Ct 19 and 7. With the longer interval provided first, a mirror image was obtained (Fig. 6b). The strength of this result was increased by computations based on a 1-h phase response curve for a parallel rhythm in *S. argyrostoma* (pupal eclosion), which predicted that the putative photoinducible phase (ϕ_i, at Ct 21.5) coincided with one of the two 1-h pulses when diapause incidence was low, but fell in the dark when diapause was high (Saunders, 1981). The results were therefore strong evidence for the involvement of circadian oscillations in photoperiodic time measurement, and also open to interpretation in terms of external coincidence.

III. BÜNNING'S HYPOTHESIS AND CIRCADIAN OSCILLATIONS IN PHOTOPERIODIC TIMING: THEORETICAL CONSIDERATIONS

Table 1 lists some of the more important models for photoperiodic time measurement, some of which (hourglass and external coincidence) have already been outlined. Some of the models, particularly those suggesting a nonclock role for the circadian system (models II and III) and the various forms of internal coincidence (models VIII to XI), will receive little attention here. Internal coincidence models which rely on mutual (i.e., internal) phase relationships within a population of oscillators have been strongly and persuasively advocated by C. S. Pittendrigh and his colleagues (Pittendrigh, 1981; Pittendrigh et al., 1984). In this chapter, in contrast, we deal more fully with the various forms of external coincidence (models IV to VII). Full details of these models cannot be included in a chapter of this length. The reader is therefore encouraged to consult the various primary publications (Table 1).

Two observations should be made at this point. One is that the various models, each proposed by authors to account for data sets obtained for their particular research animal, are not necessarily mutually exclusive. Hourglass models, for example, have much in common with external coincidence (Saunders, 1981). The second is that the most profound difference between the various propositions probably concerns the reality of the photo-inducible phase (ϕ_i): Does light have the single role of entrainment, as in internal coincidence, or a dual role, entrainment and photoinduction, as in external coincidence? So far, unequivocal experimental distinctions between these two possibilities have been hard to establish (Saunders, 1978, 1984).

Recent photoperiodic clock models have suggested that the circadian components are damping oscillators (Lewis and Saunders, 1987; Saunders and Lewis, 1987a,b; Vaz Nunes and Hardie, 1987) and that the clock may consist of a hierarchically arranged pacemaker and damping slave (Vaz Nunes et al., 1991; S. W. Gillanders and D. S. Saunders, in preparation). These models recognize the undoubted complexity of the insect circadian system, are explicit formulations consisting of coupled feedback oscillations, and are readily testable by computer simulation.

A damped circadian oscillator was introduced into clock models by Lewis and Saunders (1987) to account for some of the characteristic shapes of insect PPRCs, particularly the drop in diapause incidence at the left-hand side of the curve (Fig. 1), a feature difficult to explain by the original version of external coincidence (Saunders, 1981). The oscillation in Lewis and Saunders' model is a simple feedback control system involving the interaction of a temperature-dependent synthesis of a hypothetical oscillating chemical (c) and its temperature-independent loss. A time delay (TD) in the feedback system causes c to oscillate, and the synthesis rate (SR) of c controls the oscillation's amplitude and persistence, a high value of SR leading to a self-sustained oscillation, and a low value of SR leading to a damping oscillation. Light pulses entrain the period of the oscillation to that of the light cycle, as in all circadian rhythms, and also maintain the oscillation in high amplitude by preventing damping.

Diapause regulation by this oscillation is by a simple external coincidence mechanism in which ϕ_i falling in the dark of successive long nights (= short days) leads to a high incidence of diapause, whereas successive illuminations of ϕ_i lead to nondiapause development. In a train of "strong" light pulses the oscillation is maintained above threshold but in very short pulses, or in DD, the oscillation damps below threshold, ϕ_i is no longer apparent, and diapause incidence remains low (as in Fig. 1). The model also simulates Nanda–Hamner and night-interruption experiments. In particular it shows that a very low value of SR giving rise to an oscillation that damps below threshold within one cycle simulates those negative Nanda–Hamner results obtained with insects such as *Megoura viciae* (Fig. 2a) which are thought to have an hourglass clock. As Bünning himself suggested in 1969, therefore, an hourglass clock may merely represent a heavily damped circadian oscillator. Experimental evidence which is consistent with oscillator damping in photoperiodic phenomena has been obtained for flies (Saunders and Lewis, 1988; Vaz Nunes et al., 1990).

Where are we, then, after more than three decades of experimentation and model building? If hourglasses are regarded as heavily damped circadian oscillators (Bünning, 1969; Lewis and Saunders, 1987), all photoperiodic clocks may be regarded as devices based, one way or another, on the circadian system. With that in mind, the extraordinary perception shown by Erwin Bünning in the 1930s is largely true, with the apparent diversity of photoperiodic clocks being only a product of species diversity and natural selection.

According to Pavlidis (1971), a model is a hypothesis that has two essential properties. It must account for available experimental data, and it must be able to predict further results under new conditions. Ideally, it also ought to be capable of suggesting the nature of the physiological components involved. The first two requirements are met by most of the models listed in Table 1. The recent coupled oscillator model incorporating a pacemaker and a slave (Vaz Nunes et al., 1991) contains simple components (feedback oscillations, time delays, etc.) which might also encourage component identification.

IV. CIRCADIAN OSCILLATIONS IN GENE CONTROL AND ENDOCRINOLOGY: OPENING THE BLACK BOX?

Further progress in photoperiodism must now depend on determining concrete physiological components whose properties account for observed photoperiodic data and fit oscillatory models. Particularly important questions are: What is (are) the exact (i.e., cellular) location(s) of the photoperiodic mechanism(s) and associated photoreceptors? How complex is the mechanism (number of oscillations)? Are separate, but coupled, pacemakers and slaves involved? What is the nature of these oscillations? What is the nature of the couplings? Does light have a photo-inductive role (as in external coincidence) in addition to an entraining role? How does the accumulated experience of short/long days eventually regulate the release/retention of the neuropeptides controlling insect development? Few answers, let alone firm ones, are currently available. We will concentrate, therefore, on two: the photo-inducible phase and clock complexity.

The photo-inducible phase (ϕ_i) is one of the characteristic features of the external coincidence model (Fig. 4), yet, until recently, there has been no direct evidence for its existence. Lavialle and Dumortier (1990) have now shown that cytochrome oxidase activity in the larval optic centers of *Pieris brassicae* (reflecting neuronal activity) was greater when light pulses fell at points A and B than at a point in the middle of the night. Cytochrome oxidase activity at B, moreover, was significantly greater following a light pulse than without one. Although the authors are quick to point out that this observation might also be interpreted in terms of internal coincidence, the observation is also the first to describe physiological events occurring after illumination of a phase equivalent to ϕ_i, which might be part of the photoperiodic mechanism.

As far as clock complexity is concerned, there is now ample evidence to suggest that the insect circadian system comprises many oscillators (Saunders, 1982a). In different species and at different levels there may be hierarchically arranged pacemakers and slaves (Pittendrigh, 1981), "complex" pacemakers composed of two or more interacting oscillators (Pittendrigh and Daan, 1976; Konopka et al., 1989), and circadian pacemakers comprising a mutually coupled population of shorter-period or ultradian rhythms (Dowse and Ringo, 1987). The genetic locus *period* in *Drosophila melanogaster* may be part of (or "close to") a central pacemaker controlling the period of both pupal eclosion and adult locomotor activity rhythms (Hall and Rosbash, 1987; Hardin et al., 1990), and candidate slaves may be represented by rhythms in hormone titer produced, for example, by the prothoracic glands (Vafopoulou and Steel, 1991; B. Cymborowski et al., 1991). Somewhere among such a complex of oscillators must lie the photoperiodic phenomenon.

ACKNOWLEDGMENT

This chapter is dedicated to Erwin Bünning, who died in October 1990.

LIST OF SYMBOLS

L	duration of light, in hours
D	duration of dark, in hours
DD	continuous darkness
LL	continuous light
LD	light–dark cycle, in hours (e.g., LD 12:12 is a cycle of 12 h of light and 12 h of darkness)
PPRC	photoperiodic response curve
PPTM	photoperiodic time measurement
τ (tau)	the period of the unentrained (free running) circadian oscillation, in hours
T	period of the light–dark cycle (LD), in hours
Ct	circadian time, each circadian hour = $\tau/24$; by definition (Pittendrigh, 1966) Ct 12 is the phase of the oscillation at the beginning of the subjective night, or at transfer from extended light into darkness
$-\Delta\phi$, $+\Delta\phi$	delay and advance phase shifts of a circadian oscillation in response to single light pulses falling at different phases (Ct times)
ϕ_i	photo-inducible phase, a component of the external coincidence model of the photoperiodic clock (Pittendrigh, 1966)

REFERENCES

Beck, S. D. (1974). *J. Comp. Physiol.*, *90*: 275.
Bollenbacher, W. E. (1988). *J. Insect Physiol.*, *34*: 941.

Bowen, M. F., Saunders, D. S., Bollenbacher, W. E., and Gilbert, L. I. (1984). *Proc. Natl. Acad. Sci. USA*, *81*: 5881.
Bünning, E. (1936). *Ber. Dtsch. Bot. Ges.*, *54*: 590.
Bünning, E. (1969). *Photochem. Photobiol.*, *9*: 219.
Cymborowski, B., Muszynska-Pytel, M., Porcheron, P., and Cassier, P. (1991). *J. Insect Physiol.*, *37*: 35.
Danilevskii, A. S. (1965). *Photoperiodism and Seasonal Development of Insects*, 1st English Edition, Oliver & Boyd, Edinburgh.
Danks, H. V. (1987). *Insect Dormancy: An Ecological Perspective*, Biological Survey of Canada (Terrestrial Arthropods).
Denlinger, D. L. (1985). In *Comparative Insect Physiology, Biochemistry and Pharmacology*, Vol. 8, (G. A. Kerkut and L. I. Gilbert, eds.), Pergamon Press, Oxford, p. 353.
Dowse, H. B., and Ringo, J. M. (1987). *J. Biol. Rhythms*, *2*: 65.
Follett, B. K., and Follett, D. E. (1981). *Biological Clocks in Seasonal Reproductive Cycles*, John Wright, Bristol, England.
Gilbert, L. I., Bollenbacher, W. E., and Granger, N. A. (1980). *Annu. Rev. Physiol.*, *42*: 493.
Hall, J. C., and Rosbash, M. (1987). *J. Biol. Rhythms*, *2*: 153.
Hardin, P. E., Hall, J. C., and Rosbash, M. (1990). *Nature*, *343*: 536.
Hasegawa, K., and Shimizu, I. (1987). *J. Insect Physiol.*, *33*: 959.
Kenny, N. A. P., and Saunders, D. S. (1991). *J. Biol. Rhythms*, *6*: 217.
Konopka, R. J., Pittendrigh, C. S., and Orr, D. (1989). *J. Neurogenet.*, *6*: 1.
Lavialle, M., and Dumortier, B. (1990). *J. Comp. Physiol.*, *166*: 785.
Lees, A. D. (1965). In *Circadian Clocks* (J. Aschoff, ed.), North-Holland, Amsterdam, p. 351.
Lees, A. D. (1973). *J. Insect Physiol.*, *19*: 2279.
Lewis, R. D., and Saunders, D. S. (1987). *J. Theor. Biol.*, *128*: 47.
Nanda, K. K., and Hamner, K. C. (1958). *Bot. Gaz.*, *120*: 14.
Pavlidis, T. (1971). In *Biochronometry* (M. Menaker, ed.), National Academy of Sciences, Washington, D.C., p. 110.
Pittendrigh, C. S. (1960). *Cold Spring Harbor Symp. Quant. Biol.*, *25*: 159.
Pittendrigh, C. S. (1966). *Z. Pflanzenphysiol.*, *54*: 275.
Pittendrigh, C. S. (1972). *Proc. Natl. Acad. Sci. USA*, *69*: 2734.
Pittendrigh, C. S. (1981). In *Biological Clocks in Seasonal Reproductive Cycles* (B. K. Follett and D. E. Follett, eds.), John Wright, Bristol, England, p. 1.
Pittendrigh, C. S., and Daan, S. (1976). *J. Comp. Physiol.*, *106*: 333.
Pittendrigh, C. S., and Minis, D. H. (1964). *Am. Nat.*, *98*: 261.
Pittendrigh, C. S., Elliott, J., and Takamura, T. (1984). In *Photoperiodic Regulation of Insect and Molluscan Hormones* (R. Porter and G. M. Collins, eds.), Ciba Foundation Symposia 104, Pitman, London, p. 26.
Saunders, D. S. (1973). *J. Insect. Physiol.*, *19*: 1941.
Saunders, D. S. (1975). *J. Comp. Physiol.*, *97*: 97.
Saunders, D. S. (1978). *J. Comp. Physiol.*, *127*: 197.

Saunders, D. S. (1981). In *Biological Clocks in Seasonal Reproductive Cycles* (B. K. Follett and D. E. Follett, eds.), John Wright, Bristol, England, p. 67.
Saunders, D. S. (1982a). *Insect Clocks*, 2nd ed., Pergamon Press, Oxford.
Saunders, D. S. (1982b). *J. Insect Physiol.*, *28*: 305.
Saunders, D. S. (1984). *J. Comp. Physiol.*, *154*: 789.
Saunders, D. S. (1986). *Chronobiol. Int.*, *3*: 71.
Saunders, D. S., and Lewis, R. D. (1987a). *J. Theor. Biol.*, *128*: 47.
Saunders, D. S., and Lewis, R. D. (1987b). *J. Theor. Biol.*, *128*: 73.
Saunders, D. S., and Lewis, R. D. (1988). *J. Comp. Physiol.*, *163*: 365.
Truman, J. W., and Riddiford, L. (1970). *Science*, *167*: 1624.
Tyshchenko, V. P. (1966). *Zh. Obshch. Biol.*, *27*: 209 (in Russian).
Vafopoulou, X., and Steel, C. G. H. (1991). *Gen. Comp. Endocrinol.*, *83*: 27.
Vaz Nunes, M., and Hardie, J. (1987). *J. Insect Physiol.*, *33*: 831.
Vaz Nunes, M., and Veerman, A. (1982). *J. Insect Physiol.*, *28*: 1041.
Vaz Nunes, M., Kenny, N. A. P., and Saunders, D. S. (1990). *J. Insect Physiol.*, *36*: 61.
Vaz Nunes, M., Lewis, R. D., and Saunders, D. S. (1991). *J. Theor. Biol.*, *152*: 287.

IIIB
Sequential Formation of Periodic Patterns

17
Concentric Rings in *Escherichia coli* Colonies

James A. Shapiro *University of Chicago, Chicago, Illinois*

I. INTRODUCTION: THE SIGNIFICANCE OF BACTERIAL COLONY MORPHOLOGY

Bacterial colonies are easily studied biological self-organizing systems. Bacterial taxonomists and clinical microbiologists make daily use of the reproducibility and specificity of colony morphologies in identifying cultures. Since each taxonomic group displays a characteristic colonial shape under standard conditions, hereditary components must operate in controlling morphogenesis. The bacteria in colonies proliferate under very crowded conditions, and colony morphology responds to alterations of the growth medium (Shapiro, 1984, 1986). In the earliest stages of *E. coli* colony development, the bacterial cells can be very tightly packed, and nearby cells actually grow toward each other to maximize cell-to-cell contact (Shapiro and Hsu, 1989). Nonetheless, the intimately associated bacteria in these young microcolonies grow and divide just as rapidly as well-isolated cells in a liquid medium (Fig. 1). In effect, each colony structure represents a specific adaptation to a particular set of surface growth conditions (Legroux and Magrou, 1920). In certain natural situations, we understand the functional significance of specific colonial morphologies. One of the most dramatic examples is the trapping of cyanobacterial prey in ponds within spherical colonies of the microbial predator *Myxococcus xanthus* (Burnham et al., 1981).

Bacterial colonies on agar plates in the laboratory can assume several possible geometric forms (Carlile, 1976; Matshushita, 1989). This means that the geometry of bacterial growth is not just a matter of statistics—the accumulation of very many very small objects (the cells) on an artificially smooth surface—but involves specific interactions between the growing

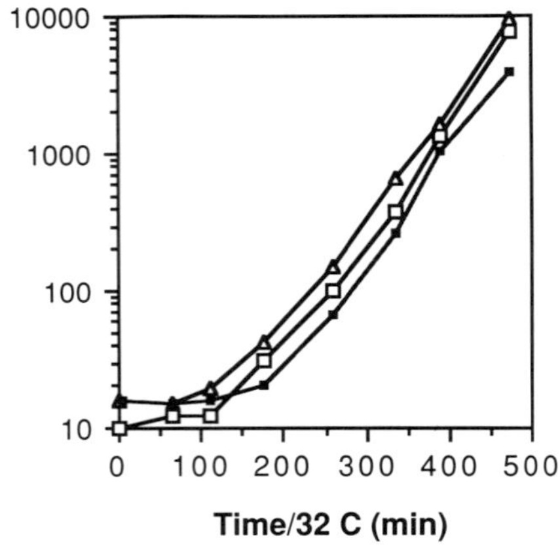

Fig. 1 Growth curves of *E. coli* MS2168 on liquid (open squares and triangles) and solid (closed squares) tryptone-yeast extract medium. Equal aliquots of a culture were inoculated into 5 mL of liquid medium (duplicate tubes) and onto a series of 1.5% agar petri dishes. All inocula were incubated at 32°C, the liquid cultures being aerated on an inclined roller drum. At intervals, samples from the liquid cultures were diluted and plated for colony-forming units (CFUs), while the bacteria on one of the petri dishes were washed off in 5 mL of liquid medium and similarly diluted and plated. The bacteria on solid medium took longer to begin dividing (*E. coli* cells grow larger in the first couple of hours on solid medium) but then continued multiplication at the same rate as the cells in liquid medium, even when there were over 100 cells in each microcolony.

cells (Rosenberg, 1984; Shimkets, 1990; Kim and Kaiser, 1990; Shapiro and Hsu, 1989). Many species produce circularly symmetrical colonies, but even these highly regular forms display considerable structural differentiation and provide evidence for biological regulation. For example, the expansive swarm colonies of *Proteus mirabilis* have beautiful concentric terracing patterns, resulting from repeated cycles of migration and multiplication. Nonetheless, *Proteus* mutants can be isolated which have lost circular symmetry and grow in branching patterns, thus showing that circularity resulted from the operation of control systems at work in morphogenesis of the nonmutant strain (Shapiro and Trubatch, 1991). This mutant also sent out its initial growth extensions along stress lines in the agar surface, thereby demonstrating that the surface was not really as smooth as might be thought from casual inspection.

II. SECTORS AND RINGS: CLONAL AND NONCLONAL PATTERNS

In this chapter we focus on certain aspects of pattern formation in colonies of the standard laboratory strain *E. coli* K-12. The heuristic value of *E. coli* is that it has long been accepted as the prototypical "single-celled" organism. Consequently, morphogenetic controls at work in this species will be seen as reflecting a general ability of bacteria to organize themselves into higher-order communities and will not be viewed as some exceptional adaptive specialization. *E. coli* colonies have a basically circular symmetry with differentiation into concentric ridges and terraces (Fig. 2). In *E. coli* on normal agar plates, expansion occurs by cell growth and division at a zone near the edge (Pirt, 1967; Shapiro, 1984).

Scanning electron microscopy of developing *E. coli* colonies has revealed various concentric zones that are composed of differentiated cell populations, each one having its own characteristic array of cell morphologies and patterns of multicellular aggregation (Shapiro, 1987). The differentiated concentric populations change over time and may be thought of as distinct tissues that appear sequentially during development. The tissue concept implies that each concentric zone also has its own composition of cells with distinct patterns of gene expression and biochemical phenotypes. Using genetic engineering methods and vital staining, such distinct biochemical zones can readily be visualized as concentric rings expressing different levels of enzyme activity (Shapiro, 1984).

In the experiments described below, a particular *E. coli* strain was used carrying the *dev1891::lacZ* genetic fusion, whose structure is such that the *lacZ* cistron encoding β-galactosidase was stably connected to regulatory signals between the *bio* and *glnP* loci at about 18 min on the *E. coli* chromosome. The isolation of the fusion has been described (Shapiro and Hsu, 1989). The *dev1891::lacZ* fusion gave rise to a characteristic pattern of light and dark rings in colonies growing on medium containing a chromogenic β-galactosidase substrate (Fig. 3). The ring patterns have been used as reporters for sequential differentiation events in colony development. Note in Fig. 3 that formation of the ring patterns was not disturbed by scratches in the agar surface.

In addition to concentric rings, *E. coli* colonies display radially oriented sectors of various shapes. These sectors can be visualized as distinct wedges of growth emerging from the circular colony, as in some of the colonies in Fig. 2, or as zones expressing new β-galactosidase and growth phenotypes (Fig. 4). Sectors are familiar to bacteriologists, and

Fig. 2 Surface structure of *E. coli* colonies. Each colony developed from a small spot of culture containing about 10^5 bacteria and grew for 5 days at 37°C on minimal salts–glucose–casamino acids agar. Some of the colonies produced sectors expressing novel surface structures that extended beyond the normal colony perimeter. Four cultures were mutants that produced larger colonies than the two parental strain cultures in the upper right. This photograph was made with a 35 mm camera fitted with a 100 mm Macro lens using a fluorescent ceiling fixture as the light source. By placing the petri dish at the appropriate angle to reflect the light from the ceiling fixture into the lens and focusing on the colonies, surface structure could be recorded in great detail.

each is generally assumed to be composed of a clone of related cells descended from a common ancestor that had undergone some kind of hereditary alteration. This assumption can frequently be verified by isolating individual bacteria from a sector and demonstrating that they carry a genetic change from the parent culture. Thus the phenotypic difference of a sector can be attributed to a genetic switch.

The outlines of sectors contain a record of the relative growth of the sector and the parent colony. When the border was straight, the populations inside and outside the sector expanded over the agar at the same rate. When the sector widened, the inside sectorial population expanded more rapidly, and when the sector narrowed, the sectorial population lost ground to the parental population. Examples of both widening and nar-

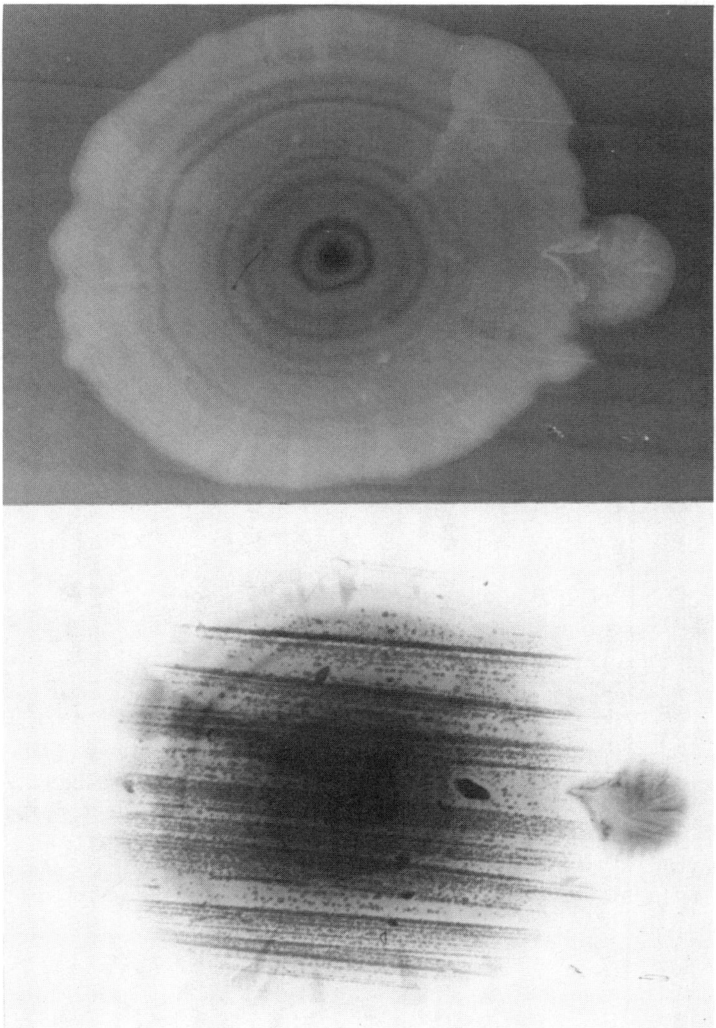

(a)

(b)

Fig. 3 A colony of a *dev1891::lacZ* strain on agar medium containing a chromogenic substrate for β-galactosidase. This colony began development from a single cell and was incubated for 6 days at 37°C. (a) Illuminated with light from the sides and from above; this illumination reveals the concentric rings of enzyme activity as well as scratches in the agar substrate outside the colony. These scratches were produced by streaking the bacterial culture with the tip of a glass pipette. (b) Illuminated with transmitted light from below and shows the lines of granules that formed inside the colony over the scratches.

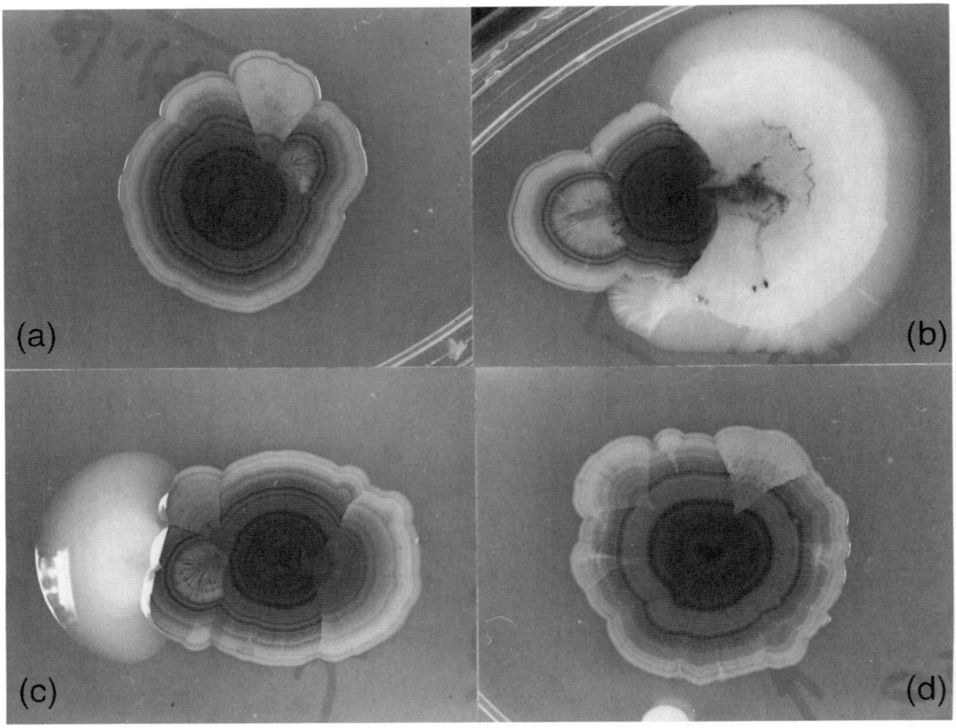

Fig. 4 Colonies with prominent sectors. These colonies began development from toothpick inocula on β-galactosidase indicator agar and were incubated 7 days at 37°C. Note how the concentric rings matched up at the edges of the various sectors. Some sectors only expressed a subset of the concentric rings. Colonies (b) and (c) produced large mucoid sectors that were very white because of excess polysaccharide synthesis. Note the "growth spurt" of the sector at 11 o'clock on colony (d).

rowing are seen, and some sectors have boundaries displaying short bursts of differential expansion. These phenotypes tell us that hereditary factors affect the growth of cell populations on the agar surface in an interesting way: Relative growth of distinct sectorial populations is not always simply faster or slower but is subject to regulation by changes that occur as colony expansion proceeds.

Concentric rings are fundamentally different from sectors. The bacteria in a ring are not related to each other by descent but by position. The cells in a dark ring, for example, are related by descent to the cells in the preceding and succeeding light rings, not to each other. The changes that occur as one ring succeeds another represent transitory phenotypic differentiations rather than permanent genetic switches. The difference between rings and sectors can be seen by isolating bacteria from rings of different phenotypes and showing that they all form colonies of the same parental phenotype. This means that the differentiated bacteria in one ring are either not genetically distinct from the bacteria in another ring, or, if there are genetically distinct bacteria in each ring, that these bacteria have become terminally differentiated and lost the capacity to form colonies when replated (Shapiro and Higgins, 1989). Rings are thus nonclonal populations. The cells in a concentric ring share a common relative position in the history of colony development, and the common phenotype expressed by each ring population must arise from some nongenetic regulatory connection between the component cells. A patterning system must coordinate the periodically changing activities of cells around the entire colony. Such a system could work either by internal timing mechanisms that synchronize cellular activities at specific chronological moments during colony development or though intercellular communication mechanisms that trigger similar responses in cells at corresponding positions around the colony.

One illustration of communication in a concentric ring between cells from different clones is seen in large colonies displaying sectors with dramatically different growth phenotypes like the ones in Fig. 4. Although expression of the ring patterns can change in some sectors, most sectors continue to express some or all of the rings seen in the main body of the colony. Remarkably, the rings line up at the boundaries where distinct sectorial and parental populations meet. Thus, even cells in genetically distinct populations with different growth regimes can be subject to a common periodic regulation affecting gene expression. A similar alignment of concentric rings was observed when colonies of the parental type and of sectorial mutants were inoculated near each other and grew together, as in the experiments described below. Thus ongoing growth of the two populations adjacent to each other is not required for coordination.

III. ALIGNMENT OF RINGS BETWEEN COLONIES GROWING ON DIFFERENT SCHEDULES

One could possibly explain alignment of concentric rings in populations growing contemporaneously with each other by some kind of morphogenetic clock hypothesis according to which cells of a similar age undergo synchronous changes in gene expression. In discussing how to distinguish clocks from ongoing communication as models for control of concentric ring formation, one of our graduate students, Marc Lavenant, came up with the idea of testing what happens when colonies inoculated at different times grow into each other. If alignment resulted from synchronous populations independently following a common chronology, populations out of phase with each other would not have coincident ring patterns. On the other hand, if alignment resulted from ongoing communication, colonies inoculated at different times might come into phase with each other and align their ring patterns. The experiment was performed by inoculating one colony, incubating it for some time, and then taking a toothpick sample and inoculating a transplant a short distance from the edge, followed by further incubation. As a control, the same toothpick was also used to inoculate a transplant on a distant zone of the agar surface. When various intervals between inoculations of adjacent colonies were tested, the result was always the same. The rings in adjacent colonies merged (Fig. 5). The adjacent transplants displayed a different pattern of gene expression from the distant transplants, and these differences appeared in the center of the transplanted colony. Thus it appears that the first colony influenced the adjacent transplant even before the two populations came into contact, presumably via some kind of signal transmissible through the agar medium.

IV. EFFECTS OF PHYSICAL OBSTACLES ON RING PATTERNS

A powerful approach to studying morphogenesis in biological systems has been to disrupt the normal process of development by physical manipulation and then observe how the developing system adjusts to the disruption. While some disruptions are so severe that they cannot be overcome, it has been known for some time that the ability of biological systems to regulate their development and compensate for perturbations is often astonishing (e.g., Driesch, 1901). The formation of symmetrical ring patterns in colonies growing on grooved agar is one example of adjustment to perturbations of the "normal" (i.e., smooth agar) development process (Fig. 3). Another way to perform disruption experiments with *E. coli* colonies is to place obstacles on the agar surface. These obstacles upset the normal sequence of colony expansion and lead to the formation of populations with a distinct growth history. By examining the developmental patterns formed by the perturbed and undisturbed populations, we can begin to understand the relative importance of "hard-wired" processes, which follow automatically from bacterial growth and metabolism,

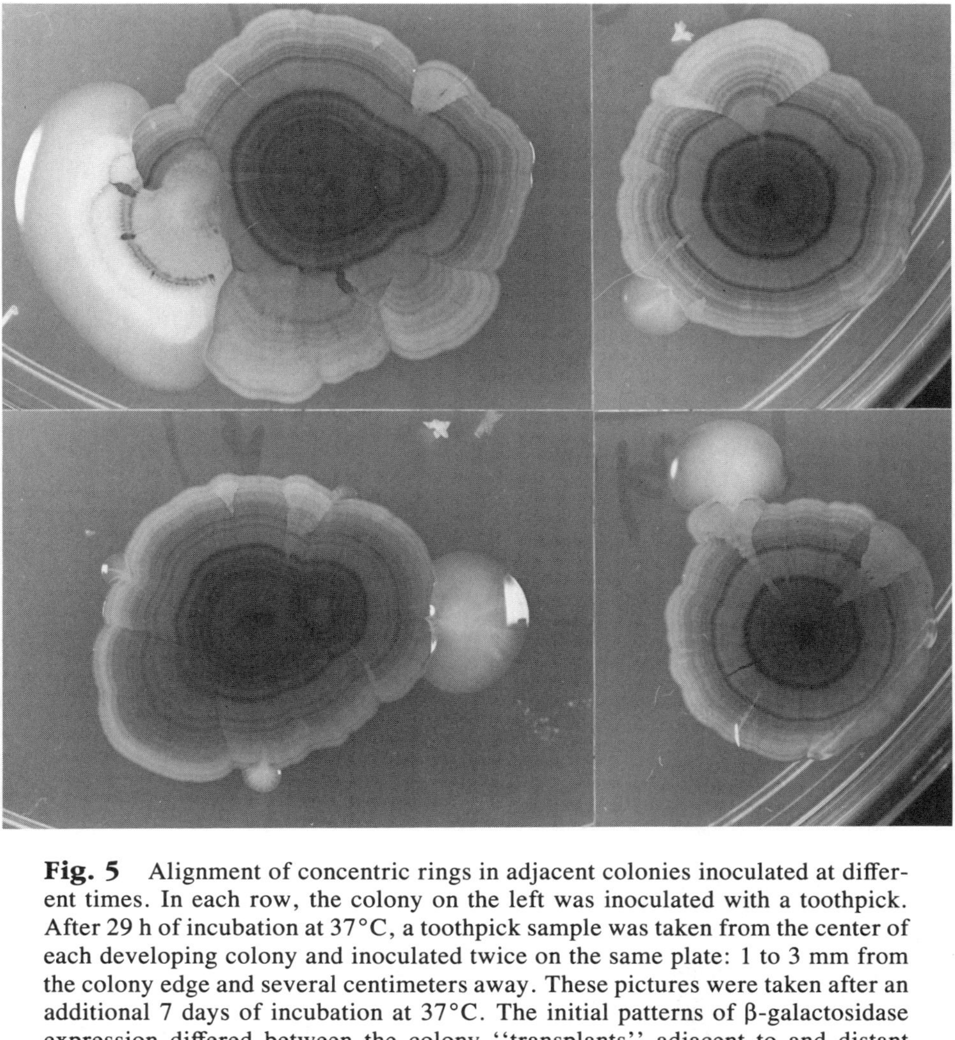

Fig. 5 Alignment of concentric rings in adjacent colonies inoculated at different times. In each row, the colony on the left was inoculated with a toothpick. After 29 h of incubation at 37°C, a toothpick sample was taken from the center of each developing colony and inoculated twice on the same plate: 1 to 3 mm from the colony edge and several centimeters away. These pictures were taken after an additional 7 days of incubation at 37°C. The initial patterns of β-galactosidase expression differed between the colony "transplants" adjacent to and distant from the first colony.

and of regulative processes, which are subject to monitoring and adjustment by feedback systems.

Fibers of glass wool have proved to be very useful obstacles for studying bacterial pattern formation. On the agar surface, the fibers are surrounded by a meniscus of liquid. When the edge of the expanding colony encounters this meniscus, motile bacteria are released and swim around the fibers, giving rise to a bacterial population that coats the fiber. This population has a very different physiological history from the densely packed bacteria that remain at the colony perimeter; its cells have undergone free swimming, settled down on the agar near the fiber, and reorganized to grow around the fiber. The bacteria around the fiber multiply but do not begin to spread over the agar surface for many hours. (*E. coli* populations derived from multicellular inocula need to achieve a certain density before they expand.) Thus if the fiber is relatively short, the spreading edge of the colony engulfs it, and there is no discernible effect on the concentric rings of β-galactosidase expression. However, if the fiber is more than a few millimeters in length, the surrounding population eventually serves as a new source of outward expansion. Figure 6 illustrates two such colonies at 2 and 6 days after inoculation; the colony on the left encountered the glass fibers at day 2, and the colony on the right encountered the fibers at day 4. The concentric ring patterns established themselves around the fibers beginning at the point where the initial contact was made, not at some earlier stage of development. In some way, therefore, the cells received the appropriate cues needed to recommence development at the appropriate stage. When extremely long fibers, 2 cm or more, were used, the populations at the most distal portions of the fibers began development in a different way, expressing higher β-galactosidase levels typical of a fresh inoculum; however, the ring patterns of these distal populations quickly came into register with the ring patterns along the rest of the fiber and around the parent colony (Fig. 7). These patterns indicate that the communication between the main body of the colony and the fiber-coating populations attenuates with distance.

V. DISCUSSION: GLOBAL CONTROL OF CELLULAR EVENTS

Like other self-organizing systems, bacterial colonies display characteristic macroscopic patterns based on the cumulative behavior of microscopic components. Previously published data indicate that the earliest stages of *E. coli* colony development involve cell–cell interactions at distances of a few micrometers (Shapiro and Hsu, 1989). The experiments described in this chapter indicate that *E. coli* in colonies are also subject to long-range controls that coordinate their activities over distances several orders of magnitude greater than the dimensions of an individual cell. These long-range controls could be seen to align the concentric ring patterns in colonies that began growth at different times a few millimeters apart or along a glass fiber several millimeters in length (Figs. 5 and 6). The

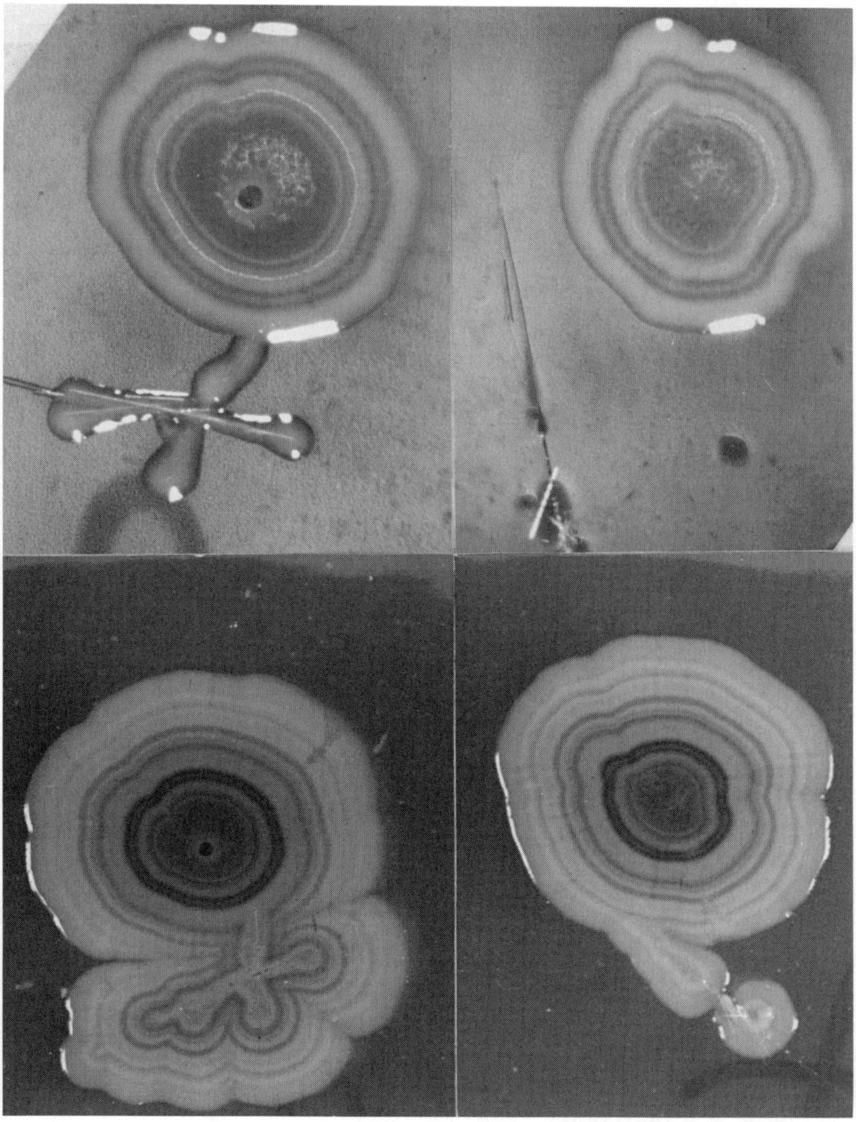

(a)

(b)

Fig. 6 Encounters of colonies with glass fibers. Duplicate colonies were inoculated with drops of a single *dev1891::lacZ* culture. After 1 day of incubation at 37°C, glass fibers were placed on the agar next to the colonies. (a) Photographed on day 2, when the colony on the left had already contacted and coated the fibers; (b) photographed on day 6. The colony on the right only produced the last rings along the fiber.

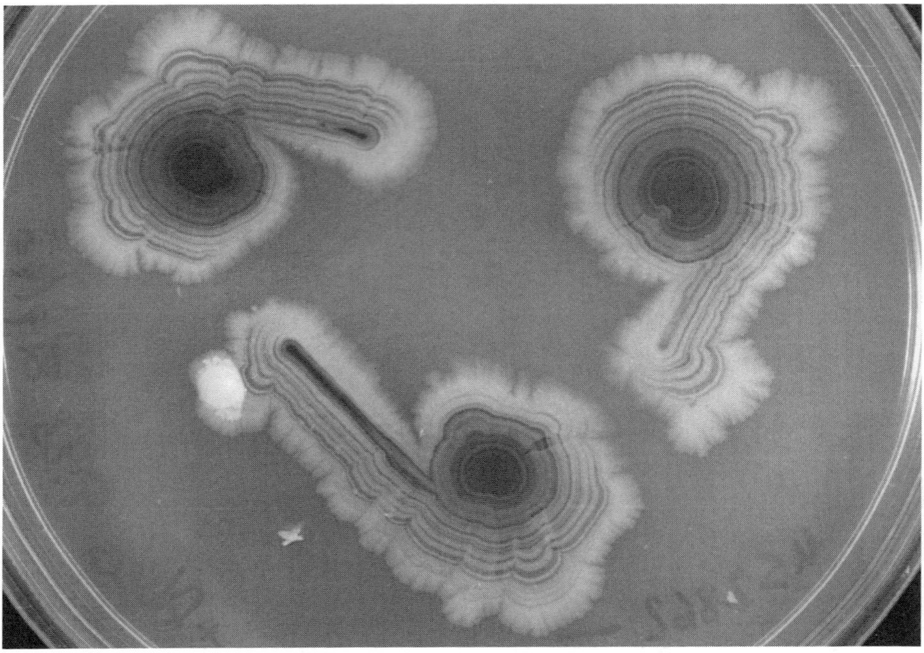

Fig. 7 Colonies after encountering very long fibers. The bacteria farthest out along the fibers began development expressing a different β-galactosidase phenotype from the bacteria closer to the main colony. This photograph also illustrates inhibition of colony spread and ring development toward the center of the dish, where gradients created by the colonies were most intense. Photographed after 14 days of incubation at 37°C. The petri dish was 9 cm in diameter.

observation that the coordination attenuated with distance along a glass fiber (Fig. 7) indicates further that the developing bacterial populations responded to signals emanating from the colonies themselves and not from the general environment (e.g., periodic fluctuations in temperature or other variables). Similar observations have been published on the operation of short- and long-range control systems in bacteria where group motility is important in morphogenesis, such as *Myxococcus xanthus* (e.g., Rosenberg, 1984; Kim and Kaiser, 1990; Shimkets, 1990) and *Proteus mirabilis* (Shapiro and Trubatch, 1991).

A basic challenge in understanding bacterial colony morphogenesis is to work out the relationship between pattern execution and pattern heredity. How does each colony express a geometrically regular pattern despite the microscopic heterogeneity of the growth substrate and despite the inevitable mistakes and disruptions that accompany any natural process? Why does a particular strain or species reproducibly produce a recognizable pattern under defined conditions? Part of the answer clearly lies in the use of the genetic apparatus to encode the molecular tools employed in morphogenesis. The intensive application of molecular biology will test the reductionist hypothesis that comprehensive knowledge of the individual molecules and their mutual interactions can explain the global developmental process. Progress in identifying these tools is accelerating in well-studied bacterial systems, such as *M. xanthus*, where it is easy to isolate mutants whose colonial structures have been altered by mutations eliminating or modifying specific components of the morphogenetic toolbox (Shimkets, 1990).

It is important to remember that certain regulatory phenomena present a special challenge to the reductionist approach. For example, concentric rings align between initially separate colonies (Fig. 5) and between adjacent sectors expressing distinct hereditary growth phenotypes (Fig. 4). These alignments suggest that there is global control imposed on the execution of individual molecular events, such as gene expression, and that independent developing systems are capable of modifying their morphogenetic programs in response to each other, as they were previously observed to do at the microscopic level (Shapiro and Hsu, 1989). Attention focused on sector–sector and colony–colony interactions may thus provide insights into the operation of higher-level control systems difficult to analyze by the reductionist approach. Perhaps the results will fit into more general explanations, and we will come to think of self-organization and pattern formation as expressions of whole-system phenomena rather than as emergent properties of large numbers of atomistic interactions.

ACKNOWLEDGMENTS

I thank Nancy Cole for excellent technical assistance, Marc Lavenant for intriguing discussions of the ring patterns, and Dan Gottschling for reading the manuscript. This research was supported by a grant from the National Science Foundation (DCB-8816274).

REFERENCES

Burnham, J. C., Collart, S. A., and Highison, B. W. (1981). *Arch. Microbiol.*, *129*: 285.

Carlile, M. J. (1976). In *Biology and Systematics of Colonial Organisms* (G. Larwood and B. R. Rosen, eds.), Academic Press, London, p. 3.

Driesch, H. (1901). *Die organischen Regulationen*, Leipzig.

Kim, S. K., and Kaiser, A. D. (1990). *Science*, *249*: 926.

Legroux, R., and Magrou, J. (1920). *Ann. Inst. Pasteur*, *34*: 417.

Matsushita, M. (1989). *J. Phys. Soc. Jpn.*, *58*: 3875.

Pirt, S. J. (1967). *J. Gen. Microbiol.*, *47*: 181.

Rosenberg, E. (1984). *Myxobacteria: Development and Cell Interactions*, Springer-Verlag, New York.

Shapiro, J. A. (1984). *J. Gen. Microbiol.*, *130*: 1169.

Shapiro, J. A. (1986). In *The Bacteria*, Vol. X (J. R. Sokatch, ed.), Academic Press, New York, p. 27.

Shapiro, J. A. (1987). *J. Bacteriol.*, *169*: 142.

Shapiro, J. A., and Higgins, N. P. (1989). *J. Bacteriol.*, *171*: 5975.

Shapiro, J. A., and Hsu, C. (1989). *J. Bacteriol.*, *171*: 5963.

Shapiro, J. A., and Trubatch, D. (1991). *Physica D*, *49*: 214.

Shimkets, L. J. (1990). *Microbiol. Rev.*, *54*: 473.

18
Oscillators and Morphogenesis in Fungi

Stéphan Jerebzoff and Simonne Jerebzoff-Quintin
Paul-Sabatier University of Toulouse, Toulouse, France

I. INTRODUCTION

Among fungi, the most common periodicity involves the development of a series of regularly spaced concentric circles of growth and/or fructification on solid media, which is called zonation. Noncircadian endogenous rhythms of hyphal branching are often observed (well documented in Nguyen Van, 1967; Jerebzoff, 1976; Lysek, 1984). The rhythmic production of fructifications can be expressed as a true circadian rhythm, with a temperature-compensated period and entrainment by light–dark cycles. It can also be exhibited as a noncircadian rhythm, light-dependent or non-light dependent, with a period modulated by temperature and different nutrients (for a review of earlier studies, see Jerebzoff, 1965a) Here we examine the different approaches of the organization of noncircadian oscillating systems, expressed by zonation rhythms of sporulation. The expression and the regulation of sporulation rhythms, the metabolic process involved, and the oscillating function of the asparagine–pyruvate pathway, which controls rhythmicity in some cases, are analyzed using a few experimental models.

II. MANIPULATION OF SPORULATION RHYTHMS BY CHEMICALS

Effects of several sugars, different nitrogen sources, and mineral salts on rhythm expression and period have been observed in free-running conditions.

1. In *Alternaria tenuis* a 50-h rhythm was well expressed when both maltose/phosphate and nitrate/phosphate ratios were low, ≤ 1; it disap-

peared when these ratios reached a value of ≥ 10. On malt-extract agar, the period length was very much dependent on temperature (its Q_{10} value was equal to 0.14). If increased quantities of water-potato extract were added to the malt extract, the Q_{10} value of the period rose and may be equal to 1 (Jerebzoff, 1965a).

2. In *Aspergillus niger* a 300-h rhythm was observed in an agar medium containing glucose. To evoke the rhythm, glucose may be replaced by arabinose, xylose, fructose, mannose, sucrose, or maltose, but not by sorbose, cellobiose, trehalose, galactose, or the galactose-containing disaccharides raffinose and melibiose, despite the positive effects of these compounds on the growth and the sporulation of *A. niger*. Zonation appeared when nitrate, ammonium salts, or urea, but not asparagine, was used as the sole nitrogen source; the level of nitrogen supplied acted on the amplitude of oscillations but not on the period. The rhythm expression and its maintenance depended on the glucose/K^+ balance in the medium; the inducing effect of K^+ could be partially replaced by Rb^+. The amplitude of the oscillations depended on the Mn^{2+} level. Isoleucine induced zonation in *A. niger* specifically by counteracting an excess of K^+ ions. In fact, isoleucamine, *N*-methylisoleucine, leucine, norleucine, valine, norvaline, and isovaline, the precursors of isoleucine, proved to be incapable of duplicating this effect. Otherwise, the inducing effect of isoleucine could be replaced by two synthetic compounds, α-hydroxy-β-methylvalerate and sodium β-methylvalerate (Jerebzoff and Jerebzoff-Quintin, 1980).

3. In *Penicillium expansum* supplied with 2.5 m*M* nitrate, a 25-h rhythm was observed in darkness. The addition of increasing levels of nitrate provoked a more and more accentuated phenomenon of period increase from one cycle to the next (i.e., period instability), which can be corrected by increasing concentrations of the sodium phosphate-citric acid buffer, but not by K^+ ions and/or the phosphate (Jerebzoff et al., 1970).

4. In *Leptosphaeria michotii*, a thiamine-heterotrophic fungus, the same phenomenon of period instability can be observed. In darkness and on glucose-agar, the fungus expressed a 4-day rhythm; the supply of increasing levels of thiamine or/and asparagine caused a rapid period instability. Moreover, when phosphates and Cu^{2+} ions were added to the glucose-agar medium, asparagine had the opposite effect. At 2.6 m*M* asparagine the period was unstable, but the phenomenon of period increase became less and less pronounced as the asparagine level was increased. At a level above 6.6 m*M*, the period was stable and equal to 28 h (Jerebzoff, 1965b). Asparagine could be only partly replaced by aspartate

in its effects on the sporulation rhythm (Jerebzoff-Quintin and Jerebzoff, 1980b). The part played by thiamine indicates a role of oxidative decarboxylation of several α-keto acids (pyruvate, α-oxoglutarate, α-oxo-β-methylvalerate) in the oscillating system of the fungus.

When the level of asparagine was not sufficient to stabilize the period in darkness, continuous dim light (3.7 to 900 nmol $m^{-2} \cdot s$) permitted a stable period, the length of which ranged between 16 and 40 h according to the energy level. Blue light was most effective, but near UV (350 to 400 nm) and wavelengths up to 595 nm had an influence, too. Strong correlations between rhythm regulation and carotenogenesis in *L. michotii* were observed. Continuous dim light, which stabilized the period, increased the synthesis of β-carotene and caused the formation of lycopene. Wavelengths from 400 to 730 nm were acting on carotenogenesis, suggesting the presence of porphyrins as photoreceptors. [$U^{14}C$]asparagine was incorporated into the two carotenes. Carotenoids must be present in relatively strict proportions to obtain a stable sporulation rhythm in darkness or in light; such a balance perhaps acts as a hydrogen pool in redox reactions (Jerebzoff-Quintin and Jerebzoff, 1980a).

5. In *Penicillium claviforme* the rhythm was elicited in the dark by very little light; the stronger the light, the longer the rhythm expression. The period remained short (30 to 36 h) if a sufficient level of irradiation (3 $\mu W\ m^{-2}$ for blue light in the range 360 to 520 nm) was continuously maintained. A very rapid desynchronization caused by irradiation at 445, 479, and 495 nm was observed; the existence of two photosensitive reactions modulating the rhythm is postulated. The stable periodicity, which became desynchronized over a few days in continuous white light, was affected by supplying asparagine (1 to 25 m*M*). (a) The rhythm remained synchronized for at least 3 weeks for all the asparagine concentrations used. (b) A strong rhythm instability was observed at 1 m*M* asparagine; it was corrected by increasing concentrations of this substance. Otherwise, asparagine could not elicit rhythmicity in darkness or in very dim light (Jerebzoff and Piskorz-Binczycka, 1987; Piskorz-Binczycka et al., 1989).

In short, analysis of five different noncircadian models shows that some of their fundamental characteristics (rhythm expression, period length, its stability or dependence on temperature, amplitude of oscillations) can be manipulated by chemicals, in the absence of external signals, indicating that metabolic processes of carbon and nitrogen metabolism are involved in the oscillating processes and can be identified. Moreover, each model presents substantial particularities. Are there biochemical bases common to these models, or not?

III. METABOLIC STEPS INVOLVED IN NONCIRCADIAN SYSTEMS

What is the role of the amino acids participating in protein synthesis in the regulation of oscillating systems in fungi? Arginine evoked a circadian rhythm in *Neurospora crassa* mutants *patch* (Brandt, 1953), *bd* (Dharmananda and Feldman, 1979), and associated with C5 amino acids in arginine-10 (Gillie, 1965). It also induced a rhythm of hyphal branching in three mutants of *Ascobolus immersus* (Berliner and Neurath, 1965). Short (1- to 2-h) rhythms were evoked by tryptophan in *N. crassa* and by a mixture of histidine, phenylalanine, tryptophan, and tyrosine in *N. tetrasperma* (Nysterakis et al., 1969). Amino acids of yeast extract evoked a circadian rhythm in *Sclerotinia fructicola* and a 4-day rhythm in *Aspergillus ochraceus*; a 4-day rhythm was evoked in *Colletotrichum lindemuthianum* by serine (in Jerebzoff, 1965b). *A. niger* rhythm was induced by isoleucine and inhibited by asparagine; asparagine can also act on the period stability in *L. michotii* and *P. claviforme* (mentioned above). Consequently, the effects of isoleucine and asparagine but also of β-methylvalerate and galactose were carefully analyzed.

A. Roles of Isoleucine and β-Methylvalerate in *A. niger*

Isoleucine permits the expression of a 150-h rhythm in *A. niger* by counteracting an excess of K^+ ions.

1. In a liquid medium, newly synthesized isoleucine, first accumulated in the medium, disappeared on the third day. It was not directly incorporated into proteins; among the labeled substances were fumaric, succinic, malic, citric, glycolic, lactic, and malonic acids and amino acids such as aspartic and glutamic acids, α-alanine, and valine (in Jerebzoff, 1976).

2. A K^+-dependent enzymatic complex actively degraded isoleucine into α-oxo-β-methylvalerate, 2-methylbutyrate, propionate, and acetate; only the first three compounds were capable of inducing a rhythm. Otherwise, β-methylvalerate and α-hydroxy-β-methylvalerate were never identified as products of isoleucine degradation (Jerebzoff and Jerebzoff-Quintin, 1980). Thus rhythm expression in *A. niger* might be correlated with the rapid release of catabolic products of isoleucine.

3. In its inducing effect on the sporulation rhythm, isoleucine can be replaced by α-hydroxy-β-methylvalerate or β-methylvalerate; the period was equal to 220 and 360 h, respectively (Jerebzoff and Lambert, 1969). Strong similarities between the degradation of β-methylvalerate and isoleucine in *A. niger* were observed. After decarboxylation of the

Na[I^{14}C]β-methylvalerate, the labeled carbon was fixed within the acids of the Krebs cycle, glutamic and aspartic acids, proline, glycine, and valine (Jerebzoff and Jerebzoff-Quintin, 1980).

The products of isoleucine degradation might play other roles in *A. niger*.

1. The 270-h rhythm expressed in the presence of 12.5 m*M* glucose and 0.2 m*M* K^+ became shorter and shorter with increasing levels of D_2O in the medium, reaching 125 h for 40% D_2O; the addition of 10 m*M* propionate and 2 m*M* K^+ nullified this effect (Jerebzoff and Jerebzoff-Quintin, 1977). Processes associated with membrane permeability and ion transport might be implicated in the oscillating system of the fungus, as in the clocks of *Euglena* (Bruce and Pittendrigh, 1960) and *Phaseolus* (Bünning and Baltes, 1963).

2. [^{14}C]Isoleucine catabolism produced radiolabeled propionate but also Krebs cycle acids. It could then be linked to the Krebs cycle, via the carboxylation of propionyl CoA, into methyl-malonyl CoA (by a biotin-dependent carboxylase), then converted into succinyl CoA by a mutase having a cobamide as coenzyme (Stadtman et al., 1960; Wegener et al., 1968). In *A. niger* receiving 10 m*M* propionate counteracting 2 m*M* K^+, the supply of avidin, a specific inhibitor of biotin-dependent carboxylases, lengthened the period and the rhythm became damped (Jerebzoff and Jerebzoff-Quintin, 1980). Propionyl CoA, pyruvate, and acetyl CoA carboxylases could then participate in the oscillating system of *A. niger*. Moreover, 3 μ*M* intrinsic factor, a specific inhibitor of B_{12}-dependent mutases, supplied in the same culture medium, lengthened the period without damping of the rhythm. Finally, in medium supplemented with [^{14}C]isoleucine, the labeling of succinate was multiplied by 10 in the presence of 5 m*M* malonate, which inhibits the conversion of succinate to fumarate in the Krebs cycle (unpublished results).

B. Role of Galactose in *A. niger*

In the presence of a concentration of glucose which permits a self-sustained rhythm of sporulation, a more and more rapid damping of oscillations was obtained with increasing levels of supplied galactose; the period was not modified. Total inhibition of the rhythm was already effective when the galactose concentration was 10 to 20 times less than glucose. [U^{14}C]galactose was slowly metabolized by the fungus. The main result concerned the changes in the asparagine–aspartate balance. In mycelium cultivated on galactose, a 55% decrease of the asparagine pool and a 130% increase in the aspartate pool were observed in comparison with cultures on glucose. Galactose was converted into glucose; labeling of aspartate, glutamate, and α-alanine shows that galactose gives rise to these compounds via oxalacetate and pyruvate crossways. Finally, aspartate supply (≥ 0.8 mM) was sufficient to replace galactose in its inhibiting effects on the sporulation rhythm, whereas glutamate was inactive (Jerebzoff and Jerebzoff-Quintin, 1980).

In short, all the data confirm the important role of the aspartate crossway in the oscillating system of *A. niger*. Moreover, manipulation of the asparagine–aspartate balance by galactose cannot be explained by the sole conversion of this sugar into glucose. In fact, the expression of a noncircadian rhythm of hyphal branching in *Podospora anserina* is under the control of *zn*, *un*, and *circ* genes; in the presence of all three genes the growth was uniform and periodic if one of these genes was absent (Nguyen Van, 1967). The wild strain expressed a rhythmic growth pattern similar to that of the *zn* mutant if the fructose medium was supplemented with galactose, not metabolized by the fungus (Lysek and Esser, 1971). These facts suggest that the *zn* gene could become silent in the presence of galactose in *P. anserina*.

C. Role of Asparagine Catabolism in *L. michotii*

In *L. michotii*, cultivated in darkness in a liquid medium, a rhythmic development of the whole mycelial pad was observed. [U^{14}C]asparagine (2.6 mM), supplied as the sole nitrogen source, produced aspartate, glutamate, threonine, α-alanine, and proline, in order of their decreasing specific labeling, the acids of the Krebs cycle also being well labeled. The other amino acids were never ^{14}C-labeled. The activities of asparaginase (EC 3.5.1.1) and that of aspartate aminotransferase (EC 2.6.1.1) and alanine aminotransferase (EC 2.6.1.2) showed a 24-h rhythm (Jerebzoff-Quintin and Jerebzoff, 1980b). The pools of aspartate, glutamate, arginine, α-alanine, and γ-aminobutyrate oscillated, with a period identical to that of the three enzymes (Fig. 1). The pools of the Krebs cycle acids oscillated, with a periodicity that could be equal to, or two times shorter than those of the amino acids (unpublished results).

All these facts demonstrate the central role of the *asparagine–pyruvate pathway* (Fig. 2), linked to glycolysis, as well as the participation of the Krebs cycle, in the oscillating system of *L. michotii*, in which the balance between two colored carotenoids, derived from acetyl-CoA condensation, also plays a role (Jerebzoff-Quintin and Jerebzoff, 1980a). This metabolic network controls 80% of the carbon flux and all of the nitrogen flux; therefore, it is able to determine the development of the entire fungus. The oscillating network in *A. niger* presents strong similarities with that of *L. michotii*; it differs mainly by the absence of carotenoids and the participation of the isoleucine degradation pathway.

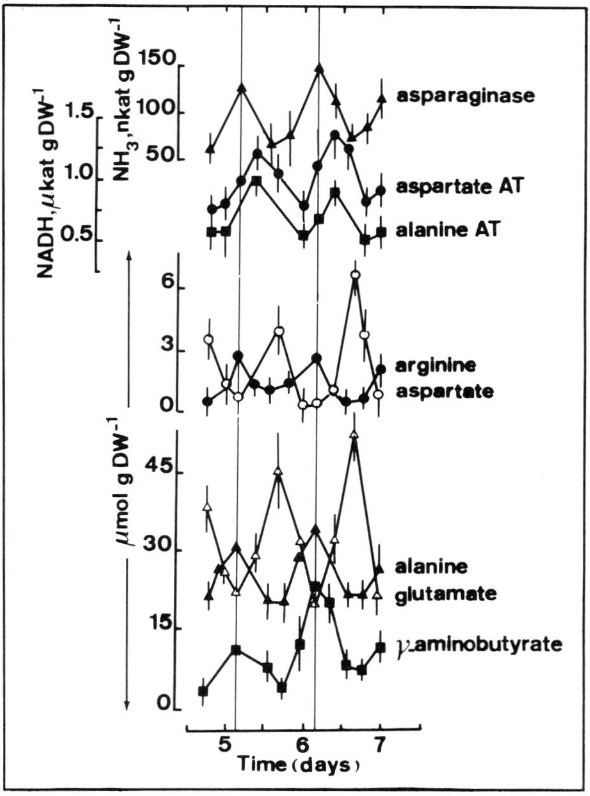

Fig. 1 Activities of three enzymes of the asparagine–pyruvate pathway and the pools of five amino acids oscillate with the same periodicity in *Leptosphaeria michotii*. The other amino acid pools remain constant. AT, aminotransferase.

Two questions arise: What is the type of regulation that determines the rhythmic activity of the asparagine–pyruvate pathway and the rhythm of morphogenesis? Is the sporulation rhythm controlled by the oscillation pattern of this metabolic pathway? Consequently, the effects of different inhibitors of transcription and translation and those of a suicide substrate for transaminases, such as propargylglycine, on the expression and control, both of the sporulation rhythm and the enzyme activity rhythms of the asparagine–pyruvate pathway, were analyzed in *L. michotii*.

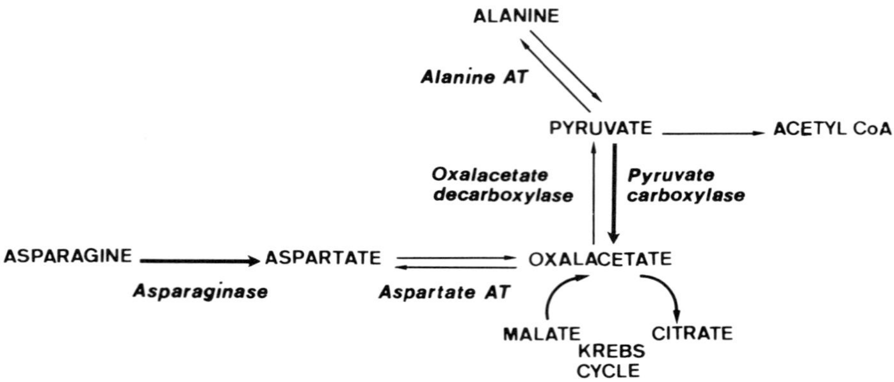

Fig. 2 Asparagine–pyruvate pathway. AT, aminotransferase.

IV. CONTROL LEVELS OF SPORULATION AND ENZYME ACTIVITY RHYTHMS IN *L. MICHOTII*

In *L. michotii* receiving 6.6 m*M* asparagine in darkness, regulation of the sporulation rhythm was inhibited by actinomycin D, cycloheximide, or puromycin (rifampicin and chloramphenicol being inactive). In particular, 10 μ*M* cycloheximide, given at the inoculation of cultures, suppressed the sporulation rhythm, whereas concentrations up to 50 μ*M* on day 2 or day 5 of growth only destabilized the period. The two first days of development are crucial for the synthesis of proteins (synthesized on 80S ribosomes), essential for the construction of the oscillating system (Jerebzoff-Quintin and Jerebzoff, 1978).

Cycloheximide and puromycin (10 to 50 μ*M*) were given at different times in the cycle of the activity rhythms of three enzymes of the asparagine–pyruvate pathway. Thus, in the asparaginase rhythm, phase advances and delays both occurred. The period was always shortened and the amplitude decreased; nevertheless, the rhythm persisted despite a complete inhibition of protein synthesis. Rhythms of aspartate and alanine aminotransferases were phase shifted at 10 μ*M* cycloheximide (period and amplitude remained unchanged and both rhythms stayed synchronous, as in the control); their periods were irregular fluctuating at a concentration of 20 μ*M* and with 50 μ*M* transaminase activities declined and rhythms disappeared completely (Jerebzoff and Jerebzoff-Quintin, 1983a). In the presence of 2.6 m*M* asparagine and in darkness, the sporulation rhythm was unstable, whereas at 6.6 m*M* it was stable and equal to 30 h; 30 to 300 μ*M* propargylglycine stabilized the period in the presence of 2.6 m*M* asparagine and shortened it at the higher concentration. Propargylglycine (100 μ*M*) was presented at different times in the cycle of the activity rhythms mentioned above. For the asparaginase and aspartate aminotransferase rhythms, both phase advances and delays occurred, which were followed by extended period transients, phase shifts, and period lengths, each being different for the two enzymes. Alanine aminotransferase activity was inhibited by 80% and became nonrhythmic, propargylglycine acting as a suicide substrate for only this transaminase in *L. michotii*; it had no effects on [$U^{14}C$]asparagine incorporation into proteins (Jerebzoff and Jerebzoff-Quintin, 1983b).

These results indicate that:

1. Cycloheximide and propargylglycine may act on different parts of the *L. michotii* pacemaker.
2. The asparagine–pyruvate pathway, considered as the main part of the oscillating system at the phenotype level, might include two oscillating subunits, one controlling the asparaginase activity, as least in part by a posttranslational process, the other the transaminase activities, where protein synthesis may play an important role.
3. The sporulation rhythm and rhythmic pattern of the asparagine–pyruvate pathway are strongly correlated in *L. michotii*. The sporulation rhythm is regulated by propargylglycine; it is destabilized by cycloheximide. When asparaginase is the only enzyme of the pathway whose activity remains rhythmic, the sporulation rhythm pursues its course, but it is no longer regulated. When both the activities of asparaginase and aspartate aminotransferase are rhythmic, the sporulation rhythm is regulated. Alanine aminotransferase activity also has an influence on the function of the sporulation rhythm, but its activity could still be nonrhythmic (Fig. 3).

Thus the synchronizing work of the foregoing three components of the asparagine–pyruvate pathway is necessary for the normal operation of the sporulation rhythm in *L. michotii*. This rhythm might be controlled through changes in the main metabolic pools driven by the asparagine–pyruvate pathway.

V. REGULATORY PROCESSES INVOLVED IN ENZYME ACTIVITY RHYTHMS

A. Regulation of Asparaginase Rhythm by a Reversible Phosphorylation Process

The asparaginase rhythm in *L. michotii* is regulated by a reversible phosphorylation process, the purified asparaginase being obtained as a cytosolic protein complex having protein kinase and protein phosphatase activi-

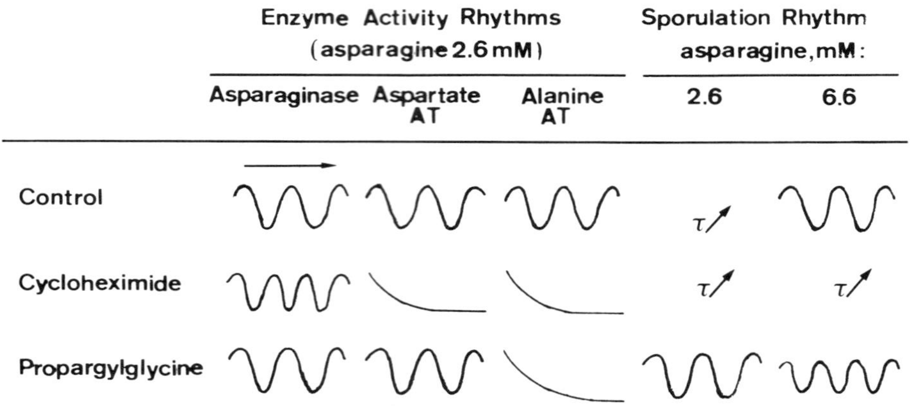

Fig. 3 Diagram depicting the effects of both cycloheximide and propargylglycine on the sporulation rhythm and on the activity rhythms of three enzymes of the asparagine–pyruvate pathway in *Leptosphaeria michotii*, to illustrate the control of the morphogenesis rhythm by this pathway. →, Stable periodicity; $\tau\nearrow$, period increased from one cycle to the next.

ties. This complex was phosphorylated by [γ-^{32}P]ATP at threonine residues. Labeling was located in an M_r 60,000 protein which could be dissociated from the complex; this protein is a Mg^{2+}–Zn^{2+}-dependent protein phosphatase, inhibited by molybdate, which can also dephosphorylate acyl phosphates. The asparaginase molecule was never phosphorylated. The role of this phosphorylation process on the asparaginase rhythm was confirmed in vivo, the time courses of asparaginase activity and ^{32}Pi incorporation into the complex being correlated. In the phase of increasing asparaginase activity, in vitro–enhanced phosphorylation of the complex strongly stimulated the asparaginase; activation of protein phosphatase by Mg^{2+} and Zn^{2+} caused a dephosphorylation and inhibited the asparaginase. When asparaginase and kinase activities were low, dephosphorylation of ^{32}P-labeled complex did not occur when Mg^{2+} and Zn^{2+} were supplied; on the contrary, a strong labeling was observed and asparaginase activity was not affected. The alkaline phosphatase activity of the complex reached a peak at this moment. Thus the same molecule exhibits a predominant protein phosphatase or an alkaline phosphatase activity, according to the phase of the asparaginase cycle; moreover, this molecule can be phosphorylated. All these facts underline the regulatory role of protein phosphatase in this oscillating system (Jerebzoff and Jerebzoff-Quintin, 1984; Jerebzoff-Quintin and Jerebzoff, 1986). When the complex was purified in the presence of chelating agents or a mixture of protein phosphatase inhibitors, two protein kinase activities were detected at different phases of the asparaginase rhythm, a casein-kinase-2 and a calmodulin-activated kinase (unpublished results).

In short, our hypothesis assumes that the oscillating activity of asparaginase, a tetrameric enzyme, is regulated by conformational changes of its molecule, due to interactions between two kinases, a protein phosphatase having autophosphorylation capacities, a regulatory protein, and the adenylate pools.

B. Regulation of Aspartate Aminotransferase Activity Rhythm at the Translational Level

Two isoforms of aspartate aminotransferase were purified to a state of apparent homogeneity; they differed in their kinetic properties, their ability to use substrates, and their in vitro capability to be associated with proteins, the isoform A with a monomer of malate dehydrogenase of M_r 25,000, the isoform B with a polypeptide of M_r 42,000. Antisera obtained were specific for each form. Isoform A was cytosolic and isoform B mitochondrial, as shown by immunocytochemical detection. Regulation of transaminase activity rhythm was analyzed. The level of isoform A remained constant. The rhythm of transaminase activity arose solely from the cyclic activity of form B, which resulted from the oscillating level of this isoenzyme. The time courses of the mRNA levels of both isoforms remained constant. The aspartate aminotransferase activity rhythm, in this species, may be caused by some mechanism controlling the efficiency of translation of mitochondrial isoform mRNA. The role of isoform A in the oscillating system is still to be defined (Romestant et al., 1989, 1990). Nevertheless, as in the case of the luciferin binding protein in *Gonyaulax* (Morse et al., 1989) the cyclic activity of a specific protease cannot be excluded; accordingly, the time course of isoform B turnover has to be determined. The mechanisms responsible for periodic translational regulation remain unknown in any system. If the degradation of isoform B transaminase proves constant in time, a plausible approach of its translational mechanism could be made by analyzing the roles of the 3′ and 5′ untranslated regions of its mRNA and identifying the cytoplasmic regulatory factors involved (Guan and Weiner, 1989; Zhang and Deutscher, 1989; Casey et al., 1989). Moreover, the part played by low-M_r RNAs and ribonucleoprotein complexes, able to modulate protein synthesis in vitro (Sarkar et al., 1981), should be investigated.

C. Regulation of Alanine Aminotransferase Activity Rhythm

Alanine aminotransferase was obtained as a purified isologous dimer, associated with a polypeptide of M_r 25,000 and having a strict substrate specificity. The enzyme level was quantified along the activity rhythm, using a monospecific antibody against the purified enzyme. The cyclic variations of alanine aminotransferase activity were correlated with those of the enzyme level (Abou-Zeid et al., 1991). All the above-mentioned data and those presented here underline the strong similarities existing between the two transaminases in *L. michotii* and suggest a similar mechanism of rhythm regulation for the two enzymes.

D. Pyruvate Carboxylase Activity in *L. michotii*

Pyruvate carboxylase activity in *L. michotii* oscillated with the same periodicity as that of the three other enzymes of the asparagine–pyruvate pathway; this rhythmicity might arise from the oscillating level of the enzyme, as measured by [^{35}S]streptavidin precipitation (Tozo et al., 1992).

In short, an essential part of the oscillating system in *L. michotii* appears to be located in one main metabolic pathway; the activities of all the enzymes of this pathway oscillate with the same periodicity and cause the oscillations of several substrate pools, which could determine the morphogenesis rhythm. Cytosol, mitochondria, and 80S ribosomes are concerned; convincing evidence for the regulatory role of membranes, the plasmalemma in particular, is still lacking. The fact that two completely distinct mechanisms control the rhythmicity of asparaginase on the one hand and that of transaminases on the other hand, imply the participation of synchronizing factors. Such a role could be played by proteins (e.g., polypeptides associated in vitro with the transaminase forms) and by some small molecules. In fact, in addition to their roles as substrates, several small molecules (amino acids, organic acids, sugars) can play various regulatory roles. Morphogenetic effects were observed for asparagine in the short-day plant *Lemna paucicostata* 6746, where 200 μM asparagine was able to promote the flower-inductive dark process (Tanaka and Takimoto, 1977) and to favor the flower producing process, at least in part by promoting iron uptake (Tanaka et al., 1987). Isoleucine constitutes a source of propionyl-CoA used for the biogenesis of ethyl-branched sesquiterpenoid juvenile hormones in insects (Brindle et al., 1987). In *Nicotiana tabacum*, nitrate reductase expression is under the control of a circadian rhythm at the transcriptional level; glutamine might play an important role in the regulation of periodicity (Deng, 1989). The presence of L-glutamate, which indicates the availability of citrate, plays a controlling role in the regulation of citrate metabolism in bacteria by influencing the enzymes that act in a complex cascade system. In particular, L-glutamate directly interacts with citrate lyase by altering its conformation. Moreover, addition of citrate is accompanied by the complete conversion of citrate lyase (sulfhydryl form) into its acetyl form (Antranikian and Gottschalk, 1989). The regulation of transcription of *gln* A, the structural gene for glutamine synthetase, which results from the phosphorylation and dephosphorylation of nitrogen regulator I, is modulated by the glutamine/2-oxoglutarate ratio, acting as a signal (Magasanik, 1989). Finally, galactose might control the rhythm of hyphal branching in *Podospora anserina* by an effect on *zn* gene expression (see Section III.B).

VI. CONCLUSIONS

The master element of the noncircadian models analyzed here is composed of a metabolic pathway, or a network, able to control the whole metabolism and, consequently, to determine the rhythm of morphogenesis (i.e., the asparagine–pyruvate pathway in the case of *L. michotii*). It appears that at least two tightly coupled or interconnected oscillators control the oscillation pattern of all the enzymes in this pathway. One regulates the rhythm of asparaginase, a cytosolic enzyme, by a reversible phosphorylation process, the other controls the rhythm of the aspartate aminotransferase isoform present in the mitochondria, through translation of its mRNA. A major challenge is to identify the proteins and small molecules, which are either participating in determining oscillations or able to synchronize the oscillators. Identification of the genes coding for such regulatory factors and obtaining their mutants might offer a successful genetic approach to the understanding of primary cellular oscillation.

Strong similarities in functional organization between the different noncircadian oscillations present in fungi can be observed; this does not imply strong similarities in the concrete mechanisms involved. For example, a similar control of period stability was obtained in darkness by asparagine in *L. michotii*, by potassium nitrate in *Penicillium expansum*, and by asparagine (but in continuous light) in *P. claviforme*. Moreover, although the network participating in the oscillating system includes aspartate and pyruvate pathways and the Krebs cycle in both *A. niger* and *L. michotii*, it must be noted that the isoleucine catabolic pathway plays a regulatory role for the *A. niger* alone, the supply of asparagine has opposite effects in the two models. Carotenoids are absent in *A. niger*.

A comparison between circadian and noncircadian metabolic systems reveals several points of interest. A key enzyme of carbon metabolism in *Bryophyllum fedtschenkoi*, a Crassulacean plant, is phosphoenolpyruvate carboxylase, whose circadian activity rhythm is regulated by a reversible phosphorylation process (Nimmo et al., 1987). Cycloheximide inhibits this rhythm indirectly, perhaps by acting on the transport of malate (an allosteric inhibitor of the enzyme) between the vacuolar and mitochondrial compartments (Bollig and Wilkins, 1979). Phase shifts induced by a step of the auxin indole-3-acetic acid reveal the presence of two different oscillators tightly coupled in *Acetabularia*, one controlling the photosynthesis and polysaccharide content rhythms and the other, the fatty acid content rhythm (Vanden Driessche et al., 1988). The reduction of nitrate to nitrite is a main point in the regulation of nitrate assimilation. The level of the mRNA coding for the tobacco nitrate reductase, as well as the protein level and activity of this enzyme, showed circadian variations. Inactivation of nitrate reductase by tungstate suppressed all these rhythmic variations; the levels of the mRNA and of the inactive nitrate reductase protein then remained very high. The same result was obtained when nitrate assimilation was blocked by glutamine synthase inhibition. Such effects could be due to overexpression of the structural gene of nitrate reductase (Deng, 1989; Deng et al., 1990). Mutants at the *frq* locus in *Neurospora crassa* exhibited not only changes in the free-running period, but also various patterns of temperature compensation of the period; moreover, the period of *frq*-7 was manipulated by both temperature and nutritional conditions (Feldman and Dunlap, 1983). Thus the mechanisms of temperature compensation of the period length, observed in circadian rhythms, can be dissociated from the oscillating mechanism of the circadian clock. The oli^r point mutation shortened the period of the circadian rhythm in *N. crassa*; it affected the primary structure of the dicyclohexyl carbodiimide binding protein, a subunit of the mitochondrial ATP synthase (Brody, 1981).

It appears that:

1. In all the cases, the master element of the circadian rhythm regulation seems to include a pathway or a network, which can control a large range of metabolic processes. These elements are specific for each organism; their modes of regulation suggest large differences in the concrete mechanisms involved, just as in the case of noncircadian systems. Consequently, it is not surprising that universal participation of a particular element (such as RNA transcription, protein synthesis, energy metabolism including the control of energy charge and nicotinamide nucleotide ratio, and second messengers; see Hardeland and Balzer, 1988) has never been demonstrated in circadian systems. Consequently, such elements were often considered as the hands of the clock, even when they might be components of the clock mechanism itself in the organisms under consideration.
2. A noncircadian system can express a temperature-compensated period by changing nutritional conditions, as in *Alternaria tenuis*. Moreover, a circadian system that has lost the mechanisms of temperature and

nutrition compensation of the period, such as the mutant *frq*-7 of *N. crassa*, can be considered as a very promising noncircadian model. Consequently, the noncircadian models present in fungi may constitute archetypes of some biological clocks. Assembly of the various parts of a noncircadian system in a structure able to oscillate may be a good, simplified model for the analysis of clock mechanisms.

REFERENCES

Abou-Zeid, A., Jerebzoff, S., and Jerebzoff-Quintin, S. (1991). *Physiol. Plant.*, *82*: 401.

Antranikian, G., and Gottschalk, G. (1989). *Biochimie*, *71*: 1029.

Berliner, M. D., and Neurath, P. W. (1965). *Mycologia*, *57*: 809.

Bollig, I. C., and Wilkins, M. B. (1979). *Planta*, *145*: 105.

Brandt, W. H. (1953). *Mycologia*, *45*: 194.

Brindle, P. A., Baker, F. C., Tsai, L. W., Reuter, C. C., and Schooley, D. A. (1987). *Proc. Natl. Acad. Sci. USA*, *84*: 7906.

Brody, S. (1981). *Fed. Proc.*, *40*: 1734.

Bruce, V. G., and Pittendrigh, C. S. (1960). *J. Cell. Comp. Physiol.*, *56*: 25.

Bünning, E., and Baltes, J. (1963). *Naturwissenschaften*, *50*: 622.

Casey, J. L., Koeller, D. M., Ramin, V. C., Klausner, R. D., and Harford, J. B. (1989). *EMBO J.*, *8*: 3693.

Deng, M.-D. (1989). *Thèse Doctorat sciences de la vie*, Université P.-M. Curie, Paris.

Deng, M.-D., Moureaux, T., Leydecker, M.-T., and Caboche, M. (1990). *Planta*, *180*: 257.

Dharmananda, S., and Feldman, J. F. (1979). *Plant Physiol.*, *63*: 1049.

Feldman, J. F., and Dunlap, J. C. (1983). *Photochem. Photobiol. Rev.*, *7*: 319.

Gillie, O. T. (1965). Thesis Sciences, Edinburgh University.

Guan, K. L., and Weiner, H. (1989). *J. Biol. Chem.*, *264*: 17764.

Hardeland, R., and Balzer, Y. (1988). *Int. J. Biometeorol.*, *32*: 149.

Jerebzoff, S. (1965a). In *The Fungi*, Vol. I, *The Fungal Cell* (G. C. Ainsworth and A. S. Sussman, eds.), Academic Press, New York, p. 625.

Jerebzoff, S. (1965b). In *Circadian Clocks* (J. Aschoff, ed.), North-Holland, Amsterdam, p. 181.

Jerebzoff, S. (1976). In *Molecular Basis of Circadian Rhythms* (J. W. Hastings and H. G. Schweiger, eds.), Dahlem Konferenzen, Berlin, p. 193.

Jerebzoff, S., and Jerebzoff-Quintin, S. (1977). *C.R. Acad. Sci. Paris Ser. D*, *284*: 631.

Jerebzoff, S., and Jerebzoff-Quintin, S. (1980). *Physiol. Plant.*, *48*: 278.

Jerebzoff, S., and Jerebzoff-Quintin, S. (1983a). *Physiol. Veg.*, *21*: 233.

Jerebzoff, S., and Jerebzoff-Quintin, S. (1983b). *Physiol. Veg.*, *21*: 247.

Jerebzoff, S., and Jerebzoff-Quintin, S. (1984). *FEBS Lett.*, *171*: 67.

Jerebzoff, S., and Lambert, E. (1969). *Physiol. Veg.*, *7*: 181.

Jerebzoff, S., and Piskorz-Binczycka, B. (1987). *Physiol. Plant.*, *70*: 163.
Jerebzoff, S., Jerebzoff-Quintin, S., and Nysterakis, F. (1970). *C.R. Acad. Sci. Paris Ser. D*, *270*: 1247.
Jerebzoff-Quintin, S., and Jerebzoff, S. (1978). *Physiol. Veg.*, *16*: 105.
Jerebzoff-Quintin, S., and Jerebzoff, S. (1980a). *Protoplasma*, *104*: 43.
Jerebzoff-Quintin, S., and Jerebzoff, S. (1980b). *Physiol. Veg.*, *18*: 147.
Jerebzoff-Quintin, S., and Jerebzoff, S. (1986). *Biochem. Biophys. Res. Commun.*, *140*: 1135.
Lysek, G. (1984). In *The Ecology and Physiology of the Fungal Mycelium* (D. H. Tennings and A. D. M. Rayner, eds.), Cambridge University Press, Cambridge, Vol. 14, p. 323.
Lysek, G., and Esser, K. (1971). *Arch. Mikrobiol.*, *75*: 360.
Magasanik, B. (1989). *Biochimie*, *71*: 1005.
Morse, D., Milos, P. M., Roux, E., and Hastings, J. W. (1989). *Proc. Natl. Acad. Sci. USA*, *86*: 172.
Nguyen Van, H. (1967). *Ann. Sci. Nat. Bot. Paris*, *9*: 257.
Nimmo, G. A., Wilkins, M. B., Fewson, C. A., and Nimmo, H. G. (1987). *Planta*, *170*: 408.
Nysterakis, F., Jerebzoff, S., and Jerebzoff-Quintin, S. (1969). *C.R. Acad. Sci. Paris Ser. D*, *268*: 1498.
Piskorz-Binczycka, B., Jerebzoff, S., and Jerebzoff-Quintin, S. (1989). *Physiol. Plant.*, *76*: 315.
Romestant, M., Jerebzoff, S., Noaillac-Depeyre, J., Gas, N., and Dargent, R. (1989). *Eur. J. Biochem.*, *180*: 153.
Romestant, M., Jerebzoff-Quintin, S., and Jerebzoff, S. (1990). *FEBS Lett.*, *260*: 285.
Sarkar, S., Mukherjee, A. K., and Guha, C. (1981). *J. Biol. Chem.*, *256*: 5077.
Stadtman, E. R., Overath, P., Eggerer, H., and Lynen, F. (1960). *Biochem. Biophys. Res. Commun.*, *2*: 1.
Tanaka, O., and Takimoto, A. (1977). *Plant Cell Physiol.*, *18*: 27.
Tanaka, O., Nasu, Y., Sonoyama, A., Maehara, Y., Kobayashi, T., Nawafune, H., and Kugimoto, M. (1987). *Plant Cell Physiol.*, *26*: 697.
Tozo, K., Séjalon-Delmas, N., Jerebzoff-Quintin, S., and Jerebzoff, S. (1992). *Physiol. Plant.* (in press).
Vanden Driessche, T., Jerebzoff, S., and Jerebzoff-Quintin, S. (1988). *J. Interdiscip. Cycle Res.*, *19*: 81.
Wegener, W. S., Reeves, H. C., Rabin, R., and Ajl, S. J. (1968). *Bacteriol. Rev.*, *32*: 1.
Zhang, J., and Deutscher, M. P. (1989). *J. Biol. Chem.*, *264*: 18228.

19
Morphogenesis of Periodic Conidiation Patterns in *Neurospora crassa*: Its Control by Circadian Rhythm and Daily Light and Temperature Signals

Ludger Rensing *University of Bremen, Bremen, Germany*

I. INTRODUCTION

Among the different aspects of morphogenesis in *Neurospora crassa*, I will concentrate on the formation of concentric rings of aerial hyphae and conidia in a radially growing mycelium (Fig. 1a). These concentric rings appear as a sequential (diachronous) periodic pattern (see Chapter 1, this volume) developing on a two-dimensional layer of branching hyphae (Russo and Pandit, 1992). Especially in the band (bd) mutant of *Neurospora*, morphogenesis of a periodic pattern in constant darkness is controlled by a circadian clock, which induces differentiation at the growth front of the mycelium during a defined phase. The clocks in individual hyphae are mutually entrained—by means of intracellular signals, anastomoses, or intercellular signals (Nakashima and Hastings, 1989), and the mycelium thus forms a "morphogenetic field." Alternatively, a periodic pattern can be induced by periodic external light or temperature signals acting on a growing mycelium.

The exact form of the pattern depends on endogenous and exogenous conditions; the growth and differentiation rate, for example, are strain specific and depend on the ambient temperature, light intensity, nutrients, pH, salts, and physical properties of the substrate. The properties of the circadian clock (e.g., period length) are strain specific as well [frequency (frq) mutants] and influenced by various internal and external factors.

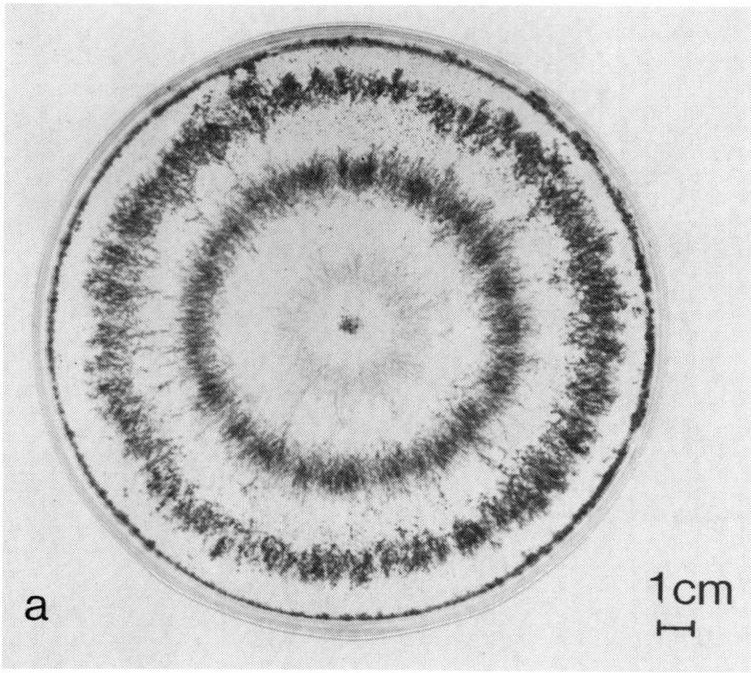

Fig. 1 Morphogenesis of concentric ring formation (dark: aerial hyphae and conidia) in a radially growing mycelium of *Neurospora crassa* (band [bd]) due to endogenous circadian oscillators. (a) The oscillator population is synchronized by the transition from constant light to constant darkness before inoculating a piece of mycelium into the middle of the petri dish. (b) Schematic section through a petri dish at different times of growth and differentiation. (After Deutsch et al., 1991.)

The clock-controlled differentiation of aerial hyphae and conidia during a defined phase of the oscillator (the "subjective" night phase in constant conditions) can be viewed as an evolutionary adaptation to the periodic day–night changes in the environment: The clock serves as a timer to elicit differentiation at night when damage by light, heat, or low humidity is minimal. Light and temperature signals, on the other hand, can directly induce and suppress differentiation and conidia formation after defined lag times, which also leads to periodic morphogenesis during the dark/cold phase of an exogenous periodicity. At the same time, these external periodic signals entrain the endogenous circadian clock. The circadian

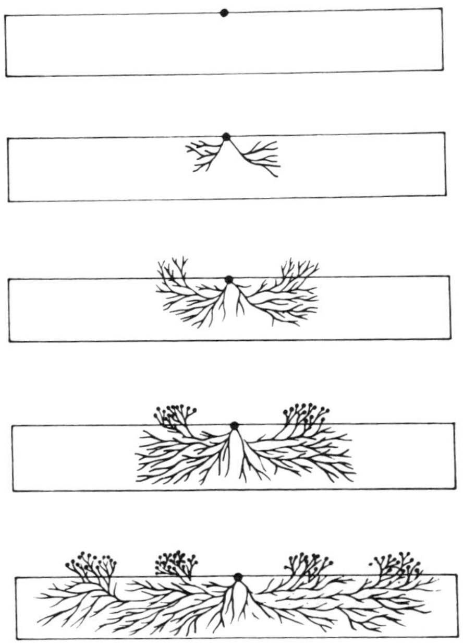

b

clock (endogenous control), light, or high-temperature signals (exogenous control) probably affect differentiation at different stages.

II. PERIODIC MORPHOGENESIS (CONIDIATION) CONTROLLED BY A CIRCADIAN OSCILLATOR

In constant darkness, the periodic formation of aerial hyphae and conidia is controlled by an endogenous circadian oscillator—particularly evident in the ''band'' (bd) mutant of *Neurospora crassa* (reviews: Feldman and Dunlap, 1983; Lakin-Thomas et al., 1990). When a petri dish containing agar and medium is inoculated in the center with a small piece of mycelium, which has been transferred from constant light to constant darkness in order to start (or synchronize) the oscillations of the circadian clock, every 21 h (at 25°C) a new ring of aerial hyphae and conidia appears. A schematic section through a petri dish shows the periodic branching of vegetative into aerial hyphae (Fig. 1b). Such a sequentially generated spatial periodicity, which is caused by an oscillating process, can be defined as ''diachronous'' periodic structure (see Chapter 1).

The basic processes involved in the formation of these periodic structures in constant conditions are the differentiation of aerial hyphae and conidia, on the one hand, and the circadian oscillation and its differentiation inducing signals, on the other hand. Both processes are principally independent on each other, as shown by the fact that conidiation occurs when no rhythm is detectable (e.g., at 40°C) and that a rhythm can exist without differentiation (Feldman and Dunlap, 1983). The processes involved in the two phenomena are summarized briefly in the following section.

A. Differentiation of Aerial Hyphae and Conidia

Some vegetative hyphae branch and differentiate to aerial hyphae. The latter grow up to and above the agar surface, stop growth (stall), and give rise to branches, which differentiate into vegetative spores (conidia) in a repeated process of apical budding (reviews: Turian and Bianchi, 1972; Nelson et al., 1975; Lysek, 1984). The entire process lasts about 10 to 12 h. When vegetative hyphae differentiate into aerial hyphae, few morphological changes can be observed, apart from a lack of the ''Spitzenkörper'' in the tips of the hyphae. Conidiophores appear more highly branched and develop cross-wall septa (Springer and Yanofsky, 1989).

A multitude of molecular processes change during differentiation: for example, oxidative metabolism (Turian and Bianchi, 1971; Hochberg and Sargent, 1974), ion transport (Jennings, 1979), amino acid contents (Brody, 1981), enzyme activity (Urey, 1971), and gene expression (Berlin and Yanofsky, 1985a,b; Roberts et al., 1988; Roberts and Yanofsky, 1989). In aerial hyphae DNA synthesis and mitosis are transiently stimulated and the conidia are then shifted into stationary phase. Stationary-phase cells are often characterized by a set of proteins (called stress proteins) which make these cells more resistent to different kinds of stress (Werner-Washburne et al., 1989; D. Zoeger et al., 1992). Conidia seem to contain a higher level of heat-shock proteins (Berlin and Yanofsky, 1985a).

External factors influencing these changes in metabolism and gene expression may enhance or suppress differentiation and morphogenesis. An important factor is the composition of the medium: differentiation is enhanced, for example, by low phosphate (Hasunuma, 1985), a pH of 6.5 to 7.0 and low biotin (West, 1975), nitrogen starvation (Brody, 1981), general starvation (Klemm and Ninnemann, 1978), and acetate as the carbon source (Turian and Bianchi, 1971); higher glucose levels inhibit sporulation in yeast (Olempska-Beer and Freese, 1987). In some fungi, high Ca^{2+} is reported to stimulate sporulation (Pitt and Ugalde, 1984).

Aeration (high oxygen and low carbon dioxide concentrations) and humidity also favor sporulation (Nelson et al., 1975). The inducing effects of light and high temperature on differentiation are discussed below.

Not yet resolved is the question of how the differentiation process, which is ultimately the result of altered gene expression, is controlled by these external factors. It is conceivable that their effects—and those of light and elevated temperature—are mediated by second messenger systems: cyclic nucleotides, inositol phosphates (IP_3), and Ca^{2+}. A decrease of the cAMP level has been demonstrated after low phosphate (Hasunuma, 1985), light (Hasunuma et al., 1987), and nitrogen deprivation (Dumbrava and Pall, 1987). Ca^{2+}- and cAMP-concentration changes each affect the pattern of protein phosphorylation (Techel et al., 1990). Addition of cAMP was reported to repress differentiation (Mishra, 1976). Differentiation and a state of cell-cycle arrest (stationary state), which often goes along with it, are not solely defined by a monotonic change of cAMP but rather by a dynamic alteration which may include a transient increase (Terenzi et al., 1976). The exact time course of these changes is not known yet. In summary, the external factors mentioned above influence processes leading ultimately to changes of gene expression and differentiation (Fig. 2).

B. The Circadian Oscillator and Its Differentiation-Inducing Signals

Circadian rhythms are self-sustained endogenous oscillations which exist in most eukaryotic (perhaps also in prokaryotic) organisms under certain conditions and which are probably based on a conserved intracellular mechanism (review: Edmunds, 1988). The molecular processes of this mechanism are not known in any detail, even though "clock" genes in *Drosophila* have been cloned and sequenced (review: Rosbash and Hall, 1989). Phase shifts of the circadian oscillation induced by perturbing pulses of various agents and treatments have been used as an indicator for the underlying mechanism and the results indicate that proteins, membrane processes, calcium calmodulin, and protein phosphorylation may be parts of the oscillator (review: Rensing and Hardeland, 1990). The oscillator can be entrained by external periodic signals ("Zeitgeber") especially by periodic light or temperature signals. These signals may be called "controlling inputs" as distinguished from "sustaining inputs," such as energy or other prerequisites of the oscillator (Fig. 2). The oscillator produces several periodic signals, which control a variety of cellular processes: In *Neurospora*, for example, many metabolic processes (Lakin-Thomas et al., 1990) as well as differentiation outputs (Fig. 2). These

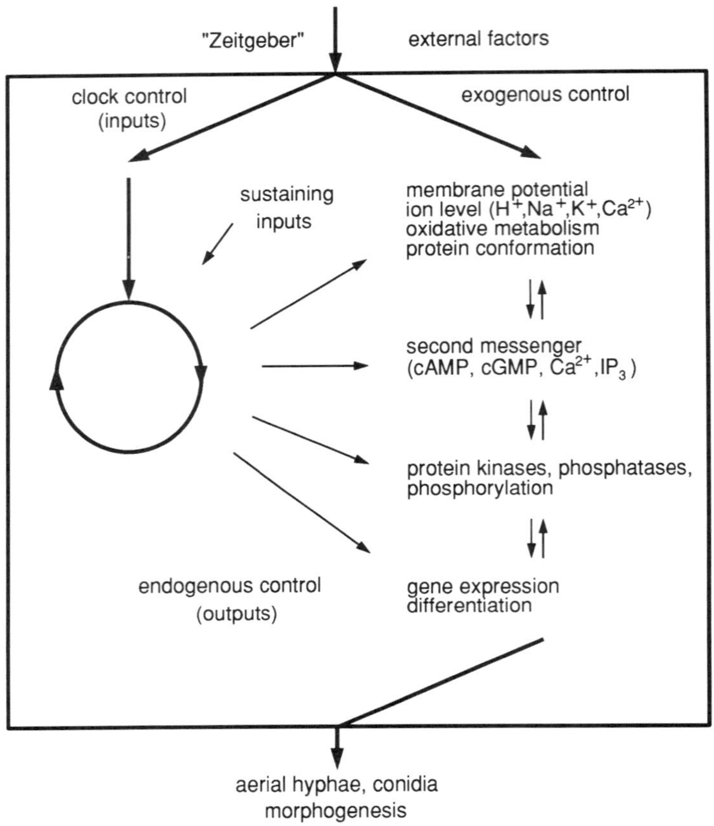

Fig. 2 Scheme for the effects of Zeitgeber and external factors on aerial hyphae and conidia formation (morphogenesis). Like light and temperature signals, Zeitgeber function as control inputs and can reset the endogenous circadian clock. The clock controls a number of processes, some of which lead to differentiation (endogenous control). Light and temperature changes—as well as other external factors—are able to induce differentiation processes (exogenous control) directly, sometimes also called ''masking'' (see Rensing, 1989).

output oscillations rarely feed back into the mechanism. Suppression of conidiation, for example, by growing mycelia in liquid medium, does not affect the functioning of the clock (Feldman and Dunlap, 1983).

The signals by which the clock controls metabolic and differentiation processes are not known. A circadian rhythm of the level of cAMP (Hasunuma et al., 1987) based on rhythmic activities of the adenylate cyclase or cAMP phosphodiesterase (Techel et al., 1990) may function as a triggering signal. The activity changes of cAMP-dependent and other protein kinases may in turn affect gene expression in a rhythmic fashion. Rhythmic transcription in *Neurospora* was shown to exist in at least two constitutionally active genes (Loros et al., 1989). Apart from circadian changes in the constitutive activity there seem to be circadian differences in the response to external signals: The induction of heat-shock mRNA

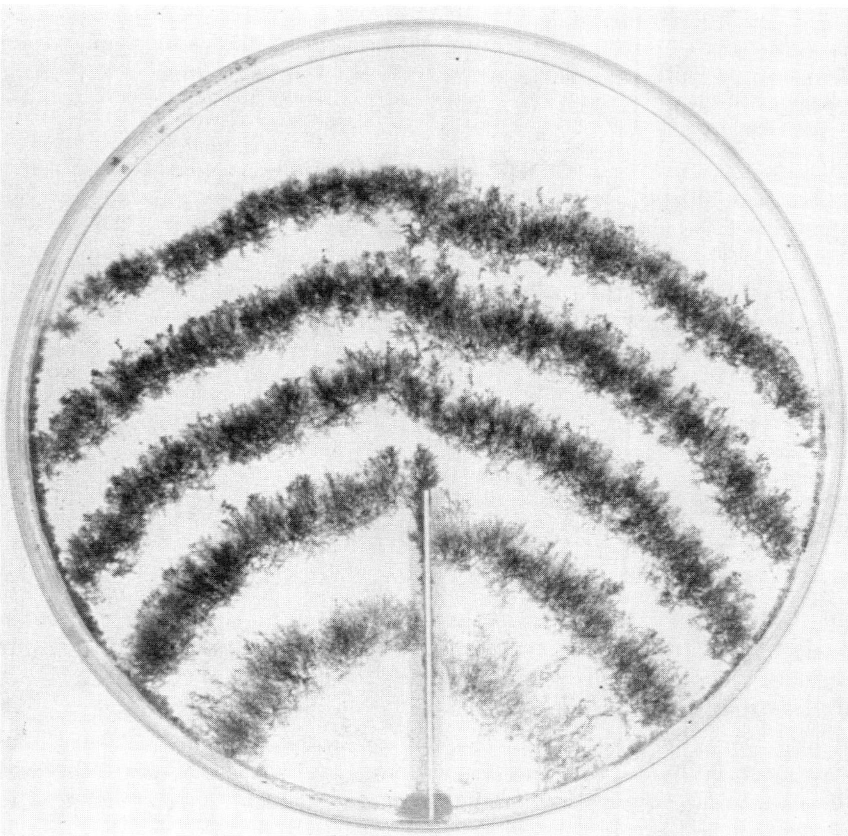

Fig. 3 Mutual interactions between mycelia of different circadian phase. Two pieces of mycelium with inversed phases (180° difference) were inoculated on a petri dish at the same site, but separated by a glass wall. They were kept in constant darkness. When the differently phased mycelia meet at the end of the wall an intermediate phase is formed. (From Deutsch, unpublished.)

and protein synthesis by elevated temperatures is influenced by the phase of the circadian oscillation (Cornelius and Rensing, 1986; Rensing et al., 1991). The possible involvement of heat-shock in differentiation is discussed below.

C. Morphogenesis at the Macroscopic Level

Morphogenetic processes leading to highly regular concentric rings of aerial hyphae and conidia in a growing mycelium in constant darkness suggest the existence of a communication system between hyphae. Experimental evidence indicates that mycelia with different circadian phases influence each other upon coming into contact on the same agar surface (Deutsch, unpublished results; Fig. 3). Similar experiments have been reported by Winfree and Twaddle (1981) and Nakashima and Hastings (1989). The regular concentric rings that develop even when conidia and mycelia were grown in constant darkness without any light/dark transition also support the interpretation of mycelia as self-organizing morphogenetic field based on interhyphal signaling.

Interactions between hyphae may occur by way of intracellular communication within hyphae and their branches, by anastomoses between hyphae of the same mating type and with the same alleles of two genes (C,c,D,d) controlling the fusion of hyphae (see Ross, 1979) or by an intercellular exchange of signals, for example, by molecules such as H^+, cAMP, or proteins.

III. PERIODIC MORPHOGENESIS INDUCED BY PERIODIC LIGHT SIGNALS

Periodic conidiation is not only controlled by an endogenous oscillation, but external light intensity changes, like the daily light–dark cycle, directly induce differentiation of spores in many fungi and thus give rise to periodic morphogenesis (review: Lysek, 1984). In *Neurospora crassa*, a pulse of light elicits a pulse of differentiation shortly after the light signal, in addition to the endogenously controlled differentiation (Fig. 4b). This was observed by Klemm and Ninnemann (1978). Furthermore, the next circadian clock–controlled band of conidia appears denser compared to the control, indicating a light effect superimposed on the circadian control (Fig. 4a,b). In a 12 : 12-h light–dark cycle (which at the same time synchronizes the circadian rhythm) a smaller peak of conidia can be observed shortly after the beginning of the light period and a larger peak of conidia in the dark period (Fig. 5a). The light-induced morphogenetic pattern can be shortened to a period of 2 : 2 light–dark in the csp-1-bd mutant (Lysek,

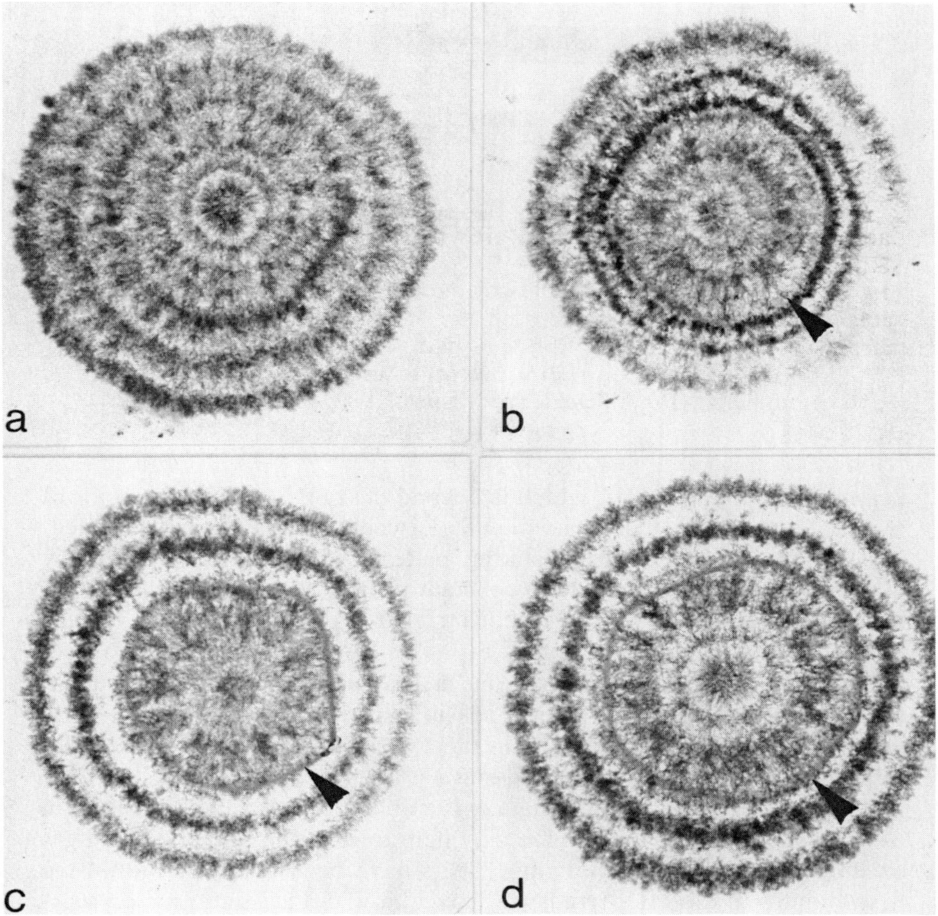

Fig. 4 Effects of short light and heat-shock pulses on differentiation. Mycelia were transferred from constant light to constant darkness just before inoculation (at constant temperature of 25°C): (a) control (in constant darkness); (b,c,d) at a defined phase of the oscillator (ct 24, i.e., shortly after the second endogenous conidiation maximum, arrow) mycelia were exposed to white light of 2.6 W/m^2 (b), or to 1 h 40°C (c), or to 1 h white light and 40°C (d).

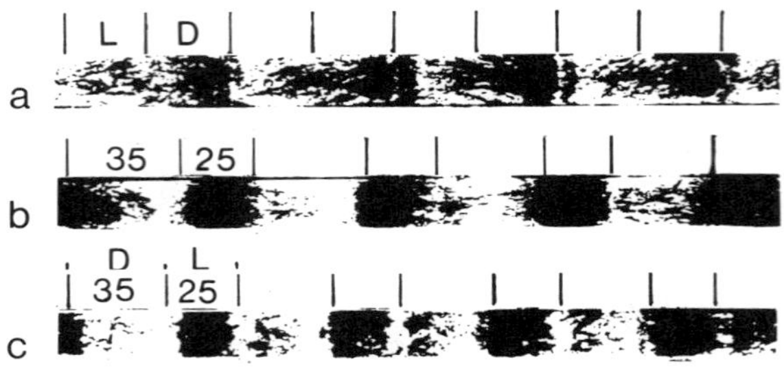

Fig. 5 Effects of 24-h periods of light–dark or high–low temperature on differentiation and entrainment of the oscillator. Mycelia were inoculated in "race tubes" and exposed to (a) 12:12-h light–dark periodicity, (b) 12:12-h high–low (35°/25°C) temperature periodicity, and (c) to a 12:12-h light (25°C) and dark (35°C) periodicity. (From Schmidt and Steier, 1987; see Rensing, 1989.)

personal communication), which is beyond the range of period lengths of external light–dark cycles by which the circadian clock can be entrained. In constant light, the periodic banding pattern is leveled off or completely overridden, which thus leads to constant sporulation. These results support the view of light as an exogenous trigger of differentiation in *Neurospora*. A possible interpretation of Fig. 4b is the assumption of two light effects on conidial differentiation: one at a rather late stage of morphogenesis, another at an earlier stage, possibly at the stage of aerial hyphae determination. This would explain the two induced rings of differentiation: an immediate and a delayed ring.

A considerable amount of data exists on the cellular effects of light in *Neurospora*, and this knowledge may help to elucidate the effect of light on differentiation. The following effects have been observed within the first minutes to 1 h after irradiating mycelia with blue light: hyperpolarization of the membrane potential (Potapova et al., 1984), a transient drop in the cAMP level and an increase in phosphodiesterase activity (Kritsky et al., 1982), changes of protein phosphorylation (Lauter and Russo, 1990; W. Kohler and L. Rensing, unpublished results), carotenoid synthesis (Zalokar, 1954), and a sequential induction of translatable mRNA species showing lag times between 2 and 45 min after the beginning of illumination (Chambers et al., 1985; Nawrath and Russo, 1990). Furthermore, within 30 min, as many as 60 to 80 genes seem to alter their activity (Sommer et al., 1989). The relationships between the light-inducible genes and genes associated with differentiation (Roberts et al., 1988; Roberts and Yanofsky, 1989) are not yet clear. The same is true for the role of genes activated after starvation (D. Zoeger et al., 1992).

IV. PERIODIC MORPHOGENESIS INDUCED BY PERIODIC TEMPERATURE SIGNALS

Temperature changes are highly effective Zeitgeber signals of the clock, similar to light signals (Rensing and Schill, 1987; Fig. 5). Whether or not conidiation in *Neurospora* can be elicited by elevated temperature (''heat shock''), similar to the effect of light, was investigated in the following series of experiments.

A 1-h pulse of 40°C suppresses conidiation at the time of application (shortly after an endogenous maximum), but it enhances conidiation in the next clock-controlled band (Fig. 4c). When applied together with 1-h light, 40°C suppresses the immediate induction of conidiation by light (Fig. 4d). In a 12:12-h temperature cycle (35° to 25°C) the circadian oscillators are apparently strongly entrained and conidiation expressed during the cold (25°C) period (Fig. 5b). High temperature again suppresses conidiation. This is also observed when a 12:12-h light–dark cycle is applied together with an inverse temperature cycle (i.e., 35°C during dark time and 25°C during light time) (Fig. 5c). Conidiation occurs only during the low-temperature period. Under normal day–night changes of light and temperature, this will lead to a more or less complete inhibition of conidia formation during daytime and a (delayed) induction of conidia formation during nighttime.

What is the molecular mechanism of immediate suppression and delayed induction of conidiation by elevated temperature? Much is known about the effects of heat shock on cellular processes (reviews: Lindquist and Craig, 1988; Morimoto et al., 1990). In *Neurospora*, for example, an increase of Ca^{2+} and H^{+} levels, changes in second messenger concentration, activity of protein kinases and phosphatases, protein phosphorylation (Cornelius et al., 1989; Techel et al., 1990), protein degradation and protein ubiquitination (Mohsenzadeh and L. Rensing, unpublished results), and gene expression (Kapoor and Chow, 1984; C. Scholle and L. Rensing, unpublished results) can be observed. These primary responses seem to repress the differentiation and conidiation process. After the primary response to heat shock—even if the elevated temperature is maintained, but especially when cells are shifted back to normal temperatures—an adaptive response is observed. This recovery process may not lead to complete restoration of the unperturbed state but may trigger a switch toward differentiation. The adaptation process includes the recovery of ion concentration, total protein synthesis and degradation, and the inhibition of heat-shock protein synthesis (Rensing et al., 1991). Similar long-term induction of differentiation by a short heat-shock pulse was observed in cultured neuroblastoma cells (Stoklosinski et al., 1992).

On the level of protein synthesis, the transient induction of a 38-kDa protein by light is suppressed by a heat-shock pulse of 45°C (45 min) given after irradiation, whereas the recovery from heat shock induces a protein similar to this light-inducible protein (Kohler and Rensing, 1992). We are completely ignorant, however, about the possible similarities between light and heat shock as early inducing signals.

V. CONCLUSIONS

The induction of differentiation by the circadian clock or by an external periodicity of light–dark or high–low temperature leads to periodic differentiation and conidiation (Fig. 6). The endogenous oscillation elicits dif-

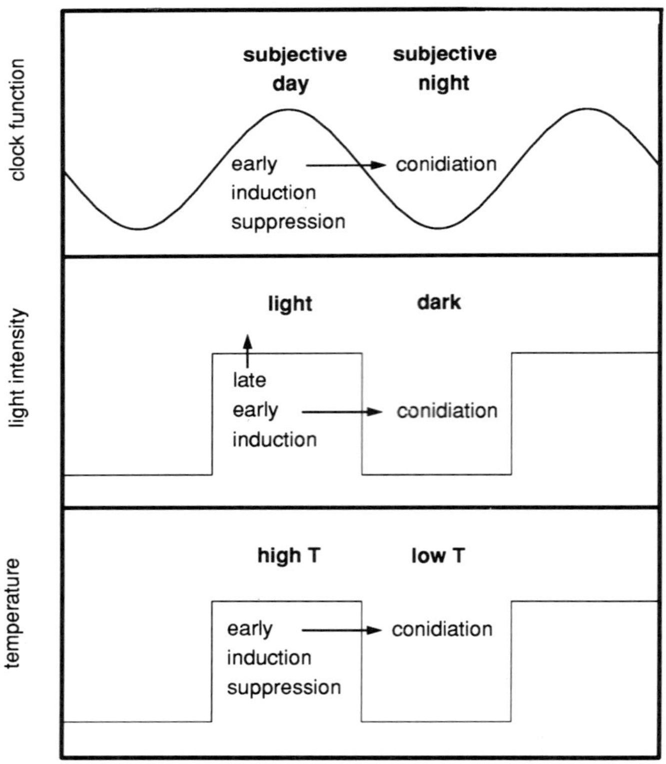

Fig. 6 Hypothetical way endogenous circadian oscillator and the exogenous periodicity of light–dark and high–low temperature might act on conidiation. For a further explanation, see the text.

ferentiation at a defined phase but may also suppress this process at other phases. Light seems to stimulate differentiation at a rather late stage of the differentiation process immediately after irradiation and at the same time trigger early events. It may even suppress differentiation during some intermediate steps of differentiation as concluded from the short periodic patterns of 6:6 and 2:2 (G. Lysek, personal communication). High temperature suppresses late differentiation steps which were enhanced by light and possibly also intermediate processes, but elicits early processes in the differentiation sequence.

In all three inducing processes, cyclic nucleotides may be involved as well as changes of protein phosphorylation. Since differentiation ultimately involves changes in gene transcription, perhaps initiated by a global switch, which starts a sequential activation of differentiation genes, some periodic control of this switch and perhaps of the subsequent activation chain must exist. It is conceivable that this primary switch is affected by the circadian clock, light, and high temperature through the same mechanism.

This periodic switch occurs exclusively at the growth fronts (even though the clock also functions in older hyphae, Feldman and Dunlap, 1983) and thereby creates a periodic spatial pattern. This pattern is quite regular, because an intercellular synchroniation mechanism exists, which, however, is not well understood.

REFERENCES

Berlin, V., and Yanofsky, C. (1985a). *Mol. Cell. Biol.*, *5*: 839.

Berlin, V., and Yanofsky, C. (1985b). *Mol. Cell. Biol.*, *5*: 849.

Brody, S. (1981). In *The Fungal Spore*: *Morphogenetic Controls* (G. Turian and H. R. Hohl, ed.), Academic Press, New York, pp. 605–626.

Chambers, J. A. A., Hinkelammert, K., and Russo, V. E. A. (1985). *EMBO J.*, *4*: 3649.

Cornelius, G., and Rensing, L. (1986). *Eur. J. Cell Biol.*, *40*: 130.

Cornelius, G., Gebauer, G., and Techel, D. (1989). *Biochem. Biophys. Res. Commun.*, *162*: 852.

Deutsch, A., Rensing, L., and Dress, A. (1991). In *Nonlinear Wave Processes in Exitable Media* (A. V. Holden, M. Markus, H. G. Othmer, ed.) Plenum Press, New York, pp. 259–267.

Dumbrava, V. A., and Pall, M. (1987). *Biochim. Biophys. Acta*, *926*: 331.

Edmunds, L. N. Jr. (1988). *Cellular and Molecular Bases of Biological Clocks*, Springer-Verlag, Heidelberg.

Feldman, J. F., and Dunlap, J. C. (1983). *Photochem. Photobiol. Rev.*, *7*: 319.

Hasunuma, K. (1985). *Bot. Mag. Tokyo*, *98*: 203.

Hasunuma, K., Funadera, F., Shinokara, Y., Furukawa, K., and Watanabe, H. (1987). *Curr. Genet.*, *12*: 127.

Hochberg, M. L., and Sargent, M. L. (1974). *J. Bacteriol.*, *120*: 1164.

Jennings, D. H. (1979). In *Fungal Walls and Hyphal Growth* (H. Burnett and A. P. J. Trinci, eds.), Cambridge University Press, Cambridge, pp. 279–294.

Kapoor, M., and Chow, A. L. W. (1984). *Can. J. Biochem. Cell Biol.*, *62*: 691.

Klemm, E., and Ninnemann, H. (1978). *Photochem. Photobiol.*, *28*: 227.

Kohler, W., and Rensing, L. (1992). *Arch. Microbiol.*, *157*, in press.

Kritsky, M. S., Sokolovsky, V. Y., Belozerskaya, T. A., and Chernysheva, E. K. (1982). *Arch. Microbiol.*, *133*: 206.

Lakin-Thomas, P. L., Coté, G. G., and Brody, S. (1990). *Crit. Rev. Microbiol.*, *7*: 365.

Lauter, F., and Russo, V. E. (1990). *J. Photochem. Photobiol. B Biol.*, *5*: 95.

Lindquist, S., and Craig, E. (1988). *Annu. Rev. Genet.*, *22*: 631.

Loros, J. J., Denome, S. A., and Dunlap, J. C. (1989). *Science*, *243*: 385.

Lysek, G. (1984). In *The Ecology and Physiology of the Fungal Mycelium* (D. H. Jennings and A. D. M. Rayner, eds.), Cambridge University Press, Cambridge, pp. 321–343.

Mishra, N. C. (1976). *Naturwissenschaften*, *63*: 485.

Morimoto, R. J., Tissières, A., and Georgopoulos, C. (eds.) (1990). *Stress Proteins in Biology and Medicine*, Cold Spring Harbor Laboratory Press, Cold Spring Harbor, N.Y.

Nakashima, H., and Hastings, J. W. (1989). *J. Biol. Rhythms*, *4*: 377.

Nawrath, C., and Russo, V. E. A. (1990). *J. Photochem. Photobiol.*, *4*: 261.

Nelson, R. E., Selitrennikoff, C. P., and Siegel, W. (1975). In *Cell Cycle and Cell Differentiation* (J. Reinert and H. Holtzer, eds.), Springer-Verlag, Heidelberg, pp. 291–310.

Olempska-Beer, Z., and Freese, E. (1987). *Mol. Cell. Biol.*, *7*: 2141.

Pitt, D., and Ugalde, U. O. (1984). *Plant, Cell and Environment*, *7*: 467.

Potapova, T. V., Levina, N. N., Belozerskaya, T. A., Kritsky, M. S., and Chailakhian, L. M. (1984). *Arch. Microbiol.*, *137*: 262.

Rensing, L. (1989). *Chronobiol. Int.*, *6*: 297.

Rensing, L., and Hardeland, R. (1990). *Chronobiol. Int.*, *7*: 353.

Rensing, L., and Schill, W. (1987). In *Temporal Disorder in Human Oscillatory Systems* (L. Rensing, U. an der Heiden, and M. Mackey, eds.), Springer-Verlag, Heidelberg, pp. 233–245.

Rensing, L., Scholle, C., and Mohsenzadeh, S. (1991). In *Chronobiology and Chronomedicine* (A. Diez-Noguera and T. Cambras, eds.), P. Lang, Frankfurt.

Roberts, A. N., and Yanofsky, C. (1989). *Nucleic Acids Res.*, *17*: 197.

Roberts, A. N., Berlin, V., Hager, K. M., and Yanofsky, C. (1988). *Mol. Cell. Biol.*, *6*: 2411.

Rosbash, M., and Hall, J. (1989). *Neuron*, *3*: 387.

Ross, J. K. (ed.) (1979). *Biology of Fungi*, McGraw-Hill, New York, p. 249.

Russo, V. E. A., and Pandit, N. N. (1992). In *Development* (V. E. A. Russo, S. Brody, D. Cove, and S. Ottolenghi, ed.) Springer, Heidelberg (in press).

Schmidt, A., and Steier, G. (1987). Diploma thesis, University of Bremen.
Sommer, I., Chambers, J. A. A., Eberle, J., Lauter, F. R., and Russo, V. E. A. (1989). *Nucleic Acids Res.*, *17*: 5713.
Springer, M. L., and Yanofsky, C. (1989). *Genes Dev.*, *3*: 539.
Stoklosinksi, A., Richter-Landsberg, C., and Rensing, L. (1992). *Exp. Cell Res.*, *200*: 89.
Techel, D., Gebauer, G., Kohler, W., Braumann, T., Jastorff, B., and Rensing, L. (1990). *J. Comp. Physiol. B*, *159*: 695.
Terenzi, H. F., Flawiá, M. M., Tellez-Inón, M. T., and Torres, H. N. (1976). *J. Bacteriol.*, *126*: 91.
Turian, G., and Bianchi, D. E. (1971). *Arch. Microbiol.*, *77*: 262.
Turian, G., and Bianchi, D. E. (1972). *Bot. Rev.*, *38*: 119.
Urey, J. C. (1971). *Dev. Biol.*, *26*: 17.
Werner-Washburne, M., J. Becker, Kosic-Smithers, J., and Craig, E. (1989). *J. Bacteriol.*, *171*: 2680.
West, D. (1975). *J. Bacteriol.*, *23*: 387.
Winfree, A., and Twaddle, G. M. (1981). In *Mathematical Biology* (T. A. Burton, ed.), Pergamon Press, New York, p. 237.
Zalokar, M. (1954). *Arch. Biochem. Biophys.*, *50*: 71.
Zoeger, D., Scholle, C., Techel, D., and Rensing, L. (1992). *Exp. Mycol.* (in press).

20
Formation of Annual Rings in Trees

Rodney Arthur Savidge *University of New Brunswick, Fredericton, New Brunswick, Canada*

I. INTRODUCTION

A. Overview

Annual rings of "secondary" xylem, or wood, arise through cell division activity in the vascular cambium (Fahn, 1974; Panshin and de Zeeuw, 1980; Philipson et al., 1971). During seasonal growth of shoots and roots, the first woody tissue to be produced in these elongating organs is the "primary" xylem, minute strands of woody tissue derived from the procambium (Larson, 1982; Philipson et al., 1971). The primary xylem mirrors the procambium, which itself consists of localized strands of axially elongated radially narrow-diameter meristematic cells. During elongation of shoot or root, interfascicular cambia form basal to shoot and root apices to join procambial strands and produce the complete ring of meristematic tissue known as the vascular cambium, or simply as cambium (Larson, 1982).

Leonardo de Vinci has been credited with the discovery that growth rings provide a measure of tree age (Glock, 1941). Seen as "rings" in stumps or disks (Fig. 1), when a stem or root is also observed in median longitudinal section, the fact that wood is produced in layers is plainly evident. The cambium ensheathes the wood of the tree from root tip to shoot tip. The layers of new wood generated by the cambium are laid down one upon the next, rather like building up a series of inverted cones (Fig. 2).

Growth and dormancy oscillations having a periodicity of one year are common but not invariable phenomena in temperate-zone trees. Although normally only a single period of wood formation occurs tempo-

Fig. 1 Photomicrograph of a transverse section of conifer wood (*Picea glauca* var. *albertiana*) from northern Canada showing annual rings. (Magnification: 67×).

rally associated with shoot growth in late spring and early summer, a second surge of cambial growth, resulting in formation of false or doublet "annual rings," not uncommonly occurs when new buds flush and elongate a second time in late summer instead of remaining dormant (Kozlowski, 1971). Moreover, annual ring formation cannot be assumed to occur throughout the tree as an automatic consequence of height growth. The cambium near bases of older shaded branches and near ground level in main stems of mature trees, particularly tall conifers, predictably remains inactive throughout the growing season, consequently producing no new annual rings despite occurrence of shoot

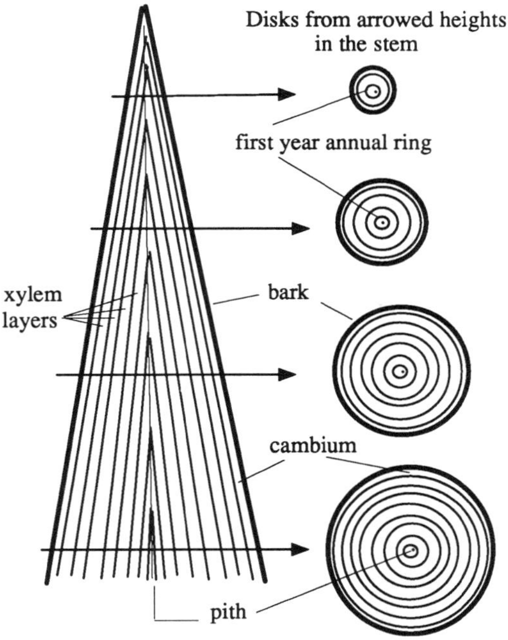

Fig. 2 Diagrammatic representation of a tree stem in median longitudinal section showing how one layer of wood is formed upon the next. The disks on the right indicate how many annual rings there would be at each arrowed location.

growth (Kozlowski, 1971). Conducting phloem and xylem, once in place, can function for a number of years without addition of new vascular tissues (Connor and Lanner, 1990; R. A. Savidge, unpublished observations).

B. Evolutionary Perspective

Cambial formation and differentiation of the cambium's derivative cells into components of long-distance transport networks (i.e., photosynthate-conducting sieve cells of phloem and lignified water-conducting tracheids of xylem are unquestionably key developments that enabled primitive plants to become established on land. Fossilized early land plants having primary xylem tracheids are known from rocks of the Silurian period (Gensel and Andrews, 1987), suggesting that genetic competence for procambial development existed then. Vascular cambium development occurred at least -387×10^6 years Before Present (BP) (i.e., middle

Devonian period), and during the Upper Devonian (-375×10^6 y BP) progymnosperms (e.g., *Archaeopteris*) growing near the equator produced woody stems greater than 5 ft in diameter; however, annual rings are lacking in these specimens (Beck, 1970; Foster and Gifford, 1974; Gensel and Andrews, 1987). Definite annual rings were being produced by trees growing in high southern latitudes at least as early as the Gondwana glaciation (-325×10^6 y BP) (Creber and Chaloner, 1984).

By the time woody plants acquired genes to produce distinct growth rings, they must also have acquired the means to become dormant and to tolerate seasonal climatic extremes. There undoubtedly were survival advantages associated with having genetic competence to tolerate seasons of moisture stress and low temperatures. The ability to produce latewood may have conferred some additional fitness. Latewood fibers and tracheids have thicker walls and smaller radial diameters than those of earlywood, and high-density latewood is stronger than low-density earlywood (Denne, 1989; Panshin and de Zeeuw, 1980). Production of stronger wood may have enabled tree stems to withstand heavy snow or wind loads. But latewood also has drawbacks; its small-diameter cells have substantially reduced numbers of bordered pits (conifers and hardwoods), and vessels (hardwoods only) are scarce. Consequently, latewood is much inferior to earlywood in hydraulic conductivity (i.e., in its ability to deliver water from the root to the shoot system in support of photosynthesis and general metabolism). A mixture of conducting earlywood and strengthening latewood represents an optimal compromise in terms of achieving fitness in two potentially conflicting ways, possibly the explanation for trees not being composed entirely of high-density wood.

C. Cambial Determination

Wood is the most abundant form of biomass on terrestrial earth, but our understanding of how trees make wood, in terms of physiology and genetics, remains far from clear. Underlying our ignorance is the central unanswered question of whether the cambial genome is committed to vascular development such that its centrifugal derivatives are predetermined to become phloem and its centripetal derivatives to become xylem or, alternatively, whether the cambium is a nondetermined meristem such that its derivatives are free to adopt any cellular differentiation pathway inherent to the plant (Savidge, 1985).

An argument supporting cambial determination for wood formation can be developed from the finding with conifers that their dominant lignin precursor, coniferin, begins to be synthesized in the cambium soon after springtime dormancy break and well before the first cambial derivatives are produced, much less begin differentiating (Savidge, 1989). Coniferin biosynthesis in conifers appears to occur exclusively in the cambium, and lignin not only is a primary constituent of wood (along with cellulose and hemicelluloses) but is the one that makes wood woody. Precocious coniferin production in springtime suggests that the cambial genome "knows" that its derivatives will be lignifying even before it has produced them, suggesting cambial determination for lignification and thus for wood formation.

There have been other compelling arguments advanced both for and against cambial commitment, but in the writer's mind more research is needed before this important question can be finally resolved (Savidge, 1985, 1990b). It is certain, however, that at some point in their evolution plants developed the ability for formation of procambium and cambium and that this became spatially coupled to cambial derivatives having genetic competence to follow cellular differentiation pathways leading to phloem and xylem development. Primitive plants such as algae (e.g., *Fucus* spp.) produce phloem but not xylem, presumably because they lack genes for xylogenesis (Fielding et al., 1987). The phellogen, a meristem in the bark of trees, consists of cambial-like fusiform-shaped cells; these divide and function similarly to those of the vascular cambium, but the phellogen's derivatives do not differentiate into phloem or xylem (Fahn, 1974; Lev-Yadun and Liphschitz, 1989). Cells such as transfusion tracheids in conifer needles undergo tracheid differentiation independently of procambium or vascular cambium (Foster and Gifford, 1974). Differentiation of "tracheary elements" in vitro occurs in single cells in suspension cultures in the absence of preceding cell division (Fukuda, 1989). In view of these and additional observations (Savidge, 1985), it appears that gene expression for cambial growth and development occurs independently of that for tracheid differentiation. The conclusion that loose if any coupling exists between genes controlling formation, activity, and perpetuation of the cambium and those underlying differentiation of cambial derivatives suggests that genetic competence for cambial development was attained independently of that for phloem- and xylem-cell differentiation during evolution. These independently regulated sets of genes nevertheless must have in common something enabling spatial and temporal coordination of their expression during diameter growth.

II. THE CONCEPT OF CAMBIAL DORMANCY

A. Rest–Quiescence

The idea that the cambium has "rest" and "quiescent" stages of dormancy comes originally from studies with buds and seeds. Rest is the stage where buds will not grow and seeds will not germinate although all extrinsic factors be optimal for growth (Wareing, 1969). At least two studies have presented evidence against the existence of rest in conifer cambia (Barnett, 1971; Worrall, 1971). On the other hand, seedling trees of numerous other conifer as well as hardwood species have been reported unable to resume diameter growth immediately after cessation of cambial growth in late summer or early autumn when transferred into optimal growing conditions (Reinders-Gouwentak, 1965; Wareing, 1949; Wodzicki, 1961, 1964).

Increasing evidence suggests that the cambium in its first year (i.e., twig cambium) is physiologically and epigenetically distinct from older cambia (Little and Savidge, 1987; Mellerowicz et al., 1989). Using 20-year-old tamarack (*Larix laricina*) trees, stem ''chips'' approximately 2 cm (*L*) by 2 cm (*W*) by 0.5 cm (*D*) containing 7- or 8-year-old intact cambium sandwiched between mature xylem and phloem were explanted onto a completely defined nutrient medium supplemented with auxin as growth promoter. When explanted immediately after the onset of cambial dormancy, very little cambial growth and callus formation occurred in these chips. The extent of cambial growth, including production of new tracheids, increased progressively as autumn graded into winter. The ability for chips to produce wood in vitro was lot by midwinter; however, these nonxylogenic chips continued to generate substantial callus around their margins (R. A. Savidge, unpublished data). Sussex and Clutter (1959) also noted seasonal variation in growth of stem explants of hardwoods cultured in vitro. My observations on explants of tamarack suggest the existence of a narrow temporal window in autumn when stem tissues associated with merchantable wood are in the rest stage of dormancy and a second, wider time span in late winter when the cambium again cannot grow despite optimal conditions. The fact that callus continued to grow during the second period suggests the existence of distinct mechanisms for regulating cell division in parenchyma cells of phloem (from which the callus is primarily derived) as opposed to cambium. It is noteworthy that a firm, compact callus was invariably associated with the chip cambium's in vitro competence for wood formation, and that with loss of xylogenic competence the callus becomes granular and friable. Whether these changes can be explained physiologically [i.e., in terms of transmissible factors (e.g., derived from leaves or roots) stored in stem tissues having become limiting] or have their explanation in epigenetic changes remains to be determined.

Following a period at temperatures slightly above freezing, rest in buds or seeds changes into quiescence provided sufficient water is available to support ongoing metabolism (Kramer, 1958). Quiescence is that stage of dormancy when shoot growth occurs after limiting extrinsic factors are removed (Wareing, 1969). For example, when potted trees are transferred from outdoor conditions in midwinter to a warm greenhouse and given water and adequate light, they grow and produce a new layer of wood, demonstrating that it was the environment and not the intrinsic status of the tree which was preventing growth at that time.

Many investigators have considered the cambium to be in the rest stage of dormancy if 1-year-old cuttings taken from outdoor are found to have produced no new tracheids in response to apically applied exogenous auxin after being incubated upright in a warm lighted environment with adequate water for several weeks. When tracheids are produced, the cambium is interpreted to have been in the quiescent stage at the moment the experiment was begun. On the other hand, Savidge and Wareing (1981a,b) reported that cambial growth was inhibited in young disbudded and defoliated stem segments of lodgepole pine (*Pinus contorta* var. *latifolia*) when exogenous auxin was applied to their apical ends. In contrast to auxin-treated cuttings, cambia in disbudded and defoliated control segments (not provided exogenous auxin) had actively dividing cambia, but despite this cell-division activity no tracheid differentiation occurred (Savidge and Wareing, 1981b; Savidge, 1983). Clearly, use of differing methods and criteria to evaluate cambial dormancy can lead to contrasting interpretations about its nature and regulation.

B. Endo-, Para-, and Ecodormancy

Dormancy recently was redefined as "the temporary suspension of visible growth of any plant structure containing a meristem" (Lang, 1987). Lang (1987) subdivided dormancy into "endo," "para," and "eco" forms, depending on whether the inductive basis has its origin within the affected structure, is generated within the plant but outside the affected structure, or comprises one or more limiting extrinsic factors, respectively. For the cambium, Lang's requirement of "visible" growth must be interpreted as "microscopically visible" because this is the only certain way a dormant cambium can be distinguished from one that is active with existing technology. It has been suggested that the presence or absence of coniferin in conifer stems could be useful as a more rapid nonmicroscopically based test to distinguish active and dormant cambia, respectively (Savidge, 1989, 1991).

C. Cambial Dormancy Defined

Cambial dormancy is defined here simply as the suspension of cell-division activity anywhere within the cambium (i.e., in cells sandwiched between mature xylem and phloem). It is this aspect of dormancy that is central to annual ring formation. Cambial dormancy unquestionably occurs in many if not all tree species, but insufficient research data are yet available to interpret how to perceive it, much less explain in definitive physiological or biochemical terms how it comes about.

III. PHENOLOGY

A. Latewood Formation and Dormancy Onset

Following the late-spring through early-summer period of cambial growth, dormancy in trees indigenous to temperate zones predictably begins in late summer or early fall. Cessation of cambial cell-division activity in aboveground parts (the cambium of the root system remains little studied) rarely is in phase with, rather usually follows, termination of springtime root and early-summer shoot elongation. An autumn surge of root growth normally follows onset of cambial dormancy in the stem (Kozlowski, 1971). In hardwoods, particularly diffuse-porous species, the onset of dormancy is more closely synchronized with shoot and leaf maturation than is the case in conifers (Wareing and Roberts, 1956; Evert, 1963; Derr and Evert, 1967; Kozlowski, 1971). Latewood formation begins at the stem base and progresses upward to the tree crown (Larson, 1964; Kozlowski, 1971); however, in sapling and larger conifers, twig cambia (which produce little or no latewood during their first year of diameter growth) nevertheless become dormant considerably before older cambia (R. A. Savidge, unpublished data).

Common stages of dormancy onset in conifers are diagrammed in Fig. 3. By late summer, cell division activity slows markedly. Whereas earlier in the growth season divisions in the tangential plane (''periclinal'' divisions) are common and divisions in the radial plane (''anticlinal'' divisions) are scarce, the frequency of anticlinal divisions usually increases as dormancy approaches (Bannan, 1967). In many species of Pinaceae, centripetal fusiform derivatives differentiate as axial resin canals instead of as latewood tracheids at this time (Panshin and de Zeeuw, 1980). Cells on the inner periphery of the cambium enlarge little if at all, but these cells nevertheless are initiated into the tracheid-differentiation pathway (Fig. 3, stage 1). Enlargement of centrifugal cambial derivatives (i.e., cells on the outer periphery of the cambium) and differentiation of these centrifugal cells into new phloem cells also occurs.

Stage 2, the complete cessation of cell-division activity, follows and constitutes cambial dormancy (Fig. 3). The onset of dormancy occurs when the extrinsic environment of the tree appears suitable for growth. The ability for ongoing metabolism in the cambial region at this time is

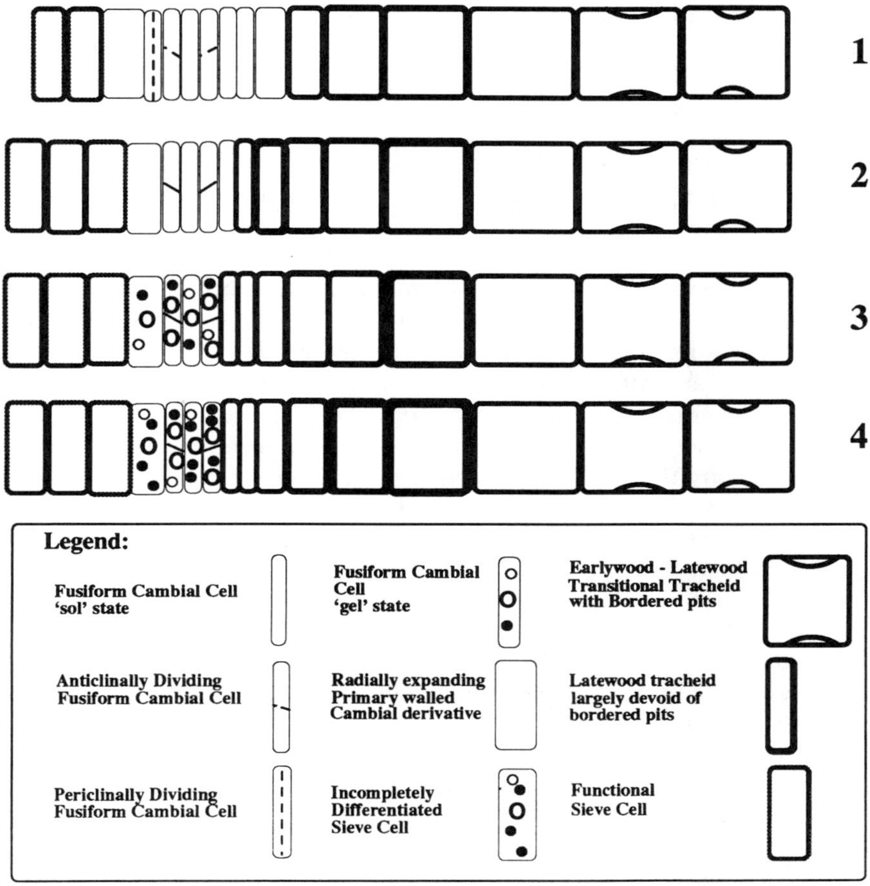

Fig. 3 Diagrammatic representation of dormancy onset in radial files of the cambium of a conifer as seen in transverse sections. The stages from active growth into dormancy, numbered 1 to 4, are explained in the text (see Section III).

evident by the fact that latewood formation and phloem development usually continue after the cambium becomes dormant.

During stage 3 (Fig. 3), the protoplasm of fusiform cambial cells begins transforming; most conspicuously, the single large vacuole breaks up into a multivacuolated condition (Bailey, 1954; Rao and Catesson, 1987). Differentiation of centrifugal and centripetal cambial derivatives into sieve cells and latewood tracheids, respectively, proceeds resulting in the radial-file cell-number width of the cambium being reduced. In conifers, the final tier of differentiating latewood tracheids typically undergoes less wall thickening than had occurred in preceding tiers, suggesting that these last differentiating tracheids are metabolically inactivated prematurely. In hardwoods the final derivatives may develop, in autumn or the following spring, as a tier of thin-walled radially narrow parenchyma rather than as lignified tracheids or fibers (Panshin and de Zeeuw, 1980).

At stage 4 (Fig. 3) the protoplasmic gel transformation within the cambium of most tree species is complete, and the differentiation of phloem and xylem elements has halted. One to two enlarging cells per radial file on the centrifugal periphery of the cambium typically do not completely differentiate into sieve cells at this time; these join the cambium in transforming from a ''sol'' to a ''gel'' protoplasmic state. Although cell-division activity and vascular development have both ceased, metabolic activities within the cambium (including increases in nuclear volume, accumulation of storage reserves, acquisition of resistance to freezing, membrane trafficking, and thickening of radial walls) may still occur (Bailey, 1954; Timell, 1973; Catesson, 1981; Sennerby-Forsse, 1986; Rao and Catesson, 1987; Mellerowicz et al., 1990).

The cambium of *Pinus radiata*, growing as an exotic species in New Zealand, differs from other conifer species (in their native habitats) in the above-described stages by remaining in the sol state throughout the winter months. Moreover, some of its centrifugal derivatives continue to differentiate as phloem during winter (Barnett, 1971, 1973). How the cambium of *P. radiata* behaves in its native habitat of California is unknown to me.

B. Dormancy Break and Earlywood Formation

After overwintering in the gel state, when soil and air temperatures, soil moisture availability, photoperiod, and light intensity all are on the increase in spring, root growth predictably resumes before aboveground parts begin to grow. At this time, the stem's dormant cambium (Fig. 4, stage 1) begins to rehydrate, changing from a dense ''gel'' to a highly vacuolated ''sol'' state (Fig. 4, stage 2) (Bailey, 1954; Timell, 1973). At about the same time incompletely differentiated cells on the centrifugal periphery of the cambium resume their differentiation into sieve cells, and radial walls of fusiform cambial cells begin thinning to dimensions about one-half of their dormant width (Evert, 1963; Derr and Evert, 1967; Timell, 1973; Rao and Catesson, 1987).

In conifers cell-division activity first becomes evident in fusiform cambial cells near and often adjoining the last differentiated tracheids of

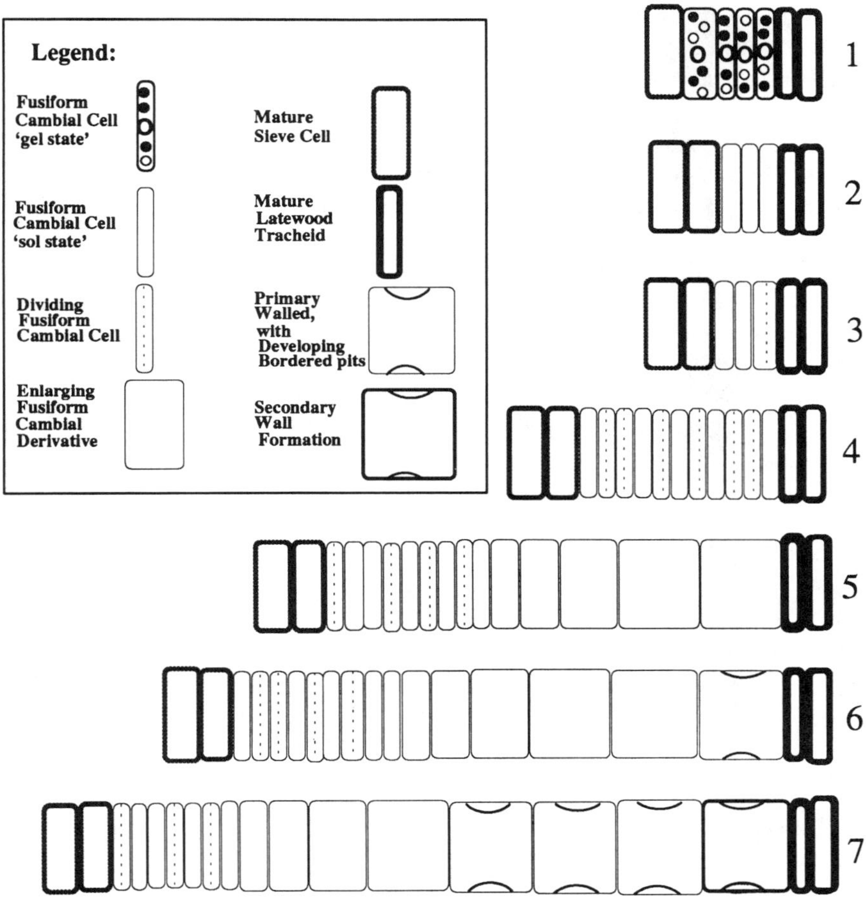

Fig. 4 Diagrammatic representation of springtime transition from dormancy to cell division activity and first earlywood formation for radial files of 2-year-old conifer cambium as seen in transverse sections. The stages from dormancy into active growth, numbered 1 to 7, are explained in the text (see Section III).

the previous year's growth ring (Fig. 4, stage 3), whereas in hardwoods the first mitotic activity occurs more generally within cells of the cambium. In conifers and diffuse-porous hardwoods, cell-division activity precedes initiation of radial expansion; however, in ring-porous hardwoods the first radial expansion accompanies or precedes the resumption of cell-division activity (Atkinson and Denne, 1988). In sapling or larger trees, these reactivation events typically occur well before bud break in both conifers and ring-porous hardwoods; however, they are concomitant with or follow bud break in diffuse-porous species (Atkinson and Denne, 1988; Savidge and Wareing, 1981b). Beginning at or near the base of the opening bud, resumption of cell-division activity shows a basipetal progression down stems of hardwoods. In contrast, it appears to begin at the crown base in conifers (Savidge and Wareing, 1981b; Savidge, 1985; Atkinson and Denne, 1988).

Once the cambium is active (Fig. 4, stage 3), repeated cell divisions in the periclinal plane increase the number of radially narrow fusiform cells in each radial file by as much as threefold, although in young twigs a lesser response occurs (Fig. 4, stage 4). Growth occurs exponentially thereafter. As many as 90% of the cells that will make up the new growth ring can be generated before the longest day of the year arrives. Following closely upon this springtime surge of meristematic activity, cambial cells adjoining mature latewood cease dividing and begin enlarging to achieve transverse diameters many times that of their original radial dimensions (Fig. 4, stage 5). In diffuse-porous hardwoods this diameter growth by radial expansion occurs after bud break, but as already noted, massive enlargement of the first vessel elements in ring-porous hardwoods can occur well before budbreak. In sapling and more mature conifers, as many as 20 radially expanded primary-walled cambial derivatives per radial file may accumulate in the main stem near the base of the crown before xylem development per se is initiated anywhere in the stem (Savidge and Wareing, 1981b, 1984; Savidge, 1988, 1989).

Depending on the tree species, there are a number of distinct xylem-cell differentiation pathways that can be pursued (Panshin and de Zeeuw, 1980), and at the stage (Fig. 4, stage 5) of radial expansion the genome of centripetal cambial derivatives apparently is uncommitted to any particular pathway. For example, if the cambial region is injured at this time the enlarged cells predictably differentiate as traumatic resin canals and axial parenchyma rather than as tracheids (Savidge and Farrar, 1984). Excepting the more or less concomitant development of first earlywood vessels that occurs over the full length of stem and branches in ring-porous species, the initiation of general xylem development (as detected by the presence of developing bordered pits, secondary walls, and lignin) begins first in foliated stem regions. From there earlywood development exhibits a basipetal progression down the stem toward the root system. As the first enlarged derivatives are differentiating into earlywood, centrifugally adjoining cambial derivatives within each radial file undergo primary-wall radial expansion and then initiate their pathway of terminal differentiation

(Fig. 4, stage 6). Through cell-division activity more and more cambial derivatives join the expansion and differentiation queues (Fig. 4, stage 7).

IV. REGULATION OF CAMBIAL GROWTH AND DEVELOPMENT

A. Extrinsic Factors

On first principles alone, there can be no doubt that ecodormancy (Lang, 1987), mediated through changes in temperature, photoperiod, and water availability, affects stems of trees. Temperate-zone trees adapt readily to the yearly cycles of temperature, photoperiod, and wet and dry seasons. Temperature and photoperiod variations are slight for trees growing in the tropics; here moisture cycles are the more important (Breitsprecher and Bethel, 1990).

Direct actions of water, temperature, and photoperiod on the cambium are considered in the comments below. It should be remembered that these same extrinsic factors also act indirectly on the cambium by controlling the timing and extent of metabolism and primary growth in shoot and root systems, thereby influencing the availability of substrate and regulatory factors to the cambium.

Photoperiod

In an experimental greenhouse, trees exhibit the capacity for continuous growth both in height and diameter when maintained in either long-day or continuous-light photoperiodic conditions where none of water, temperature, and nutrient is limiting. When the day length is artificially shortened to 10 h or less, trees enter rest and will not resume growth until a chilling requirement has been met (Wareing, 1969; Hänninen, 1987; Vince-Prue, 1984).

Bands of earlywood and latewood can be induced in the absence of induction of cambial dormancy by manipulating photoperiod during cambial growth (Larson, 1962; Wodzicki, 1961, 1964). A brief exposure to short days gives rise to thick-walled latewood-like tracheids, whereas long days yield wide-diameter earlywood-like tracheids. Leaves are thought to be the site of photoperiod perception (Wareing and Roberts, 1956). In addition, there is evidence that a factor from leaves controls the extent of secondary-wall material deposition during wood formation (Savidge and Wareing, 1981a,b; Savidge, 1983).

When maintained in continuous darkness the dormant cambium within chips of *L. laricina* grown in vitro reactivates, divides, and produces new xylem (R. A. Savidge, unpublished observations), and similar observations have been made in vivo (Savidge, 1983). In vivo, it appears that the growing cambium is indirectly sensitive to decreasing photoperiod due to biochemical changes in foliage or shoot apices being transmitted to the cambium (Wareing and Roberts, 1956; Wodzicki, 1961); however, the molecules or signals involved remain to be identified. Once dormant the cambium in foliated regions of evergreens is maintained dormant more or less synchronously with buds, in agreement with the finding that photoperiod perception by the cambium is mediated indirectly through buds and leaves (Wareing and Roberts, 1956; Wodzicki, 1961).

Temperature

Cambial dormancy in the shoot system usually begins when soil temperatures are still sufficiently high to support autumn root growth and when above ground, daytime temperatures are also adequate for growth. This suggests that low temperature probably does not induce cambial dormancy. Following onset, cool to subfreezing temperatures characterize the greater period of cambial dormancy, and plant growth, whether by cell division or expansion, or both, occurs slowly if at all when temperatures hover at or below 10°C (i.e., under usual late autumn and early winter conditions). The catalytic activity of enzymes decreases exponentially as a function of decreasing temperature; moreover, cytoskeletal microtubules in plant cells destabilize at low temperatures, assuming a soluble protomeric form below 4°C (Akashi and Shibaoka, 1987; Voet and Voet, 1990). The cytoskeleton, considered essential for xylogenesis, when unstable presumably limits numerous metabolic processes related to cambial growth and wood formation (Hepler and Fosket, 1971; Kobayashi et al., 1988; Lloyd, 1984; Thorsch and Esau, 1982; Murmanis, 1971).

Both Worrall (1971) and Savidge and Wareing (1981b) provided evidence that the cambium of conifers could be maintained in ecodormancy due to cool temperatures. Upon warming localized regions of the stem, while chilling other regions in the tree, cambial activity and wood formation were promoted beneath the warmed site but not elsewhere.

It is well established that cool temperatures prepare buds and seeds in the resting stage of dormancy for ensuing growth once temperatures become elevated, and related observations have been made on the cambium (Reinders-Gouwentak, 1965; Savidge and Wareing, 1981b). Cool (nonfreezing) temperatures are known to favor hydrolysis of starch to sugar, and it is likely that they fulfil additional metabolic roles associated with progression from one stage of dormancy to the next.

Water

Water makes up a high proportion of the fresh weight of actively dividing meristems such as the cambium (Savidge et al., 1982). In summer a mature evergreen conifer can readily absorb 40 L of water per day, and hardwoods still more, but of this less than 5% may be used for metabolism, the rest being evaporated unavoidably to air through open stomata as CO_2 enters leaves. Water moves up to the leaves in older sapwood, and in addition to maintaining turgidity, water serves as a solvent and carrier of solutes and also as reactant or substrate in innumerable enzymatically catalyzed transformations. When water is limiting, respiration and associated energy (e.g., ATP) production decrease. The ability for cell enlargement is also reduced due to decreased turgor. Stomata close, and photosynthesis is greatly reduced. Annual ring width and its distribution along the bole, seasonal duration of cambial growth, proportion of xylem to phloem increment, time of latewood initiation, and duration of latewood production are all influenced by water availability during the growth season (Kozlowski, 1979).

Water moves passively from cell to cell, whether it be a meristematic cambial cell, dead xylem element, or conducting sieve cell, by diffusion along gradients of decreasing water potential (ψ_w).* The xylem transpiration stream typically has a low solute content, consequently little osmotic strength ($\psi_o \sim 0$). The xylem sap's weak osmotic potential is compensated for by its very negative pressure potential (ψ_p). Excepting periods when the soil is at field capacity and the air at 100% relative humidity, several processes work together to generate tension (i.e., $\psi_p \ll 0$) in the xylem sap. These are extremely negative atmospheric ψ_w values leading to rapid daytime transpirational losses of water from stomata during photosynthesis, strongly negative ψ_o values in photosynthetic cells which result in water flowing up the tree into the leaves, and relatively slow rates of daytime replenishment of stem water reserves through root absorption of soil water.

The phloem, in contrast to the xylem, contains approximately 20% sucrose and, consequently, it attracts water quite strongly through osmosis. The osmotic forces ($\psi_o \ll 0$) in the phloem in combination with stoppage of transpirational losses from leaves by stomatal closure ensures that the phloem maintains positive (turgor) pressure ($\psi_p > 0$) by drawing on reserves of stem water even during periods of drought.

---

*Water potential comprises the algebraic sum of ψ_o (osmotic potential), ψ_p (pressure potential), ψ_g (gravitational potential), and ψ_m (matric potential). ψ_w never exceeds zero and usually is negative.

Whenever water potential in the stem is low, cambial water diffuses into xylem or phloem, depending on which has the more negative ψ_w. Maintenance of an adequate water content in the cambium probably is a key factor controlling when it grows and does not grow. Although all other conditions may be conducive for growth, the cambium cannot divide and the dormant cambium will not reactivate if water is limiting (R. A. Savidge, unpublished observations).

B. Intrinsic Factors

Nutriment

In the absence of carbohydrates and inorganic nutrients the cambium does not grow in vitro (R. A. Savidge, unpublished observations); however, tree stems harbor considerable quantities of protein, starch, and lipid reserves, and it is unlikely that nutritional factors are normally in such short supply that they limit seasonal activity of the cambium (Savidge and Wareing, 1981b; Savidge, 1991).

Phytohormones

Under in vitro conditions where neither water, temperature, or light period is limiting, unorganized plant cell cultures require not only the inorganic substances acquired from the soil and carbohydrates generated through photosynthesis but also small but definite amounts of phytohormones (e.g., auxin and cytokinin) for cell-division activity. Without these hormonal supplements, although all other nutritional and environmental factors are nonlimiting, there is negligible growth. Other phytohormones known to affect cell-division and differentiation activities include gibberellins, abscisic acid, and ethylene.

Auxin. Indol-3-ylacetic acid (IAA), otherwise known as auxin, appears to be ubiquitous in tree cambia (Savidge, 1990a). Exogenous IAA promotes primary-wall growth in cambial derivatives (Savidge and Wareing, 1981a). Investigations into seasonal variation of IAA have repeatedly shown that the cambium's endogenous IAA content is higher during growth than during dormancy, and higher IAA concentrations are associated with earlywood formation than with latewood formation (Savidge, 1988; Little and Savidge, 1987). When applied to the surface of a population of parenchymatous isodiametric cells, the cells in the auxin path elongate to form procambial-like strands (Savidge and Wareing, 1981b).

Considerable evidence supports the theory that IAA, derived from leaves and also from shoot and root apices (as opposed to being synthesized in the cambium itself), is the common basis for cambial development being spatially associated with phloem and wood formation (Savidge and Wareing, 1981b; Little and Savidge, 1987; Savidge, 1988; 1990b). IAA promotes both xylem and phloem development, and when auxin sources such as leaves and buds are removed, fusiform cambial cells lose their stability (Savidge and Wareing, 1981b; Little and Savidge, 1987; Savidge, 1988, 1990b). They enlarge in diameter, their nuclei change in morphology from being fusiform-shaped to being spherical, and concomitantly the cells undergo repetitive true transverse divisions resulting in the cambium de-differentiating into axial parenchyma. The cambium's normal appearance is maintained if exogenous IAA is applied in place of leaves and buds (Savidge and Wareing, 1981b; Savidge, 1983).

Cytokinins. The little research that has been done into seasonal variation of endogenous cytokinins suggests that the cambium's content becomes substantially elevated in association with late-spring to early-summer cambial cell-division activity (Little and Savidge, 1987; Savidge, 1988, 1990b). Cytokinins are derived from growing roots and are transported acropetally to the tree crown in the transpiration stream, the transport process characteristically beginning in early spring before bud break (Savidge, 1989, 1990b). There is little doubt that cytokinins participate with auxin in regulating lignification (Savidge, 1989, 1991). It is possible that there exists an obligate requirement for the presence of both auxin and cytokinin before cambial cell-division activity can occur, that the surge of root growth in spring and the surge of shoot growth in early summer contribute sufficient cytokinin and auxin, respectively, to the cambium to support initially rapid followed by declining cell-division activity as observed in nature. Contradicting this, however, is the observation that auxin is the only phytohormone supplement that has been found essential for promotion of growth in *Larix laricina* chips and stem-tissue cultures of other conifer and hardwood species in vitro (Sussex and Clutter, 1959; R. A. Savidge, unpublished data). Addition of minute quantities of synthetic cytokinin to the culture medium results in substantial inhibition of cell-division activity (R. A. Savidge, unpublished data). Possibly this is because the exogenous cytokinin levels, although minute, are together with endogenous levels supraoptimal and therefore inhibitory.

Gibberellins. Gibberellic acid (GA3), well known for its ability to break dormancy and promote both cell division and cell elongation in higher plant cells, strongly promotes cambial cell-division activity in hardwoods (Savidge and Wareing, 1981b; Savidge, 1985). GA3 and other gibberellins have been identified in conifer cambia (Pharis et al., 1981; R. Pharis, personal communication) but none has been reported to promote cambial activity when applied to stem segments of conifers (Little and Savidge, 1987). The cambium's content of gibberellins in relation to cambial growth and dormancy remains to be investigated. A combination of auxin, cytokinin, and GA3 was reported to promote more cambial growth in hardwoods than auxin with GA3 (Zakrewski, 1983). Recently it was observed that this triplet combination of phytohormones was more active on conifer cambia than any doublet combination (R. A. Savidge, unpublished data).

Abscisic Acid. The phytohormone *S*-abscisic acid (ABA) occurs abundantly in leaves and also is present in the cambium (Little and Savidge, 1987). ABA was considered to have roles in inducing both latewood formation and cambial dormancy; however, a series of investigations involving applications of varied auxin/ABA ratios to apical ends of dormant conifer cuttings showed that ABA had little or no inhibitory effect on cell-division activity (M. Leitch and R. A. Savidge, unpublished data). Pharis et al. (1981) reported that ABA enhanced rather than inhibited diameter growth in *Pinus radiata*; concomitantly, ABA promoted gibberellin metabolism. Measurements of endogenous ABA have yielded no evidence for an ABA increase in relation to cambial dormancy (Little and Savidge, 1987; Lachaud, 1989). On the other hand, the overall inhibitor content of the cambium, when measured by bioassays, does increase in autumn (Wodzicki and Wodzicki, 1980). Recent findings with noncambial higher plant tissues suggest that ABA may have roles in increasing the cambium's frost hardiness, freezing tolerance, and unloading of phloem-transported solutes into the cambial zone.

Ethylene. The role of ethylene in cambial growth remains uncertain despite it having been shown many years ago that ethylene transforms the cambium from rest to quiescence (Reinders-Gouwentak, 1965; Savidge, 1988). The growing cambium generates considerable volumes of ethylene (Savidge, 1988) and also contains the water-soluble immediate precursor of ethylene (i.e., 1-aminocyclopropane-1-carboxylic acid (ACC)] (Savidge et al., 1983). In vitro, it has been reported that the dormant cambium's ACC content can be so low as to be limiting for ethylene production (Savidge, 1988). High ethylene levels, generally associated with meristematic activity in higher plants, give rise to heightened respiratory activity and resultant ATP production in support of growth. It is possible that auxin promotion of meristematic activity with consequent ethylene generation aids the cambium in resisting entry into rest. In addition, ethylene is produced in response to most stresses, including water deficiency and chilling, suggesting a possible mechanism by which the cambium's rest stage of dormancy is changed into quiescence.

V. CONCLUSIONS

Annual layers of wood are formed because the cambium enters into and awakens from dormancy each fall and spring, respectively. Through trial and error over geological time, trees able to produce latewood and enter dormancy presumably were favored in temperate zones by the process of natural selection, with annual rings arising as a consequence of this selection pressure. Both dormancy onset and release are complex processes involving numerous subcellular changes the intrinsic regulation of which remain poorly understood.

The triplet combination of high soil water potential, warm temperatures, and long-day photoperiods that typically occurs in temperate zones in late spring and early summer promotes annual-ring formation by providing the cambium, as well as the shoot and root systems, with sufficient water and energy for growth. As they grow, leaves and roots produce essential hormonal regulatory factors, such as auxins and cytokinins, respectively, some of which are exported to the cambium (Savidge and Wareing, 1981b; Little and Savidge, 1987; Savidge, 1988). The phenology of development in shoot and root systems overlaps but imperfectly in spring and is out of phase in autumn when root growth is active and the shoot system is dormant. These differences in phenology undoubtedly affect the timing of cambial growth, but more research is needed before the molecular basis for growth and dormancy will be understood. At the end of each growth period, the termination of cambial activity occurs when temperatures remain conducive to growth. Decreasing photoperiod at this time is believed to lead indirectly to cambial dormancy through leaf production of unknown physiological factors which are transmitted to the cambium. Low water potential is hypothesized to be another major factor in the initiation of cambial dormancy; the positioning of the cambium between highly osmotic phloem and tensed water columns in the xylem is suggested to make it difficult for the cambium to retain sufficient water in support of growth whenever water is limiting. Once dormancy has begun, cool temperatures contribute to maintaining dormancy, but they also prepare the cambium, metabolically, for the next spring's surge of diameter growth.

REFERENCES

Akashi, T., and Shibaoka, H. (1987). *Plant Cell Physiol.*, *28*: 339.

Atkinson, C. J., and Denne, M. P. (1988). *Ann. Bot.*, *61*: 679.

Bailey, I. W. (1954). *Contributions to Plant Anatomy*, Chronica Botanica, Waltham, Mass.

Bannan, M. W. (1967). *For. Prod. J.*, *17*: 63.
Barnett, J. R. (1971). *N.Z. J. For. Sci.*, *1*: 208.
Barnett, J. R. (1973). *Ann. Bot.*, *37*: 1005.
Beck, C. B. (1970). *Biol. Rev. Cambridge Philos. Soc.*, *45*: 379.
Breitsprecher, A., and Bethel, J. S. (1990). *Ecology*, *71*: 1156.
Catesson, A.-M. (1981). *Bull. Soc. Bot. Fr. Actual. Bot.*, *128*: 43.
Connor, K. F., and Lanner, R. M. (1990). *Am. J. Bot.*, *77*: 1070.
Creber, G. T., and Chaloner, W. G. (1984). *Bot. Rev.*, *50*: 357.
Denne, M. P. (1989). *IAWA Bull. n.s.*, *10*: 59.
Derr, W. F., and Evert, R. F. (1967). *Am. J. Bot.*, *54*: 147.
Evert, R. F. (1963). *Am. J. Bot.*, *50*: 149.
Fahn, A. (1974). *Plant Anatomy*, 2nd ed., Pergamon Press, Elmsford, N.Y.
Fielding, A. H., Carter, P. L., and Smith, C. A. (1987). *Phycologia*, *26*: 501.
Foster, A. S., and Gifford, E. M., Jr. (1974). *Comparative Morphology of Vascular Plants*, 2nd ed., W. H. Freeman, New York.
Fukuda, H. (1989). *Bot. Mag. Tokyo*, *102*: 491.
Gensel, P. G., and Andrews, H. N. (1987). *Am. Sci.*, *75*: 478.
Glock, W. S. (1941). *Bot. Rev.*, *7*: 649.
Hänninen, H. (1987). *Silva Fenn.*, *21*: 279.
Hepler, P. K., and Fosket, D. E. (1971). *Protoplasma*, *72*: 213.
Kobayashi, H., Fukuda, H., and Shibaoka, H. (1988). *Protoplasma*, *143*: 29.
Kozlowski, T. (1971). *Growth and Development of Trees*, Vol. 2, Academic Press, New York.
Kozlowski, T. (1979). *Tree Growth and Environmental Stresses*, University of Washington Press, Seattle, Wash.
Kramer, P. J. (1958). In *The Physiology of Forest Trees* (K. V. Thimann, ed.), Ronald Press, New York, pp. 573–580.
Lachaud, S. (1989). *Trees Struct. Funct.*, *3*: 125.
Lang, G. A. (1987). *Hort. Sci.*, *22*: 817.
Larson, P. R. (1962). *Am. J. Bot.*, *49*: 132.
Larson, P. R. (1964). In *The Formation of Wood in Forest Trees* (M. H. Zimmermann, ed.), Academic Press, New York, pp. 345–365.
Larson, P. R. (1982). In *New Perspectives in Wood Anatomy* (P. Baas, ed.), Nijhoff, Dordrecht, The Netherlands, pp. 85–122.
Lev-Yadum, S., and Liphschitz, N. (1989). *IAWA Bull. n.s.*, *10*: 43.
Little, C. H. A., and Savidge, R. A. (1987). *Plant Growth Regul.*, *6*: 137.
Lloyd, C. W. (1984). *Int. Rev. Cytol.*, *86*: 1.
Mellerowicz, E. J., Riding, R. T., and Little, C. H. A. (1989). *Can. J. Bot.*, *67*: 990.
Mellerowicz, E. J., Riding, R. T., and Little, C. H. A. (1990). *Can. J. Bot.*, *68*: 1857.
Murmanis, L. (1971). *New Phytol.*, *70*: 1089.
Panshin, A. J., and de Zeeuw, C. (1980). *Textbook of Wood Technology*, 4th ed., McGraw-Hill, New York.

Pharis, R. P., Jenkins, P. A., Aoka, H., and Sassa, T. (1981). *Aust. J. Plant Physiol.*, *8*: 559.
Philipson, W. R., Ward, J. M., and Butterfield, B. G. (1971). *The Vascular Cambium: Its Development and Activity*, Chapman & Hall, London.
Rao, K. S., and Catesson, A.-M. (1987). *Can. J. Bot.*, *65*: 246.
Reinders-Gouwentak, C. (1965). In *Encyclopedia of Plant Physiology*, Vol. 15 (W. Ruhland, ed.), Springer-Verlag, New York, pp. 1076–1105.
Savidge, R. A. (1983). *Histochem. J.*, *15*: 447.
Savidge, R. A. (1985). In *Attributes of Trees as Crop Plants* (M. G. R. Cannell and J. E. Jackson, eds.), NERC Press, ITE, Abbots Ripton PE17 2LS, England, pp. 208–227.
Savidge, R. A. (1988). *Tree Physiol.*, *4*: 401.
Savidge, R. A. (1989). *Can. J. Bot.*, *67*: 2663.
Savidge, R. A. (1990a). *Can. J. Bot.*, *68*: 521.
Savidge, R. A. (1990b). In *Fast Growing Trees and Nitrogen Fixing Trees* (D. Werner and P. Müller, eds.), Gustav Fischer, Jena, Germany, pp. 142–151.
Savidge, R. A. (1991). *For. Sci.*, *37*: 953.
Savidge, R. A., and Farrar, J. L. (1984). *Can. J. Bot.*, *62*: 2872.
Savidge, R. A., and Wareing, P. F. (1981a). *Planta*, 153: 395.
Savidge, R. A., and Wareing, P. F. (1981b). In *Xylem Cell Development* (J. R. Barnett, ed.), Castle House, Tunbridge Wells, England, pp. 192–235.
Savidge, R. A., and Wareing, P. F. (1984). *Can. J. For. Res.*, *14*: 676.
Savidge, R. A., Heald, J. K., and Wareing, P. F. (1982). *Planta*, *155*: 89.
Savidge, R. A., Mutumba, G. M. C., Heald, J. K., and Wareing, P. F. (1983). *Plant Physiol.*, *71*: 434.
Sennerby-Forsee, L. (1986). *Physiol. Plant.*, *67*: 529.
Sussex, I. M., and Clutter, M. E. (1959). *Science*, *129*: 836.
Thorsch, J., and Esau, K. (1982). *J. Ultrastruct. Res.*, *78*: 73.
Timell, T. E. (1973). *Ultrastructure of the Dormant and Active Cambial Zones and the Dormant-Phloem Associated with Formation of Normal and Compression Woods in* Picea abies *(L.) Karst*, State Univ. Coll. Environ. Sci. For., Syracuse Univ. Tech. Publ. 96.
Vince-Prue, D. (1984). *Plant Cell Environ.*, *7*: 507.
Voet, D., and Voet, J. G. (1990). *Biochemistry*, Wiley, New York.
Wareing, P. F. (1949). *Nature*, *163*: 770.
Wareing, P. F. (1969). *Symp. Soc. Exp. Biol.*, *23*: 241.
Wareing, P. F., and Roberts, D. L. (1956). *New Phytol.*, *55*: 356.
Wodzicki, T. J. (1961). *Acta Soc. Bot. Polon.*, *30*: 293.
Wodzicki, T. J. (1964). *J. Exp. Bot.*, *15*: 584.
Wodzicki, T. J., and Wodzicki, A. B. (1980). *Physiol. Plant.*, *48*: 443.
Worrall, J. (1971). *Can. J. For. Res.*, *1*: 84.
Zakrewski, J. (1983). *Physiol. Plant.*, *57*: 537.

21
Periodic Factors Influencing the Spatial Patterns in Bivalve Mollusk Shells

Christopher A. Richardson *School of Ocean Sciences, Marine Science Laboratories, Menai Bridge, Anglesey, Wales*

I. INTRODUCTION

Many marine bivalve mollusk shells contain a wealth of information about the way in which environmental factors affect shell deposition. Numerous growth rings, spines, and ridges adorn the shell surface, while internally the shell structure is constructed of an intricate crystalline and organic matrix. Within the internal structure of the shell are observed a series of growth lines and bands that result from periodic changes in the environment surrounding the organism. The periodic factors influencing the formation of internal and surface growth patterns have been studied, not only from the biological standpoint, but also from their importance in understanding the palaeoenvironmental conditions for growth in fossil species. Palaeontologists have not been slow to recognize that patterns within the structure of fossil species could be used to determine palaeoecological conditions and to obtain evidence for changes in the earth's rotation. In view of the extensive literature concerning various aspects of growth rhythms in molluscan shells this review is necessarily selective in its coverage. I have concentrated particularly on the important factors that govern the patterns in bivalve mollusks and have avoided reference to other molluscan species, especially any reference to the complex and poorly understood biochemical and physiological mechanisms by which the patterns are produced.

II. SURFACE GROWTH PATTERNS

A. Annual Growth Checks

Surface growth rings or checks commonly observed on the shells of bivalve mollusks have long been observed and widely used in age determinations. These rings, however, are not always annual, and even when they are present their appearance may be confused by the presence of other nonannual checks or disturbance rings.

Regional and latitudinal variations in linear growth rates can produce annual rings of different clarity. In temperate waters seasonally induced variations in shell growth rates often produce clear surface rings, while in warmer or tropical waters where seasonal variations are less pronounced, rings may be ill defined or absent. In intertidal bivalves the rings may be more pronounced than those grown under more uniform conditions subtidally. Surface growth rings are produced during periods of suspended shell growth and may be associated with changes in seawater temperatures (Richardson et al., 1980c), environmental disturbances such as storms (Farrow, 1971), predation (Richardson et al., 1980a), or with the reproductive cycle (Seed, 1976).

During the early years of shell growth the rings are widely spaced, but in later years, as linear growth rates slow down, they progressively become deposited closer together and individual rings may become more difficult to identify. The shell surface of long-lived species such as ocean quahogs (*Arctica islandica*) and horse mussels (*Modiolus modiolus*) are covered with many growth rings, and the identity of individual rings is difficult to ascertain toward the shell margin.

B. Patterns of Growth Ridges and Spines

In addition to the annual surface growth rings there are frequently undulations, ridges, or spines on the surface of mollusk shells. Many of these structures form complex sculpturing on the surface, which may be purely ornamental or may help to anchor the animal in the sediment during burrowing (Stanley, 1970). The shell surface of the scallops *Pecten diegensis*) (Clark, 1968, 1975) and *P. maximus* (Antoine, 1978) and the queen scallop *Chlamys opercularis* (Broom and Mason, 1978) are covered with small, fine, curled growth ridges or striae. Panel 1 in Fig. 1 shows the appearance of the growth ridges of *P. maximus* as seen in shell section. It has been suggested that these structures have a daily periodicity of formation, although there is contrary evidence. Clark (1968, 1975) claimed that the ridges observed on the shell surface of *P. diegensis* grown in field and laboratory experiments were deposited with a daily

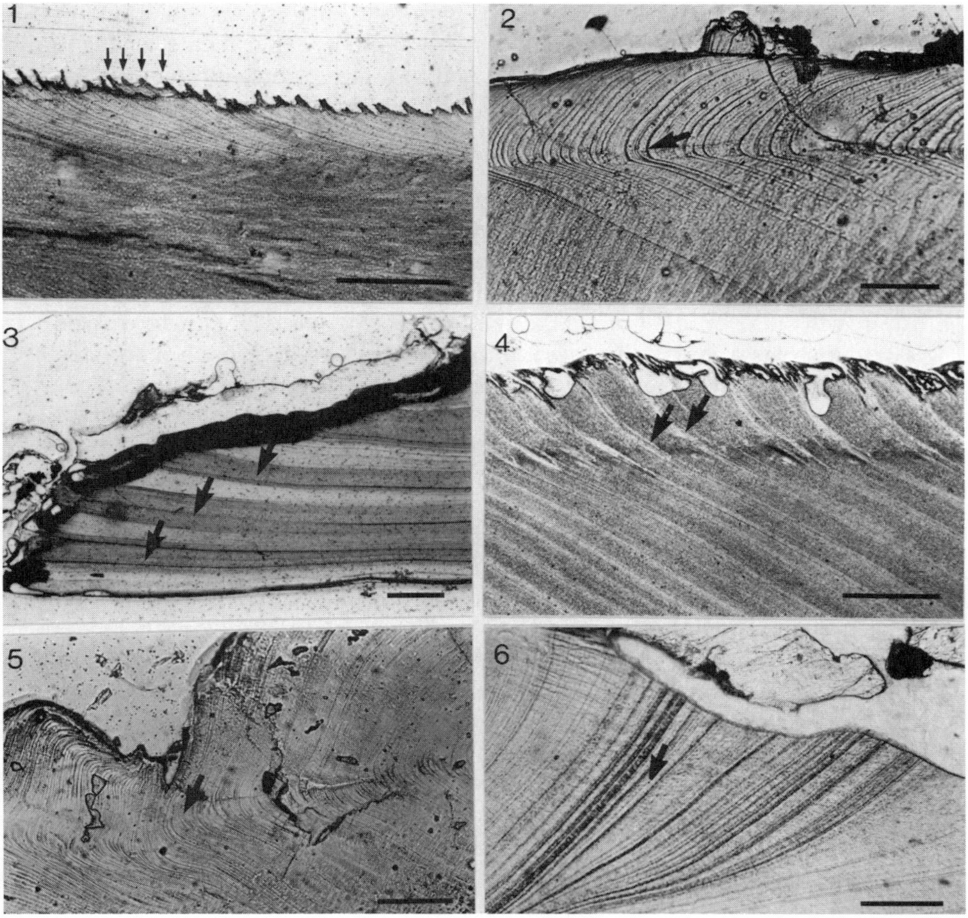

Fig. 1 Photomicrographs of acetate peel replicas of shell sections of bivalve shells. Growth from right to left, except figure 6, where growth is from left to right. 1, Ridges or striae (arrows) on the shell surface of *Pecten maximus*, scale bar = 500 μm; 2, section of a cockle shell, *Cerastoderma edule*, to show the alternating pattern of dark growth bands (arrow) and light growth increments between, scale bar = 100 μm; 3, annual growth lines (arrows) in the shell of the horse mussel *Modiolus modiolus*, scale bar = 500 μm; 4, annual growth lines (arrows) in the shell of the ocean quahog *Arctica islandica*, scale bar = 500 μm; 5, narrowing of the microgrowth bands (arrow) during the winter in the shell of *C. edule*, scale bar = 100 μm; 6, salinity stress lines (arrow) in the shell of the Malaysian cockle *Anadara granosa*, scale bar = 200 μm.

periodicity, although the number of ridges on marked shells did not coincide exactly with the number of days. In other experiments, using a variety of pectinid species, he obtained similar results, although occasionally some species produced more ridges than there were days. In the case of *Argopecten gibbus* some shells formed two lines per day, suggesting that the ridges might not be daily. Similarly, Broom and Mason (1978) found that on the shell of *C. opercularis* the ridges were laid down daily but only when the animals were growing rapidly; during less favorable conditions either side of the summer, no ridge was formed. Gruffydd (1981) has resolved the controversy by clearly demonstrating that the rate of growth ridge production on the shell surface of small laboratory-reared *P. maximus* is related to the size of the shell and to the growth increment of shell deposited during a given period, not to a daily rhythm. The larger the shell, the greater is the growth increment and the number of ridges produced in a given time. Furthermore, Gruffydd was able to demonstrate that the rate of ridge production is influenced by water temperature, ridges forming at less than half the rate at 10°C than at 14.5°C. It is more likely that the frequency of ridge formation is related to the physiology of the animal and the ability to transport calcium to the growing edge of the shell (Gruffydd, 1981) and will be closely regulated by environmental factors such as temperature and food supply. The formation of other, less pronounced surface ridges has been shown not to be related to daily or lunar cycles. Evans (1975) found that the formation of spines, which occur as external rows radiating from the umbo on the surface of the basket cockle (*Clinocardium nuttalli*) were not correlated to a lunar periodicity. Thus it would appear that the coincidence between the external ridges on the surface of bivalve shells and diurnal, lunar, and daily cycles is spurious.

III. INTERNAL GROWTH PATTERNS

Internal microgrowth patterns in the shells of bivalve mollusks can be observed by examining acetate peel replicas of polished and etched radial sections of resin-embedded shell valves. These replicas are prepared by allowing small pieces of acetate sheet to become almost molten in ethyl acetate before placing them onto the dry cut shell surface. The etching procedure, which involves exposing the shell surface to a dilute acid, normally 0.01 *M* HCl for 30 min, differentially etches the crystals and organic matrix, which together make up the shell structure, results in differences in surface relief which are revealed when the acetate peel replicas are examined microscopically. The appearance of a typical acetate peel of a section of a cockle (*Cerastoderma edule*) shell reveals a pattern of alternating dark growth bands and light growth increments (Fig. 1, panel 2). All bivalve shells so far examined contain patterns of microgrowth bands.

Barker (1964) was one of the first to identify and describe a hierarchy of five cyclic groupings of growth layers in the shells of several species of marine bivalves. He suggested that these layers reflected the annual changes of temperature and salinity (first order), equinoctial storms and tides (second order), the fortnightly tidal cycle (third order), day and night cycles (fourth order), and daily tidal rhythms (fifth order). These hierarchies are discussed below.

A. Annual Growth Patterns

In temperate waters shell growth is highly seasonal, with maximum growth occurring during the warmer spring and summer months and minimum growth occurring during the colder winter months. In many subtidal bivalve species the periods of winter growth are represented by clearly defined annual growth lines which coincide with the surface growth rings. Panels 3 and 4 in Fig. 1 show the appearance of annual growth lines in acetate peels of sections of the horse mussel (*Modiolus modiolus*) and the ocean quahog (*Arctica islandica*). The alternating pattern of dark and light lines which are deposited throughout the length of the shell represent periods of winter and summer growth, respectively. Many of these species are long-lived. The age of *M. modiolus* has been estimated from the growth lines to live up to 48 years (Anwar et al., 1990), whereas *A. islandica* can reach an age up to and in excess of 100 years old (Murawski et al., 1982; C. A. Richardson and H. Austin, unpublished data; Witbaard and Duinveld, 1990). During the first 5 to 10 years of growth the lines are widely spaced, but at an age of about 10 years there is an abrupt decline in growth with progressive narrowing of the growth lines during later life.

In many intertidal bivalve species the period of winter growth is represented by a characteristic narrowing of the microgrowth bands during the autumn and winter and an increase in width the following spring and summer (Fig. 1, panel 5). Similar patterns have been described for *M. mercenaria* (e.g., Pannella and MacClintock, 1968; Kennish, 1980) and *Phacosoma japonicum* (Tanabe and Oba, 1988). Richardson et al. (1980c) measured the width of groups of bands in *C. edule* and compared them with the seawater temperature during shell deposition. Shell growth increased rapidly during the spring (April to June) and slowed during the autumn and winter (October to March). However, the correlation between water temperature and growth rate, although obvious, may not be causal. Water temperature, for instance, may strongly covary with seasonal phytoplankton abundance, making it difficult to generalize about the importance of temperature in regulating growth rate without appropriate measurements of food availability (Seed and Richardson, 1990). The effect of winter temperatures on the formation of growth patterns can be highly variable. Mussels (*Mytilus edulis*), for example, show a narrowing of growth increments during the winter (Richardson et al., 1990), but shell growth does not slow down to the same extent as in cockles of similar age. Similarly, razor clams (*Ensis siliqua*) show almost no interuptions to the summer growth patterns, making it difficult to determine the position of winter growth (Richardson, 1992). By contrast, cockles (*C. edule*) from Norway display prominent growth lines within the shell structure (Richardson et al., 1980c), when shell deposition becomes almost completely curtailed. Cockles are able to grow for only 5 to 6 months during the summer, when seawater temperatures and food supply are favorable for shell deposition. During the remaining months little or no growth takes place at temperatures close to freezing point at a time when food availability is similarly low. In tropical waters there may only be small seasonal fluctuations in water temperatures. For example, in the shells of the cockle *Anadara granosa* from Penang, Malaysia, there is no temperature-related growth pattern (Richardson, 1987b). Although the periodic influx of low-salinity seawater over the cockle beds during the intermonsoon period causes the formation of characteristic growth lines (Fig. 1, panel 6). These lines are formed during periods of lowered salinity and presumably result from long periods of shell valve closure under these unfavorable conditions.

B. Influence of the Neap-Spring Lunar Cycle

An examination of the microgrowth patterns in the shells of cockles, mussels, and clams growing in the intertidal zone has shown that they are accurate recorders of the conditions of growth experienced by the animal at a particular tidal level. The position on the shore of the bivalve and the precise form of the neap-spring lunar tidal cycle are important in influencing the appearance of the banding patterns. Although evidence of equinoctial storms and tides (Barker's second-order hierarchy) have not been observed in cockles (*C. edule*), the influence of the fortnightly tidal cycle is well documented (Barker's third-order hierarchy). Richardson et al. (1980a) studied the effect of tidal level on microgrowth pattern formation in *C. edule*. Cockles collected from the high water of neap tides (HWN) grew significantly more during spring tides, when they were immersed for longer periods than during neap tides, when they were emersed for all but a few minutes of the tidal cycle. The observed pattern of increments reflects the lunar cycle of emersion and immersion (Fig. 2, panel 7). The reverse was found for cockles living near low water, where they grew faster during the neap periods when the shells were continuously immersed. During spring tide emersion, cockles have a shorter period of time for feeding resulting in a pattern of wide increments at neaps and narrow ones at spring tides. Similar findings have been recorded for *C. edule* growing elsewhere (Ohno, 1983, 1985; Lønne and Gray, 1988). This pattern of increased growth at neap tides relative to spring tides has also been observed in other species such as clams (*T. philippinarum*) (Fig. 2, panel 8) (Richardson, 1987a) and mussels (*M. edulis*) (Richardson et al., 1990)

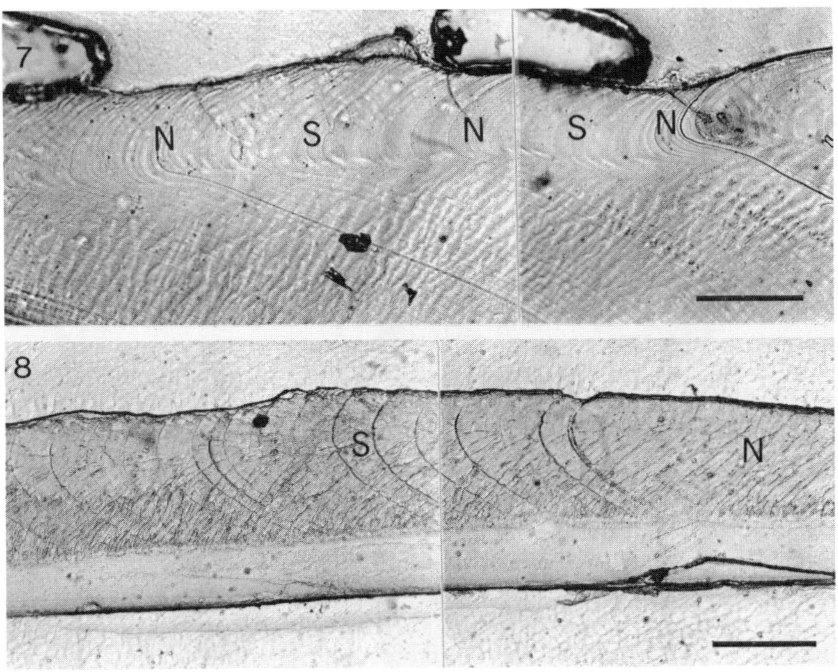

Fig. 2 Photomicrographs of acetate peel replicas of shell sections of bivalve shells. Growth from right to left. Scale bars = 200 μm. 7, Neap-spring lunar pattern of microgrowth bands in the shell of *C. edule* from high water of neap tides (HWN) to show clustering of growth bands during neap (N) tides and wide growth increments during spring (S) tides; 8, neap-spring lunar pattern of microgrowth bands in the clam *Tapes philippinarum* from low water of spring tides. Spring (S) and neap (N) tides.

and produces a clear fortnightly lunar cycle of patterns of bands. Cockles living at or near mean tide level (MTL) showed a 5% increase in shell growth during spring tides over neap tides (Richardson et al., 1980a) even though the animals did not experience any difference in the proportion of time emersed between neaps or spring tides, the difference in growth being attributed to an increase in water flow at spring tides.

The form of the neap-spring lunar cycle has been shown to influence the appearance of the microgrowth patterns. In his study of the basket cockle *C. nuttalli* growing in a mixed diurnal–semidiurnal tidal regime on the coast of Oregon, in which the cockles experienced twice-daily emersion at neap tides and daily emersion at spring tides, Evans (1972, 1975) demonstrated that this lunar pattern was recorded in the patterns of bands. During neap tides a pattern of semidiurnal bands was produced, while at spring tides, when the cockles were emersed only once a day, a diurnal pattern of bands was produced. Ohno (1985) similarly demonstrated the importance of the form of the lunar cycle in his study of *Fragum unedo* growing along the coast of Ishigaki Island, Japan. He showed that the semidiurnal tides had a profound influence on the microgrowth patterns. Animals growing on the lower part of the shore were emersed only once a day during spring tides and remained immersed at neap tides. The pattern of bands in the shells reflected the cycle of emersion and immersion with clear daily bands deposited during emersion at spring tides and weak bands during continuous immersion during neap tides. By contrast, Richardson (1987b) could find no influence of the mixed semidiurnal tidal regime at Penang, Malaysia, on the appearance of the banding patterns in the Malaysian cockle (*A. granosa*). Like Ishigaki Island, the tidal pattern in Penang has two highs and two lows of unequal amplitude each day, with the greatest difference at spring tides. Cockles from the lower shore are emersed only once a day during spring tides and remain immersed at neap tides. The pattern of bands in cockle shells showed a pattern of clear bands during spring tides alternating with a few wider increments during neap tides. The wider increments were only produced for 1 to 3 days and occasionally had weaker bands between (see Richardson, 1987b, Fig. 8). It is clear from this work that tidal level and the neap-spring lunar cycle are periodic factors influencing the patterns of microgrowth bands laid down in the shells.

C. Diurnal and Semidiurnal Growth Patterns

Effects of Tidal Emersion

Examination of an acetate peel replica of a shell section of *C. edule* reveals the presence of an alternating series of dark and light growth bands (Fig. 1, panel 2). Experimental investigations into the periodicity of the microgrowth patterns has been detailed in Pannella (1975), Richardson et al. (1979, 1980a), Lutz and Rhoads (1980), and Richardson (1987a,b, 1989). Recent experiments on living bivalves has not concurred with the once-prevalent view that the bands are laid down once daily (Barker's fourth-order hierarchy) (House and Farrow, 1968; Pannella and MacClintock, 1968; Rhoads and Pannella, 1970). House and Farrow (1968) and later Whyte (1975) claimed that in *C. edule* the bands were formed once a day. Whyte (1975) claimed on the basis of samples of shells collected every 3 h during neap tides that shell deposition is a daily nocturnal event which can be interrupted by tidal or other events. Similarly, nocturnal calcification has been claimed to occur in the clam *M. mercenaria* (Pannella and MacClintock, 1968; Rhoads and Pannella, 1970; Thompson, 1975). Pannella and MacClintock (1968) marked shells of *M. mercenaria* and withdrew them 2 years later. They found one growth band per day, although they also stated that daily increments were further divided into two bands. Rhoads and Pannella (1970) reported observing strongly marked bands in intertidal transplants of the same species, which is more consistent with tidally produced bands.

Date-marking bivalve shells has been the key to interpreting the periodicity of the patterns in species such as cockles (*C. edule*) (Richardson et al., 1979, 1980a), clams (*T. philippinarum*) (Richardson, 1987a), and mussels (*M. edulis*) (Richardson, 1989). One or more bands were marked on a given date to which all subsequent growth bands could be related. Examination of the shells grown in these short-term field and laboratory experiments showed conclusively that the bands were produced at semidiurnal intervals (Barker's fifth-order hierarchy). Richardson et al. (1981) showed on the basis of samples of cockles collected every hour for 24 h that deposition of the dark band occurred during emersion, while formation of the light band took place during immersion, when the cockles were immersed and actively feeding. More recent studies using time-series analysis on widths of 200 consecutive microgrowth increments also supports the view of a tidal influence on the production of growth increments in intertidal *C. edule* from Norway (Lønne and Gray, 1988).

Observation of the semidiurnal tidal bands reveals that there is a regular changing pattern of definition of the bands with a fortnightly period. Dolman (1975) and Richardson et al. (1979, 1981) noticed that periodically, groups of strong and weak bands separated by regions in which the bands were more or less uniform in definition were deposited in the shell of *C. edule* (Fig. 3, panel 9). Dolman (1975) believed that the alternation in banding could be related to tidal emersion in light and dark conditions. Since the low-tide period advances progressively during the

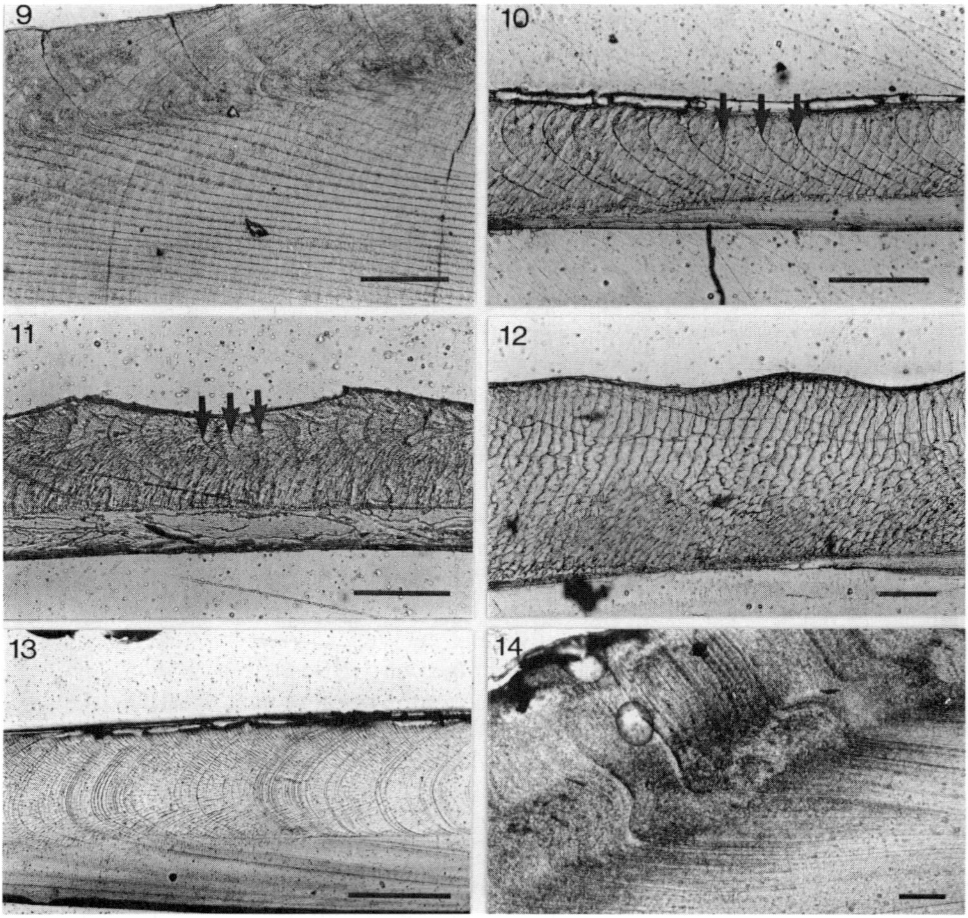

Fig. 3 Photomicrographs of acetate peel replicas of shell sections of bivalve shells. Growth from right to left. 9, Alternations in the distinctiveness of the growth bands formed during spring tides in the shell of *C. edule*, scale bar = 100 μm; 10, tidal growth bands (arrows) deposited in the shell of *T. philippinarum* grown in a semidiurnal tidal regime in the laboratory, scale bar = 200 μm; 11, weak microgrowth bands (arrows) in the shell of *T. philipinarum* grown in continuously immersed conditions in the laboratory, scale bar = 200 μm; 12, weak microgrowth bands in the shell of *T. philippinarum* grown on a raft in the Menai Strait, North Wales, scale bar = 100 μm; 13, microgrowth patterns in the shell of *Donax trunculus* from Barcelona in the Mediterranean, scale bar = 500 μm; 14, growth patterns in the shell of the hydrothermal vent clam *Calyptogena pacifica* from a depth of ca. 4000 m, scale bar = 200 μm.

lunar cycle, there would be periods when both low tides would be at twilight and others a week later, when low water was around midnight and midday, thus producing a lunar periodicity in alternating patterns. Richardson et al. (1981) were able to show on the basis of marked shells in which each band could be assigned to a particular low tide that the stronger bands were laid down during the heat of the afternoon low tide, while the less distinct bands were formed during the cool of the morning. A similar conclusion was reached by Ohno (1985) for *C. edule* and *M. mercenaria*. The occurrence of alternating bands at spring low tides has also been observed in *Spisula subtruncata* growing at extreme low water of spring tides in the Menai Strait (Richardson, 1988b).

Effects of Continuous Immersion

Microgrowth patterns in the shells of bivalves growing subtidally have been shown to be weaker in definition than those growing in the intertidal (Richardson, 1990). Rhoads and Pannella (1970) sought to explain the presence of microgrowth bands in the shells of subtidal *M. mercenaria* as the result of changes in deposition during night and day. This view was shared by Thompson (1975) and Kennish (1978, 1980); it was believed most shell growth took place at night, when the animals were actively feeding. Thompson (1975), however, showed that shell growth of *M. mercenaria* in the laboratory was similar in regimes of both continuous illumination and light–dark conditions, so it is difficult to see how a theory of nocturnal shell deposition has any credence. Kennish (1978, 1980) also studied microgrowth patterns in *M. mercenaria* in detail and recognized both daily and twice-daily growth patterns. His conclusions were based on experiments in which several thousand notched clams were transplanted into the subtidal zone (Kennish, 1980). Specimens subsequently recovered and studied contained approximately the same number of days that had elapsed since the day of transplantation. While their conclusion that the microgrowth bands in subtidal bivalves are formed at daily intervals appears to be borne out by experiments the evidence is unconvincing with little clear demonstration of daily microgrowth bands in the shells.

Richardson et al. (1979 et seq.) have made considerable progress in attempting to understand the periodicity of microgrowth patterns in bivalves growing subtidally. The patterns of bands are always weaker in definition in the shells of cockles (*C. edule*) (Richardson et al., 1980b), clams (*T. philippinarum*) (Richardson, 1987a, 1988a), and mussels (*M. edulis*) (Richardson, 1989) held under subtidal conditions when compared with their intertidal conspecifics (Fig. 3, panels 10 and 11). Richardson et al. (1980b) presented evidence to show that the bands in continuously immersed *C. edule* were of tidal origin. Counts of microgrowth bands in the shell corresponded to an approximate tidal rhythm of formation which they concluded was evidence for an endogenous rhythm of shell deposition in *C. edule*. Since the cockles had been maintained in constant conditions of continuous darkness, temperature, food supply, and water flow, there were no exogenous factors to entrain microgrowth band formation. Further support for an endogenous tidal rhythm of shell deposition has been demonstrated for *T. philippinarum* by Richardson (1987a). Here again in the absence of tidal emersion the clams, like cockles, produced a pattern of bands with an approximate tidal periodicity. However, although the average number bands laid down in the shells coincided with the number of expected tidal periods, several shells showed deviation from the expected number. Richardson (1990) expressed doubts about the early interpretation of the weak patterns in cockles and clams. Based on the interpretation of patterns in shells from later experiments an alternative explanation for the production of microgrowth patterns in both continuously immersed clams and mussels, held in constant conditions, has been put forward (Richardson, 1988a, 1989). It has been shown that the weak banding in subtidal clams (Fig. 3, panel 12) and mussels is not related to the tides or days but to the growth of the shell. In a series of laboratory experiments, it was demonstrated that the total number of microgrowth bands deposited was significantly correlated with shell growth. Further support for this explanation comes from mussels growing continuously immersed in the natural environment on a raft in the Menai Strait; in these shells there was a positive and significant correlation between shell growth and the deposition of microgrowth bands.

This explanation for the formation of microgrowth patterns in subtidal bivalves, which is based on experiments with living animals held in continuously immersed and constant conditions, provides a creditable alternative to a theory based on diurnal and/or tidal changes in shell deposition. Indeed, this would explain why shells from the Mediterranean, off the coast of Barcelona, Spain, where there is little or no tidal rise and fall (M. Ramon and C. A. Richardson, unpublished data), contain clear microgrowth bands (Fig. 3, panel 13). Bivalves living in these waters where the tidal influence is slight might not be expected to contain tidal or even daily patterns. Similarly, the shells of bivalves living in near-constant conditions at great depth (ca. 4000 m), such as the hydrothermal vent clam *Calyptogena pacifica*, contain microgrowth patterns (Fig. 3, panel 14), even though they are unlikely to be exposed to obvious tidal or daily environmental perturbations. Since the deposition of microgrowth bands in subtidal bivalves from shallow waters does not appear to be related to a tidal or daily periodicity, the question remains as to why the bands are formed. Perhaps the bands are organic sheets that are deposited at the crystal boundaries, a form of intercrystalline "cement" which is necessary to hold the growing crystals of the shell together and ensure the integrity of the shell structure (Richardson, 1989)? (Specifically, the interband distance is a feature of the animals' intrinsic morphology.) This question and others remain to be answered.

ACKNOWLEDGMENTS

This paper was written while in support of a NERC research grant (GR3/7390). I am particularly grateful to Professor A. Southward and Dr. Eve Southward for providing a specimen of the hydrothermal vent bivalve *Calyptogena pacifica* for examination.

REFERENCES

Antoine, L. (1978). *Haliotis*, *9*: 627.
Anwar, N. A., Richardson, C. A., and Seed, R. (1990). *J. Mar. Biol. Assoc. UK*, *70*: 441.
Barker, R. M. (1964). *Malacologia*, *2*: 69.
Broom, M. J., and Mason, J. (1978). *Mar. Biol.*, *47*: 277.
Clark, G. R., II (1968). *Science*, *161*: 800.
Clark, G. R., II (1975). In *Growth Rhythms and the History of the Earth's Rotation* (G. D. Rosenberg and S. K. Runcorn, eds.), Wiley, London, p. 103.
Dolman, J. (1975). In *Growth Rhythms and the History of the Earth's Rotation* (G. D. Rosenberg and S. K. Runcorn, eds.), Wiley, London, p. 191.
Evans, J. W. (1972). *Science*, *176*: 416.
Evans, J. W. (1975). In *Growth Rhythms and the History of the Earth's Rotation* (G. D. Rosenberg and S. K. Runcorn, eds.), Wiley, London, p. 119.
Farrow, G. E. (1971). *Paleontology*, *14*: 571.
Gruffydd, L. D. (1981). *J. Mar. Biol. Assoc. UK*, *61*: 401.
House, M. R., and Farrow, G. E. (1968). *Nature* (*Lond.*), *219*: 1384.
Kennish, M. J. (1978). *Environ. Geol.*, *2*: 223.
Kennish, M. J. (1980). In *Skeletal Growth of Aquatic Organisms* (D. C. Rhoads and R. A. Lutz, eds.), Plenum Press, New York, p. 255.
Lønne, O. J., and Gray, J. S. (1988). *Mar. Ecol. Prog. Ser.*, *42*: 1.
Lutz, R. A., and Rhoads, D. C. (1980). In *Skeletal Growth of Aquatic Organisms* (D. C. Rhoads and R. A. Lutz, eds.), Plenum Press, New York, p. 203.
Murawski, S. A., Ropes, J. W., and Serchuk, F. M. (1982). *Fish. Bull. NOAA USA*, *80*: 21.
Ohno, T. (1983). In *Tidal Friction and the Earth's Rotation* (P. Brosche and J. Sündermann, eds.), Springer-Verlag, Berlin, p. 222.

Ohno, T. (1985). *Palaeontographica A*, *189*: 63.
Pannella, G. (1975). In *Growth Rhythms and the History of the Earth's Rotation* (G. D. Rosenberg and S. K. Runcorn, eds.), Wiley, London, p. 253.
Pannella, G., and MacClintock, C. (1968). *J. Paleontol.*, *42*: 64.
Rhoads, D. C., and Pannella, G. (1970). *Lethaia*, *3*: 143.
Richardson, C. A. (1987a). *Proc. R. Soc. Lond. B*, *230*: 367.
Richardson, C. A. (1987b). *J. Exp. Mar. Biol. Ecol.*, *111*: 77.
Richardson, C. A. (1988a). *J. Exp. Mar. Biol. Ecol.*, *122*: 105.
Richardson, C. A. (1988b). *J. Mol. Stud.*, *54*: 71.
Richardson, C. A. (1989). *J. Mar. Biol. Assoc. UK*, *69*: 477.
Richardson, C. A. (1990). In *Earth's Rotation from Eons to Days* (P. Brosche and J. Sündermann, eds.), Springer-Verlag, Berlin, p. 215.
Richardson, C. A. (1992). *Proc. First Int. Conf. Mar. Biol. Hong Kong South China Sea '90*, Hong Kong (in press).
Richardson, C. A., Crisp, D. J., and Runham, N. W. (1979). *Malacologia*, *18*: 277.
Richardson, C. A., Crisp, D. J., and Runham, N. W. (1980a). *Proc. R. Soc. Lond. B*, *210*: 513.
Richardson, C. A., Crisp, D. J., and Runham, N. W. (1980b). *J. Mar. Biol. Assoc. UK*, *60*: 991.
Richardson, C. A., Crisp, D. J., Runham, N. W., and Gruffydd, L. D. (1980c). *J. Mar. Biol. Assoc. UK*, *60*: 997.
Richardson, C. A., Crisp, D. J., and Runham, N. W. (1981). *J. Mar. Biol. Assoc. UK*, *61*: 465.
Richardson, C. A., Seed, R., and Naylor, E. (1990). *Mar. Ecol. Prog. Ser.*, *66*: 259.
Seed, R. (1976). In *Marine Mussels: Their Ecology and Physiology* (B. L. Bayne, ed.), Cambridge University Press, Cambridge, p. 13.
Seed, R., and Richardson, C. A. (1990). In *Neurobiology of Mytilus edulis* (G. B. Stefano, ed.), Manchester University Press, Manchester, p. 1.
Stanley, S. (1970). *Geol. Soc. Am. Mem.*, *125*: 1.
Tanabe, K., and Oba, T. (1988). *Mar. Ecol. Prog. Ser.*, *47*: 75.
Thompson, I. (1975). In *Growth Rhythms and the History of the Earth's Rotation* (G. D. Rosenberg and S. K. Runcorn, eds.), Wiley, London, p. 149.
Whyte, M. A. (1975). In *Growth Rhythms and the History of the Earth's Rotation* (G. D. Rosenberg and S. K. Runcorn, eds.), Wiley, London, p. 177.
Witbaard, R., and Duineveld, G. C. A. (1990). *Basteria*, *54*: 63.

IV
Synchronous Formation of Periodic Patterns

22
Periodic Alignment of Chloroplasts in *Acetabularia*

Silvano Bonotto *University of Turin, Turin, Italy*

I. INTRODUCTION

In 1966, when I started to work with *Acetabularia* in the laboratory of Jean Brachet at the University of Brussels, I noticed in vegetative cells a periodic alignment of chloroplasts, which was more marked at the beginning and at the end of the day. However, this interesting phenomenon could be studied only 10 years later in the photobiology laboratory of Cyrille Sironval at the University of Liège, where the equipment necessary for the investigations became available. Chloroplast bands, formed at daybreak, were found to persist for several hours when "banded" cells were brought back to darkness, but to disappear rather rapidly (in about 30 min) in the light. Moreover, light microscopy and fluorescence recording, with a scanner, of frozen cells excited with a pencil of blue light (436 nm) unequivocally demonstrated that band formation was due to chloroplast accumulation (Sironval et al., 1977). The introduction of optical density recording of living cells both in Belgium and in Germany allowed more detailed investigations on chloroplast distribution along the stalk during the light–dark cycle. It was thus discovered that chloroplasts migrate according to a circadian rhythm with a period length of about 27 h (Koop et al., 1978). In addition, band formation was interpreted as a direct consequence of the circadian rhythm itself (Koop et al., 1978; Schmid and Koop, 1983). At the same time, data obtained in various experiments on the effects of a narrow beam of white or blue light prompted introduction of the "chloroplast photoaccumulation concept" (Pâques and Brouers, 1985). Moreover, these authors, by examining light action spectra, suggested that a photoreceptor containing riboflavin and carotenoids is possibly involved in chloroplast movement (Pâques and

Brouers, 1985). According to their hypothesis, the chloroplasts, under the action of light, would activate a trigger mechanism leading to their transport, thus displaying an indirect phototactic behavior. Although this interpretation requires more substantial experimental support to be proven, the idea that chloroplasts may play an active role in their transport should be taken into consideration.

To add to the complexity of chloroplast movement in *Acetabularia*, it may be noted that this alga, despite active cytoplasmic streaming, shows apicobasal morphological and physiological chloroplast gradients (Issinger et al., 1971; Puiseux-Dao et al., 1972b; Arapis et al., 1985). Moreover, *Acetabularia* shows both phototropic and geotropic reactions (Nasr, 1939; Puiseux-Dao, 1962; Bonotto et al., 1976; Bonotto and Sironval, 1977; Gibor, 1977), which, directly or indirectly, might interfere with chloroplast distribution in the cell. In this chapter it thus seems advisable to deal first with phototropism and geotropism, cell gradients and cytoplasmic streaming, and then with chloroplast migration and band formation. In fact, a more complete picture of the morphological and physiological features of *Acetabularia* may be helpful for a better understanding of chloroplast distribution and function during cell morphogenesis.

II. PHOTOTROPISM AND GEOTROPISM

Under natural conditions, *Acetabularia* cells grow toward the light (Fig. 1a,b). Moreover, their vertical habit suggests that they are subjected to the orienting action of the force of gravity (Bonotto and Sironval, 1977). The first observations on the phototropism of *Acetabularia* cells were made in 1937 on the species *A. calyculus* growing in the Red Sea (Nasr, 1939). Later, a positive phototropism of the stalk of *A. acetabulum* (= *A. mediterranea*) was observed by Puiseux-Dao (1962). More recently, experimental work on phototropism and geotropism has been undertaken on *A. acetabulum*, *A. crenulata*, and *A. major* (Bonotto et al., 1976; Bonotto and Sironval, 1977; Gibor, 1977). It was found that when *Acetabularia* are grown undisturbed in glass bottles or petri dishes, their stalk grows toward the light source, whereas the rhizoid develops in the opposite direction (Bonotto et al., 1976). In addition, vegetative cells of *A. crenulata* were capable of reorienting their stalk when light direction changed from 55° to 90° (Bonotto and Sironval, 1977). Besides, Gibor (1977) has reported that *A. acetabulum* responds to gravity while *A. major* does not.

Of particular interest is the fact that the gametes of *Acetabularia*, which contain a single elongated chloroplast (Crawley, 1966, 1970; Kellner and Werz, 1969; Werz, 1974) show a positive phototropism,

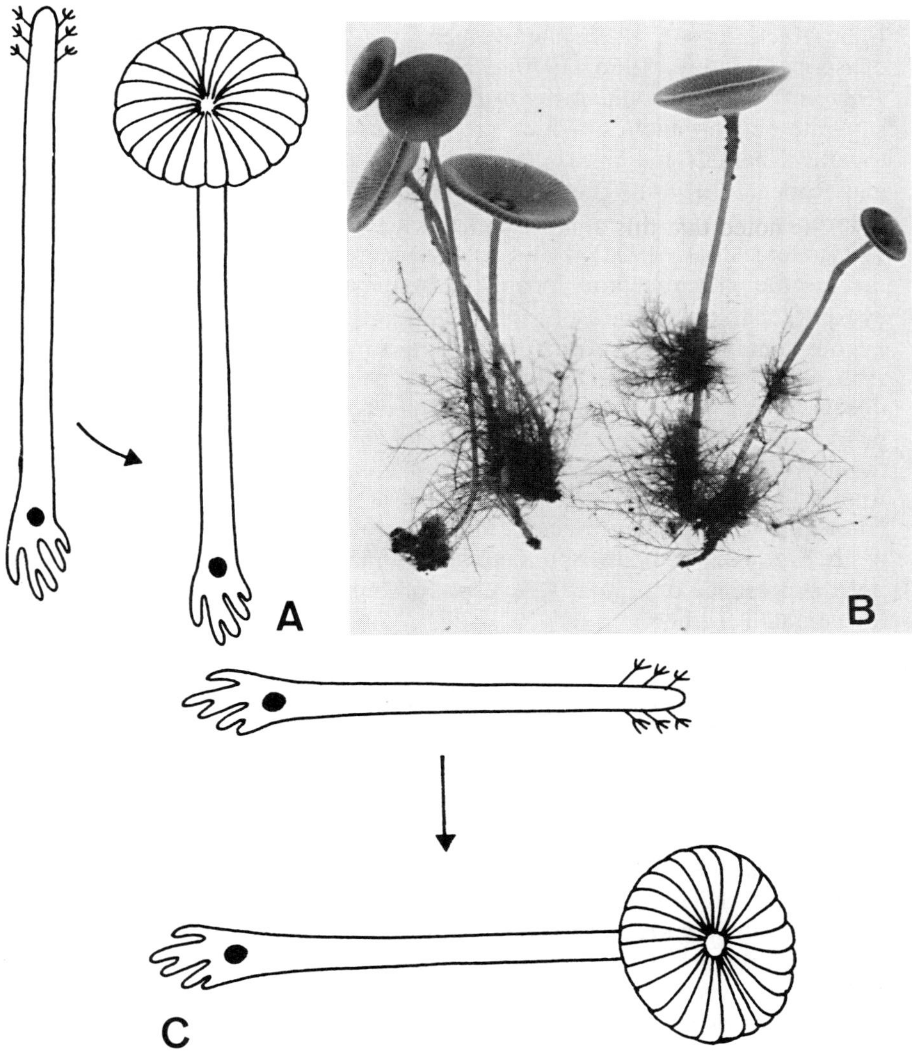

Fig. 1 Growth and cap formation in *Acetabularia acetabulum* (= *A. mediterranea*) in nature (A,B) and in the laboratory (C), where the cells are maintained horizontally. Under natural conditions, the cells are strongly calcified and covered by various epiphytes.

whereas the zygotes display a negative one, swimming away from the light (Keck, 1964). In their movement, the gametes are assisted by the eyespot (Crawley, 1966, 1970; Kellner and Werz, 1969; Herth et al., 1981). The eyespot apparatus in flagellated green algae consists of two separate cell organelles, which are structurally and functionally linked together, namely the chloroplast and the plasma membrane (Melkonian and Robenek, 1984). The layer of lipid globules present in the single chloroplast of the *Acetabularia* gametes (Crawley, 1966, 1970; Werz, 1974; Herth et al., 1981) would serve to increase the light intensity at the level of the photoreceptor, located in the plasma membrane overlying the eyespot (Melkonian and Robenek, 1984). After gamete fusion, the young zygote contains two eyespots, according to its two chloroplasts derived from each gamete cell (Kellner and Werz, 1969). These authors also observed that in later stages of zygote development, the eyespots appeared fragmented, suggesting that the fragments became distributed into the newly formed chloroplasts (Kellner and Werz, 1969). The observations of Kellner and Werz (1969) should be kept in mind when studying chloroplast migration in *Acetabularia*. Unfortunately, to our knowledge, neither the fate of the eyespot fragments in the *Acetabularia* chloroplasts nor the presence of a putative eyespot precursor (or prostigma) has been investigated further.

III. MORPHOLOGICAL AND PHYSIOLOGICAL CHLOROPLAST GRADIENTS

In the marine environment, where a pronounced vertical light gradient exists, it is not surprising that *Acetabularia* cells grow toward the light source. However, since the cell walls become strongly calcified and covered by epiphytes (Fig. 1B) (van der Ben et al., 1972) only the apical part of the cell, bearing numerous ramified whorls, would benefit from the light. It is thus beneficial for *Acetabularia* to have developed a transport system allowing the chloroplasts to go up (apex) and down (rhizoid). Taking into account that the *Acetabularia* chloroplast population (several million plastids per cell; Shephard, 1965) is heterogeneous (Puiseux-Dao et al., 1972a), one can imagine that big heavy chloroplasts would be transported with more difficulty than the light ones from the base to the apex of the cell. Consequently, a morphological chloroplast gradient may be formed in vertically growing cells. Under laboratory conditions, however, *Acetabularia* cells are maintained in horizontal position (Fig. 1C). Nevertheless, they show a marked apicobasal morphological gradient, the small young chloroplasts predominating in their apices (Puiseux-Dao et al., 1972a). Possibly, to the maintenance of the apicobasal chloroplast gradient concur both physical (light, temperature, chemical composition of the external medium, force of gravity) and various cellular factors (polar growth, currents, substances of nuclear origin transported to the apex where growth occurs, rate of chloroplast division, rate of photosynthetic activity, and others). In this context it is of interest to note that the photosynthetic capacity is higher in the apical than in the basal part of the cell (Issinger et al., 1971) and that this capability is maintained by isolated chloroplasts in vitro for at least 2 h (Arapis et al., 1985). Moreover, apical chloroplasts divide more frequently and show a higher rate of nucleic acid and protein synthesis when compared with the basal ones (Bonotto et al., 1978; Arapis et al., 1984). The experimental evidence obtained until now suggest that a vegetative cell of *Acetabularia* contains heterogeneous population of chloroplasts at various stages of differentiation, maturing and aging from the apex toward the base, despite the active cytoplasmic streaming (Puiseux-Dao et al., 1978, 1979).

IV. CYTOPLASMIC STREAMING

Five types of streaming were described in the literature (see references in Strömgren Allen, 1983) for plant cells: (1) saltation (agitation), (2) circulation, (3) fountain and reverse-fountain streaming, (4) rotational streaming, and (5) multistriate streaming. In the latter, which is typical of *Acetabularia*, the direction of a strand is either acropetal or basipetal (Kamiya, 1962, 1981; Kamiya and Kuroda, 1966; Puiseux-Dao, 1979). Moreover, cytoplasmic streaming shows a rhythm, even in continuous light, with a periodicity similar to that of the electrical polarity (Borghi et al., 1983; Puiseux-Dao, 1984). In addition, these authors reported that in darkened *Acetabularia* cells brought back to light, the timing of recovery of the electrical behavior and cellular streaming was exactly the same (Puiseux-Dao, 1984). The rhythm of cytoplasmic streaming is probably tightly related to that of chloroplast migration (see Section V.B). However, it has not been investigated whether or not the former occurs in the absence of chloroplasts (i.e., in centrifuged cells). The streaming cytoplasm appears to be composed of two interconnected networks, one made up of thin, fast-moving strands, more easily detectable in the apical part (Fig. 2a,b), the other of larger strands, which are thicker in the basal part of the cell (Fig. 2c,d; Puiseux-Dao, 1979). This author has also observed that the large basal strands contained numerous big chloroplasts having one or more starch granules (Puiseux-Dao, 1979).

Studies with various techniques [differential interference microscopy (DIC), video-enhanced contrast microscopy (AVEC-DID), TEM and SEM electron microscopy, laser Doppler spectroscopy, and microcine-

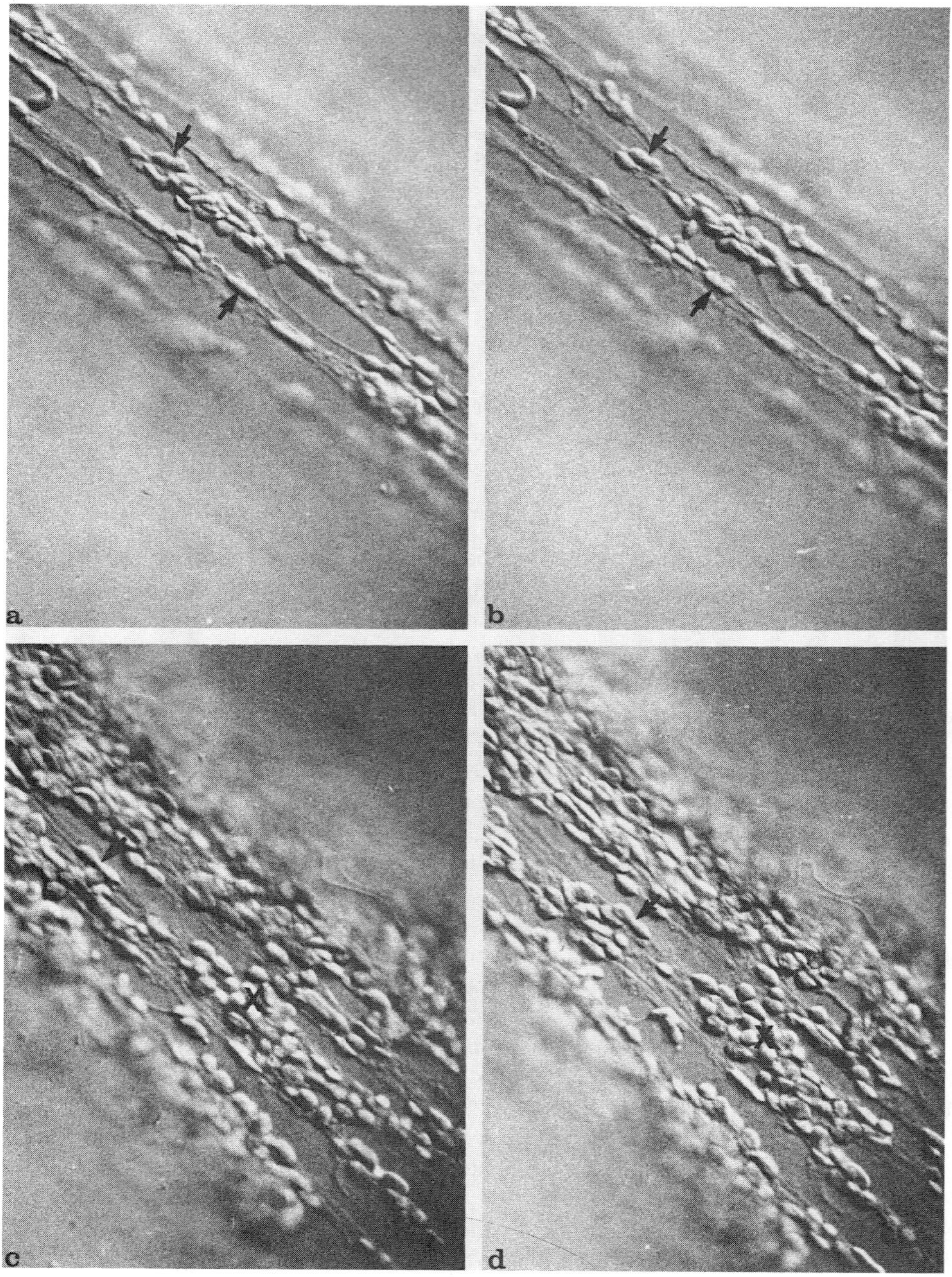

matography] have permitted a detailed analysis of cytoplasmic streaming in *Acetabularia* (Puiseux-Dao, 1979; Heunert, 1978; Koop, 1981; Nagai and Fukui, 1981; Pâques, 1982; Strömgren Allen, 1983; Fukui and Nagai, 1985). Scanning electron microscopy (SEM) has shown that the chloroplasts are connected by fibrillar structures (Pâques, 1982). Besides, the use of cell centrifugation, perfusion, and the treatment of *Acetabularia* with ATP, cytochalasin B (CB, an inhibitor of actin microfilament function), *N*-ethylmaleimide (NEM, a SH-blocking agent), colchicine (a microtubule inhibitor), and saponin supported the hypothesis that cytoplasmic streaming is due to the interaction of actin and myosin (Nagai and Fukui, 1981; Fukui and Nagai, 1985). These authors reported that "the actomyosin system generates the motive force for the cytoplasmic streaming in *Acetabularia*," in the presence of Mg-ATP and Ca^{2+}. However, according to another hypothesis, both microfilaments and microtubules would play an active role in cytoplasmic streaming (Koop, 1981; Koop and Kiermayer, 1980a,b,c). In agreement with previous work of Heunert (1978) and Puiseux-Dao (1979), Koop and Kiermayer (1980b,c) have observed thin cytoplasmic filaments, at which chloroplast movement occurs at a rate of 1 to 2 μm/s. They also noted headed streaming bands, containing small vesicles, groups of polyphosphate granules, and secondary nuclei during the reproductive phase, which moved faster, 3 to 11 μm/s, along the filament system. Both streaming systems are sensitive to CB, whereas the headed streaming bands alone are inhibited by colchicine, chloroisopropyl-*N*-phenylcarbamate (CIPC), amiprophosmethyl (APM), and vinblastine (Koop and Kiermayer, 1979, 1980b,c; Koop, 1981). According to Koop (1981), the induction of birefringent vinblastine crystals in the stalk of vegetative *Acetabularia* cells further supports the involvement of microtubules in streaming. Menzel (1986) suggests that microtubules and actin "must be coordinately involved in the key events of the developmental cycle of *Acetabularia*," such as the transport of secondary nuclei. However, microtubules alone do not seem to confer motility to the nuclei (Menzel, 1986).

Actin and tubulin have been already detected in *Acetabularia* (Weber et al., 1977; Dazy et al., 1981; Menzel, 1986), whereas the presence of

---

Fig. 2 Cytoplasmic streaming in the apical (a,b) and in the basal (c,d) region of the stalk of *Acetabularia acetabulum*. The pictures were taken at time *t* (a,c) and *t* + 10 seconds (b,d). The arrows indicate chloroplast movement, whereas the cross (c,d) shows a large strand where the plastids move slowly. (From Puiseux-Dao, 1979.)

myosin remains hypothetical (Menzel, 1986) and is inferred from the action of NEM, which is known as an inhibitor of the F-actin-activated ATPase of myosin (Shibata-Sekiya and Tonomura, 1975). According to Nagai and Fukui (1981), "microfilaments must reside in the cortex, especially in the ectoplasmic striations, while the putative myosin must reside in the moving endoplasm." Experiments with CB showed, in fact, that treatment of the cortex alone blocked streaming, while it continued at a normal rate when the endoplasm alone was supplied with the drug (Nagai and Fukui, 1981). The effect of CB was reversible, whereas that of NEM was irreversible. The use of radioactive cytochalasin B ([^{3}H]CB) has shown that the reversibility of its action is not due to intracellular destruction of the drug but to its release into the medium (Bonotto, unpublished results). This observation suggests that the complex actin-[^{3}H]CB is rather labile and may be easily dissociated.

Although the main macromolecules and structures involved in cytoplasmic streaming were detected or visualized, some findings deserve further investigation. For example, electron microscopical observations have revealed in the cytoplasm of *A. acetabulum* at least five types of filaments (Garcia, 1986), whose precise function is not known at present. In addition, Strömgren Allen (1983), by using AVEC-DIC microscopy, has observed a "thin membranous polygonal network (PN) lying against the plasma membrane." This author states that "these networks shift and attach themselves to membranes associated with the cortical ridges and thus with the streaming." The function of these structures should be investigated for obtaining more insight into the mechanisms responsible for chloroplast migration in *Acetabularia*.

V. CHLOROPLAST MIGRATION

A. Chloroplast Band Formation

Our first investigations (Sironval et al., 1977) on the formation of chloroplast bands in *A. acetabulum* showed that:

1. They formed very rapidly at daybreak, spontaneously or following gentle stirring of the cells, in the stalk (Fig. 3) as well as in the whorls (Fig. 4).
2. They could be induced at any time during the day, provided that the cells were previously maintained in darkness.
3. There was no band formation in darkness.
4. The bands persisted for several hours in darkness, but in the light they shifted, became diffuse, and then disappeared in about 30 min,

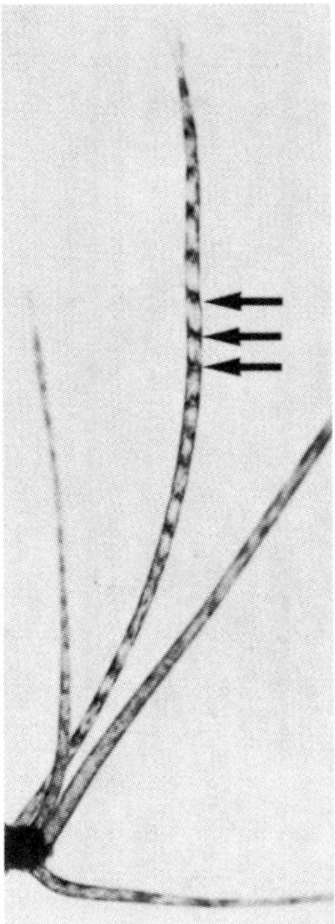

Fig. 3 Formation of 24 bands in a young vegetative *Acetabularia* cell at the dark–light transition. (From Sironval et al., 1977.)

due to a rapid transition of the ''banded'' cytoplasm to a more homogeneous state.

5. The bands were induced in both whole and enucleated cells, suggesting that the phenomenon was independent from the presence of the nucleus, located in the rhizoid.
6. Band frequency in 0.25 to 0.30 mm wide stalks (Fig. 3) was about 10 cm^{-1}.

Further observations on chloroplast distribution during the light–dark cycle showed that the cytoplasm passed from a ''homogeneous'' (light period) to a ''heterogeneous'' (dark period) state, the transverse bands appearing at the dark–light (beginning of the day) and at the light–dark (end of the day) transitions (Fig. 5, arrows) (Pâques, 1982; Pâques and Brouers, 1981, 1985). The heterogeneous state of the cytoplasm during the dark period was probably provoked by a decreased movement of the

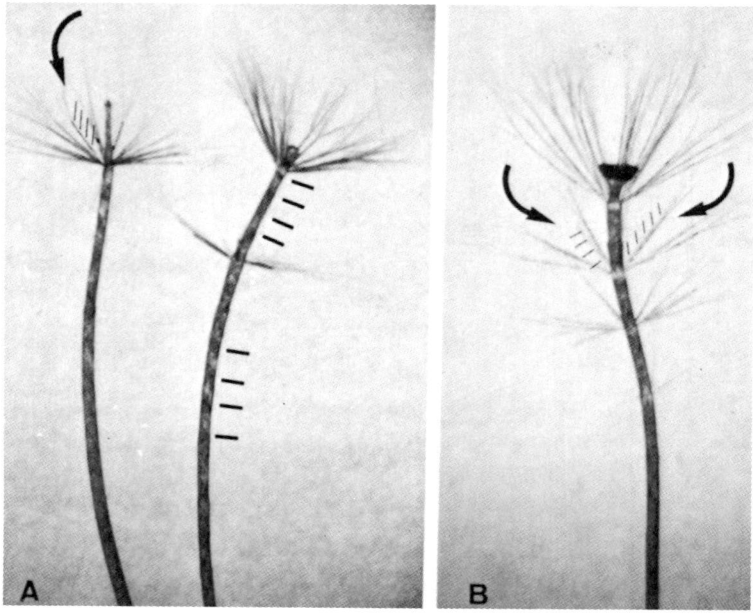

Fig. 4 Band formation in the whorls (arrows and small marks) and in the stalk (ticker marks) at day-break. (Modified from Sironval et al., 1977.)

chloroplasts, which separated into small clots. Besides, heterogeneous cells (Fig. 5, h) showed pale-green apices (Fig. 5, arrowheads) but dark-green basal regions, suggesting that a chloroplast migration toward the rhizoid had occurred at the end of the day and probably also during the night, when plastid movement continues at a reduced rate (S. Puiseux-Dao, personal communication). Recorder tracing of the optical density of the tip of the cell made it possible to discover that ''banding'' was closely related to the circadian rhythm of chloroplast migration, coinciding with phases of rapid net translocation of plastids (Schmid and Koop, 1983). In addition, serial photographs showed that the bands moved along the stalk (where streaming is of the multistriate type) in upward and downward directions at the same time (Fig. 6), with a velocity of 0.6 to 0.9 μm/s (Schmid and Koop, 1983). This finding indicated the involvement of only the slow streaming system (1 to 2 μm/s) of *Acetabularia* in band formation. Moreover, Schmid and Koop (1983) suggested that the effect of the dark-to-light transition was to phase-shift the circadian cycle and not to induce banding, an interpretation that seems quite plausible but needs further experimental support.

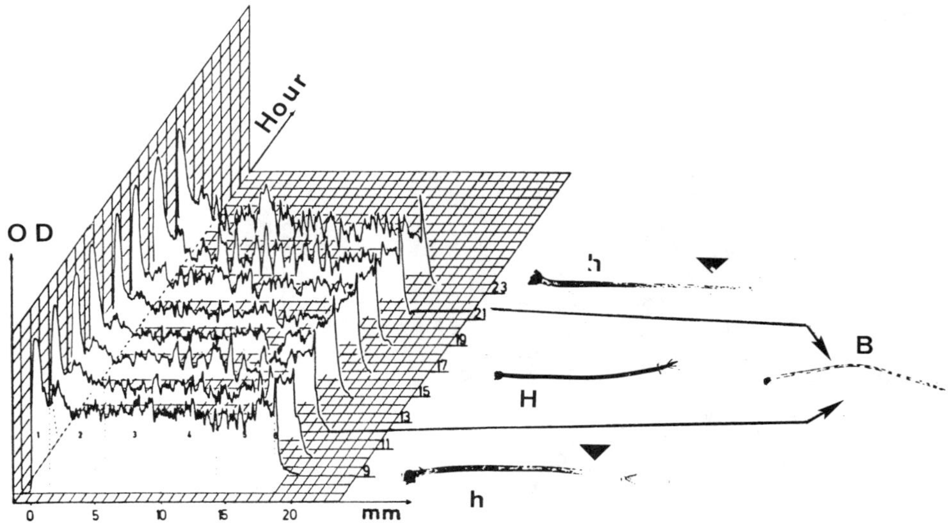

Fig. 5 Optical density recording of *Acetabularia* cells, in the light, between 9 A.M. and 23 P.M. The cytoplasm shows transitions from a "heterogeneous" (h) to a "homogeneous" (H) state, band formation (B) occurring two times (long arrows) during a day. In "heterogeneous cells" the apical regions are pale-green (arrowheads), indicating that an important proportion of chloroplasts descended toward the rhizoid. These observations are in agreement with those of Schmid and Koop (1983). (From Pâques, 1982.)

B. Circadian Rhythm of Chloroplast Migration

Time-lapse cinematography of young *Acetabularia* cells and recording of the optical transmission of the apex and the rhizoid have revealed the existence of a circadian rhythm of chloroplast migration, having a period length of about 27 h (Koop et al., 1978; Broda, 1979; Broda et al., 1979; Schmid, 1981; Schmid and Koop, 1981, 1983; Hartwig et al., 1985). The absorbance was at the maximum in the apex, when it was at the minimum in the rhizoid, and vice versa (Fig. 7). Clearly, chloroplast migration is directed "toward the apex of the cell at the beginning of the circadian day and toward the rhizoid at the beginning of the circadian night" (Koop et al., 1978). According to these authors, chloroplast bands are an expression of the rhythm itself.

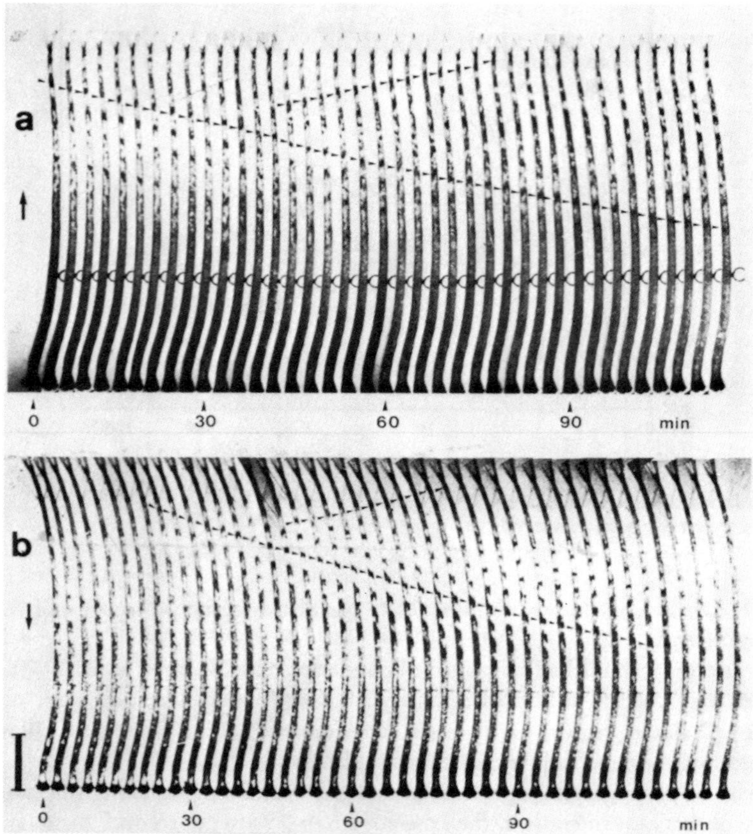

Fig. 6 Cutouts of circadian phases of chloroplast translocation: (a) at the circadian morning; (b) same cell at the circadian evening. The big arrows indicate the direction of the chloroplast migration (omission of arrowheads does not mean that the bands are lacking). Bands move in acropetal as well as in basipetal direction. This can best be seen when the series is viewed from one of the edges. The frequency of the takes was one per 3 min. The bar represents 2 mm. (From Schmid and Koop, 1983.)

Fig. 7 Time course of rhythmical changes of absorbance in apices and rhizoids of two different cells of *Acetabularia acetabulum*. Note that in both cells I and II the absorbance is at the maximum in the apex when it is at the minimum in the rhizoid. (From Koop et al., 1978.)

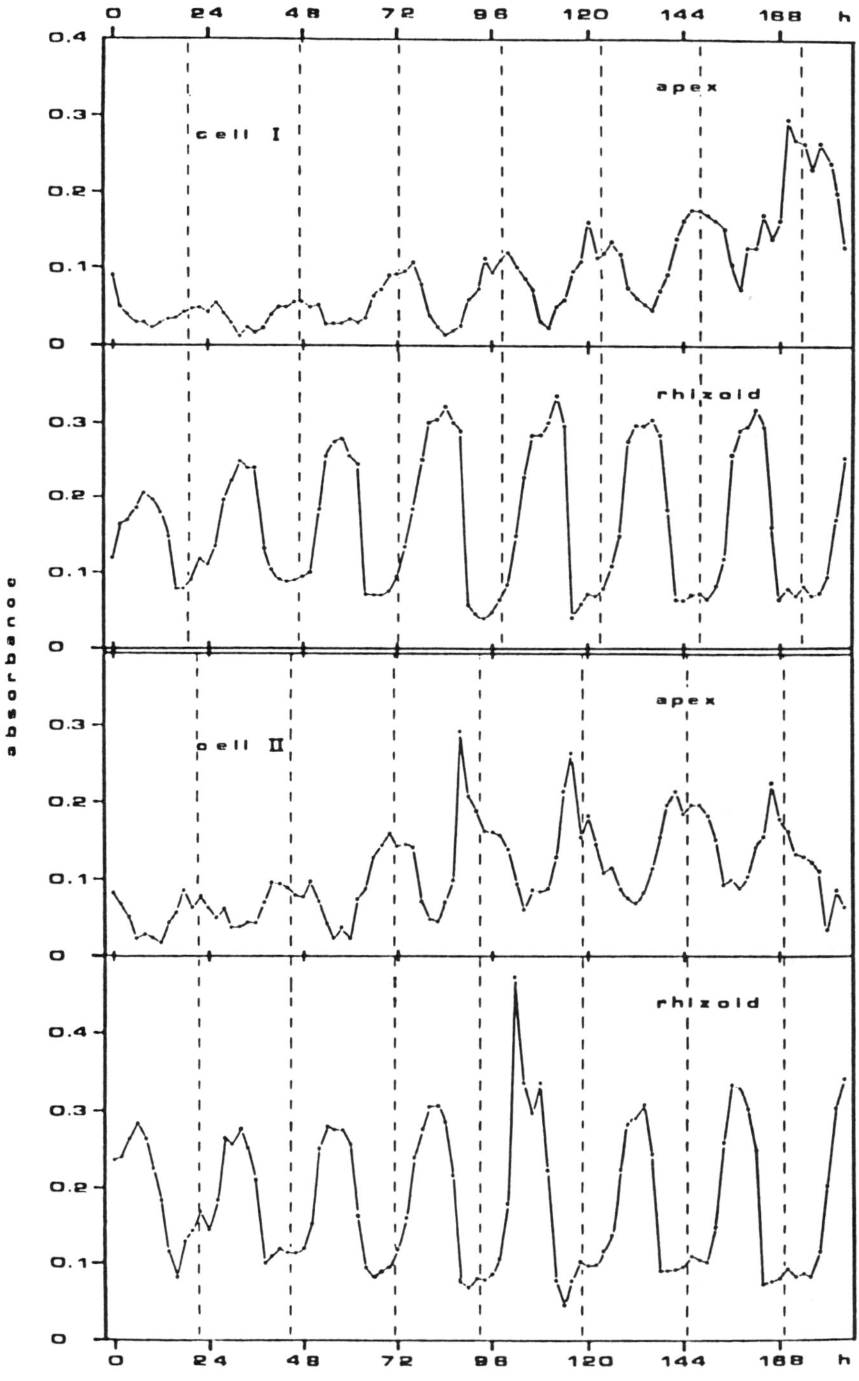
0
24
48
72
96
120
144
168
h
0.4
0.3
0.2
0.1
0
apex
cell I
rhizoid
absorbance
cell II

C. Short-Period Oscillations

By recording chloroplast accumulation in an *Acetabularia* cell irradiated with a small white light beam, Pâques and Brouers (1985) found that the optical density eventually fluctuated with a period of about 4 min. This observation suggests that part of the chloroplasts may move from the light-induced band or change their position in such a way as to affect the optical density of the stalk. It is not known whether these short-period (4 min) oscillations are related to normally occurring chloroplast migration and/or to the release of bands from the apex and the rhizoid. According to Schmid and Koop (1983), the time elapsed between the release of two chloroplast bands would be of about 13 to 20 min. Band release would provoke chloroplast density oscillations along the stalk with the same frequency.

VI. CONCLUSIONS

It is reasonable to assume that in *Acetabularia*, the complex multistriate streaming system, which obviously comprises numerous macromolecules, is the physical support for the migration of chloroplasts, including their periodic alignment. Moreover, on the basis of the results reported by Schmid and Koop (1983), it seems quite plausible that band formation in the stalk represents an expression of the circadian rhythm of chloroplast migration. If this hypothesis is correct, the presence of small regularly distributed bands in the whorls (Fig. 4) would indicate that the circadian rhythm of chloroplast migration occurs also in these tiny cell structures, which contain plastids, whose length increases with the degree of ramification (Mangenot and Nardi, 1931).

Although the velocity of band formation was not measured in our first work (Sironval et al., 1977), we were surprised by their sudden appearance, a fact which suggested the presence of a "concerted structural organization" along the stalk (and the whorls) capable of reacting very rapidly to a stimulus. Puiseux-Dao (1984) presented a working hypothesis supporting the idea that light effects, electrical activity, and cytoplasmic streaming in *Acetabularia* are linked. In addition, according to her opinion, water and ions transport play an important role in cytoplasmic streaming. The rapidity of band formation might be related to the electrical activity of the cell.

Schmid and Koop (1983) also emphasize the importance of the electrical activity, suggesting that changes in the apicobasal potential in *Acetabularia*, which follow circadian characteristics (Broda and Schweiger, 1981; Borghi et al., 1983; Puiseux-Dao, 1984), may provoke a switch in the main direction of the chloroplast stream, resulting in the release of successive chloroplast bands. A better knowledge of streaming and band formation in *Acetabularia* would be obtained by coupling physicochemical with physiological studies. For example, the rheological properties of the *Acetabularia* cytoplasm were not characterized (Strömgren Allen, 1983). Moreover, the dynamics of fluid transport in the stalk, which may compared to a vertical (in nature) tube, was not investigated. It should be added here that the rhythmic change of the shape of the chloroplasts (Vanden Driessche, 1966, 1967) would influence the velocity of their migration.

ACKNOWLEDGMENTS

I thank Simone Puiseux-Dao, Marc Pâques, and Reiner Schmid for their kind permission to reproduce their figures. I also thank the Société Française de Microscopie Electronique, the Laboratoire de Cryptogamie, and Academic Press. Mrs. Maggi Groenen has word-processed the manuscript.

REFERENCES

Arapis, G., Hoursiangou-Neubrun, D., Bonotto, S., and Puiseux-Dao, S. (1984). *Hydrobiologia*, *116/117*: 184.

Arapis, G., Bonotto, S., and Puiseux-Dao, S. (1985). In *Acetabularia 1984* (S. Bonotto, F. Cinelli, and R. Billiau, eds.), Belgian Nuclear Center, Mol, Belgium, p. 279.

Bonotto, S., and Sironval, C. (1977). In *Progress in Acetabularia Research* (C. L. F. Woodcock, ed.), Academic Press, New York, p. 241.

Bonotto, S., Lurquin, P., and Mazza, A. (1976). In *Advances in Marine Biology*, Vol. 14 (F. S. Russel and M. Yonge, eds.), Academic Press, London, p. 123.

Bonotto, S., Nuyts, G., Baugnet-Mahieu, L., Gilles, J., Dujardin, E., and Sironval, C. (1978). In *Chloroplast Development* (G. Akoyunoglou and J. H. Argyroudi-Akoyunoglou, eds.), Elsevier/North-Holland Biomedical Press, Amsterdam, p. 333.

Borghi, H., Puiseux-Dao, S., Durand, M., and Dazy, A.-C. (1983). *Plant. Sci. Lett.*, *31*: 75.

Broda, A. (1979). *Eur. J. Cell Biol.*, *20*: 127.

Broda, H., and Schweiger, H. G. (1981). *Eur. J. Cell Biol.*, *26*: 1.

Broda, H., Schweiger, G., Koop, H. V., Schmid, R., and Schweiger, H. G. (1979). In *Developmental Biology of Acetabularia* (S. Bonotto, V. Kefeli, and S. Puiseux-Dao, eds.), Elsevier/North-Holland Biomedical Press, Amsterdam, p. 163.

Crawley, J. C. W. (1966). *Planta, 69*: 365.
Crawley, J. C. W. (1970). In *Biology of Acetabularia* (J. Brachet and S. Bonotto, eds.), Academic Press, New York, p. 73.
Dazy, A.-C., Hoursiangou-Neubrun, D., and Sauron, M. E. (1981). *Biol. Cell, 41*: 235.
Fukui, S., and Nagai, R. (1985). *Plant Cell Physiol., 26*: 737.
Garcia, E. (1986). Thesis, University of Paris VII.
Gibor, A. (1977). In *Progress in Acetabularia Research* (C. L. F. Woodcock, ed.), Academic Press, New York, p. 231.
Hartwig, R., Schweiger, H., Schweiger, R., and Schweiger, H. G. (1985). *Proc. Natl. Acad. Sci. USA, 82*: 6899.
Herth, W., Heck, B., and Koop, H. V. (1981). *Protoplasma, 109*: 257.
Heunert, H. H. (1978). *Protoplasma, 93*: 477.
Issinger, O., Maass, I., and Clauss, H. (1971). *Planta, 101*: 360.
Kamiya, N. (1962). In *Handbuch der Pflanzenphysiologie*, Vol. 17, Part 2 (W. Ruhland, ed.), Springer-Verlag, Berlin, p. 979.
Kamiya, N. (1981). *Annu. Rev. Plant Physiol., 32*: 205.
Kamiya, N., and Kuroda, K. (1966). *Bot. Mag. Tokyo, 79*: 706.
Keck, K. (1964). In *Methods in Cell Physiology*, Vol. 1 (D. M. Prescott, ed.), Academic Press, New York, p. 189.
Kellner, G., and Werz, G. (1969). *Protoplasma, 67*: 117.
Koop, H. V. (1981). *Protoplasma, 109*: 143.
Koop, H. V., and Kiermayer, O. (1979). *Eur. J. Cell Biol., 20*: 123.
Koop, H. V., and Kiermayer, O. (1980a). *Protoplasma, 102*: 147.
Koop, H. V., and Kiermayer, O. (1989b). *Protoplasma, 102*: 295.
Koop, H. V., and Kiermayer, O. (1980c). *Eur. J. Cell Biol., 22*: 355.
Koop, H. V., Schmid, R., Heunert, H. H., and Milthaler, B. (1978). *Protoplasma, 97*: 301.
Mangenot, G., and Nardi, R. (1931). *Rec. Trav. Cryptogr. Dédiés à Louis Mangin, Paris, 11*: 459.
Melkonian, M., and Robenek, H. (1984). In *Progress in Phycological Research*, Vol. 3 (F. E. Round and D. J. Chapman, eds.), Biopress, Bristol, England, p. 193.
Menzel, D. (1986). *Protoplasma, 134*: 30.
Nagai, R., and Fukui, S. (1981). *Protoplasma, 109*: 79.
Nasr, A. H. (1939). *Rev. Algol., 11*: 347.
Pâques, M. (1982). Thesis, University of Liège, Belgium.
Pâques, M., and Brouers, M. (1981). *Protoplasma, 105*: 360.
Pâques, M., and Brouers, M. (1985). In *Acetabularia 1984* (S. Bonotto, F. Cinelli, and R. Billiau, eds.), Belgian Nuclear Center, Mol, Belgium, p. 303.
Puiseux-Dao, S. (1962). *Rev. Gen. Bot., 819*: 409.
Puiseux-Dao, S. (1979). *Biol. Cell., 34*: 83.
Puiseux-Dao, S. (1984). In *Cell Cycle Clocks* (L. N. Edmunds, Jr., ed.), Marcell Dekker, New York, p. 351.

Puiseux-Dao, S., Dazy, A.-C., Hoursiangou, D., and Matthys, E. (1972a). In *Biology and Radiobiology of Anucleate Systems*, Vol. 2, Plant Cells (S. Bonotto, R. Goutier, R. Kirchmann, and J. R. Maisin, eds.), Academic Press, New York, p. 101.

Puiseux-Dao, S., Dazy, A.-C., Hoursiangou-Neubrun, D., and Borghi, H. (1972b). *Protoplasma*, *75*: 484.

Puiseux-Dao, S., Hoursiangou-Neubrun, D., and Dubacq, J. P. (1978). In *Chloroplast Development* (G. Akoyunoglou and J. H. Argyroudi-Akoyunoglou, eds.), Elsevier/North-Holland Biomedical Press, Amsterdam, p. 345.

Puiseux-Dao, S., Hoursiangou-Neubrun, D., and Dubacq, J. P. (1979). In *Developmental Biology of Acetabularia* (S. Bonotto, V. Kefeli, and S. Puiseux-Dao, eds.), Elsevier/North-Holland Biomedical Press, Amsterdam, p. 141.

Schmid, R. (1981). *Protoplasma*, *105*: 364.

Schmid, R., and Koop, H. V. (1981). *Protosplasma*, *105*: 364.

Schmid, R., and Koop, H. V. (1983). *Z. Pflanzenphysiol.*, *112*: 351.

Shephard, D. (1965). *Biochim. Biophys. Acta*, *108*: 635.

Shibata-Sekiya, K., and Tonomura, Y. (1975). *J. Biochem.*, *77*: 543.

Sironval, C., Bonotto, S., Pâques, M., and Dujardin, E. (1977). In *Progress in Acetabularia Research* (C. L. F. Woodcock, ed.), Academic Press, New York, p. 207.

Strömgren Allen, N. (1983). In *The Application of Laser Light Scattering to the Study of Biological Motion* (J. C. Earnshaw and M. W. Steer, eds.), Plenum Press, New York, p. 529.

Vanden Driessche, T. (1966). *Exp. Cell Res.*, *42*: 18.

Vanden Driessche, T. (1967). *Sci. Prog.*, *55*: 293.

van der Ben, D., Felluga, B., and Bonotto, S. (1972). *Protoplasma*, *75*: 475.

Weber, K., Osborn, H., Franke, W. W., Seib, E., Scheer, U., and Herth, W. (1977). *Cytobiologie*, *15*: 285.

Werz, G. (1974). *Int. Rev. Cytol.*, *38*: 319.

23
Cell Movement and Pattern Formation in *Gonyaulax polyedra*

Till Roenneberg *University of Munich, Munich, Germany*

J. Woodland Hastings *Harvard University, Cambridge, Massachusetts*

I. INTRODUCTION

Small organisms tend to form aggregates, which often exhibit distinct patterns. This phenomenon has been known and studied for quite some time (Loefer and Mefferd, 1952; Plesset and Winet, 1974; Teodoresco and Angelescu, 1941; Wagner, 1911), and several models have been put forward to account for it (Levandowsky et al., 1975; Kessler, 1985, 1986; Harashima et al., 1988; Hill et al., 1989). Such patterns typically occur at the surface and upper regions of a culture, and range from polka dots and swarm-like structures to branched or parallel reticules and highly geometrical structures such as honeycombs. The geometry of these structures has been reported to be fractal (Harashima, personal communication); that is, the length between bifurcations maintains a systematic ratio throughout all orders of branching.

This kind of pattern formation is postulated to involve a type of nonthermal convection for which Platt (1961) coined the term *bioconvection*, by analogy with the fluid movements that occur in nonuniformly heated fluids (Rayleigh-Bénard instability; Bergé et al., 1988). In bioconvection, bulk fluid at the surface containing a high cell concentration falls toward the bottom, so that the cells are dispersed and diluted.

The cells accumulate at the surface by virtue of negative geotaxis. Once a portion of the culture reaches a critical density, it falls convectively toward the bottom. This convective bulk movement depends on the fact that the organisms have a specific gravity slightly greater than that of the medium. The effective density increase at the surface does not require or involve cell–cell aggregations or clumping; individual cells remain free swimming and generally experience no lasting contact with other cells.

We have recently reported such pattern formations in the unicellular marine dinoflagellate *Gonyaulax polyedra* (Roenneberg et al., 1989). Stability, shape, and size of aggregations depend on several factors, including container type, lighting conditions (intensity, direction and possibly color), and cell concentration. In addition, the character of the patterns exhibits a daily oscillation, which persists in cultures maintained in constant conditions. Daily variations in motility and/or vertical migration have also been previously described for numerous unicellular organisms, including *Euglena* (Brinkmann, 1966), *Gonyaulax tamarensis* (Anderson and Stolzenbach, 1985), *G. polyedra* (Hasle, 1950), *Gyrodinium* (Kamykowski et al., 1988), *Noctiluca* (Uhlig and Sahling, 1982), *Paramecium* (Hasegawa and Tanakadate, 1987), *Peridinium* (Sibley et al., 1974), and *Tetrahymena* (Wille and Ehret, 1968). However, the daily changes in pattern formation have frequently been attributed simply to direct effects of environmental factors such as light and temperature (Forward, 1976; Hand et al., 1965; Kamykowski et al., 1988), nutrient gradients (Ballek and Swift, 1986; Cullen and Horrigan, 1981), gravity (Roberts, 1970), and predation (Gliwicz, 1986). Here we describe the reasons for the differences in aggregation shapes and discuss the influence of various factors, such as vertical migration, phototaxis, and the circadian clock.

II. PATTERN FORMATION IN *GONYALAUX POLYEDRA*

Populations of the marine dinoflagellate *Gonyaulax polyedra*, grown in Fernbach flasks (1 to 1.5 liters) under a cycle of 12 h light and 12 h darkness, form clear patterns. When the lights are on, one can observe that cells aggregate in vertical bands forming parallel and branching streaks near the surface, which extend downward as finger- and curtain-like structures (Fig. 1a). During the dark phase, however, cells are distributed in a thin layer or "lawn" at the bottom of the flask. To the circadian biologist, it is not surprising that the occurrence of these day and night patterns appears to be independent of the light–dark transitions. The nightly "lawn" begins to form about 2 h before the lights go off and the population begins to rise to the surface of the culture medium also approximately 2 h before the lights in the culture room are turned on.

Clear but somewhat different patterns can also be observed with cultures in petri dishes. These also exhibit characteristic daily changes, but they tend to be locally more stable with time, making them favorable for

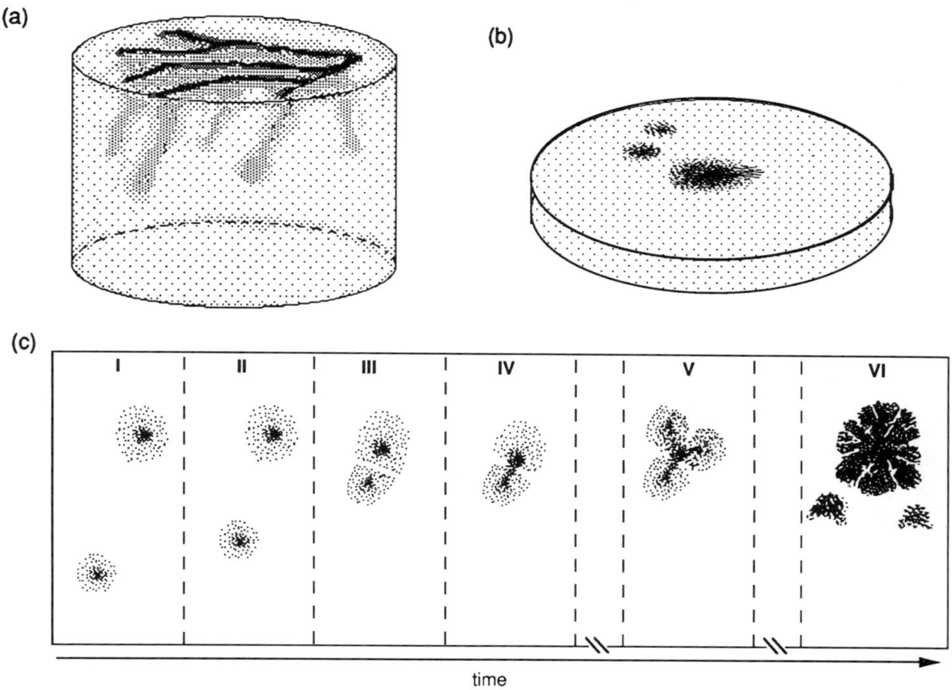

Fig. 1 Different patterns of the *Gonyaulax* aggregations: (a) band-shaped aggregations with falling fingers or curtains found in beakers; (b) round swarms as found in petri dishes (for a cross section of a swarm showing the dynamics of the aggregation, see icon below stage V in Fig. 6b); (c) when several swarms form in a petri dish, the smaller ones tend to fuse with the larger ones over the course of several hours forming complex structures (see also Fig. 2).

study. After cells in a culture have been evenly distributed by swirling, they aggregate within minutes in numerous small polka dots, which then fuse to form one or several larger swarms with dense centers (Fig. 1b). In this process, smaller swarms frequently form in the dish, commonly moving toward and fusing with the larger aggregation (Fig. 1c). Before fusion, the two are separated by a clear and practically cell-free interface (Fig. 1c, III). Thereafter, the dense centers of the two aggregates fuse into a bone-shape structure (Fig. 1c, IV). Repetitive fusions, which occur especially at high cell densities (> 8000 cells/mL), lead to the formation of star- or cartwheel-like structure (Fig, 1c, V and VI), which is also shown in the photograph in Fig. 2. Cell-free areas, probably related to the

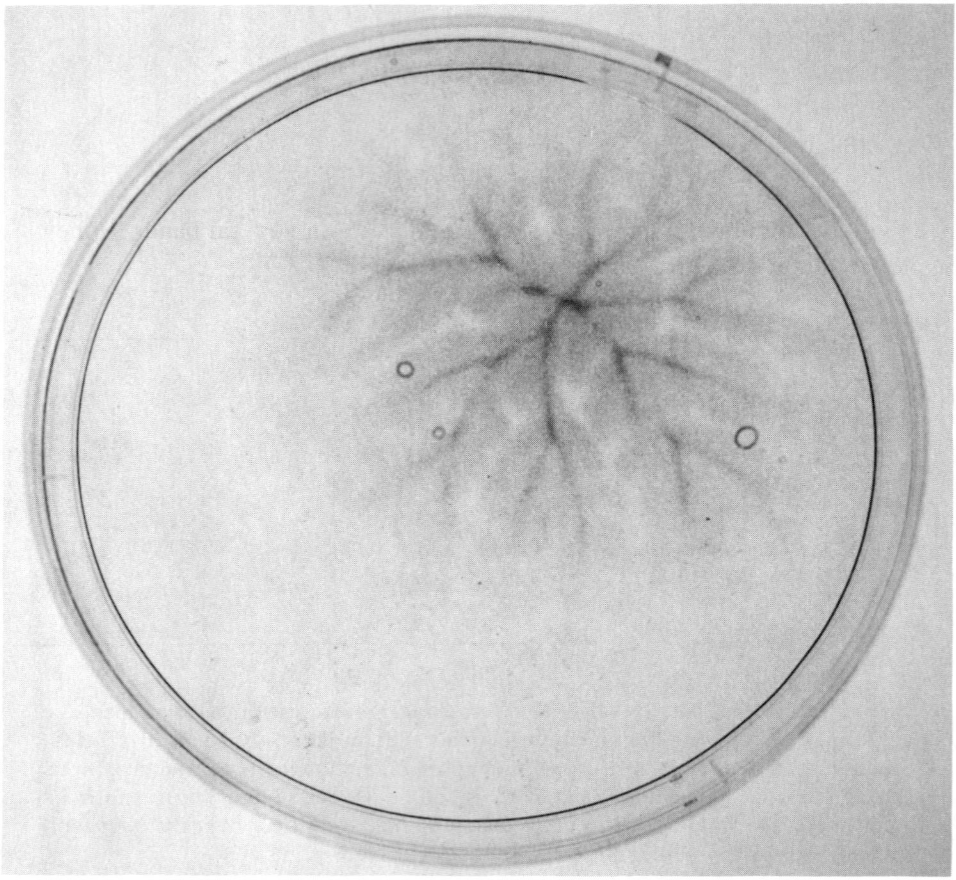

Fig. 2 Photograph illustrating the complex nature of the aggregations in petri dishes. Note the two almost separate reticules, one consisting of dark lines representing high cell concentrations and white lines representing almost cell-free regions.

cell-free interfaces seen prior to fusion (light areas in Fig. 2), commonly persist throughout these complex structures, giving the impression of two reticules, one formed by the dense streaks and centers, the other formed by streaks of cell-free areas (Fig. 2).

The dynamics of the aggregations can best be seen in the compact aggregates, which we have called swarms: Cells move downward in the dense core, and rise back to the surface in the periphery (see icon below stage V of Fig. 6b). Within these dynamic structures, cells move rapidly. As judged by microscopic observations, the velocity of the *Gonyaulax* cells is higher within the swarms than elsewhere, attributable to a passive motion due to the convection dynamics. This has also been observed in *Tetrahymena* (Plesset and Winet, 1974).

Typical day-phase swarms occur in petri dishes only over a limited range of light intensities (~500 to 5000 lux) (Roenneberg et al., 1989). In both dimmer and bright light, cells tend to form vertical bands as they do in the large culture flasks with light provided from below. At dusk, cells fall to the bottom of the dish, concentrating to one side. This vertical migration is independent of the direction of the light source and occurs in light–dark cycles as well as in constant light (independent of wavelength). In time-lapse recordings, the size of the daytime swarms appears to pulsate with a period of minutes. In contrast, the nighttime pattern does not show such high frequency pulsations.

III. CIRCADIAN ASPECTS

For the investigation of circadian aspects of the aggregation activity, we have developed an automated recording system (Fig. 3). Several petri dishes (< 10) are filmed by a video camera which is mounted below the cultures. By recording from below, the picture is less obstructed by condensation droplets, which often form on the underside of the lid. To achieve the largest contrast between cells and background, light is provided from the side so that only reflections from the cells themselves are recorded by the video camera. The video image is digitized approximately every 10 min. With the help of a computer program (developed by W. Engelmann and J. Schuster, Tübingen), the experimenter can define a number of rectangles on the computer screen, thereby selecting parts of the displayed digitized image for separate evaluation. Within these rectangles, the program records the number of pixels set (i.e., the presence of a bright object at this coordinate of the image) and stores the data in a file that can be analyzed later.

In cell cultures maintained under constant conditions of light and temperature, the daily changes in pattern formation exhibit a persistent rhythm with an endogenous period length (Fig. 4). Like other circadian rhythms in *Gonyaulax* (Johnson and Hastings, 1986), the exact period length depends on the experimental conditions: temperature (Hastings and Sweeney, 1957), as well as the intensity (Broda et al., 1986; Hastings and Sweeney, 1958) and spectral composition (Roenneberg and Hastings, 1988; Roenneberg and Hastings, 1991) of the background light. Under

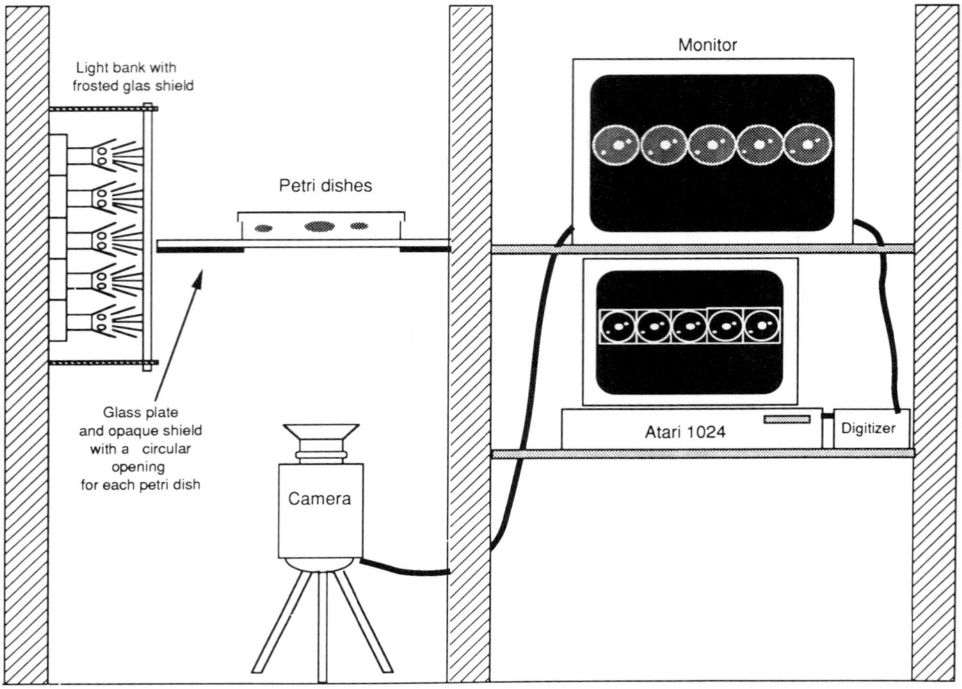

Fig. 3 Setup for automatic recordings of the circadian aggregation rhythm in *Gonyaulax polyedra* (see the text for details).

blue light, the period length is shorter than 24 h, whereas it is longer than 24 h under red and yellow light.

The experiment shown in Fig. 4 was carried out in constant yellow light. Figure 4a displays the number of pixels set within the rectangle circumscribing the video image of a petri dish over a time course of almost 10 days. The same data are ''double plotted'' in Fig. 4b: sections of 48 h are shown on the horizontal axes, and consecutive days are also drawn beneath each other. In such a raster plot, the daily events (in this case, the presence of aggregations) appear directly beneath each other if they occur exactly every 24 h. If the period of the rhythm is shorter than 24 h, peaks are located progressively more toward the left; if it is longer, to the right. In double plots, the stability of the rhythm or possible period changes and phase jumps can easily be detected. The endogenous period length of the rhythm under ''free-running'' (constant) conditions can be

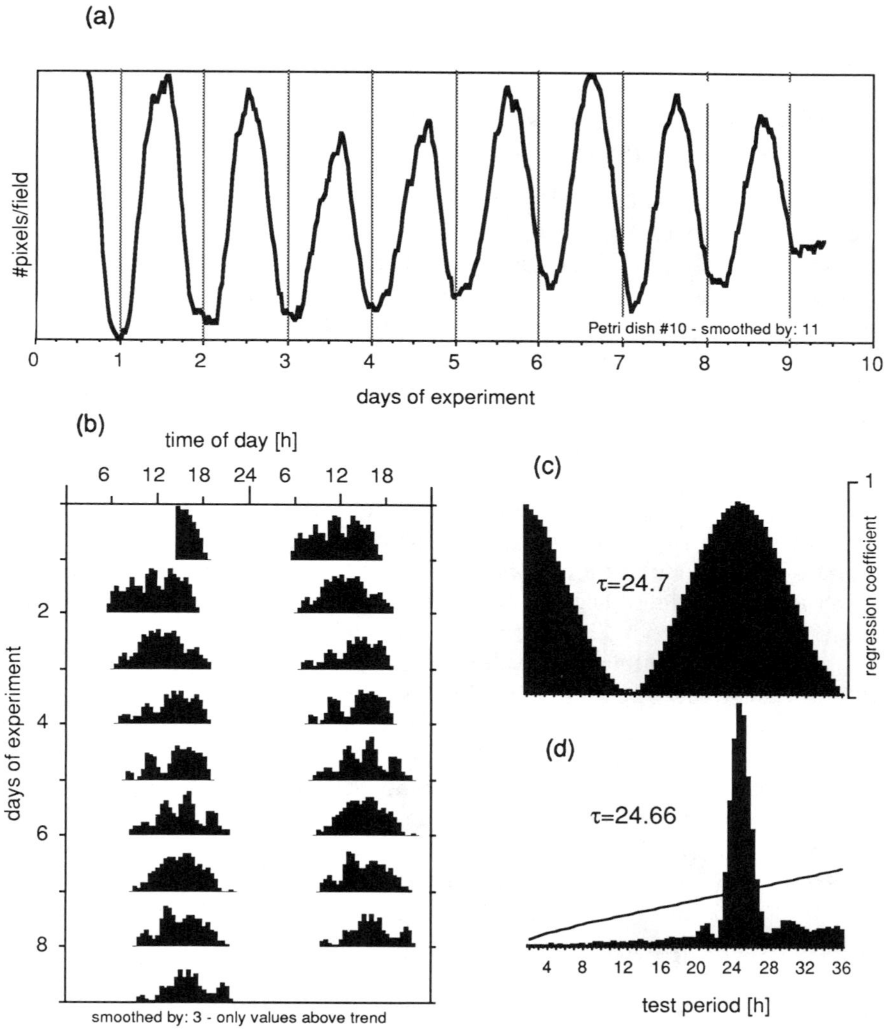

Fig. 4 The formation of dense surface-near aggregations is controlled by the circadian clock. (a) Number of pixels set within a chosen area of the digitized video image of a petri dish measured for almost 10 days. (b) "Double plot" of the same data (see the text). The period length of the "free-running" rhythm is analyzed either by autocorrelation (c) or by periodogram (d). The experiment was carried out under constant conditions of yellow light (>510 nm, ~100 $\mu E\ m^{-2}\ s^{-1}$) and temperature (19 ± 1°C).

analyzed by mathematical procedures, such as the autocorrelation (Fig. 4c) or the periodogram (Sokolove and Bushell, 1978) (Fig. 4d).

IV. OTHER FACTORS AND POSSIBLE MECHANISMS

The aggregation "behavior"* has variations other than the mere circadian alteration between presence or absence of swarms. One other feature is the exact location of the main swarm in the culture container and its changes over time. If the experimental conditions are kept constant, the daily location of the main swarm is extremely stable (Fig. 5). But if conditions (e.g., the direction of the light) are changed, the location of the main swarm, relative to the light source, can be used as a sensitive indicator for the cells' response to light. To analyze such responses, the digitized video image of a single petri dish is divided into a 10 × 10 matrix of rectangles which are analyzed separately by the computer. The location of each aggregation and its development over time can later be reconstructed in the computer by an analysis program. In the upper portion of Fig. 5, the individual pixel densities for each rectangle are shown as a function of time. The two lower graphs of Fig. 5 show the same data with the x and y coordinates plotted separately.

Aggregation patterns also depend on container shape, depth of the culture, and average cell density. For example, there is a great difference between the swarm-like aggregations found in petri dishes and the streaks and bands found in the much larger culture flasks (see Fig. 1). One difference is that the cell aggregations in petri dishes are rather stable, whereas those in flasks form and dissipate within minutes.

We have tried to analyze the mechanisms behind these intriguing differences by visual observations; these are summarized very schematically in Fig. 6. The main reason for the differences appears to relate to the depth of the medium and the consequent differences in the cells' movement within the aggregations. If the depth of the medium is less than approximately 0.4 cm, no distinct patterns form. In depths not exceeding about 2 cm, cells aggregate in round swarms, which at high cell concentrations (>8000 cells/mL) can develop complex structures (see Fig. 2). At greater depths, the structures appear and gradually elongate to form

---

*Around the turn of the century, H. S. Jennings introduced the term "behavior" for the aggregations formed by the unicellular organism *Paramecium* (Jennings, 1899). Although it may be questionable to use "behavior" for single cells, we believe that their differentiated spatial distribution, motility, and orientation (taxis) has many features comparable to what is commonly called behavior in higher organisms.

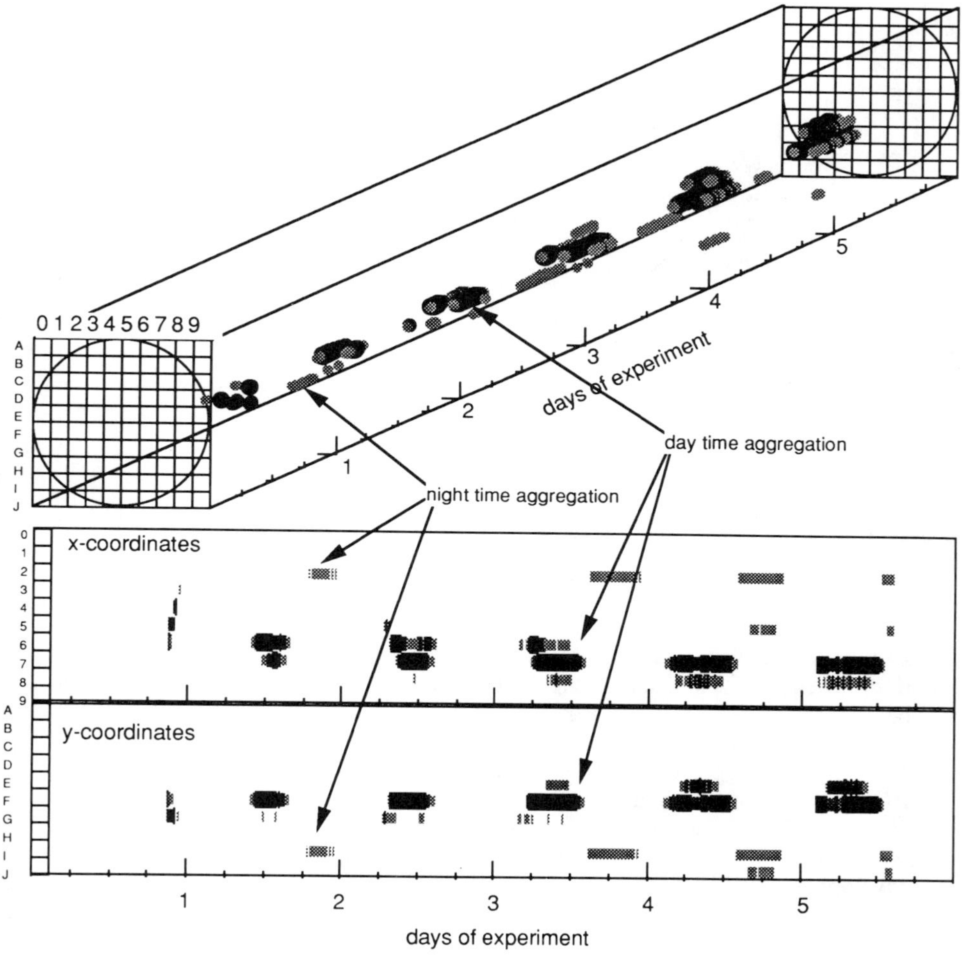

Fig. 5 To investigate the daily location of the aggregates within a petri dish and their changes over time, the digitized video image of a dish is divided into a 10 × 10 matrix of rectangles. The number of pixels set in each of these rectangles is recorded every 10 min. The developments of the aggregations can then be reconstructed by an analysis program. The top graph shows the density distributions within the petri dish at different times as a "stack." The two bottom graphs show the *x*- and *y*-distribution of aggregates at subsequent time steps.

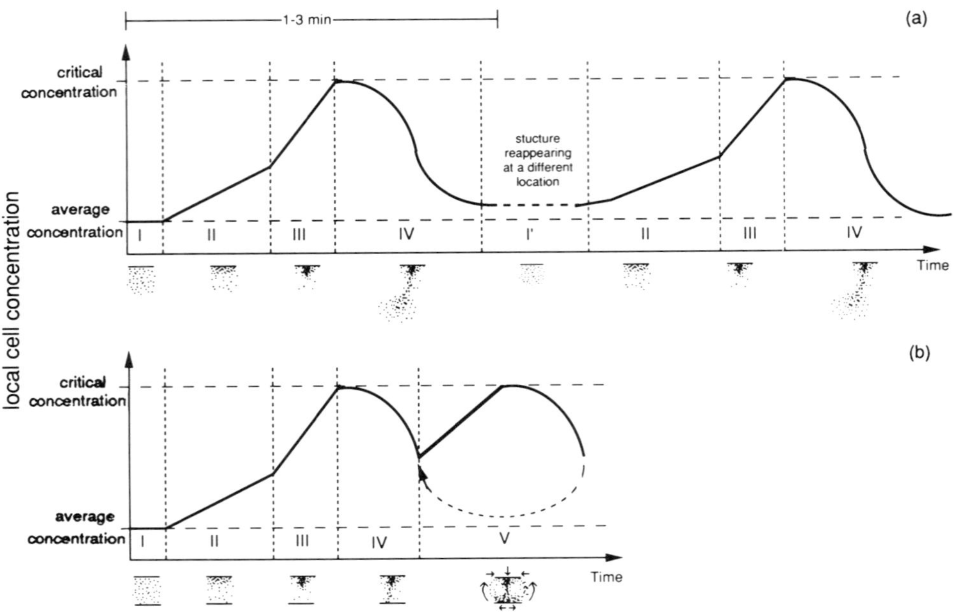

Fig. 6 Schematic description of the mechanisms leading to different aggregation shapes. The cell distributions, viewed from the side, are represented as icons below each stage. The two panels represent different water depths: (a) > 2 cm; (b) 0.4 to 2 cm). Stage I, even distribution of cells; Stage II, due to negative geotaxis, cells aggregate at the surface and, additionally, convene locally (stage III); stage IV, at a critical local cell concentration, the surface-near cell aggregates start to fall as a bulk. Cells are dispersed if water depth exceeds 2 cm (a) but rise back locally in lower depth (b). Due to their intrinsic dynamics, the swarms developing at lower depths are relatively stable in their location, (b, stage V).

streaks or bands, sometimes branching, and then dissipate, all of this within 1 to 3 min.

The prerequisites of the intrinsic movement of bioconvection patterns are that the organisms be motile and swim toward the surface (negative geotaxis) and that their specific gravity be slightly higher than that of the medium (the average specific gravity of *Gonyaulax* cells is approximately 1.069 g/cm^3; Eppley et al., 1968). These factors, combined with horizontal phototactic movements and probably with stochastic processes, lead to high local cell concentrations in the upper layer of the medium.

In the case of deep cultures, the development and dissipation of an aggregation takes only a few minutes, and a new aggregation will not usually recur in the exact same location. This is attributed to the fact that the more dense bulk fluid is well diluted as it falls through several centimeters, so that the subsequent nucleation of a dense region will occur elsewhere (Fig. 6b).

In the case of more shallow cultures, the bottom of the vessel limits cell dispersion. There is less dilution and the cells that have fallen down then return to the surface at the peripheral regions of the now quasi-stable aggregate (Fig. 6a). Its self-stabilizing nature is such that its center is most dense; as this region falls down, cells are drawn inward from the periphery and concentrated, perpetuating the structure. Two or more such structures may attract one another, first remaining distinct (while close), followed by mergers (see Fig. 1c).

V. INTERACTIONS OF MANY FACTORS

The many factors that influence the aggregation behavior either directly or indirectly are summarized in Fig. 7. The initiation of bioconvections depends on the establishment of a high local cell concentration (center of the diagram) which is influenced by several physical and biological factors. Different functions of the ''intracellular environment'' (e.g., geo- and phototaxis, swimming speed and the circadian clock) are specifically influenced by the ''physical environment'' (e.g., gravity, light, and temperature). Geotaxis, together with phototaxis, are the main factors contributing to the development of high local cell concentrations near the surface. The location of the aggregations during the subjective* day indicates that cells prefer certain light intensities. This is especially obvious in the stable daytime swarms, found in petri dishes, when the light is provided from the side: depending on light intensity, swarms form at defined loci in the petri dish, where they occur repeatedly each day (Fig. 5). Unlike this horizontal movement, the daily change in vertical migration is independent of the intensity and direction of the light (Roenneberg et al., 1989).

Both photo- and geotaxis change during the course of the day (i.e., are influenced by the circadian clock). During the subjective day, cells show an intensity-dependent phototaxis, whereas during the subjective night, their entirely positive phototaxis appears to be independent of light intensity. In addition, cells only move to the surface during the subjective day (negative geotaxis). Their swimming speed is also higher during the

*The term ''subjective'' refers to the endogenous phase of the circadian clock when the external conditions are kept constant.

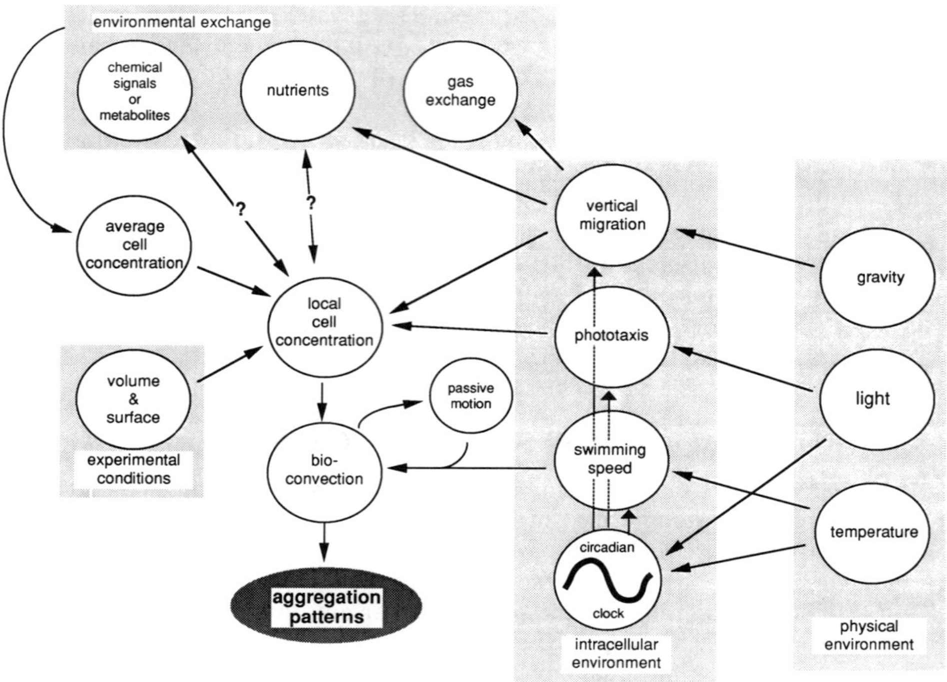

Fig. 7 Summary of the various parameters contributing to the aggregation phenomenon. The initiation of bioconvections depends on a high local cell concentration (center of the graph). Influences are shown as arrows and are separated into groups: physical and intracellular environment, environmental exchange, average cell density, and experimental conditions.

day than during the night (D. Morse and L. Fritz, personal communications). Similar diurnal changes in motility have also been reported for other dinoflagellates (Kamykowski et al., 1988).

The formation of aggregations may also depend on environmental factors, such as nutrients and gases, whose availability certainly changes due to vertical migration. We know very little about these factors, including possible chemical signals between cells within aggregations. However, we have shown that the upward movement of the cells during the daytime is independent of the location of the medium–air interface: If a culture is transferred to a small cuvette, it is possible to turn the cuvette upside down without emptying it. Under these conditions (the medium–air interface now being at the bottom) the cells still swim upward during the subjective day and sink to the bottom during the subjective night (Lee and Roenneberg, unpublished results). The composition of the seawater medium influences growth rate, so the chemical qualities of the environment certainly influence the aggregation behavior indirectly via the average cell concentration.

The differences in aggregation patterns (swarms or bands) are mainly due to water depth (Fig. 6). Because cells rise to the surface during the subjective day, they share the available surface area. Therefore, the number of cells concentrating below the surface will also depend on the volume/surface ratio of the container. Thus these ''experimental conditions'' will also influence the size and shape of the aggregation. The centripetal surface convection draws cells into the circular current, shaped much like a magnetic field, whereas they are dispersed at the lower tip of the falling central density. Therefore, volume and surface also contribute to the local cell concentration.

The factors influencing cell movement and aggregation clearly interact in complex ways. While the analysis of the phenomena will be aided by considering these factors individually, one should not lose sight of their interdependences.

ACKNOWLEDGMENTS

This work was supported by grants from the Deutsche Forschungsgemeinschaft (Ro 656/2) and the National Institutes of Health (MH-56660). We would like to thank K. Schill and H. Distel for helpful discussions and I. Neher for her excellent technical assistance.

REFERENCES

Anderson, D. M., and Stolzenbach, K. D. (1985). *Mar. Ecol. Prog. Ser.*, *25*: 39.

Ballek, R. W., and Swift, E. (1986). *J. Exp. Mar. Biol. Ecol.*, *101*: 175.

Bergé, P., Pomeau, Y., and Vidal, C. (1988). *L'ordre dans le chaos: Vers une approche déterministe de la turbulance*, Hermann, Paris.

Brinkmann, K. (1966). *Planta*, *70*: 344.

Broda, H., Gooch, V. D., Taylor, W. R., Aiuto, N., and Hastings, J. W. (1986). *J. Biol. Rhythms*, *1*: 251.

Cullen, J. J., and Horrigan, S. G. (1981). *Mar. Biol.*, *62*: 81.

Eppley, R. W., Holm-Hansen, O., and Strickland, J. D. H. (1968). *J. Phycol.*, *4*: 333.

Forward, R. B. (1976). *Photochem. Photobiol. Rev.*, *1*: 157.

Gliwicz, M. Z. (1986). *Nature*, *320*: 746.

Hand, W. G., Collard, P. A., and Davenport, D. (1965). *Biol. Bull. Mar. Biol. Lab. Woods Hole*, *128*: 90.
Harashima, A., Watanabe, M., and Fujishiro, I. (1988). *Phys. Fluids*, *31*: 764.
Hasegawa, K., and Tanakadate, A. (1987). *J. Biol. Rhythms*, *2*: 269.
Hasle, G. R. (1950). *Oikos*, *2*: 162.
Hastings, J. W., and Sweeney, B. M. (1957). *Proc. Natl. Acad. Sci.*, *43*: 804.
Hastings, J. W., and Sweeney, B. M. (1958). *Biol. Bull.*, *115*: 440.
Hill, N. A., Pedley, T. J., and Kessler, J. O. (1989). *J. Fluid Mech.*, *208*: 509.
Jennings, H. S. (1899). *Am. J. Psychol.*, *10*: 503.
Johnson, C. H., and Hastings, J. W. (1986). *Am. Sci.*, *74*: 29.
Kamykowski, D., McCollum, S. A., and Kirkpatrick, G. J. (1988). *Limnol. Oceanogr.*, *33*: 66.
Kessler, J. O. (1985). *Nature*, *213*: 218.
Kessler, J. O. (1986). In *Progress in Phycological Research* (R. Chapman, ed.), Biopress, New York, p. 191.
Levandowsky, M., Childress, W. S., Spiegel, E. A., and Hunter, S. H. (1975). *J. Protozool.*, *22*: 296.
Loefer, J. B., and Mefferd, R. B., Jr. (1952). *Am. Nat.*, *86*: 325.
Platt, J. R., (1961). *Science*, *133*: 1766.
Plesset, M. S., and Winet, H. (1974). *Nature*, *248*: 441.
Roberts, A. M. (1970). *J. Exp. Biol.*, *53*: 687.
Roenneberg, T., and Hastings, J. W. (1988). *Naturwissenschaften*, *75*: 206.
Roenneberg, T., and Hastings, J. W. (1991). *Photochem. Photobiol.*, *53*: 525.
Roenneberg, T., Colfax, G. N., and Hastings, J. W. (1989). *J. Biol. Rhythms*, *4*: 201.
Sibley, T. H., Herrgesell, P. L., and Knight, A. W. (1974). *J. Phycol.*, *10*: 475.
Sokolove, P. G., and Bushell, W. N. (1978). *J. Theor. Biol.*, *72*: 131.
Teodoresco, E. C., and Angelescu, E. (1941). *Kolloid Z.*, *97*: 216.
Uhlig, G., and Sahling, G. (1982). *Ann. Inst. Oceanogr.*, *58*: 277.
Wagner, H. (1911). *Proc. R. Soc., Lond. B*, *83*: 94.
Wille, J. J., Jr., and Ehret, C. F. (1968). *J. Protozool.*, *15*: 789.

24
Spatial and Temporal Control of Ion Transport at the Plasma Membrane of *Chara corallina*

William J. Lucas *University of California, Davis, California*

I. INTRODUCTION

The thallus of the giant freshwater alga, *Chara corallina*, is composed of a simple main axis formed by large highly vacuolated internodal cells that are cylindrical in shape, having diameters of approximately 0.7 to 1.2 mm and lengths of 2 to 20 cm. A nodal complex forms the junction between internodal cells and this is composed of small connecting cells that join the internodal cells of the main axis to a whorl of branch internodal cells that are arranged almost perpendicular to the main axis. Under normal developmental conditions, these branches are made up of four internodal cells whose lengths decrease from the basal to the apical position. Main axis and branch internodal cells appear to be identical in all cellular and physiological aspects, except for their difference in physical dimensions.

The *Chara* thallus is easily dissected into individual internodal cells that have a range of cell lengths and surface areas: Under our culture conditions, a normal internodal cell has a length of 4 cm and a total plasma membrane surface area of 1.0 cm^2. Single cells having these dimensions have proved ideal for the study of cellular aspects of control over spatial regulation of membrane transport function. In this chapter we describe the phenomenon of the acid and alkaline banding pattern that forms along the length of these cells. The characteristics of this system are then discussed in terms of our recent studies aimed at elucidating the underlying mechanism(s) responsible for the formation of this pattern.

II. DEVELOPMENT OF EXTRACELLULAR pH AND IONIC CURRENT PATTERNS

A. Light Dependence

It has long been known that many of the ecorticate species of *Chara* and *Nitella* form calcarious deposits on the surface of their internodal cells and that for a large number of species, these deposits are somewhat regularly spaced along the length of physiologically mature cells. Using acid–base indicator dyes, Spear et al. (1969) showed that these characean species establish alternating acid and alkaline bands or zones at the cell surface and the alkaline regions correlated with the areas of $CaCO_3$ deposition. An extensive analysis of this phenomenon has been provided through the use of miniature pH and extracellular vibrating probe technologies (Lucas, 1975a,b, 1982; Lucas and Nuccitelli, 1980), and the general features have been illustrated in Fig. 1. The alternating acid and alkaline regions correlate with areas of outward and inward current, respectively, and these extracellular pH and ionic current profiles are completely light dependent. An interesting feature of this phenomenon is the precision with which the cellular control system maintains the spatial positions at which the direction of the current changes from outward to inward, and vice versa, the profile being stable over many hours, provided that the cell is under suitable illumination.

Omission from the experimental medium of inorganic carbon, in the form of HCO_3^-, or the inhibition of photosynthesis by DCMU, causes the elimination of this banding pattern. This suggests that the component of the control system that imparts light dependence involves an element of the reactions occurring in association with the chloroplasts (i.e., electron transport or the reactions of carbon fixation). However, in that once the cell has been light activated, exogenous HCO_3^- can be replaced with either organic or inorganic buffers, with no effect being detected on the extracellular current profile (Lucas, 1982), any involvement of the *specific* fixation products of photosynthesis can be eliminated. In addition, these results establish that although HCO_3^- is required for the activation of the membrane transport systems, which give rise to this extracellular current pattern, the outward current in the acid bands *cannot* be the result of an electrogenic HCO_3^- influx system in the plasma membrane of *Chara*. Rather, HCO_3^- uptake within these acid regions appears to be mediated by an electrically neutral $H^+ : HCO_3^-$ cotransport system (Lucas et al., 1983), with the outward current being generated by an electrogenic H^+-ATPase.

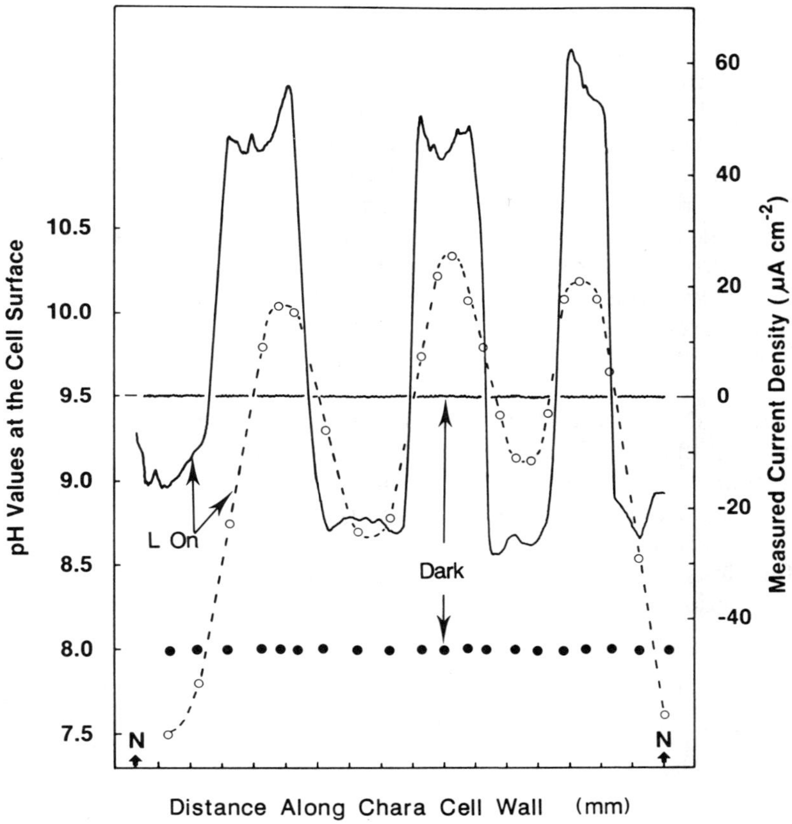

Fig. 1 Extracellular current scans (solid lines) and surface pH (●, ○) along the length of a *Chara* branch internodal cell obtained under dark and light conditions. Positions of the nodes (N) are marked by arrows. The background pH value of the experimental medium was 7.95. (Modified from Lucas and Nuccitelli, 1980.)

B. Inversion of the Extracellular Current Pattern

The identity of the membrane transport system responsible for producing the alkaline bands remains controversial. In our initial model (Lucas, 1975b) we proposed the involvement of an active (?) OH^- transport system, rather than a H^+ influx system or H^+ channel, as later proposed by Bisson and Walker (1980, 1982). Although over the past decade there seemed no way to differentiate between these two systems, our recent results offer strong support for the operation of a passive H^+ channel. As illustrated in Fig. 2, manipulations of the experimental bathing medium, in terms of pH and the form of dissolved inorganic carbon, often give rise to an inversion of the extracellular current pattern (Fisahn and Lucas, 1990a). Thus regions along the membrane that are associated with H^+ efflux or OH^- production can be converted, quite rapidly (within a few seconds), into alkaline or acid bands, respectively. In addition, pH and extracellular vibrating probe studies revealed that the current, in regions where the current changes direction, oscillates between outward and inward current with a frequency of approximately 2 min (Fisahn et al., 1989).

Based on the observation that the extracellular current pattern can be inverted, and that the direction of the current can oscillate with a single time constant, we proposed that the extracellular current pattern is generated by a single class II transporter (see Fig. 3). In this reaction kinetic model, one arm of the transporter functions as a H^+ pump, while the other functions as a H^+ channel. Activation of the transport system involves transition from the two proposed lazy states, L_1 (H^+ channel) and

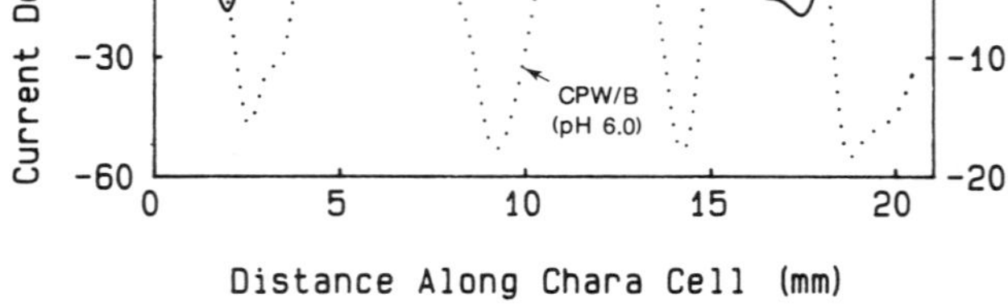

Fig. 2 Acidic pH-induced inversion of the extracellular current pattern that develops at the surface of internodal cells of *Chara*. The current profile measured in *Chara* pond water (CPW/B: 1.0 m*M* NaCl, 0.2 m*M* KCl, 0.2 m*M* $CaSO_4$, 1.0 m*M* $NaHCO_3$) represents the control pattern (solid line with scale indicated on left current axis). Changing the bathing medium to CPW/B (pH 6, buffered with 10 m*M* MES) elicited an inversion of the current pattern (dashed line, with scale indicated on the right current axis). (From Fisahn and Lucas, 1990a.)

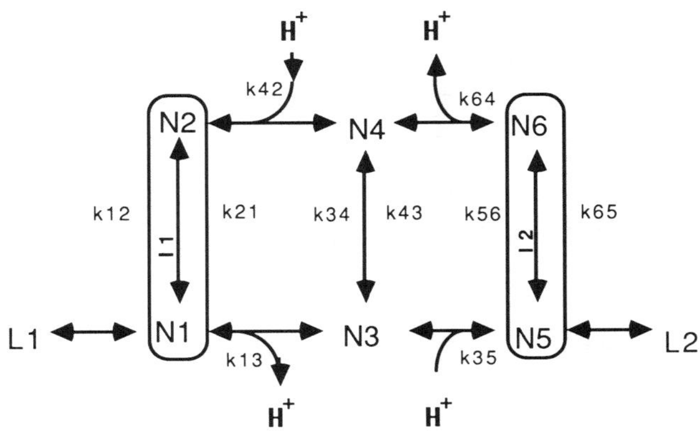

Fig. 3 Unified class II transporter model, comprising a H^+ pump (right-hand side) and H^+ channel (left-hand side). Transition from the lazy states L_1 and L_2 is thought to be regulated by an internal spatial control system. (From Lucas and Fisahn, 1989.)

L_2 (H^+-ATPase) (Lucas and Fisahn, 1989). An important element of this model is the concept that these transitions are controlled by a cellular spatial control system.

C. Hierarchy of Spatial Domains

The complexity of the control system that activates the proposed class II transporter is illustrated by the established response of *Chara* cells to activation by various light intensities (Fig. 4). Under light intensities sufficient to saturate photosynthesis, numerous alkaline bands (regions of functional H^+ channels) are present along the internodal cell length. However, as the activating light intensity is lowered, the overall H^+ channel current does not become reduced, but rather, fewer alkaline bands become activated, with an hierarchy being involved in terms of which regions will (or will not) become functional (Lucas, 1975b). It would appear that the *Chara* spatial control system must be capable of imparting special transition properties to the class II transporters that are in the L_1 state. Based on our earlier identification of *primary* (*p*), *subprimary* (*sp*), and *subsidiary* (*s*) alkaline bands (Lucas, 1975b), we suggest that at least three L_1 states exist for the *Chara* H^+ channel, L_{1p}, L_{1sp}, and L_{1s}.

In *Chara*, the presumed normal function of this class II H^+ transport system is the acquisition of inorganic carbon for photosynthesis. Irrespec-

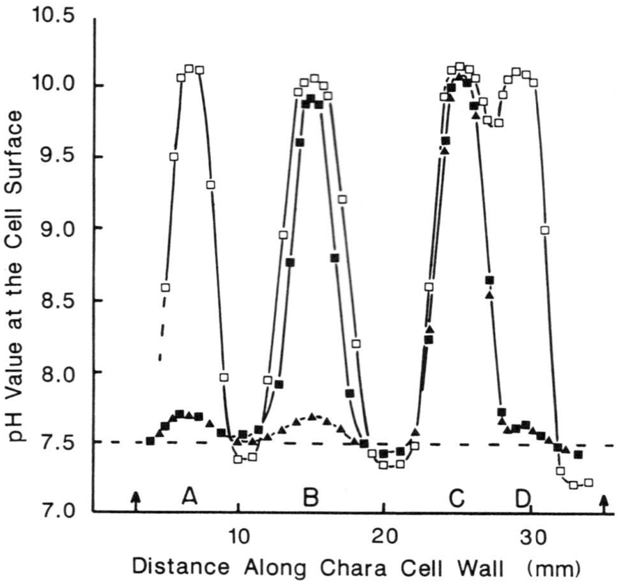

Fig. 4 Cell surface pH values measured along a *Chara* internodal cell after 2 h of illumination at the indicated light intensity. Light intensities were as follows: □, 10 W m^{-2}; ■, 3 W m^{-2}, ▲, 2 W m^{-2}. A 1-h dark period was employed between different light intensities.

tive of whether HCO_3^- enters via the proposed electroneutral $H^+:HCO_3^-$ cotransport system, photosynthetic fixation of the inorganic carbon provided by the HCO_3^- would result in the production of alkalinity within the cytoplasm. In the initial model proposed to account for the alkaline band hierarchy, we suggested that cytoplasmic pH plays an important role in activating the OH^- transporters. Although no specific details are currently available, it still seems plausible to us that changes in cytoplasmic pH are somehow involved in activating the H^+ channel component of the *Chara* class II transporter.

D. Developmental/Cytological Aspects of Spatial Specialization in *Chara*

Banding pattern studies performed on young developing internodal cells of *Chara* revealed that although acid and alkaline bands were formed, their positions were *not* stable (Lucas and Nuccitelli, 1980). Usually, the alkaline bands appear to migrate, at random, over the surface of the cell.

Only when the cell is nearing its final length does a stable pattern form, indicating the presence of an operational spatial control system. This instability in the banding pattern may relate to the acidic pH requirement for characean cell elongation (Metraux and Taiz, 1977, 1979; Metraux et al., 1980). In any event, it is a very interesting observation that warrants further careful study.

Ultrastructural studies conducted on mature *Chara* internodal cells revealed that the plasma membrane in this species undergoes a remarkable degree of amplification by forming extensive tubular anastomoses, termed charasomes (Franceschi and Lucas, 1980; Pesacreta and Lucas, 1984). Of particular importance to the present topic was the finding that the highest densities of these charasomes were found along the plasma membrane of the acid bands, with few, if any being located along the plasma membrane of the alkaline band. In addition, histochemical studies revealed that the plasma membrane that forms the charasome has a high level of ATPase activity (Franceschi and Lucas, 1982).

Although charasomes have been implicated in the acquisition of inorganic carbon (Price and Badger, 1985; Price et al., 1985), the fact that they are not present in *Nitella*, and that HCO_3^- utilization is possible in *Chara*, in which charasome development has been drastically reduced by manipulations in culture conditions (Lucas et al., 1986, 1989), suggests that they are more likely essential for the acquisition of other essential anions, like Cl^-. In that Cl^- is taken up in the acid bands (Spear et al., 1969) by a $H^+:Cl^-$ cotransport system (Sanders, 1980), it is clear that kinetic regulatory systems are required to control the recycling of H^+ through these various transport systems that are driven by the electrochemical potential gradient for protons.

III. TEMPORAL CONTROL

A. Relationship Between Light Intensity, Chlorophyll Level, and Activation Lag

Following illumination of a dark-adapted *Chara* internodal cell there is a precise lag period before the pH value begins to increase within the alkaline band. The duration of this lag period is dependent on the light intensity and the chlorophyll level within the cell (Lucas, 1975b). Over the lower range of light intensities, a linear relationship is observed between the reciprocal of the lag period and light intensity, with the critical light intensity being the condition where the lag period approaches infinity. A precise linear relationship also exists between the chlorophyll level in the cell and the critical light intensity, with this intensity being lower for cells that have higher chlorophyll content.

These results establish, unequivocally, that the chloroplasts form an integral component of the temporal control system that is responsible for the light-induced activation of the class II H^+ transport system(s). Since normal activation requires the presence of exogenous HCO_3^-, it is possible that the link between the chloroplasts and the transporters, within the plasma membrane, involves a change in any of a number of biochemical compounds in the cytoplasm that are affected by CO_2 fixation. Alternatively, the association between light intensity, critical light intensity, chlorophyll content, and the induction lag period may relate to some aspect of the rate of electron transport, per se. The latter possibility is supported by our finding that the characteristics of dark-induced inactivation of the transporter can be simulated by the introduction of DCMU (which blocks electron transport in the chloroplasts) to *Chara* cells that have had the exogenous HCO_3^- replaced by an artificial buffer (i.e., in these cells the dark reactions of photosynthesis have been blocked).

It has also been well established, using isolated chloroplasts from higher plants, that during chloroplast light activation there is an exchange of inorganic ions such as Mg^{2+}, Ca^{2+}, Cl^-, and H^+ between the cytosol and the stroma. Thus the relationship between the chloroplasts and activation of the class II transporter may be indirect, through a light-induced uptake of Ca^{2+} into the chloroplasts (see Miller and Sanders, 1987), resulting in a modulation in a second messenger signal within the cytosol. Experiments to investigate these possibilities are presently being conducted in our laboratory.

B. Electrophysiological Studies

The membrane potential of physiologically competent *Chara* internodal cells is highly sensitive to both light and the presence of exogenous HCO_3^- (Lucas, 1982). Transferring cells that exhibit strong pH banding patterns into the dark causes the membrane potential to hyperpolarize, transiently, from around -250 mV to approximately -350 mV. Reillumination results in a very rapid depolarization of the membrane potential back to the steady-state resting potential. Similar changes in the potential can be achieved simply by reducing the light intensity incident upon the cell. These results offer further support for our hypothesis that some form of signaling takes place between the chloroplasts and the class II transporter within the plasma membrane. The hyperpolarization upon darkening the cell suggests that the H^+ channels become inactivated in advance at the H^+-ATPase.

IV. POSSIBLE SPATIAL CONTROL SYSTEMS

A. Nonequilibrium Thermodynamic Model

Toko et al. (1985) adopted a nonequilibrium thermodynamic approach to develop a theoretical model of the characean banding pattern based on the concept of a dissipative structure. In their model, light intensity and the concept of a critical activating light intensity (Lucas, 1975b) was incorporated as an external parameter that produced a signal which was proposed to be responsible for shifting the cell, spontaneously, from an homogeneous state to a far-from-equilibrium self-organized state, represented by the pH banding pattern.

Diffusion equations and boundary conditions were simulated for a cylindrical cell in which H^+-ATPase activity was assumed to be distributed uniformly along the length of the cell. The activity of the H^+-ATPase was assumed to be a nonlinear function of the proton activity gradient across the plasma membrane, with external acidification acting to stimulate pump activity. A cytoplasmic inhibitory parameter (perhaps ADP [Toko et al., 1988]) was invoked to control activation of the H^+-ATPase. The steady-state boundary condition of flux continuity (zero net current) was achieved by imposing an inwardly directed H^+ leak.

A numerical analysis approach was employed to solve the various flux equations for a range of boundary conditions. Interestingly, the pH profiles generated by these simulations showed some degree of similarity to the experimentally measured banding patterns. Perhaps the most important outcome of these theoretical studies relates to the advancement of the concept that a *local potential* could give rise to the transition (bifurcation) from an unstable, homogeneous state to a stable self-organized state. Although there are real predictive limitations with this model, its major achievement is that it clearly brings into focus the concept that spatial patterns in biology could be formed by reaction–diffusion systems (see Turing, 1952).

B. Autonomous Local Area Control Model

Considering the magnitude of the currents that circulate between the acid and alkaline regions of characean internodal cells (Lucas and Nuccitelli, 1980; Lucas, 1982), it is clear that a penultimate aspect of the spatial control system will be the establishment of zero net current under steady-state conditions. To achieve this state, the spatial controller must be capable of performing some form of integration of the currents flowing through the regions of outward and inward current. In a recent study we tested the hypothesis that the spatial controller integrates the current flowing over the *entire* plasma membrane surface.

Internodal cells of *Chara* were mounted in a special chamber that allowed us to separate each cell physically into two cytoplasmically connected segments that could then be exposed to different exogenous conditions. The extracellular vibrating probe was used to characterize the control current patterns in the two segments and then a range of manipulations was performed to investigate the effects of local inhomogeneities on the current patterns. The data presented in Fig. 5 illustrate that changing the form of dissolved inorganic carbon from HCO_3^- to CO_2 (accompanied by a decrease in bathing medium pH from 8.2 to 6.0) resulted in a complete inactivation of the current pattern along the test segment, while the pattern and magnitude of the current in the control segment remained completely unaffected. We interpret these results to indicate that one element of the spatial control system of *Chara* involves the activation and operation of local autonomous regions in which out-

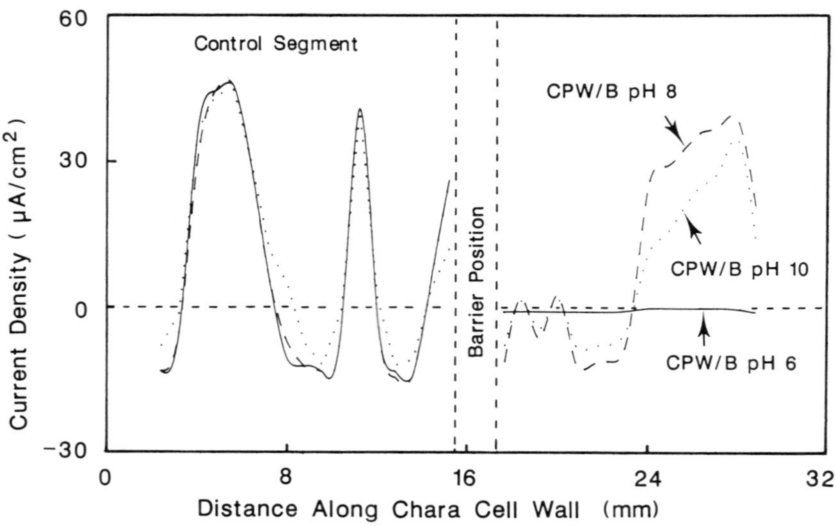

Fig. 5 Influence of changing both the bathing medium pH and the form of dissolved inorganic carbon on the extracellular current pattern of a divided *Chara* internodal cell. CPW/B-pH 6.0 was introduced into the experimental compartment and the response along the control and experimental segments was measured over a 3-h period. The experimental solution was then changed to CPW/B-pH 10 and the current profile measured within 30 s of this solution change. (From Fisahn and Lucas, 1991.)

ward and inward currents are integrated to give rise to zero net current crossing the plasma membrane in this domain (Fisahn and Lucas, 1991).

Support for this concept of local functional domains is also provided by our earlier experiments in which we showed that cessation of cytoplasmic streaming, through the application of cytochalasin B, resulted in the replacement of the normal pH pattern with small areas of transport activity (Lucas and Dainty, 1977). These results imply that a local autonomous domain may be composed of numerous smaller sectors. It will be interesting to see whether future experiments will confirm this hypothesis.

C. Functional Involvement of Cytoskeletal Elements

A microtubular network is located beneath the plasma membrane of *Chara* (Williamson et al., 1986). That these microtubules are involved in some way, in terms of establishing spatial control over membrane transport, was illustrated by exposing *Chara* cells to chemical agents known to disrupt plant microtubules (Fisahn and Lucas, 1990b). Colchicine, oryzalin, and vinblastine all caused a complete perturbation of the extracellular current pattern. However, it is interesting to note that upon removal of these agents, an extracellular current pattern equivalent to the original control was eventually reestablished. This response to microtubular disrupting agents is similar to the effects of cytochalasin B, where removal of the perturbative agent resulted in the reestablishment of the control banding pattern.

This inherent stability of the control pattern, following such perturbative manipulations, implies that the *Chara* spatial control system has some form of long-term "memory." On a speculative note, this memory component may involve discrete microtubular nucleating centers embedded within the plasma membrane. Although the microtubule disrupting agents cause the depolymerization of the microtubular network, they would have no effect on the nucleating/anchoring sites at the plasma membrane. Further, over the time course of these experiments, little longitudinal migration of these putative nucleating sites is likely to take place.

V. CONCLUSIONS

The operational features of the pH banding and extracellular current patterns developed by internodal characean cells have been well characterized. These studies establish that a complex hierarchical spatial and temporal control system(s) has evolved to orchestrate the operation of a class II H^+ transporter. Little information is currently available as to the nature of this spatial control system. The possible involvement of dissipative structures, analogous to a reaction–diffusion system, have been modeled using numerical analysis and nonequilibrium thermodynamics. Although this approach can simulate the experimental system, the predictive capabilities of these models are limited. Further studies are required to establish the putative components of such a reaction–diffusion control system that may establish the operational boundaries of individual local autonomous transport domains.

ACKNOWLEDGMENTS

Research in our laboratory on this topic has been funded by grants from the National Science Foundation (PCM-87-10474, PCM-80-03133, PCM-81-17721, DMB-85-03153, and DCB-88-16077).

REFERENCES

Bisson, M. A., and Walker, N. A. (1980). *J. Membr. Biol.*, *56*: 1.
Bisson, M. A., and Walker, N. A. (1982). *J. Exp. Bot.*, *33*: 520.
Fisahn, J., and Lucas, W. J. (1990a). *J. Membr. Biol.*, *113*: 23.
Fisahn, J., and Lucas, W. J. (1990b). *Planta*, *182*: 506.
Fisahn, J., and Lucas, W. J. (1991). *Plant Physiol.*, *95*: 1138.
Fisahn, J., McConnaughey, T., and Lucas, W. J. (1989). *J. Exp. Bot.*, *40*: 1185.
Franceschi, V. R., and Lucas, W. J. (1980). *Protoplasma*, *104*: 253.
Franceschi, V. R., and Lucas, W. J. (1982). *Planta*, *154*: 525.
Lucas, W. J. (1975a). *J. Exp. Bot.*, *26*: 271.
Lucas, W. J. (1975b). *J. Exp. Bot.*, *26*: 347.
Lucas, W. J. (1982). *Planta*, *156*: 181.
Lucas, W. J., and Dainty, J. (1977). *J. Membr. Biol.*, *32*: 75.
Lucas, W. J., and Fisahn, J. (1989). In *Plant Membrane Transport: The Current Position* (J. Dainty, M. I. De Michelis, E. Marrè, and F. Rasi-Caldogno, eds.), Elsevier, Amsterdam, pp. 25–29.
Lucas, W. J., and Nuccitelli, R. (1980). *Planta*, *150*: 120.
Lucas, W. J., Brechignac, F., Mimura, T., and Oross, J. W. (1989). *Protoplasma*, *151*: 106.
Lucas, W. J., Keifer, D. W., and Pesacreta, T. C. (1986). *Protoplasma*, *130*: 5.
Lucas, W. J., Keifer, D. W., and Sanders, D. (1983). *J. Membr. Biol.*, *73*: 263.
Metraux, J.-P., and Taiz, L. (1977). *Proc. Natl. Acad. Sci. USA*, *74*: 1565.
Metraux, J.-P., and Taiz, L. (1979). *Plant Physiol.*, *63*: 657.
Metraux, J.-P., Richmond, P. A., and Taiz, L. (1980). *Plant Physiol.*, *65*: 204.
Miller, A. J., and Sanders, D. (1987). *Nature*, *326*: 397.
Pesacreta, T. C., and Lucas, W. J. (1984). *J. Cell Biol.*, *98*: 1537.
Price, D. G., and Badger, M. R. (1985). *Aust. J. Plant Physiol.*, *12*: 257.

Price, D. G., Badger, M. R., Bassett, M. E., and Whitecross, M. I. (1985). *Aust. J. Plant Physiol.*, *12*: 241.
Sanders, D. (1980). *J. Membr. Biol.*, *53*: 129.
Spear, D. G., Barr, J. K., and Barr, C. E. (1969). *J. Gen. Physiol.*, *54*: 397.
Toko, K., Chosa, H., and Yamafuji, K. (1985). *J. Theor. Biol.*, *114*: 125.
Toko, K., Nosaka, M., Fujiyoshi, T., Yamafuji, K., and Ogata, K. (1988). *Bull. Math. Biol.*, *50*: 255.
Turing, A. M. (1952). *Philos. Trans. R. Soc. Lond. B*, *237*: 37.
Williamson, R. F., Perkin, J. L., McCurdy, D. W., Craig, S., and Hurley, U. A. (1986). *Eur. J. Cell Biol.*, *41*: 1.

25
Dissipative Patterns of Ion Channels in Membranes

Peter Fromherz *Universität Ulm, Ulm, Germany*

I. INTRODUCTION

Standard concepts of morphogenesis are based on reaction–diffusion equations as introduced by Turing (1952) and as brought into canonic form by Gierer and Meinhardt (1972). The model reviewed in the present paper (Fromherz, 1988a) is of a different nature. Pattern formation is not connected with a chemical reaction. It is driven by concentration gradients of ions across a membrane. The gradient couples to nonlinearities of membrane dynamics—diffusion and drift of ion channels interacting with the membrane potential. Physically the model is related *cum grano salis* to hydrodynamic instabilities (Bénard, 1900), where a temperature gradient couples to the nonlinear dynamics of a thin liquid film. Formally, the model is related to equations used to describe the motion of amoebae interacting with self-produced chemo-attractant (Keller and Segel, 1970).

II. SMOLUCHOWSKI–KELVIN EQUATIONS

Biomembranes are usually described by two quite different concepts (Fig. 1): the fluid-mosaic model of biochemistry (Singer and Nicolson, 1972) and the electrical-circuit model of physiology (Hodgkin and Huxley, 1952). (i) In the fluid-mosaic model the membrane is described by a sheet of lipid with floating integral proteins (Fig. 1a). (ii) In the circuit model every element of the membrane is described by a capacitor and by selective ion conductors are driven by Nernst-type batteries. The elements are connected electrically through the resistance of an electrolyte (Fig. 1b). These two aspects of membrane structure are not independent: (a) Membrane proteins may transport ions; they form discrete ion channels

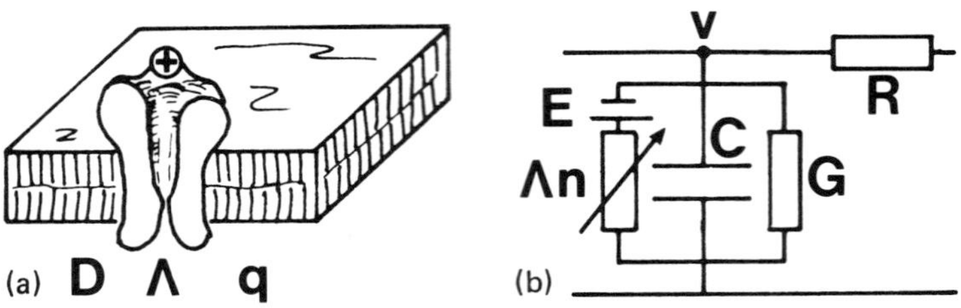

Fig. 1 (a) Fluid-mosaic model. Proteins are characterized by diffusion coefficient D, channel conductance Λ, and electrophoretic charge q. (b) Circuit model. Every element of the membrane is characterized by the voltage v, the reversal potential E, and in a one-dimensional system, by the capacitance C, the leak conductance G, the number n of proteins, and the resistance R per unit length.

(Hladky and Haydon, 1970; Neher and Sakmann, 1976). They are the structural correlates of the conductors of the circuit model; their redistribution affects the circuit. (b) Membrane proteins move in an electrical field; they bear an electrophoretic charge (Jaffe, 1977; Poo and Robinson, 1977). A voltage gradient, as described in the circuit model, gives rise to electrophoretic drift and to a redistribution of proteins. We may imagine positive feedback by combination of (i) and (ii) with (a) and (b): An accumulation of proteins (fluid mosaic) gives rise to a locally enhanced current across the membrane (channels), and a lateral voltage gradient is formed (circuit) which gives rise to further attraction of proteins (electrophoresis).

Whether self-organization of charged ion channels is possible at all and whether it occurs for a reasonable set of parameters can be decided only on the basis of a rigorous physical formulation which is simple enough to allow quantitative predictions. The circuit model is described by equation (1), which is the two-dimensional analogon of Kelvin's cable equation for voltage v across the membrane as a function of space and time. c and g are capacitance and leak conductance per unit area; n, Λ, and E are density, conductance, and reversal potential of channels; and r is the surface resistance of the electrolyte. The fluid-mosaic model is described by a Smoluchowski equation for the density n of membrane proteins according to equation (2). D and q are the diffusion coefficient and electrophoretic charge, and kT is the thermal energy.

$$c\partial_t v = \nabla(r^{-1}\nabla v) - gv - \Lambda n(v - E) \tag{1}$$

$$\partial_t n = D\nabla\left(\nabla n + \frac{q}{kT} n\nabla v\right) \tag{2}$$

Equation (1) is linear with respect to voltage v; equation (2) is linear with respect to the protein density n. It is the crucial point that both equations are bilinear in n and v such that the coupled system of the Smoluchowski–Kelvin (SK) equation is nonlinear (Fromherz, 1988a).

III. INSTABILITY

The homogeneous distribution of proteins at an average density $\bar{n}$ is a stationary solution of the SK equations with the voltage at the resting potential $v_R = \alpha E$ with $\alpha = \Lambda\bar{n}/(g + \Lambda\bar{n})$ (density parameter). This state may be stable or instable.

The threshold of instability is obtained from the linearized SK equations (Fromherz, 1988a,b, 1989a). In the limit of low channel density $\Lambda\bar{n} \ll g$, the criterion for spontaneous growth of a mode of spatial frequency k is given by the relation

$$-\frac{\Lambda g E\bar{n}}{kTg} > 1 + \left(\frac{k}{gr}\right)^2 \tag{3}$$

The instability is controlled by the product Λq of the two coupling constants of protein and voltage: the channel conductance Λ and the electrophoretic charge q. A finite Nernst-type potential E is required to trigger instability. The sign of E must be opposite to the electrophoretic charge q. The threshold is lowered by a high channel density $\bar{n}$ The threshold is enhanced with respect to growth of high spatial frequency k. The shift is suppressed by high leak conductance g and high surface resistance r. The diffusion coefficient D does not affect the threshold, in contrast to reaction–diffusion models (Gierer and Meinhardt, 1972).

In the general case the threshold is expressed by $p > 1 + K^2$. The pump parameter is defined as $p = -(qE/kT)\alpha(\Lambda - \alpha)$. The normalized spatial frequency is $K = \lambda k$, with the electronic length constant $\lambda = 1/\sqrt{(\Lambda\bar{n} + g)r}$ (Fromherz, 1988a,c).

IV. STATIONARY WAVES

There exist periodic stationary states beyond the threshold of instability of the homogeneous state. They are found by numerical solution of the stationary SK equations (Fromherz and Kaiser, 1991). They depend on the pump parameter p, on the density parameter α, and on the size of the membrane as compared to the electrotonic length constant λ. An example is shown in Fig. 2 for a one-dimensional system of length $l = 20\lambda$ with periodic boundary conditions with $p = 1.9$ and $\alpha = 0.5$. There exist three periodic stationary solutions. The normalized wavenumbers of their

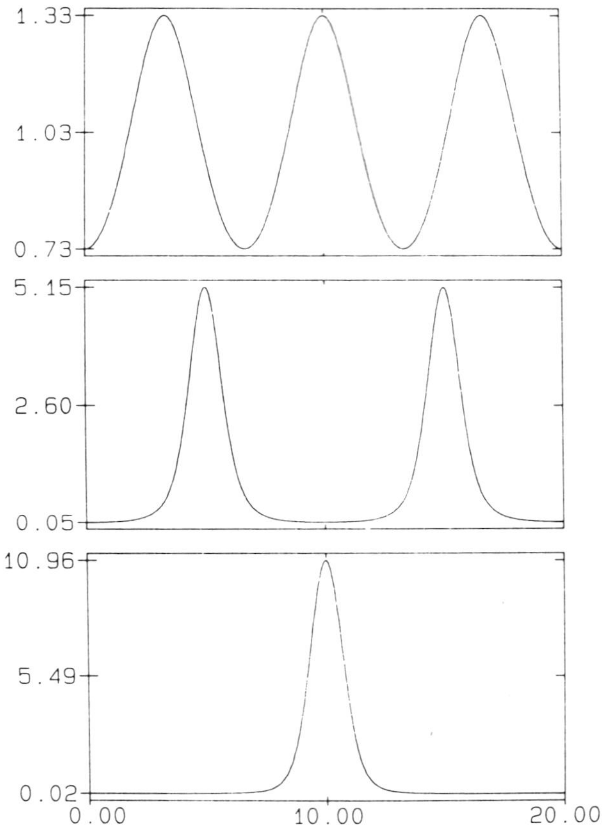

Fig. 2 Stationary patterns in a one-dimensional system. Relative density of proteins molecules versus distance. Cable length $l = 20\lambda$ (λ-electrotonic length constant). Density parameter $\alpha = 0.5$, pump parameter $p = 1.9$.

fundamentals correspond to the three instable eigenmodes within the bandwidth of instability as defined by $p = 1 + K^2$.

The stationary state of lowest spatial frequency is stable with respect to arbitrary perturbations. The higher stationary states are instable if they are perturbed by modes which are not among the spectrum of their overtones (Fromherz and Kaiser, 1991).

V. DYNAMICS OF PATTERN FORMATION

Above threshold a growth of a pattern is triggered by a perturbation of the homogeneous state. In a real membrane such perturbations are present due to the random distribution of protein molecules in the membrane which is homogeneous only on average.

An example of the dynamics of growth is shown in Fig. 3 for the same one-dimensional system as studied in Section IV. In a first phase a pattern is formed which resembles the second stationary state. This pattern corresponds to the eigenmode of fastest growth constant (Fromherz, 1988b). In a second phase a metamorphosis sets into a state of lower spatial

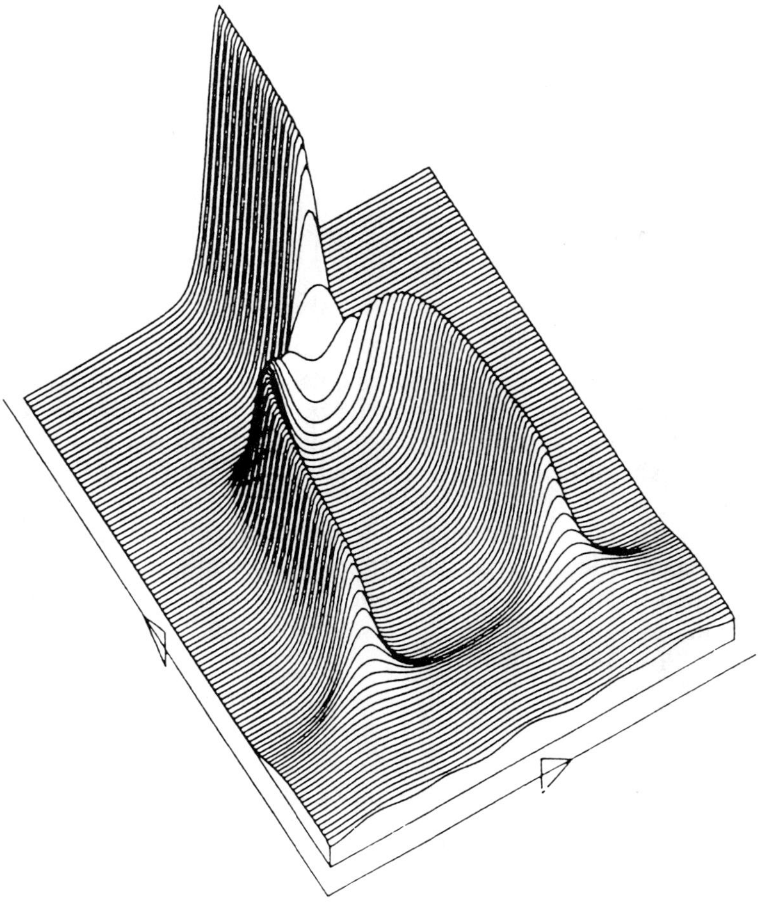

Fig. 3 Pattern formation in a one-dimensional system. Density of channels versus distance—from left to right—and versus time—from front to back. Cable length $l = 20\lambda$ (λ-length constant), density parameter $\alpha = 0.5$, pump parameter $p = 1.9$, time interval $0 < t/\tau < 200$ (time constant $\tau = \lambda^2/D$).

frequency. Eventually, the stationary state of lowest spatial frequency is formed which is stable (Fromherz and Kaiser, 1991). We conclude that the higher stationary states—although instable—play an important role as metastable transients in the dynamics of pattern formation.

VI. OSCILLATORY WAVES

Pattern formation becomes more involved if more than a single mobile protein couples to the electrical potential. Oscillatory waves may appear (Fromherz, 1989b). The simplest case is a membrane in which the homogeneous leak conductance g is replaced by mobile channels and in which conductance and charge of the two kinds of channels are chosen to be identical. The one-dimensional system is described by the following equations:

$$c\partial_t v = r^{-1}\partial_{xx} v - \Lambda n_L v - \Lambda n(v - E) \tag{4}$$

$$\partial_t n = D\partial_x \left(\partial_x n + \frac{q}{kT} n\partial_x v\right) \tag{5}$$

$$\partial_t n_L = D_L\partial_x \left(\partial_x n_L + \frac{q}{kT} n_L\partial_x v\right) \tag{6}$$

For equal average density of the two channels with $\bar{n} = \bar{n}_L$ the density parameter is $\alpha = 0.5$. For a pump parameter $p = 5$ (e.g., for $E = -100$ mV, $q = 5e_0$) the dynamics are shown in Fig. 4 in a closed system of length $l = 10\lambda$. The diffusion of the leak channels is assumed to be slow with $D_L = D/10$. For $D = 1\ \mu\text{m}^2/\text{s}$ and $\lambda = 2\ \mu\text{m}$ the total time shown in Fig. 4 is 280 s. The channels accumulate in a central patch, the leaks are attracted with some delay. The central patch spreads and splits eventually into two fragments which move at constant velocity and shape. They are reflected at the ends and collide. The system develops all features of a solitary wave (Fromherz, 1989b).

VII. SYNAPSES

Synaptic structures may be particularly prone to pattern formation:

1. The narrow cleft between pre- and postsynaptic membrane gives rise to a high surface resistance r (i.e., to a short electrotonic length constant λ): The threshold is lowered for growth of patterns with high spatial frequency.
2. Opening of channels by chemical transmitters provides a simple mechanism for triggering patterns by modulating $\bar{n}$ and α at invariant

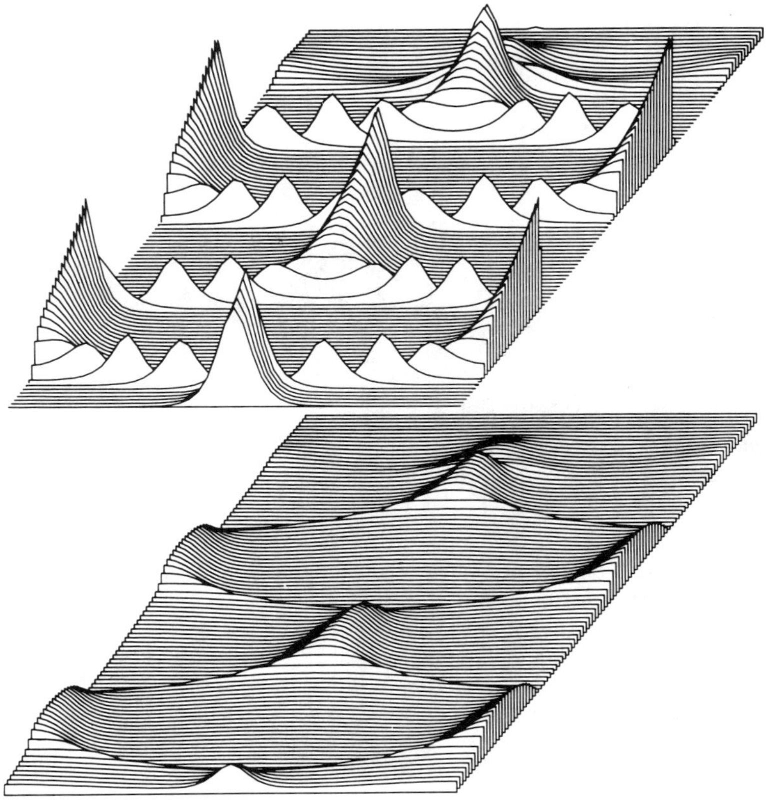

Fig. 4 Spatiotemporal dynamics in a one-dimensional system. Density of two channels versus space—from left to right—and versus time—from back to front. (a) Fast channels; (b) slow leak channels. The diffusion coefficient of the leak channels is 10 times lower $D_L = D/10$. Solitary waves appear which move at constant shape and velocity and which collide with the walls and with each other. Cable length $l = 10\lambda$ (λ-length constant), density parameter $\alpha = 0.5$, pump parameter $p = 5$. Time interval $0 < t/\tau < 70$ (time constant $\tau = \lambda^2/D$).

chemical composition and invariant reversal potential. An example is shown in Fig. 5, where the geometry is chosen to resemble a motor endplate. A periodic accumulation of channels develops which attracts further channels from the surround (Fromherz, 1989b).

The dynamics in the postsynaptic membrane is unchanged if the extended presynaptic terminal of the endplate is replaced by a set of closely

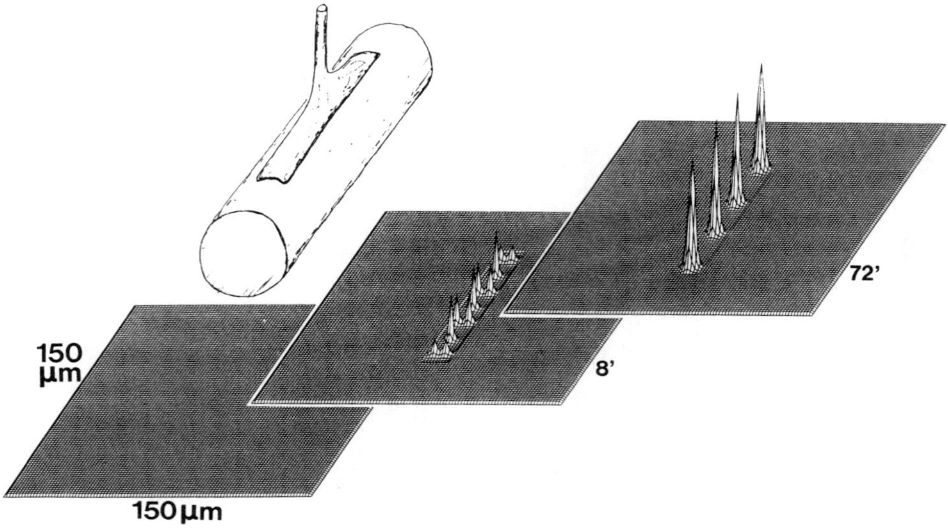

Fig. 5 Structural dynamics in a synapse. The geometry resembles a neuromuscular junction. Density of channels 0, 8, and 72 min after initiation of pattern formation. Density parameter $\alpha = 0.5$, pump parameter $p = 5$. Average channel density 200 μm^{-2}, channel conductance $\Lambda = 30$ pS, electrophoretic charge $q = 5e_0$, diffusion coefficient $D = 0.1\ \mu m^2/s$. Width of synaptic cleft 50 nm, length constant in the cleft $\lambda = 2\ \mu m$.

packed small presynaptic terminals which are active simultaneously. For such a polysynaptic assembly Fig. 5 reveals that channels are accumulated in certain synapses and are depleted in others. An autonomous modulation of synaptic strengths occurs as a result of presynaptic stimulation: Dissipative patterns of ion channels may give rise to memory effects in neural networks.

VIII. CONCLUSIONS

The coupling of the membrane potential to the electrical properties of membrane proteins—ion conductance and electrophoretic charge—gives rise to spatiotemporal patterns. The Smoluchowski–Kelvin equations are a simple approach to estimate the conditions of pattern formation. The parameters are found to be within the range of the properties of biomembranes. It is unknown whether the "tool" of pattern formation by ion channels is actually used in cellular morphogenesis or whether it is suppressed by the immobilization of membrane proteins.

Finally, the relation may be pointed out between pattern formation and action potentials. Hodgkin and Huxley (1952) have shown that a coupling of the membrane potential—as described by Kelvin's equation—to membrane proteins with voltage-dependent conductances—as described by chemical rate equations—gives rise to oscillations and action potentials. The system is driven by Nernst-type potentials. In the present approach the Kelvin equation is coupled to a modulation of conductances as caused by lateral motion. The rate equations are replaced by Smoluchowski equations. The system is driven by Nernst-type potentials again. Apparently, the two concepts are complementary to each other. The time scale of pattern formation, however, is extremely slow as compared to the time scale of action potentials: Whole protein molecules move along the membrane in the former case, whereas small "gating particles" move across the membrane in the latter case. The relevance of channel patterns lies in morphogenesis and memory not in signal transmission.

ACKNOWLEDGMENTS

The cooperation of Bernhard Kaiser was of great value in the progress of this work. The project was supported by the Fonds der Chemischen Industrie and by the Deutsche Forschungsgemeinschaft (SFB 239/D8).

REFERENCES

Bénard, H. (1900). *Rev. Gen. Sci. Pures Appl.*, *12*: 1261.
Fromherz, P. (1988a). *Proc. Natl. Acad. Sci. USA*, *85*: 6353.
Fromherz, P. (1988b). *Bunsenber. Phys. Chem.*, *92*: 1010.
Fromherz, P. (1988c). *Biochim. Biophys. Acta*, *944*: 108.
Fromherz, P. (1989a). *Chem. Phys. Lett.*, *154*: 146.
Fromherz, P. (1989b). *Biochim. Biophys. Acta*, *986*: 341.
Fromherz, P., and Kaiser, B. (1991). *Europhys. Lett.*, *15*: 313.
Gierer, A., and Meinhardt, H. (1972). *Kybernetik*, *12*: 30.
Hladky, S. B., and Haydon, D. A. (1970). *Nature*, *225*: 451.
Hodgkin, A. L., and Huxley, A. F. (1952). *J. Physiol.* (*Lond.*), *117*: 500.
Jaffe, L. F. (1977). *Nature*, *265*: 600.
Keller, E., and Segel, L. A. (1970). *J. Theor. Biol.*, *26*: 399.
Neher, E., and Sakmann, B. (1976). *Nature*, *260*: 779.
Poo, M., and Robinson, K. R. (1977). *Nature*, *265*: 602.
Singer, S. J., and Nicolson, G. L. (1972). *Science*, *175*: 720.
Turing, A. M. (1952). *Philos. Trans. R. Soc. Lond. B*, *237*: 37.

26
Periodic Spatial Patterns in the Cellular Slime Mold *Polysphondylium pallidum*

Edward C. Cox *Princeton University, Princeton, New Jersey*

I. INTRODUCTION

How organisms take on their shape is an abiding question in developmental biology. Broadly speaking, the emergence of shape can be considered as the interaction of growth, differentiation, and morphogenesis, often acting in concert. The latter stages of differentiation and morphogenesis are separated from growth in the cellular slime molds *Polysphondylium pallidum* and *Dictyostelium discoideum*, and this accounts for some of the appeal these organisms hold for developmental biologists (Bonner, 1967). Both species undergo similar well-studied life histories. Free-living amoebae in the forest detritus feed on bacteria until they exhaust their food supply. As the population starves, rare individual cells begin spontaneously signaling to their neighbors. This signal is relayed from amoeba to amoeba in a coordinate way, such that large territories, from one to several millimeters in diameter, quickly form. Signaling continues, and cells migrate toward the center of the territory, forming a rounded mass of cells which undergoes a transformation into a cooperative, phototactic, and thermotactic slug capable of migrating many centimeters across the substratum in search of a suitable surface. On this surface the terminal part of the morphogenetic cycle begins.

The mature fruiting bodies of *P. pallidum* and *D. discoideum* contain only two cell types, mature, highly vacuolated dead stalk cells, and spores. The spores are gathered together as droplets (sori) at the end of each stalk. Each sorus contains from a few to several hundred spores. Unlike *D. discoideum*, whose fruiting body consists of a single stalk terminated by a single sorocarp, the mature *P. pallidum* fruiting body carries many whorls of secondary fruiting bodies arrayed symmetrically about the primary stalk (Fig. 1). As the sorogen mass ascends, new stalk cells continuously differentiate at the tip of both the main, or primary, sorogen, and the secondary sorogen. In Fig. 1a, which is a photograph of a fruiting body caught part way through morphogenesis, the most basal mass of cells has formed a whorl with six secondary fruiting bodies; the

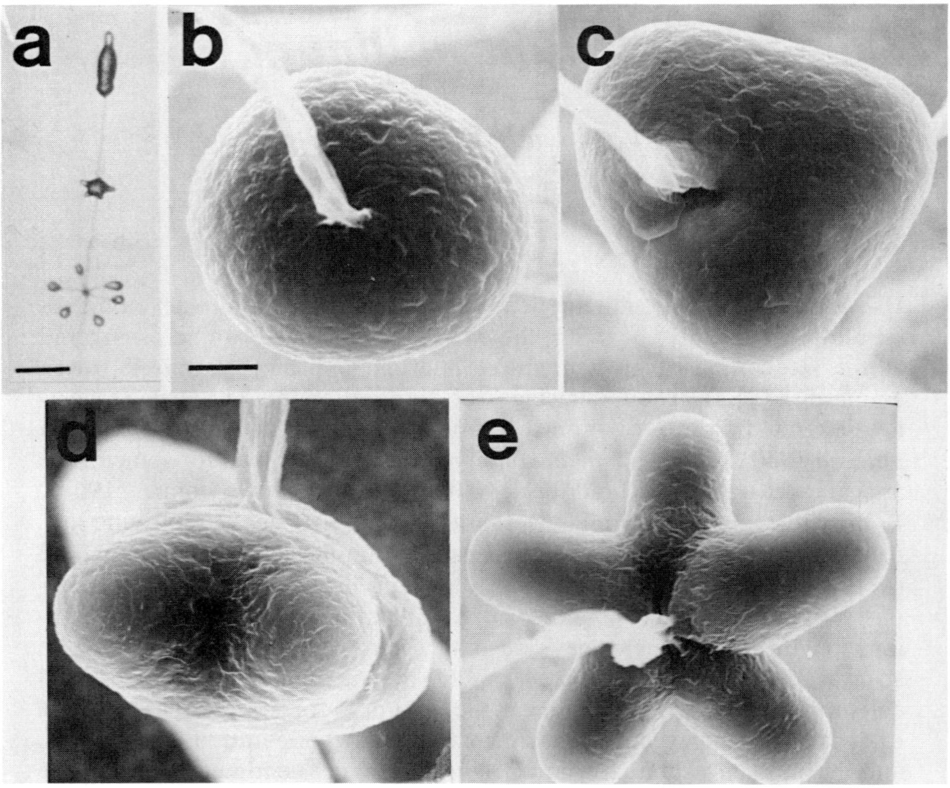

Fig. 1 Morphogenesis in *Polysphondylium pallidum*. (a) Culminating sorogen with two whorls of different developmental age. The bottom whorl with six arms is older than the middle whorl, which has just formed five tips. These tips will recruit cells to form a mature whorl of secondary fruiting bodies. Scale bar = 200 μm. (b–d) Scanning electron micrographs of developing whorls. Micrograph (b) is the youngest, (e) the oldest. The mature stalk may be seen running through the center of each whorl mass. Scale bar: b–e = 8.5, 7.3, 6.7, and 7.3 μm, respectively. [(a) From Byrne and Cox, 1987; (b–e) from Byrne and Cox, 1986.]

next most apical whorl mass has formed five tips on the equator of a mass of cells recently pinched off from the main sorogen. These will act as recruiting centers. Eventually, this whorl will form five arms, very much like the six-armed whorl below it. The main body of the sorogen is about to release a mass of cells which will become the next whorl. Whorls are spaced regularly along the central or primary stalk. The number of whorls can number from none in the tiniest fruiting body to 15 or more in the largest. Within a whorl, the angular spacing of secondary fruiting bodies shows surprisingly low variance (Cox et al., 1988).

How do these two patterns arise? In the remainder of this chapter I summarize what we know about these processes. I also propose models for both of them that, although quite speculative, seem plausible and amenable to analysis.

II. THE SPACING OF ARMS WITHIN WHORLS

One of the special fascinations of this aspect of morphogenesis in *P. pallidum* is the chance to study the change in symmetry that occurs as a spheroidal mass of cells, numbering from several hundred to a few thousand, undergoes a transition from a spherical to a radial symmetry. Morphological changes accompanying this transition are shown in Fig. 1b to e. These scanning electron micrographs demonstrate that the spherical cell mass freshly deposited from the base of the culminating sorogen undergoes a series of shape changes that occur simultaneously about the equator of the whorl (Byrne and Cox, 1986).

These observations, and others discussed below, immediately suggest a mechanism whereby all points on the equator communicate simultaneously with all others. That is, they argue against a model in which cells that will become tip organizers fire randomly; or one in which a traveling wave positions tips as it moves around the equator, since these mechanisms lead to the random or sequential appearance of tips, respectively.

In seeking an answer to the cellular mechanism underlying tip patterning, we studied the spatial distribution during morphogenesis of antigens revealed by a panel of monoclonal antibodies raised against tip cells or tip precursor cells (Byrne and Cox, 1986). This approach showed that the first detectable event in whorl morphogenesis was the uniform expression of tip-specific antigens over the surface of the nascent whorl. In thin sections these antigens can be shown to be present mainly in the cells on the whorl surface.

Following the initial synthesis of tip-specific antigens everywhere on the surface, further synthesis becomes restricted to the equator (Fig. 2c), then to randomly distributed equatorial patches (Fig. 2d), and finally to

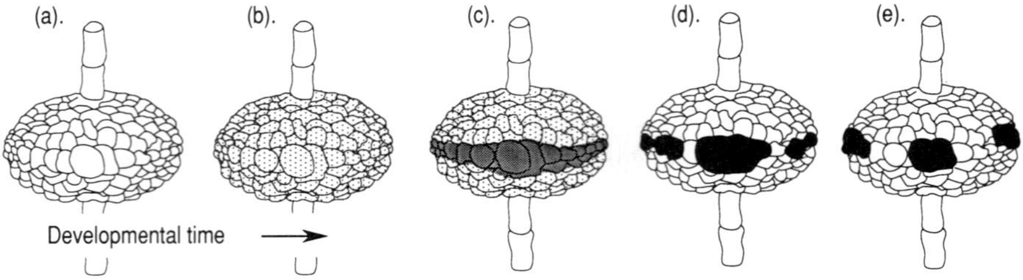

Fig. 2 Origin of a periodic prepattern in a whorl mass. This figure summarizes results obtained with a panel of monoclonal antibodies raised against *P. pallidum* cells. (a) Whorl mass just released from the base of the culminating sorogen. (b) Tip-specific antigens first appear over the entire surface of the whorl mass. (c) With time, tip-specific antigens become restricted to the equatorial surface of the whorl. (d) Additional antigens appear in randomly distributed equatorial patches. (e) Periodic patches of tip-specific antigens grow while nonperiodic patches fail to grow. All of the changes above occur before the appearance of tips or bulges. (Adapted from Byrne and Cox, 1986.)

evenly distributed patches which presumably act as organizers for secondary tips (Fig. 2e). With time, the concentration of antigen increases.

The steps illustrated in Fig. 2 are defined in another sense as well. The antigen detected by one antibody (anti-Tp423) is on from the first to the last step shown in Fig. 2b to e; Pg101 can first be detected as an equatorial band (Fig. 2c) and persists during subsequent morphogenetic events, localized finally to the tips of secondary sorogens; and anti-Tp200 stains only when regularly spaced equatorial patterns are already established (Fig. 2e). These three antibodies thus define a biochemical pathway intimately associated with the genesis of the radial pattern so apparent in Fig. 1. Throughout the later stages of morphogenesis, all three antibodies also stain primary sorogen tips. Because Tp423, Pg101, and Tp200 are expressed first at the tip of the secondary sorogen, and then later on secondary tips, it is reasonable to believe that at least some aspects of secondary sorogen morphogenesis recapitulate early events leading to the establishment of the primary sorogen.

III. MOLECULAR CHARACTERIZATION OF TIP-SPECIFIC ANTIGENS

Cell extracts prepared from synchronous populations of cells were used by Vocke to study the appearance and subcellular distribution of the antigens recognized by several of the monoclonal antibodies described above (Table 1; Vocke and Cox, 1990). The distribution of antigens is in broad agreement with data obtained using stained whole mounts: Tp423, Pg101, and Tp200 all stain prestalk and stalk cells, and are regulated during the developmental cycle as expected.

When Tp200 and Tp423 are used to stain populations of starved cells, rare cells can be detected which appear to be founder cells, the cells that establish territories early during morphogenesis (McNally, personal communication). This leads us to believe that founder cells play similar roles during both the establishment of aggregates and the establishment of primary and secondary sorogen tips.

Two features are of special interest for the purposes of this discussion. Both Pg101 and Tp200 undergo a change in apparent molecular weight just as the sorogen begins to form. Whether or not this is because the protein is cleaved or modified by, for example, glycosylation, is not known. The change is suggestive, however, because both antigens are closely associated with morphogenetic movement, and it is possible that a change at the molecule level of a protein, especially a membrane protein, underlies the transition from one spatial symmetry to another.

3D10Pnk is interesting because it stains all but the tips of primary sorogens, which is why it is called a prespore antigen in Table 1. However, it stains all cells in secondary sorogens, and this suggests that whorl genesis may be more than simply the recapitulation of tip formation in the primary sorogen.

IV. MODELING SECONDARY TIP SPACING

With the foregoing results in hand we modeled the emergence of the tip prepattern using nonlinear reaction diffusion theory (Byrne and Cox, 1987). For technical reasons we restricted our modeling to the emergence

Table 1 Antigen Characterization

Antigen	Localized to:	Molecular mass	Properties	Location during development
Tp423	Prestalk, stalk cells	Heterodisperse	Membrane bound	Early aggregates to sorogens
Pg101	Prestalk, stalk cells	66 kDa → 63 kDa	Soluble	Sorogens
Tp200	Prestalk, stalk cells	77 kDa → 74 kDa	Integral membrane	Late mounds to sorogens
3D10Pnk	Prespore cells	215 kDa	Soluble	Tight aggregates to sorogens

of a tip prepattern from the time Pg101 first appears distributed evenly in the equatorial cells (Fig. 2c). Since this subject is covered elsewhere in this volume by Meinhardt, and in detail by McNally and Cox (1988), I will not discuss the modeling assumptions we employed. However, a semiquantitative analysis of the distribution of one antigen, Pg101, revealed that a two-morphogen model accurately described the initial, intermediate, and final prepattern in realistic detail. The experimental results and model were analyzed by comparing the Fourier transform of the data to the Fourier transform of the simulation results, particularly with respect to the wavelengths of both as a function of developmental time. In model and experiment, peaks in morphogen (antigen) concentration shifted from a characteristic higher frequency distribution to a lower, equilibrium distribution, as the final pattern emerged. Moreover, since the peaks in a Fourier transform can be assigned statistical weight, we could demonstrate that initially (Fig. 2c) Pg101 was distributed in each equatorial cell in approximately equal concentrations. This too agreed well with the assumptions of the model, and is strong evidence that tip patterns emerge from an initial uniform state.

However, as first pointed out by Gmitro and Scriven (1966), reaction diffusion mechanisms are formally similar to models that describe shape changes by, say, mechanical forces: surface tension, tissue buckling, haptotaxis, and so on (Oster et al., 1983).

To make this point clearly I discuss the equations below, chosen from a paper by Xu et al., (1983), but characteristic of a large literature. These authors demonstrated that a thin vibrating plate successfully mimics animal coat patterns, patterns first modeled by Turing (1952) and later in realistic detail by Murray (1981). It is worth understanding why this is so.

The generalized reaction diffusion model taken from Xu et al. is shown in (1) and (2):

$$\frac{\partial s}{\partial t} = g(s,a) + \nabla^2 s \tag{1}$$

$$\frac{\partial a}{\partial t} = f(s,a) + \beta\nabla^2 a \tag{2}$$

s and a are two diffusible morphogens proposed by Turing. Their rates of synthesis are interdependent. β is the ratio of the diffusion coefficients for a and s, D_a and D_s. All of the spatial information is contained in the operator ∇^2 (familiar to most biologists as the operator $\partial^2/\partial x^2$ for diffusion in a capillary along the capillary axis x). With appropriate choices for the various parameters contained in $g(s,a)$ and $f(s,a)$, certain harmonic modes will grow, driven by intrinsic instabilities in the homogeneous initial distribution of a and s.

The harmonic patterns that evolve from these equations for the appropriate parameter choices are solutions to an equation that takes the general form

$$\nabla^2 w + k^2 w = 0 \tag{3}$$

Equations like (1) and (2) will always have solutions like those of (3). Xu et al. point out that their model describing the growth of vibrational nodes on an excited plate are also solutions to this equation. This will be true for virtually all models that invoke short-range activating and long-range inhibiting forces—reacting and diffusing chemicals, a combination of mechanical forces, or nerve impulse as the long-range inhibiting force. Such models will invariably contain the Laplacian operator ∇^2, solutions to which will be solutions to equation (3). The time required to establish the equilibrium pattern may of course vary, depending on the model, but the outcome will look the same.

Thus agreement between simulation and experiment for a pattern that has achieved equilibrium does not allow us to infer very much about *mechanism*. In one important way, however, reaction–diffusion models and models based on mechanical forces differ: Reaction–diffusion mechanisms can in principal set up a true prepattern—a system of positional coordinates preceding morphogenetic change—whereas there is no prepattern in mechanical models—visible changes in shape accompany the growth of one spatial mode over another (see Oster, 1988 and Murray and Maini, 1988 for an extended discussion of these ideas).

In our studies of *P. pallidum* we see evidence for the establishment of a well-defined prepattern in the absence of shape change, at least at the resolution of transmission light microscopy. Also, because the slime molds use cyclic AMP and ammonia for cell–cell communication and sorogen movement (reviewed in Odell and Bonner, 1986), a reaction–diffusion model is not unreasonable, although this is not a strong argument. It remains possible therefore that curvature of the whorl surface, and changing surface tension in the slime sheath, for example, contribute to the forces establishing the tip prepattern.

V. TIP PATTERNS ON MONSTER WHORLS

When *P. pallidum* is grown under conditions favoring massive sorogens, monster whorls are released at frequent intervals (McNally and Cox, 1989). The shapes vary from exceptionally large spherical masses, capable of producing 10 or more arms, to various ellipsoids of revolution. The latter figures produce densely packed arrays of arms that look like bottle brushes. The prepattern for all three classes, revealed by anti-Pg101 staining, is quite striking. This antibody reveals staggered rows of reasonably well-spaced tip-specific antigen. Very long ellipsoids often stain in stripes orthogonal to the central stalk, and these stripes are unstable, breaking up into patches. McNally found that a linear reaction diffusion mechanism is fully capable of modeling these experimental results.

Solutions to (3) take the form $e^{\lambda t}H$, where H is a harmonic function, t is time, and λ a constant, the eigenvalue. In a single dimension (e.g., the equator of a newly formed whorl) H is a sine or cosine function. However, for spheroidal surfaces, the functions are spheroidal harmonic functions, which take the form of an equatorial band of spots (corresponding to the standard pattern), parallel rows of stripes, and checkerboard arrays of spots (patterns on monsters) (McNally and Cox, 1989).

This agreement between experiment and theory is very suggestive. Both the standard patterns and the monsters are explained by the same geometry-sensitive mechanism—one that uses short-range activating and long-range inhibiting forces to produce surface harmonics. Because there appears to be a true prepattern, fully established in the absence of detectable shape changes, mechanical models can probably be ruled out. Because the prepattern appears globally and simultaneously on the surface of both the standard and monster whorls, traveling waves, especially for the largest spheroids, also seem unlikely.

Finally, although stripes are found on large ellipsoids, they appear to be unstable, breaking into spots. In reaction–diffusion theory it is easy to see why stripes are unstable, since at the leading edge of a stripe, synthesis of the activating morphogen can fire. Further synthesis quickly outruns inhibition by neighboring cells. The result is a spot, and numerous examples of this may be found in the literature (Meinhardt and Gierer, 1974). This raises an interesting problem. What stabilizes striped prepatterns in segmented animals, for example, the *ftz* stripes in *Drosophila*? There does not seem to be biological evidence bearing on this problem, although McNally and Cox (1989) have pointed out that one might predict that mutations in the *ftz* gene, or genes interacting with *ftz*, would produce *ftz* spots at the cellular blastoderm stage. Lacalli and co-workers have shown that it is possible to stabilize stripes by varying the rate constant for activator synthesis in an anterior to posterior gradient (Lacalli et al., 1988).

VI. CELL MOVEMENT AND TIP FORMATION

The evidence summarized above seems clear on at least one central point: The organizing centers for tip placement are established as a prepattern, in the absence of detectable change in whorl geometry. How might the prepattern be translated into a morphological change, the emergence of visible tips (Fig. 1)?

A clue to an answer can be found in ultrastructural changes that accompany, and may even precede, tip formation. These are shown in Fig. 3. Both scanning and transmission electron microscopy reveal characteristic changes in regions that will become whorl tips. During early tip

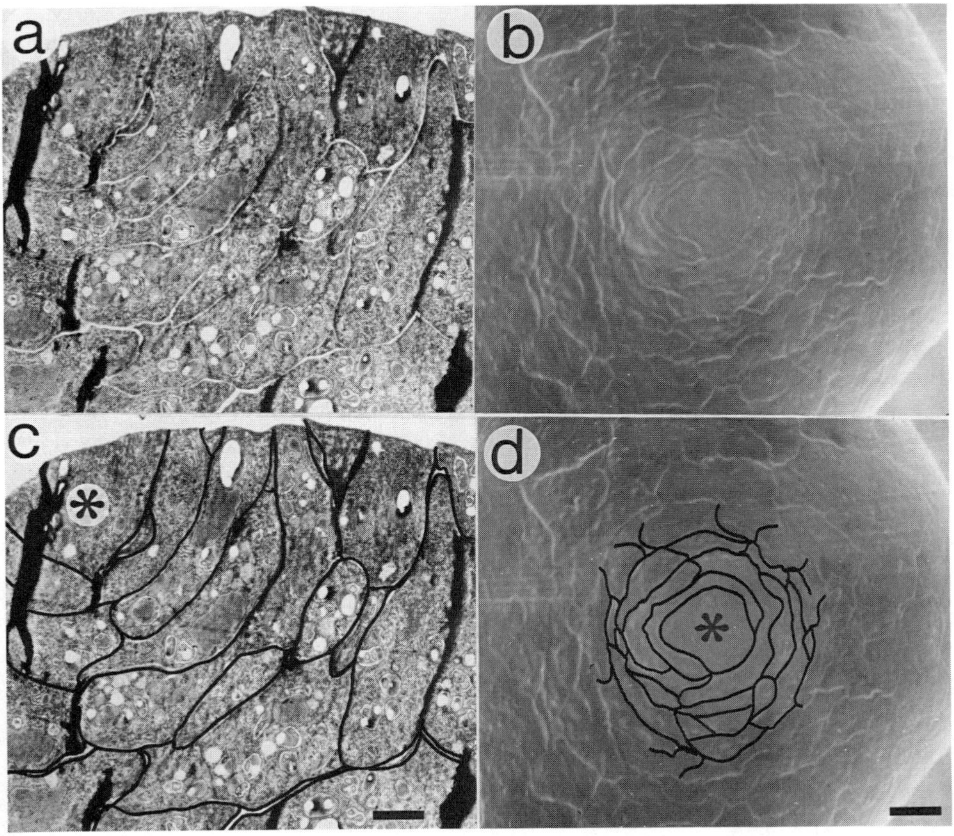

Fig. 3 Arrangement of cells in pretips prior to shape change. (a) Transmission electron micrograph of a thin section orthogonal to the central stalk of a whorl and passing through a pretip region. (b) Scanning electron micrograph of the surface of a secondary whorl in a region corresponding to (a). In (c) and (d) cell borders are outlined and the putative tip founder or cap cell marked with an asterisk. Scale bars: (c) = 1 μm, (d) = 5 μm. (From Spiegel and Cox, unpublished.)

formation, a single cell can be seen "capping" the tip (marked with an asterisk in Fig. 3c and d). Cells surrounding the cap cell are arranged symmetrically about it and partially under it.

The micrograph shown here is from a developmental stage corresponding approximately to Fig. 1c. This can be seen from the curvature of the right edge of the whorl in Fig. 3b and d. Scanning electron micrographs of whorls at the developmental stage corresponding to Fig. 1b also have cells similarly arranged about the equator of the whorl, and in about the correct number. However, when the whorl mass is completely symmetrical, it cannot be claimed that these cells are associated with pretips.

The inferences drawn from scanning electron micrographs are borne out by thin sections (Fig. 3a,c). The cap cell, marked again with an asterisk, lies over other cells which are cupped under the cap cell. Note that cell–cell contact is extensive in the pretip region, as extensive in the pretip region as elsewhere in the sorogen.

I wish to use these observations to suggest a mechanism for the transition from the smooth convex surface of the recently formed whorl to a structure with a second convex surface, the tip, bulging from the surface of the first. The mechanism makes two assumptions: first, that cap cells are equivalent to founder cells, known to found aggregation territories (Shaffer, 1961); and second, that cap cells and their recruits are less adhesive than noncap cells.

There is good evidence for the first assumption. As discussed above, the antigens that appear sequentially during the establishment of the tip prepattern also stain the tip of the primary sorogen, and thus are tip and pretip specific. Both anti-Tp200 and anti-Tp423 stain cells during aggregation that appear by their frequency and morphology to be founder cells.

There is no evidence in *P. pallidum* bearing on the relative adhesivity of tip cells compared to the adhesivity of prespore cells (the majority of the cells in a *P. pallidum* sorogen). Adhesive molecules and their receptors have been isolated in *P. pallidum* (Bozzaro et al., 1981; Chang et al., 1977; Rosen et al., 1976; Rosen et al., 1974), and while mutants lacking either one or the other have not been isolated, very similar molecules have been well characterized in *D. discoideum* (Bozzaro et al., 1987; Gerisch, 1976; Noegel et al., 1986; Loomis, 1988; Murray et al., 1983). In this slime mold mutational analysis has shown that they are required for cell–cell adhesion and normal morphogenesis (Noegel et al., 1985). Furthermore, *D. discoideum* membranes prepared from prestalk cells are known to be less adhesive than membranes prepared from prespore cells (Lam et al., 1981).

The results with live cell-sorting experiments are not as straightforward as those with isolated membranes, since chemotaxis and cell adhesion complete with each other. Nonetheless, the balance of the evidence strongly favors the view that prestalk cells are less adhesive than prespore, in agreement with the cell membrane work. For example, when the influence of chemotaxis is offset by a low concentration of cAMP in the aggregation medium, balls of cells are observed, with prestalk cells forming a shell around a core of prespore (Sternfeld and David, 1981). More telling, cells mutant for a glycoantigen known to confer adhesivity during aggregation sort to the tip of chimeric sorogens containing both mutant and wild-type cells (Houle et al., 1989).

This information is incorporated in the model outlined in Fig. 4. The cap cell, whose position is determined by the prepattern described in Fig. 2, signals to its neighbors (panel a), either with the *P. pallidum* chemoattractant glorin (Shimomura et al., 1982) or with cAMP. Cells respond by moving toward the chemotactic source. As cells respond to the chemo-

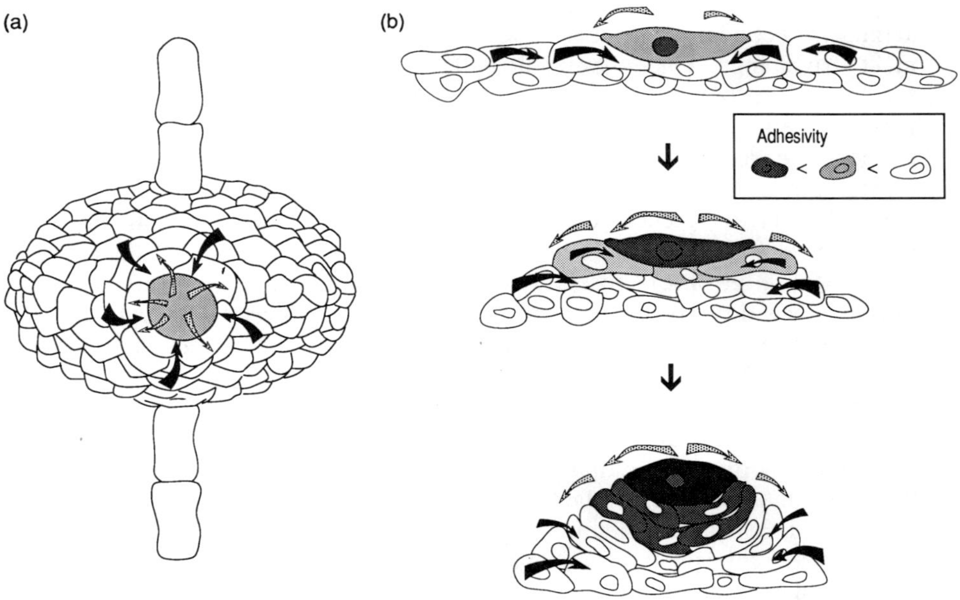

Fig. 4 Model for tip formation in the whorl. (a) The founder cell secretes a chemoattractant (stippled arrows). Nearby cells respond by moving toward the source of the signal (bold arrows). (b) Cross section of a whorl illustrating how graded cell–cell affinities can be used to cause shape changes in the vicinity of the founder cell. The relative magnitude of the adhesive force between cells is indicated by cell shading (box).

attractant, they begin to differentiate into prestalk cells, losing a fraction of their adhesive sites, and signaling to others. The consequences of this are shown in Fig. 4b.

According to this idea, the forces responsible for dimpling the surface are chemoattractive and adhesive. The chemoattractant force recruits cells, while the change in adhesion between prestalk cells, and between prestalk cells are prespore, acts as the organizing force. I postulate a *reduction* in the adhesivity of prestalk cells for the experimental reasons noted above, and because an increase in adhesivity would drive the shape change in the wrong direction. If prestalk cells were more adhesive, they would invade the whorl mass as the number of contacts between adhesive sites was maximized. I assume also that the change in adhesivity as prestalk cells begin to differentiate is graded, as shown in the figure. This is reasonable because our experiments with stalk specific antigens show graded staining, both from anterior to posterior in the primary sorogen, and radially within a patch positioned on a whorl (Byrne and Cox, 1986).

In making the transition from a spherical mass to a series of radially arranged secondary sorogens (Fig. 1c to e), the surface/volume ratio must also change. Since there is no cell division throughout morphogenesis, there must be either a substantial rearrangement of cells within sorogens, or cells must increase their surface/volume ratio. Both appear to happen. Figure 3 shows that cells farthest away from the cap cell are the most cuboidal, while cells closest to the cap cell are flattened and elongated. Similar differences between tip proximal and tip distal cells have been recorded by others (Hohl et al., 1977). This change in cell shape is quite local (Fig. 3a,c), possibly a reflection of cytoskeletal changes induced either directly by the chemoattractant or indirectly by cell–cell interaction. Similar mechanisms may operate early during development in *P. pallidum* and *D. discoideum*, perhaps providing a mechanism for tip formation at the early mound stage.

An interesting parallel to these results occurs when embryonic tissue from the flank of the developing amphibian is transplanted from a posterior to a more anterior region of the flank. The cells quickly form a convex bump on the surface while maintaining good adhesive contacts between transplanted and recipient tissues (Steinberg and Poole, 1982). This has been interpreted as evidence for an anterior–posterior adhesive gradient along the flank of the animal, with the most adhesive region anterior.

VII. SPACING OF WHORLS ALONG THE CENTRAL STALK

The exquisite spacing of whorls on the central stalk of the mature fruiting body is a striking feature of *P. pallidum*. Clearly, some periodic force is at work. At first glance, a wave propagated from the tip, or perhaps dividing the sorogen up into whorl sized packets long before they are released basally, is an attractive way to think of this aspect of morphogenesis.

For a variety of reasons, however, I favor a different model, based on the following observations:

1. The primary tip of the culminating sorogen acts to repress secondary tip formation—when a tip is excised, secondary tips break out over the surface of the primary sorogen (Byrne et al., 1982).
2. Staining patterns with monoclonal antibodies reveal that some tip-specific antigens are graded from the apex of the sorogen to the base (Byrne and Cox, 1986). Cell masses released from the base of the sorogen contain few if any cells synthesizing tip-specific antigens (Fig. 2).
3. The velocity of sorogen culmination is constant throughout morphogenesis and independent of sorogen size (Cox et al., 1988).
4. Cells at the base of the primary sorogen are not systematically organized, as they are in the remainder of the sorogen [see Fig. 3a of Vocke and Cox (1990) for a good example],
5. The tip of the mature fruiting body always has a sorus. The organism knows how many cells it has left to commit to this task. (Not all members of the genus are this smart. *P. filamentosum* and *P. candidum* often form fruiting bodies without this terminal structure.)

These features of *P. pallidum* development are combined in the qualitative model shown in Fig. 5. The vectors represent the combined effects of chemotaxis and chemokinesis on the motion of individual cells [see Odell and Bonner (1986) for the likely necessity of both]. A regulated gradient of tip-repressing activity is shown to the right of each panel. This activity must also control the sensitivity of cells to cAMP, the likely chemotactic substances in *P. pallidum* sorogens. As the basal concentration of this gradient drops during culmination (panel b), basal cells begin to move randomly. With time, more and more cells lose their orientation. Adhesive forces combine to round up the whorl mass. Because there is no longer directed motion toward the tip, the rounded mass of cells is left behind, free of the tip inhibitory substance, and the synthesis of tip-specific antigens begins. This sets in train the events described in Fig. 2, and cells begin to reorient toward cap cells (Fig. 5d). By this time the organization of secondary tips is well under way.

This hypothesis argues that whorl deposition is a consequence of the sum of two sets of opposing forces. Chemotaxis and chemokinesis act together, directing cells apically; retarding forces—cell–cell adhesion

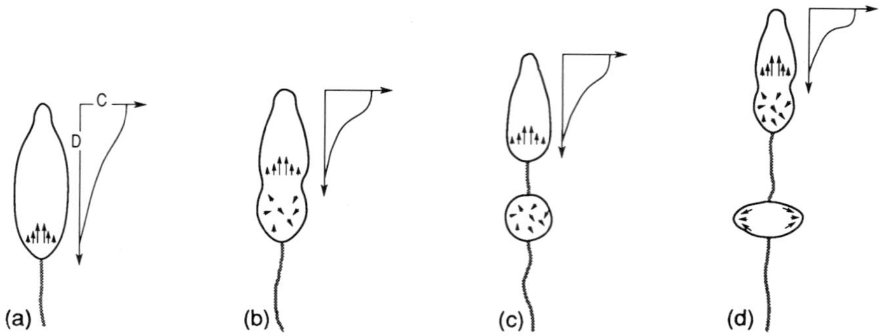

Fig. 5 Periodic release of whorl masses from an ascending sorogen. Each panel is a highly schematic view of the sorogen and a plot of a morphogen gradient that represses secondary tip formation within the sorogen. The vectors in each case represent the magnitude and direction of cell motion. The gradient is regulated and its source density constant. *D* is distance and *C* is concentration of the proposed morphogen.

and random cell movement—act to oppose chemokinesis and chemotaxis. The constant velocity of culmination and the linearly decreasing basal activity of the inhibitory gradient then guarantee even spacing. The mass of the primary sorogen determines the number of whorls that will form; and the number of cells released in the whorl mass will determine the number of arms per whorl, since the whorl prepattern has a fixed wavelength (Byrne and Cox, 1987).

The gradient described here could also be responsible for sensing the number of tip cells. It is illustrated as a fixed source density in Fig. 5 to account for the fact that there is always a terminal sorus in *P. pallidum*.

VIII. PATTERNING IN OTHER *POLYSPHONDYLIA*

P. filamentosum is "withal a very striking slime mold" (Rapier, 1984). The mature fruiting body sports a second set of regularly spaced whorls on the arms of the first set, giving rise to tertiary as well as secondary branches. Laboratory cultures show that whorl masses released at the base of the ascending sorogen contain many more cells than the typical *P. pallidum* whorl. When partitioned into radially arranged secondary sorogens, the secondary sorogens culminate, and in the process release basal whorls which then form a second set of whorls orthogonal to the first. Although this is occasionally observed under the conditions used to produce *P. pallidum* monster whorls, it is a genetically programmed and stable part of the *P. filamentosum* life cycle.

The reasons for elaborate branches, and branches on branches, is a mystery. Perhaps selection operates to fix these patterns because it is an efficient way to disperse spores (Bonner, 1982). Another explanation might lie in the nature of pattern generating mechanisms themselves. I have pointed out in this essay that most patterning models can be reduced to equation (3). A corollary of this is that once a signaling mechanism for the establishment of aggregates evolved in the slime molds, the mechanism itself is not necessarily easy to suppress later during morphogenesis. The whorl patterns we admire in *P. pallidum* may be inevitable consequences of this fact, not a response to natural selection beyond the obvious need for signaling early in the life cycle.

ACKNOWLEDGMENTS

This work was supported by the NIH and NSF. I thank John Bonner, Cathy Vocke, and Jim McNally for commenting on early versions of this manuscript, and Jim McNally for pointing out the importance of the Houle et al. (1989) reference.

REFERENCES

Bonner, J. T. (1967). *The Cellular Slime Molds*, Princeton University Press, Princeton, N.J.

Bonner, J. (1982). *Am. Natl.*, *119*: 530.

Bozzaro, S., Tsugita, A., Janku, M., Monok, G., Opatz, K., and Gerisch, G. (1981). *Exp. Cell Res.*, *134*: 181.

Bozzaro, S., Merkl, R., and Gerisch, G. (1987). *Methods Cell Biol.*, *28*: 359.

Byrne, G., and Cox, E. C. (1986). *Dev. Biol.*, *117*: 442.

Byrne, G., and Cox, E. (1987). *Proc. Natl. Acad. Sci. USA*, *84*: 4140.

Byrne, G., Trujillo, J., and Cox, E. (1982). *Differentiation*, *23*: 103.

Chang, C., Rosen, S., and Barondes, S. (1977). *Exp. Cell Res.*, *104*: 101.

Cox, E., Spiegel, F., Byrne, G., McNally, J., and Eisenbud, L. (1988). *Differentiation*, *38*: 73.

Gerisch, G. (1976). *Surface Membrane Receptors*, Plenum Press, New York, p. 67.

Gmitro, J. I., and Scriven, L. E. (1966). In *Intracellular Transport* (K. B. Warren, ed.), Academic Press, New York, p. 221.

Hohl, H. R., Honegger, R., Traub, F., and Markwalder, M. (1977). In *Development and Differentiation in the Cellular Slime Molds* (P. Cappuccinelli and J. M. Ashworth, eds.), Elsevier/North-Holland Biomedical Press, Amsterdam, p. 149.

Houle, J., Balthazar, J., and West, C. M. (1989). *Proc. Natl. Acad. Sci. USA*, *86*: 3679.
Lacalli, T. C., Wilkinson, D. A., and Harrison, L. G. (1988). *Development*, *103*: 105.
Lam, T., Pickering, G., Geltosky, J., and Siu, C. (1981). *Differentiation*, *20*: 22.
Loomis, W. (1988). *Dev. Genet.*, *9*: 549.
McNally, J. G., and Cox, E. C. (1988). *Dev. Genet.*, *9*: 663.
McNally, J. G., and Cox, E. C. (1989). *Development*, *105*: 323.
Meinhardt, H., and Gierer, A. (1974). *J. Cell Sci.*, *15*: 321.
Murray, J. D. (1981). *Philos. Trans. R. Soc. Lond. B*, *295*: 473.
Murray, J. D., and Maini, P. K. (1988). *Phys. Rep.*, *171*: 59.
Murray, B. A., Niman, H. L., and Loomis, W. F. (1983). *Mol. Cell. Biol.*, *3*: 863.
Noegel, A., Harloff, C., Hirth, P., Merkl, R., Modersitzki, M., Stadler, J., Weinhart, U., Westphal, M., and Gerisch, G. (1985). *EMBO J.*, *4*: 3805.
Noegel, A., Gerisch, G., Stadler, J., and Westphal, M. (1986). *EMBO J.*, *5*: 1473.
Odell, G., and Bonner, J. (1986). *Philos. Trans. R. Soc. Lond. B*, *312*, *487.*
Oster, G. F. (1988). *Math. Biosci.*, *90*: 265.
Oster, G. F., Murray, J. D., and Harris, A. K. (1983). *J. Embryol. Exptl. Morph.*, *78*: 83.
Raper, K. B. (1984). *The Dictyostelids*, Princeton University Press, Princeton, N.J., p. 384.
Rosen, S., Simpson, D., Rose, J., and Barondes, S. (1974). *Nature*, *252*: 128.
Rosen, S., Haywood, P., and Barondes, S. (1976). *J. Cell Biol.*, *70*: 133A.
Shaffer, B. M. (1961). *J. Exp. Biol.*, *38*: 833.
Shimomura, O., Suthers, H., and Bonner, J. (1982). *Proc. Natl. Acad. Sci. USA*, *79*: 7376.
Steinberg, M. S., and Poole, T. J. (1982). In *Developmental Order: Its Origin and Regulation* (S. Subtelny, ed.), Alan R. Liss, New York, p. 351.
Sternfeld, J., and David, C. (1981). *Differentiation*, *20*: 10.
Turing, A. M. (1952). *Philos. Trans. R. Soc. Lond. B*, *237*: 37.
Vocke, C. D., and Cox, E. C. (1990). *Dev. Genet.*, *11*: 427.
Xu, Y., Vest, C. M., and Murray, J. D. (1983). *Appl. Opt.*, *22*: 3479.

27
Pigment Cell Migration and Pattern Formation in Salamander Larvae

Lennart Olsson and Jan Löfberg *Uppsala University, Uppsala, Sweden*

I. INTRODUCTION

The migration and differentiation of pigment cells, which create the pigment patterns of vertebrates, have been studied extensively from both experimental and theoretical aspects. Pigment cell development has been used as a model system for investigations into questions of fundamental importance for the developmental biologist. How does a cell know how to move to the right place in the embryo (cell migration), and how does it know what to become (cell differentiation)? Are cells preprogrammed to migrate and differentiate in certain ways, or is their fate determined by their environment? In this brief review we give a general background and then concentrate on work done in the senior author's laboratory, in collaboration with Dr. Epperlein (Freiburg), on salamander pigment cells.

II. THE NEURAL CREST

All pigment cells in vertebrates, except for those in the retina, are derived from a transient embryonic structure called the neural crest (NC). The NC cells migrate from an original position on top of the neural tube to many different regions in the embryo, giving rise to a variety of tissues in diverse structures, such as the peripheral nervous system, the head, and the adrenal gland, besides giving rise to pigment cells (for reviews, see LeDouarin, 1982, and Hall and Hörstadius, 1988). The elegant transplantation experiments by LeDouarin and her co-workers, where they showed that NC cells could change their fate according to a changed position, gave inspiration to investigations into the role of the extracellular environment in cell differentiation and migration.

III. NEURAL CREST CELL MIGRATION

Cell *migration* implies that the cells move from one place to another following a specific pathway. Cell locomotion can be random, but cell migration is directional. How does the extensive migration of NC cells come about? Although the word "migration" implies that the cells are moving actively, it should be kept in mind that there is also a passive component in NC cell migration. Newgreen and Erickson (1986) distinguish passive carriage and passive displacement. Passive carriage occurs when NC cells are moved as a result of morphogenetic changes of the tissues with which they are associated but do not themselves move relative to their environment. Passive displacement implies that NC cells move relative to their environment without contributing actively to the process. If latex beads are injected into neural crest pathways, they will be displaced ventrally along these pathways, in the direction of NC cell migration (Bronner-Fraser, 1982, 1984). The impetus for bead movement is not known but is probably connected to morphogenetic movements of the surrounding tissues.

The relative importance of passive and active mechanisms has not been determined, but active migration has been considered more important. Neural crest cells move actively in culture and, as indicated by time-lapse recordings of pigment cells (Keller and Spieth, 1984), marker systems (LeDouarin, 1982; Bronner-Fraser and Fraser, 1988), and shape and ultrastructure, it is well established that the cells also do so in vivo (birds: Tosney, 1978, 1982; Newgreen et al., 1982; mice: Erickson and Weston, 1983; salamanders: Löfberg and Ahlfors, 1978, and Löfberg et al., 1980).

In the embryo, NC cells migrate through extracellular spaces filled with a three-dimensional fibril network, the extracellular matrix (ECM). The ECM can initiate NC cell migration in vivo (Löfberg et al., 1985). Several types of extracellular matrix molecules that might influence cell migration have been localized in the NC migratory pathways. These include collagens (van der Mark et al., 1976; Löfberg et al., 1980; Duband and Thiery, 1987); glycoproteins such as fibronectins, laminin, and tenascin (Newgreen and Thiery, 1980; Duband and Thiery, 1987; Epperlein et al., 1988; Mackie et al., 1988); and proteoglycans/glycosaminoglycans (Derby, 1978, 1982; Pintar, 1978; Perris et al., 1990).

Fibronectins (FNs) have received much attention, since NC cells in vitro migrate faster and are more directionally persistent on FNs than on other ECM molecules that have been tested (Erickson and Turley, 1983). Many perturbation experiments performed with synthetic peptides containing the cell-binding region of FNs, and with antibodies directed against FNs or FN receptors, demonstrated that FNs are required for NC cell adhesion and motility (Boucaut et al., 1984; Bronner-Fraser, 1985, 1986).

The cells in the neural crest are held together by intercellular junctions and cell adhesion molecules (CAMs; Tucker et al., 1988). The initiation of NC cell migration is correlated with loss of intercellular junctions (Revel and Brown, 1975) and the CAMs *N*-CAM (Edelman et al., 1983) and *N*-cadherin (Hatta et al., 1987; Duband et al., 1988). NC cell migration is probably initiated when the crest cells shift from cell–cell adhesion to cell–substrate adhesion. Perhaps the FN receptors are the most important of the substrate adhesion molecules (SAMs) involved. When neural crest cell migration has been initiated, it proceeds rapidly. The only model of the molecular events underlying this migration that has been proposed (Duband et al., 1986; Dufour et al., 1988; Tucker et al., 1988) assumes that FNs are the only ECM molecules of importance for NC cell migration. In this model a motile cell is characterized by a disorganized cytoskeleton, a diffuse distribution of mobile FN receptors, many FN receptors that do not bind to extracellular FNs, and a diffuse, cytoplasmic distribution of the molecules talin and vinculin (Fig. 1a). The bonds between FNs and their receptors can easily be made and broken. A stationary cell, on the other hand, displays clusters of FN receptors concentrated close to focal adhesions and microfilament bundles. The bonds that form cannot be broken. Almost all FN receptors are bound to extracellular FNs and are poorly mobile. The FN receptors build a transmembrane link between extracellular FNs and the cytoskeleton via talin and vinculin. Many actin microfilaments are cross-linked by α-actinin (Fig. 1b).

Is there any evidence for this model? The correlation between focal adhesions and stationary cells has been observed (see Trinkaus, 1984 for a review). A higher lateral mobility of FN receptors on mobile cells than on stationary cells has been shown (Duband et al., 1988), and talin, vinculin, and α-actinin have been demonstrated to participate in the attachment of actin microfilaments to the internal side of the membrane at focal adhesions (see Burridge, 1986 for a review). Talin binds to the FN receptors (Horwitz et al., 1986), and the difference in the organization of the microfilaments has been shown by immunolabeling (Duband et al., 1986). Some critical points are that the model does not explain how NC cells cultured on other ECM molecules than FNs [e.g., laminin or collagen; see, e.g., Bilozur and Hay (1988)] can move. The *directionality* of neural crest cell migration cannot be explained by this model. It only

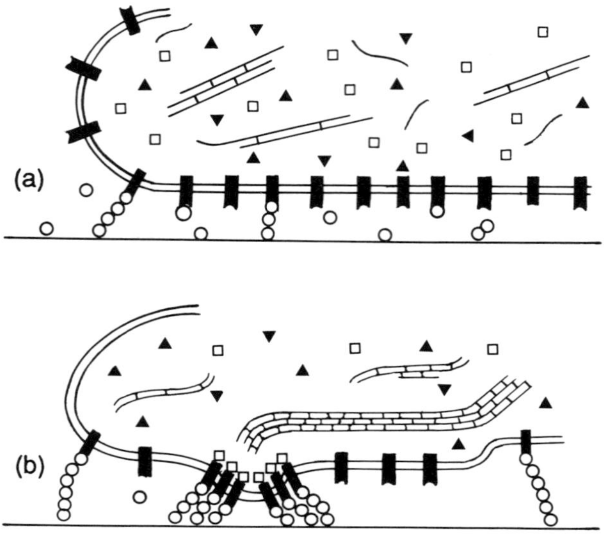

Fig. 1 Model at the molecular level of a motile (a) and a stationary (b) cell. The circles are symbols for FNs, the black rectangles for FN receptors, the open squares for vinculin, and the black triangles for talin. In the microfilament bundles the long lines are actin and the short, cross-linking lines α-actinin. The model is explained in the text. (Redrawn from Tucker et al., 1988.)

predicts that cells do move, but it does not explain the precision of NC cell migration. Therefore, the in vivo situation is probably more complex than the model predicts.

Many different mechanisms have been proposed to explain the directionality of NC cell migration (see Erickson, 1988, 1990 for a review). One group of explanations is based on long-range taxisms, when cells are responding to a gradient of either a substance in solution (chemotaxis) or an electric potential (galvanotaxis). Another group is based on cell–cell interactions, including contact inhibition (cells that collide change direction away from each other) and gradients in cell–cell adhesiveness (cells move from less to more adhesive neighbors), and a third group is based on cell–ECM interactions including haptotaxis [cells move up (or down) a gradient in adhesiveness of the substratum] and contact guidance (cells follow discontinuities in the substratum, e.g., oriented fibrils). Finally, the creation of spaces in the ECM, perhaps by the migrating cells themselves, could provide pathways of least resistance. While there is evidence that many of these mechanisms are active in vivo as well as in vitro, there is at present no consensus as to what mechanisms are most important in the embryo. This is due primarily to the lack of experiments that clearly distinguish between the mechanisms by testing one at a time. We also lack an understanding of many of the proposed mechanisms at the molecular level, although data are now accumulating from studies of both the neural crest and other systems (see, e.g., Erickson, 1990 for a review).

In our research we have concentrated on the role of the ECM in NC cell migration (and differentiation) in a salamander, the axolotl (*Ambystoma mexicanum*), paying special attention to the so-called white (dd) mutant. In this mutant, the migration of the pigment cells is defective. The pigment cells never leave the neural crest, but they differentiate into the two types of pigment cells—xanthophores (yellow) and melanophores (black)—that embryos of *A. mexicanum* normally possess. Transplantations of epidermis with attached ECM (Keller et al., 1982) and of ECM adsorbed to microcarriers (Löfberg et al., 1989) between mutant and wild-type embryos showed that the NC cells of the white mutant were normal but the ECM defective (Fig. 2). Interestingly, the defect was found to be transient. ECM from older white embryos was permissive to pigment cell migration in younger embryos. These results were interpreted to indicate that the "maturation" of the ECM in the white mutant is delayed. When the ECM is finally "mature" (i.e., permissive to NC cell migration), the cells are no longer able to respond to it and to start migration. If correct, this interpretation emphasizes the importance of precise timing in cell–ECM interactions for proper pigment cell patterning. The restricted pigmentation of the white axolotl larva could, therefore, result from heterochrony (Gould, 1977; Raff and Kaufman, 1983). Heterochrony, which implies changes in the relative timing of developmental processes, is recognized as an important ontogenetic basis for phylogenetic change and therefore of great interest in attempts to tie together developmental and evolutionary biology.

What is the nature of the "defect" in the ECM of the white mutant? It is divergent in ultrastructure as well as molecular composition (Perris et al., 1990; Olsson and Perris, unpublished observations). Immunohistochemistry revealed only minor differences in the distribution of ECM molecules permissive to cell migration [e.g., fibronectin, laminin, collagen I), whereas some molecules on which cells migrate poorly in vitro (collagen III, some chondroitinsulfate proteoglycan (CSPG) and keratansulfate proteoglycan (KSPG) epitopes] were more abundant in the white mutant. This suggests that the prospective pigment cells in the white mutant are unable to migrate not because the ECM is lacking a molecule

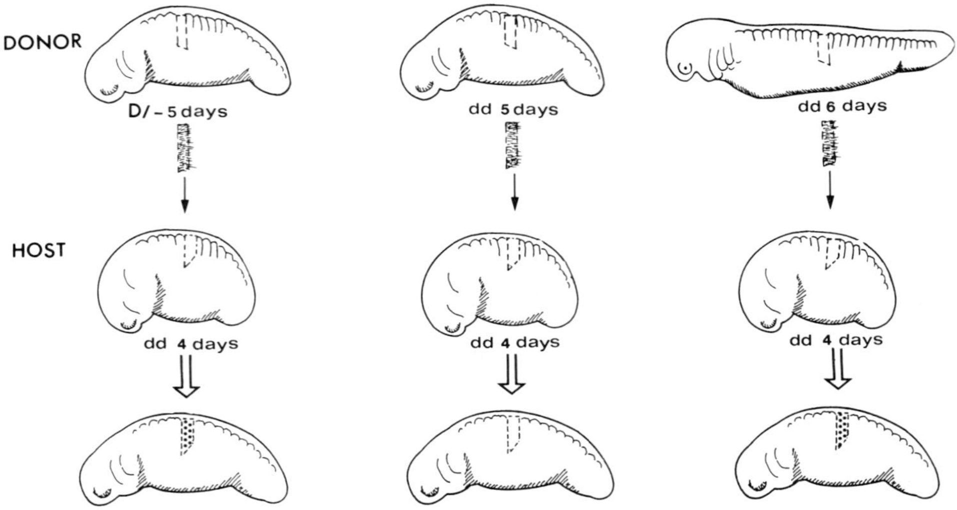

Fig. 2 Transplantation of ECM using microcarriers reveals that the ECM of the white mutant is transiently defective. ECM from 5-day-old wild-type (D/-) embryos transplanted from a region where pigment cell migration has not yet started into 4-day-old white (dd) embryos trigger pigment cell migration (left column, black dots). This does not happen if the donor is a 5-day-old white embryo (middle column). But if the white donor is older (6 days), pigment cells migrate out (right column). An interpretation of these results is given in the text. (Based on data from Löfberg et al., 1989.)

that promotes migration, but because of an overabundance of molecules that inhibit migration.

IV. PIGMENT PATTERN FORMATION

Pigment cell migration leads to the creation of pigment patterns. The axolotl larva has a pigment pattern consisting of alternating *vertical* bars of melanophores and xanthophores. Many other species, including the alpine newt (*Triturus alpestris*), have two *horizontal* melanophore stripes with xanthophores in between. What mechanisms are at work here? In a comparative study (Epperlein and Löfberg, 1990) based on both earlier work (Epperlein, 1982; Epperlein and Claviez, 1982a,b; Epperlein and Löfberg, 1984) and new data, the pigment pattern formation in the alpine newt was compared to that in the axolotl. In the alpine newt, the two types of pigment cells leave the neural crest at the same time, migrate out on the

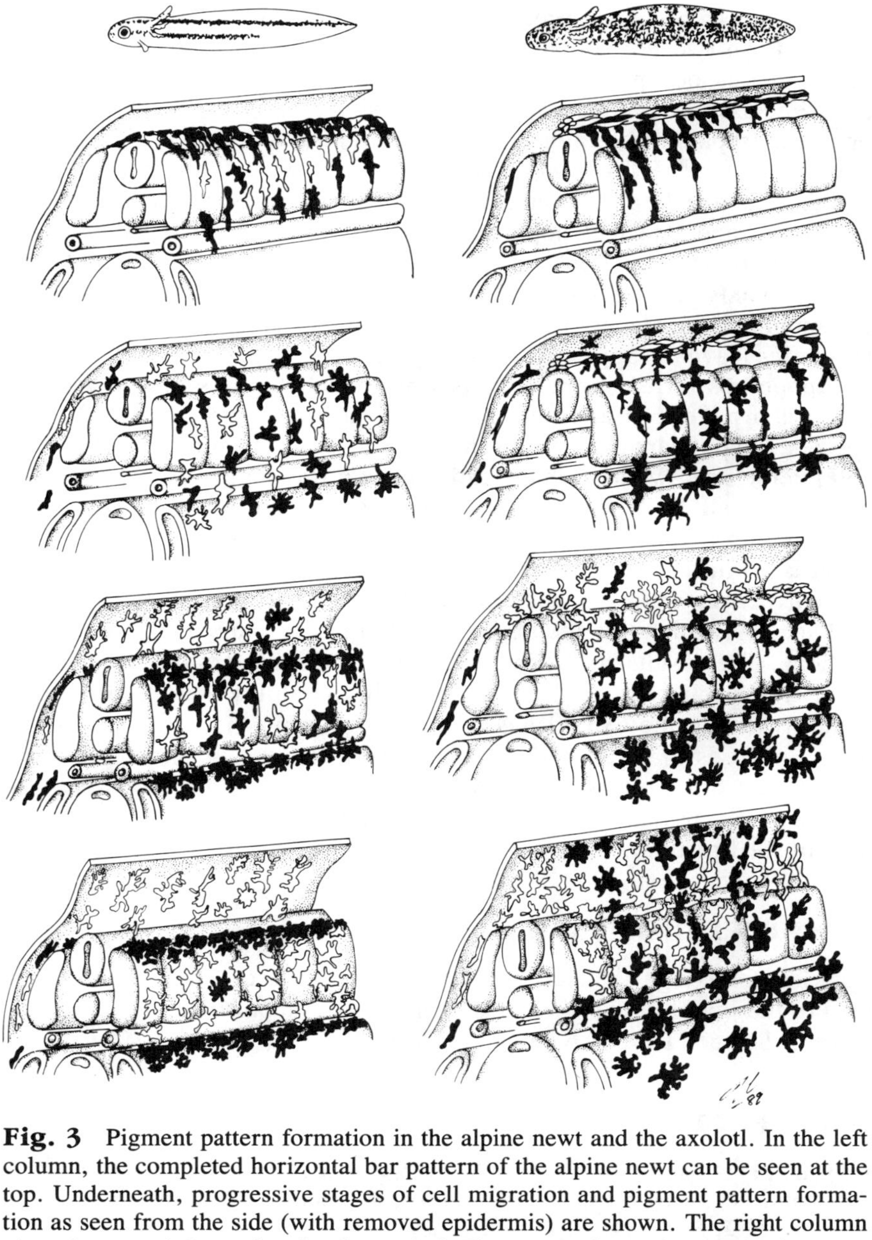

Fig. 3 Pigment pattern formation in the alpine newt and the axolotl. In the left column, the completed horizontal bar pattern of the alpine newt can be seen at the top. Underneath, progressive stages of cell migration and pigment pattern formation as seen from the side (with removed epidermis) are shown. The right column gives the same information for the axolotl. The pattern formation mechanisms are explained in the text. Black cells, melanophores; white cells, xanthophores. (Based on data from Epperlein and Löfberg, 1990.)

flank of the larva, and spread out evenly (Fig. 3). Later, the melanophores form two horizontal stripes along the body of the larva and the xanthophores occupy the interstripe area. From various analyses and experiments, the authors reach the conclusion that the melanophores recognize a prepattern of environmental cues laid down in the ECM as two horizontal stripes. The xanthophores show a lower affinity for the cues and therefore end up between the melanophore stripes. In the axolotl, on the other hand, the melanophores start to migrate out from the neural crest before the xanthophores. While melanophores migrate out and spread evenly on the flank of the larva, xanthophores and remaining melanophores form aggregates in the crest (Fig. 3). Later, the xanthophores leave the aggregates and migrate out to form vertical bars. The melanophores retreat from regions where xanthophores invade and the result is a pattern of alternating yellow and black vertical bars. In this species, the pigment pattern was indicated to be formed primarily by cell–cell interactions, whereas in the alpine newt it was suggested to be formed mainly by cell–ECM interactions (Epperlein and Löfberg, 1990).

How do pigment patterns and their formative mechanisms evolve? What is primitive and what is derived? To answer such questions, we are currently investigating pigment patterns and how they are formed in several species, which like the axolotl are members of the family Ambystomatidae. The results so far indicate that a transition from a pattern and formative mechanism of the type found in the alpine newt to a derived state present in the axolotl has taken place during the evolution of the ambystomatids. A pattern intermediate between that in the alpine newt and in the axolotl exists in certain ambystomatid species, which is initiated in a way similar to that seen in the axolotl (L. Olsson and J. Löfberg, unpublished results).

ACKNOWLEDGMENTS

Inga Thomasson is thanked for technical assistance. The research that lead to the invitation to write this chapter was supported by the Swedish Natural Science Research Council (Grant No. B-BU 3810-104 to JL).

REFERENCES

Bilozur, M. E., and Hay, E. D. (1988). *Dev. Biol.*, *125*: 19.

Boucaut, J.-C., Darribère, T., Poole, T. J., Aoyama, H., Yamada, K. M., and Thiery, J. P. (1984). *J. Cell. Biol.*, *99*: 1822.

Bronner-Fraser, M. (1982). *Dev. Biol.*, *91*: 50.

Bronner-Fraser, M. (1984). *J. Cell Biol.*, *98*: 1947.

Bronner-Fraser, M. (1985). *J. Cell Biol.*, *101*: 610.
Bronner-Fraser, M. (1986). *Dev. Biol.*, *115*: 44.
Bronner-Fraser, M., and Fraser, S. E. (1988). *Nature (Lond.)*, *335*: 161.
Burridge, K. (1986). *Cancer Rev.*, *4*: 18.
Derby, M. A. (1978). *Dev. Biol.*, *66*: 321.
Derby, M. A. (1982). *Cell Tissue Res.*, *225*: 379.
Duband, J.-L., and Thiery, J. P. (1987). *Development*, *101*: 461.
Duband, J.-L., Rocher, S., Chen, W.-T., Yamada, K. M., and Thiery, J. P. (1986). *J. Cell Biol.*, *102*: 160.
Duband, J.-L., Nuckolls, G. H., Ishihara, A., Hasegawa, T., Yamada, K. M., Thiery, J. P., and Jacobson, K. (1988). *J. Cell Biol.*, *107*: 1385.
Dufour, S., Duband, J.-L., Kornblihtt, A. R., and Thiery, J. P. (1988). *Trends Genet.*, *4*(7): 198.
Edelman, G. M., Gallin, W. J., Delouvée, A., Cunningham, B. A., and Thiery, J. P. (1983). *Proc. Natl. Acad. Sci. USA*, *80*: 4334.
Epperlein, H. H. (1982). *Roux's Arch. Dev. Biol.*, *191*: 19.
Epperlein, H. H., and Claviez, M. (1982a). *Roux's Arch. Dev. Biol.*, *191*: 5.
Epperlein, H. H., and Claviez, M. (1982b). *Dev. Biol.*, *91*: 497.
Epperlein, H. H., and Löfberg, J. (1984). *Roux's Arch. Dev. Biol.*, *193*: 357.
Epperlein, H. H., and Löfberg, J. (1990). *Adv. Anat. Embryol. Cell Biol.*, *118*: 1.
Epperlein, H. H., Halfter, W., and Tucker, R. P. (1988). *Development*, *103*: 743.
Erickson, C. A. (1988). *Development*, *103(Suppl.)*: 63.
Erickson, C. A. (1990). *Curr. Opin. Cell Biol.*, *2*: 67.
Erickson, C. A., and Turley, E. A. (1983). *J. Cell Sci.*, *61*: 299.
Erickson, C. A., and Weston, J. A. (1983). *J. Embryol. Exp. Morphol.*, *74*: 97.
Gould, S. J. (1977). *Ontogeny and Phylogeny*, The Belknap Press of Harvard University Press, Cambridge, Mass., p. 501.
Hall, B. K., and Hörstadius, S. (1988). *The Neural Crest*, Oxford University Press, New York, p. 303.
Hatta, K., Takagi, S., Fujisawa, H., and Takeichi, M. (1987). *Dev. Biol.*, *120*: 215.
Horwitz, A., Duggan, K., Buck, C., Beckerle, M. C., and Burridge, K. (1986). *Nature (Lond.)*, *320*: 531.
Keller, R. E., and Spieth, J. (1984). *J. Exp. Zool.*, *229*: 109.
Keller, R. E., Löfberg, J., and Spieth, J. (1982). *Dev. Biol.*, *89*: 179.
Le Douarin, N. M. (1982). *The Neural Crest*, Cambridge University Press, Cambridge, p. 259.
Löfberg, J., and Ahlfors, K. (1978). *Zoon*, *6*: 87.
Löfberg, J., Ahlfors, K., and Fällström, C. (1980). *Dev. Biol.*, *75*: 148.
Löfberg, J., Nynäs-McCoy, A., Olsson, C., Jönsson, L., and Perris, R. (1985). *Dev. Biol.*, *107*: 442.
Löfberg, J., Perris, R., and Epperlein, H. H. (1989). *Dev. Biol.*, *131*: 168.

Mackie, E. J., Tucker, R. P., Halfter, W., Chiquet-Ehrismann, R., and Epperlein, H. H. (1988). *Development*, *102*: 237.

Newgreen, D. F., and Erickson, C. A. (1986). *Int. Rev. Cytol.*, *103*: 89.

Newgreen, D. F., and Thiery, J. P. (1980). *Cell Tissue Res.*, *211*: 269.

Newgreen, D. F., Gibbins, I. L., Sauter, J., Wallenfels, B., and Wütz, R. (1982). *Cell Tissue Res.*, *221*: 521.

Perris, R., Löfberg, J., Fällström, C., von Boxberg, Y., Olsson, L., and Newgreen, D. F. (1990). *Development*, *109*: 533.

Pintar, J. E. (1978). *Dev. Biol.*, *67*: 444.

Raff, R. A., and Kaufman, T. C. (1983). *Embryos, Genes and Evolution*, Macmillan, New York, p. 543.

Revel, J.-P., and Brown, S. S. (1975). *Cold Spring Harbor Symp. Quant. Biol.*, *40*: 443.

Tosney, K. (1978). *Dev. Biol.*, *62*: 317.

Tosney, K. (1982). *Dev. Biol.*, *89*: 13.

Trinkaus, J. P. (1984). *Cells into Organs*, Prentice Hall, Englewood Cliffs, N.J., p. 543.

Tucker, G. C., Duband, J.-L., Dufour, S., and Thiery, J. P. (1988). *Development*, *103*(*Suppl.*): 82.

van der Mark, H., van der Mark, K., and Gay, S. (1976). *Dev. Biol.*, *48*: 237.

28
A Novel Cellular Automaton Approach to Pattern Formation by Filamentous Fungi

Andreas Deutsch *University of Bremen, Bremen, Germany*

I. INTRODUCTION

Pattern formation in filamentous fungi occurs on at least two different scales. A microscopic pattern is created by successive growth and ramification of rather uniform structure elements: namely, vegetative hyphae (Fig. 1). Differentiation of vegetative hyphae to aerial hyphae and conidia causes secondary structures which may manifest themselves as macroscopic patterns (Fig. 2). How can branching patterns, such as the ones shown in Fig. 1, give rise to the concentric rings, spirals, and synchronous banding patterns that have been reported for a great number of filamentous fungi (e.g., in Winfree, 1973; Deutsch et al., 1991)?

Branching structures occur not only in filamentous fungi but also in plants, hydroids, or corals and as cellular networks of nerve cells and capillaries, for example. Their generation is one of the central problems of theoretical biology. For modeling growth patterns, branched filamentous organisms such as fungal mycelia, algal thalli, or moss protonemata have the advantage, compared to higher organisms, of having a relatively simple organization. One model of branching patterns of bryozoan cell clusters is based primarily on mechanical constraints (e.g., convection) (Goldwasser et al., 1989). Another model analyzes branching networks with applications to fungal growth patterns (Edelstein, 1989). Meinhardt considered separate activator–inhibitor models, which produced branching networks and stripe patterns, respectively (Meinhardt, 1984; see also Chapter 2, this volume). Branching models of the Gierer–Meinhardt type were also proposed for fungal mycelia (Briére and Buis, 1985; Hyver et al., 1979). These and an additional fungal morphogenesis

(a)

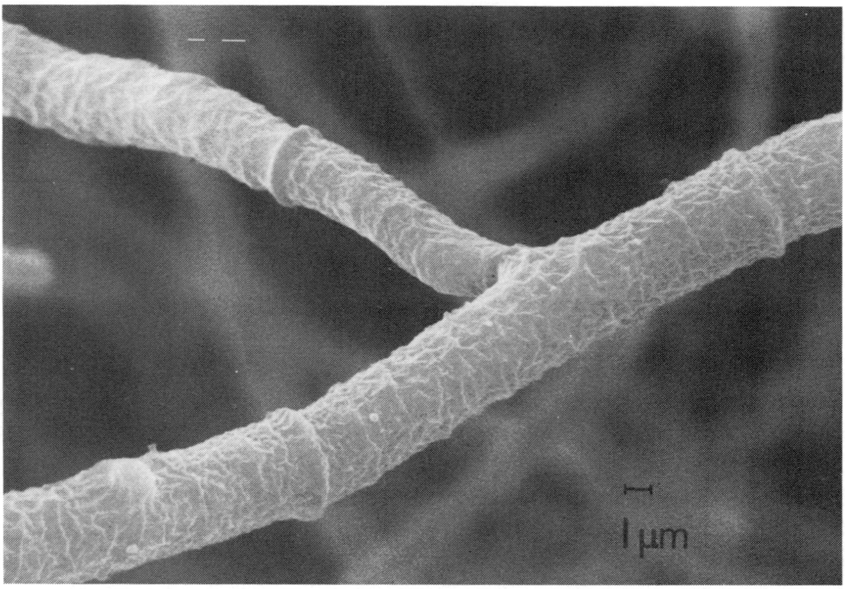

(b)

Fig. 1 (a) Light micrograph of *Neurospora crassa* growth front (magnification factor 100); (b) scanning electron micrograph of *Neurospora* hyphal ramification.

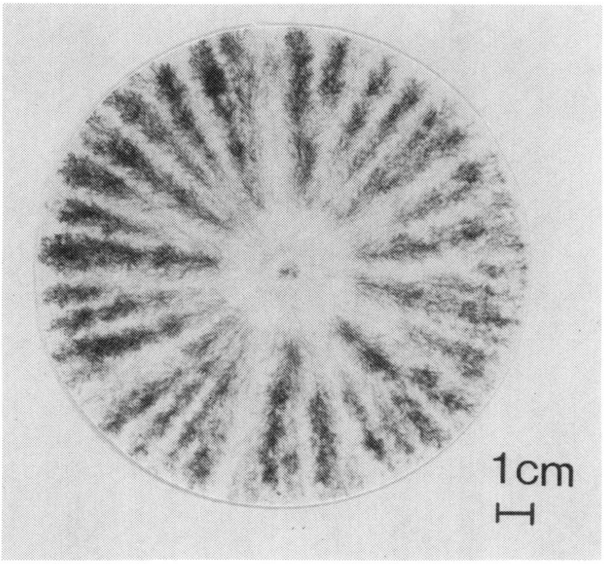

Fig. 2 Radial zonation pattern in the spore distribution of *Neurospora*. The fungus was grown 3 days in constant darkness in a petri dish containing a semi-solid mixture of agar and nutrients (amino acids, salts, and a carbon source). See Deutsch et al. (1991) for experimental details.

model (Bartnicki-Garcia et al., 1989) are concerned with the developmental fate of individual cells and try to explain the mechanisms of growth and ramification of single cells.

In Section II I introduce the ascomycete *Neurospora crassa* as an example of a filamentous fungus. Not much is known about the synergetic effects of branching and differentiation of single hyphae, which lead to the aforementioned microscopic and macroscopic patterns of this fungus. Mathematical modeling may help to get a deeper understanding of the relevant processes. The idea is to relate complex dynamical system behavior to the behavior and the rules of interaction between the elementary parts of the system. For this purpose we present in Section III a model in the language of cellular automata that connects the levels of microscopic and macroscopic pattern formation in filamentous fungi. In particular, computer simulations (Fig. 3) suggest a hypothesis for a control mechanism of fungal differentiation, leading to radial spore distributions. Upon introduction of an (intracellular) time-dependent switching device, the automaton is also able to produce concentric ring and spiral patterns. The

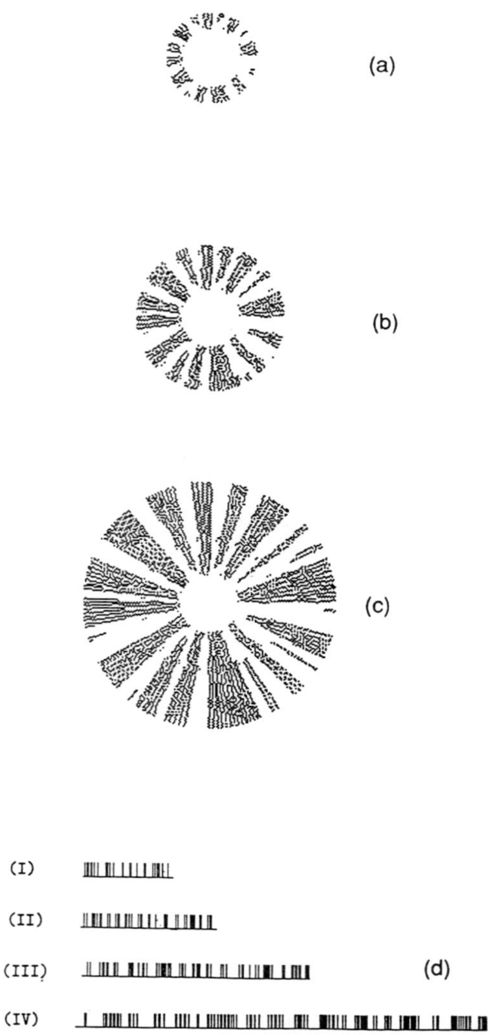

Fig. 3 Computer simulation leading to a radial zonation pattern (''macroscopic'' pattern). The distribution of phenomenological states $p = 1$ is shown. Differentiation rule g (see the text for an explanation). (a) After 10 time steps; (b) after 30 time steps; (c) after 60 time steps. (d) Phenomenological state distribution along growth fronts for times $t = 1$ (I), $t = 10$ (II), $t = 30$ (III), $t = 60$ (IV), respectively. (Cells with phenomenological states $p = 1$ are indicated by strokes.) Parameters: $|B_1| = 50$, $r_1 = 2$, $r_2 = 5.5$, $k_1 = 1$, $k_2 = 0.4$, $k_0 = 20$, $\delta = 3$, $\varepsilon = 1$.

model incorporates dynamic properties of growing cellular networks and stresses geometrical implications of radial growth. While fixed cellular networks have been studied extensively, in my model it is essential that the (filamentous) network develops successively. Constraints of geometry and scale (i.e., of the topology of the system) on development were analyzed already in continuous models (Murray, 1981). The model presented here is a modified cellular automaton. "Classical" cellular automata are discrete dynamical systems defined on single physical spaces (termed base spaces) (Dress et al., 1988) and well-established tools to model self-organized processes (Wolfram, 1986). A classical (two-dimensional) cellular automaton was used earlier to explain pattern formation of the ascomycete *Neurospora crassa* as a growing field of coupled oscillators (Deutsch et al., 1991). This automaton, however, proved unable to create radial zonation patterns, because the geometry of the base space is rectangular. One possible way to overcome this difficulty is to assume successive circular growth fronts as separate physical spaces. The elements of a particular base space then represent the hyphae constituting the corresponding growth front. This idea leads to the extension of the classical cellular automaton definition to sequences of base spaces (Deutsch, 1991). Differentiation is incorporated in the model by assuming an activator–inhibitor mechanism. A similar (discrete) activator–inhibitor mechanism, but in a fixed rectangular cellular network, was introduced to model vertebrate skin patterns (Young, 1984).

II. PATTERN FORMATION BY THE ASCOMYCETE *NEUROSPORA CRASSA*

Among the most striking macroscopic patterns produced by the ascomycete *Neurospora crassa* are concentric ring and radial zonation patterns, in particular. Similar patterns have also been reported for other filamentous fungi [e.g., *Nectria cinnabarina* in Winfree (1973)], but so far no hypothesis for their formation has been proposed in either case. Therefore, as a model system I examine branching and macroconidial pattern formation in the (filamentous) ascomycete *Neurospora crassa*. Macroconidia are spores produced during the asexual life cycle. The cycle is initiated experimentally by a spore (or a small portion of mycelium) in the center of a nutrient-containing agar plate. Vegetative hyphae grow and branch in radial directions from this center (Fig. 1). Some hyphae differentiate into aerial hyphae, which give rise to spores (macroconidia).

Differentiation is triggered by various signals. The concentric ring pattern of spore distribution in the band mutant is due to an endogenous clock mechanism (circadian rhythm), producing temporal periodic signals (see Chapter 19, this volume). Under certain growth conditions, especially with a defined medium composition, *Neurospora* exhibits radial zonations of spore distribution apart from the concentric ring pattern (Fig. 2). Since the underlying mycelial pattern remains unchanged, this pattern suggests spatial interactions during differentiation. It is well known that differentiation can be induced by various (external) stress signals: for example, starvation or heat shock (see Chapter 19). Possibly, endogenous signals comparable to stress signals are produced in the growing mycelium, leading to spontaneous differentiation. The molecular basis of stress-induced differentiation processes is unknown. Neither is it known whether the different processes are perhaps only variations of one basic mechanism. Thus the radial zonation pattern may be the result of synergetic effects of stress responses.

Knowledge of the molecular mechanisms of branching in *Neurospora* is fragmentary. Probably, ramification is initiated by enzymes (e.g., proteases and glucanases) that have a lytic influence on components of the cell wall. This weakening of the cell wall may in turn cause budding of the cell, initiating a branch (Mahadevan and Mahadkar, 1970). Production of these enzymes may be triggered by certain nutritional conditions or by a loss of inhibitory influence of adjacent hyphae around the growing hyphae. Branching can occur either apically (i.e., by splitting the growing apex into two or more apices) or laterally (i.e., by the emergence of new apices along the lateral hyphal walls). The study of profusely branched (i.e., colonial) mutants of *Neurospora* provided evidence for the involvement of cyclic AMP (cAMP) in the regulation of branching, profuse branching being dependent on a low intracellular level of cAMP (Scott and Solomon, 1975). An antagonistic role of Ca^{2+} and cAMP is known for many organisms and may also be important for the *Neurospora* branching process. The Ca^{2+}-ionophore A23187 induces profuse branching in *Neurospora* after calcium starvation, whereas starvation involving other divalent cations had no detectable effect (Reissig and Kinney, 1983). The precise role of calcium ions in *Neurospora* morphogenesis remains unclear, the published results often being contradictory (Schmid and Harold, 1988). Nothing whatsoever is known biochemically about synergetic effects caused by the coupling of individual branching events with secondary differentiation processes that eventually lead to highly synchronized patterns, such as concentric rings or radial zonations. Mathematical modeling may help to get a better understanding of these effects.

III. THE MODEL

Common tools to describe spatiotemporal organization in dynamic systems are differential equations, difference equations, or cellular automata. A cellular automaton—itself being a discrete dynamical system—is a theoretical concept that can be applied to certain spatially distributed systems in which macroscopic behavior results from small-scale relationships. John von Neumann was one of the pioneers of this automaton approach (Neumann, 1966). Dress et al. (1988) introduced a formal definition which fits all cellular automata considered so far in the literature. My model is based on an extended form of this definition (see Deutsch, 1991, for a formal introduction to this approach).

A. A Cellular Automaton on a Sequence of Base Spaces

The idea that led to the development of the automaton presented here is sketched in Fig. 4a. The automaton simulates growth, branching, and differentiation along a circular growth front that is typical of filamentous fungi growing from a "point inoculum." The successively constructed sequence of base spaces in the model corresponds to the moving growth fronts in the biological system. Thus a particular (discrete) base space B_t consists of all filaments (hyphae) constituting the growth front at time step t ($t = 1, 2. 3, \ldots$). The "mycelium" at time $k \in N$ is therefore given as

$$\bigcup_{t=1}^{k} B_t$$

Cells (i.e., vegetative hyphae) can differentiate (this means development of aerial hyphae/spores in the biological system). The model allows one to test various differentiation hypotheses, among them those that are based on an activator–inhibitor mechanism.

Formal Definition. According to the general definition of cellular automata based on a sequence of base spaces (Deutsch, 1991), base spaces, state space, neighborhoods, and transferring maps have to be specified in particular.

Base Spaces

A sequence of base spaces $(B_t)_{t=1,2,\ldots}$ is (successively) defined giving rise to a microscopic pattern (Fig. 5). B_1 defines the "inoculum" (i.e., the hyphae positions of the initial circular growth front).

$$B_1 := \{\alpha_i \in [0, 1), i \in I_1\} \qquad I_1 \subset N$$

$$B_{t+1} := \bigcup_{\alpha \in B_t} b_t(\alpha) \qquad t = 2, 3, \ldots$$

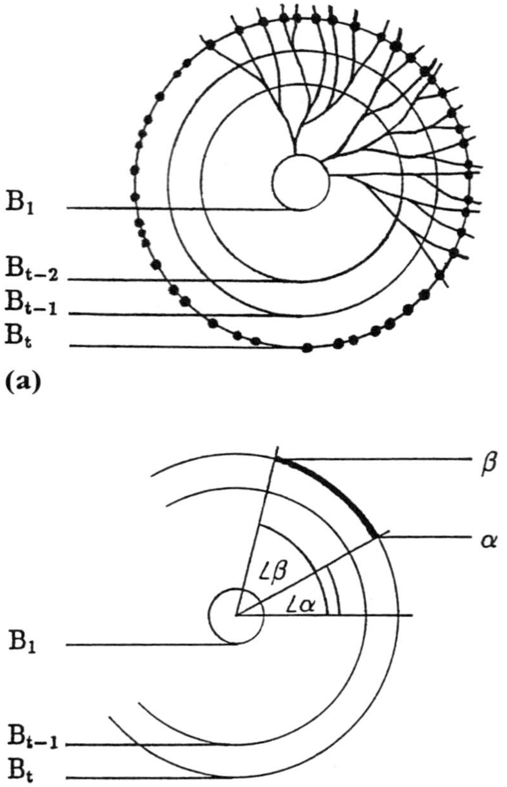

Fig. 4 Geometrical interpretation of a cellular automaton on a series of base spaces. (a) Schematic view of a series of base spaces that correspond to successive growth fronts of a growing fungal mycelium (only parts of the mycelium are sketched). The element positions of a particular base space (B_t) are marked by dots. B_1 corresponds to the inoculum. (b) "Distance measure" $A_t(\alpha, \beta)$ within a particular base space B_t of the cellular automaton. The distance $A_t(\alpha, \beta)$ is given

The correspondence between a particular base space B_t and the growth front at time t is as follows: The position of a particular element $\alpha \in B_t$ at the growth front is given by the section of the angle $2\pi\alpha$ and the circle with radius $k_0 t$ around the center of the inoculum (Fig. 4b). Here k_0 is the radius of the inoculum. Growth and branching of cells at one particular growth front generate the next growth front. These processes are implicitly defined by keeping some minimum distance (δ) between neighboring

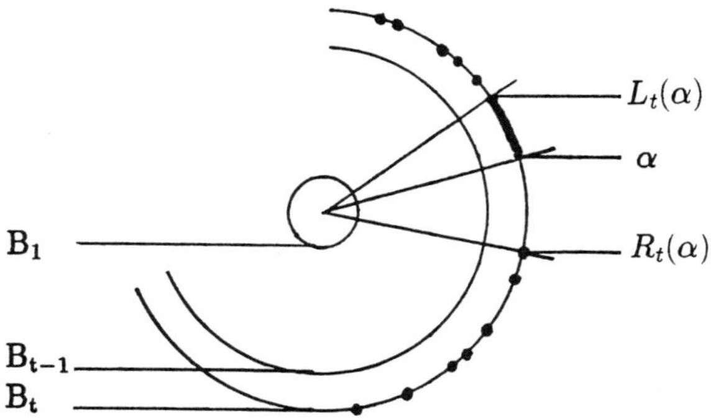

(c)

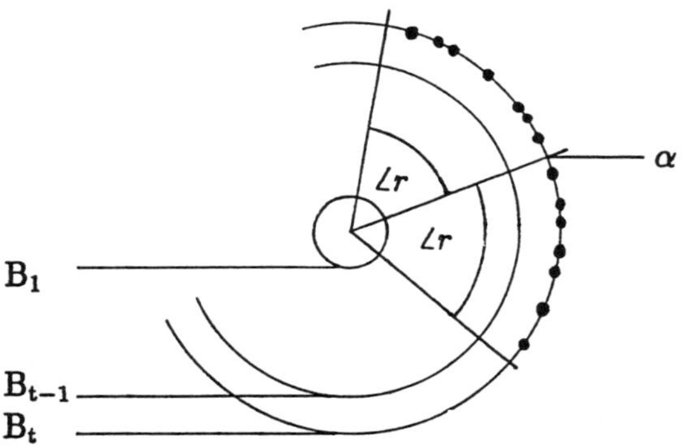

(d)

by the length of the thick line ($\angle\alpha = 2 \cdot \pi \cdot \alpha$, $\angle\beta = 2 \cdot \pi a$). (c) Left $L(\alpha)$ and right neighbor $R(\alpha)$ of a particular element $\alpha \in B_t$. $A_t(\alpha, L(\alpha))$ is marked by a thick line. If $A_t(\alpha, L(\alpha)) > \delta$ or $A_t(\alpha, R(\alpha)) > \delta$ new elements are created according to the definition of function b_t. Some element positions of (B_t) are marked by dots (see the text for details). (d) Neighborhood $\varphi^r_\alpha(N_t)$ of a particular element $\alpha \in B_t$. The elements of $\varphi^r_\alpha(N_t)$ are marked by dots ($\angle r = 2\pi r$).

filaments. The distance measure $A_t(\alpha, \beta)$ defines the circular length between two cells $\alpha, \beta \in B_t$ at the growth front at time t (Fig. 4b). As soon as the threshold distance (δ) between two adjacent cells is superseded, cells at the growth front start to branch: A new cell is created at a certain distance from the old cell (determined by the parameter ε). This means, biologically, that only apical growth and branching occur. $L_t(\alpha)$ and $R_t(\alpha)$

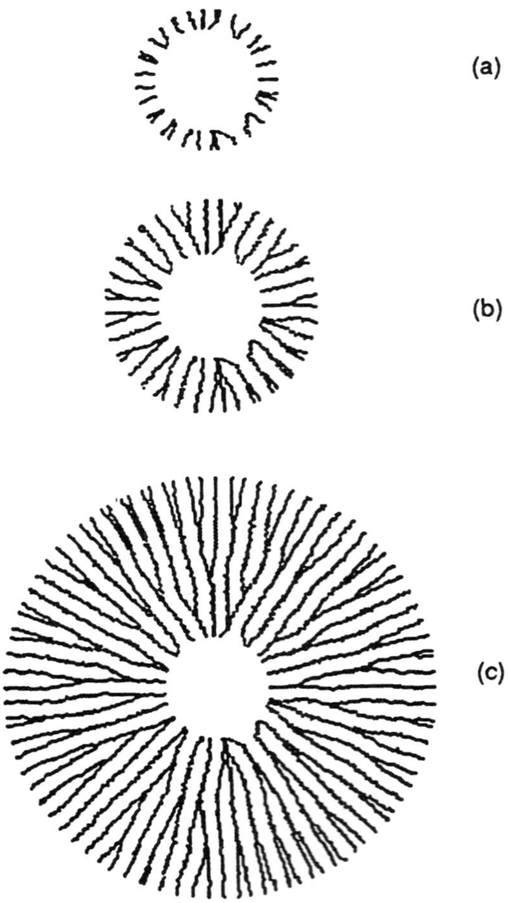

Fig. 5 Computer simulation of a "microscopic" automaton pattern: (a) $\bigcup_{t=1}^{10} B_t$, (b) $\bigcup_{t=1}^{30} B_t$, (c) $\bigcup_{t=1}^{97} B_t$. Parameters: $\delta = 15$, $\varepsilon = 1$, $k_0 = 30$.

define for each cell its left and right neighbor at the growth front at time t, respectively (Fig. 4c). Thus we have*

$$b_t : B_t \to \mathcal{P}([0, 1))$$

$$b_t(\alpha) := \begin{cases} \{\alpha\} \\ \quad \text{if } A_{t+1}(\alpha, L_{t+1}(\alpha)) < \delta \text{ and } A_{t+1}(\alpha, R_{t+1}(\alpha)) < \delta \\ \left\{\alpha, \dfrac{(2\pi k_0(t+1)\alpha + \varepsilon) \bmod (2\pi k_0(t+1))}{2\pi k_0(t+1)}\right\} \\ \quad \text{if } A_{t+1}(\alpha, L_{t+1}(\alpha)) < \delta \text{ and } A_{t+1}(\alpha, R_{t+1}(\alpha)) \geq \delta \\ \left\{\alpha, \dfrac{(2\pi k_0(t+1)\alpha - \varepsilon) \bmod (2\pi k_0(t+1))}{2\pi k_0(t+1)}\right\} \\ \quad \text{if } A_{t+1}(\alpha, L_{t+1}(\alpha)) \geq \delta \text{ and } A_{t+1}(\alpha, R_{t+1}(\alpha)) < \delta \\ \left\{\alpha, \dfrac{(2\pi k_0(t+1)\alpha + \varepsilon) \bmod (2\pi k_0(t+1))}{2\pi k_0(t+1)}, \right. \\ \quad \left. \dfrac{(2\pi k_0(t+1)\alpha - \varepsilon) \bmod (2\pi k_0(t+1))}{2\pi k_0(t+1)}\right\} \\ \quad \text{if } A_{t+1}(\alpha, L_{t+1}(\alpha)) \geq \delta \text{ and } A_{t+1}(\alpha, R_{t+1}(\alpha)) \geq \delta \end{cases}$$

where $\alpha \in B_t$ and $\delta, \varepsilon > 0$

Distance

$$A_t : B_t \times B_t \to R$$

$$A_t(\alpha, \beta) := 2\pi k_0 t \min\{|\alpha - \beta|, \alpha + 1 - \beta, \beta + 1 - \alpha\}$$

where $\alpha, \beta \in B_t$ and $k_0 \in \mathbb{R}_0^+$

(Fig. 4b).

Left and Right Neighbors

$$L_t, R_t : B_t \to B_t$$

$$L_t(\alpha) := \begin{cases} \min\{\beta \in B_t/\beta > \alpha\} & \text{if } \{\beta \in B_t/\beta > \alpha\} \neq \emptyset \\ \min\{\beta \in B_t/\beta < \alpha\} & \text{if } \{\beta \in B_t/\beta > \alpha\} = \emptyset \end{cases}$$

$$R_t(\alpha) := \begin{cases} \max\{\beta \in B_t/\beta < \alpha\} & \text{if } \{\beta \in B_t/\beta < \alpha\} \neq \emptyset \\ \max\{\beta \in B_t/\beta > \alpha\} & \text{if } \{\beta \in B_t/\beta < \alpha\} = \emptyset \end{cases}$$

where $\alpha \in B_t$ and $|B_t| \geq 2$, $t = 1, 2, \ldots$

(Fig. 4c).

---

*$\mathcal{P}(M) := \{S/S \subset M\}$ for any set M.

State Space

The local state space F_α of a cell α (total space) is defined by

$$F = F_\alpha = \{(p_\alpha, \phi_\alpha, m_\alpha) \mid p_\alpha \in Z_2,\ \phi_\alpha \in Z_\nu,\ m_\alpha \in Z\}$$

The elements are vectors:

p_α marks the "phenomenological state" of cell α. $p_\alpha = 0$ corresponds to the vegetative (undifferentiated) state (i.e., a vegetative hypha in the biological system), while $p_\alpha = 1$ marks an aerial hypha/conidium.

ϕ_α is the phase of the endogenous (temporal) switching device of cell α, and ν is the period length of the switching device ($\nu \in \mathbb{N}$).

m_α describes some "morphogenetical influence" on cell α. Morphogenetical influence is a value m_α that can be interpreted as an evaluation of all processes that have any influence on the differentiation of cell α. Thus m_α is not necessarily simply the concentration of a particular morphogen. The processes determining differentiation are defined by the transferring map below as well as the meaning of positive and negative m_α values.

Neighborhood

The neighborhood set $\varphi^r_\alpha(N_t)$ of a cell α at time t contains all cells along the growth front B_t that have smaller distance than r from α (Fig. 4d).

$$\left.\begin{aligned} N_t &:= B_t \\ \varphi^r_\alpha(N_t) &:= \{\beta \in N_t / A_t(\alpha, \beta) < r\} \end{aligned}\right\} \quad t = 1, 2, \ldots$$

where $\alpha \in B_t,\ r > 0$

Extension Maps

The extension maps ζ^t, $t = 1, 2, \ldots$, guarantee that all "newly born" cells α (i.e., all $\alpha \in B_{t+1} - B_t$) inherit the state of their "mother cell."

$$\zeta^t : F^{B_t} \to F^{B_{t+1}},$$

$$s(t) \mapsto s(t+1), \quad s(t) \in F^{B_t}, \quad s(t+1) \in F^{B_{t+1}}$$

$$s(t+1)(\alpha) := \begin{cases} s(t)(\alpha) \\ \quad \text{if } \alpha \in B_t \\ (\max\{p_\beta/\beta \in b^{-1}(\alpha),\ p_\beta = s(t)(\beta)/_p\}, \\ \max\{\phi_\beta/\beta \in b^{-1}(\alpha),\ \phi_\beta = s(t)(\beta)/_\phi\}, \\ \max\{m_\beta/\beta \in b^{-1}(\alpha),\ m_\beta = s(t)(\beta)/_m\}) \\ \quad \text{if } \alpha \in B_{t+1} - B_t \end{cases}$$

where $\alpha \in B_{t+1},\quad b_\alpha^{-1} := \{\beta \in B_t / b(\beta) = \alpha\}$

The choice of the maximum function is not biologically motivated. It is simply one (of many possibilities) to introduce a proper map (the maximum of a finite set of real numbers is properly determined).

Transferring Maps

The transferring maps ψ^t describe the global state change

$$s(t) \mapsto s(t+1), \qquad s(t) \in F^{B_t}, \qquad s(t+1) \in F^{B_{t+1}}$$

by

$$\psi^t : F^{N_{t+1}} \to F$$

$$\psi^t(\zeta^t(s(t)) \circ \varphi^r_\alpha(N_{t+1}))(\alpha) := (p_\alpha(t+1), \varphi_\alpha(t+1), m_\alpha(t+1))$$

$$\text{where} \quad \alpha \in B_t, \quad t = 1, 2, \ldots$$

ψ^t is defined by "rules" f, g, and h for each component of the elements of the (local) state space, respectively:

$$p_\alpha(t+1) := g(\zeta^t(s(t)) \circ \varphi^r_\alpha(N_{t+1}))$$

$$\phi_\alpha(t+1) := f(\zeta^t(s(t)) \circ \varphi^r_\alpha(N_{t+1}))$$

$$m_\alpha(t+1) := h(\zeta^t(s(t)) \circ \varphi^r_\alpha(N_{t+1}))$$

It is essential that the state development of a cell α is not dependent on the complete global state $\zeta^t(s(t))$, but only on some neighborhood $\zeta^t(s(t)) \circ \varphi^r_\alpha(N_{t+1})$. It can easily be shown that under the stated assumptions a cellular automaton on the sequence of base spaces $(B_t)_{t=1,2,\ldots}$ is defined in a natural way (Deutsch, 1991).

B. Computer Simulations

The complexity of the automaton does not allow analytical solutions. Cellular automata defined on a sequence of base spaces have not been investigated experimentally or theoretically so far. To get an impression of the dynamics of the automaton, typical state developments are shown calculated with the help of a digital computer.* To reproduce radial zonation patterns as observed in *Neurospora* spore distributions, I tested several hypotheses concerning the differentiation rules (maps f, g, and h). The following hypothesis yields a phenomenological similarity between fungal and simulation patterns (Fig. 3): The basic assumption is that grow-

---

*Simulations were performed on an IRIS 4D-Workstation. Programs are written in C (B. W. Kermighan and D. M. Ritchie, New Jersey, 1978). Source code (on 360-kilobyte $5\frac{1}{4}$-in. floppy disks) can be ordered from the author.

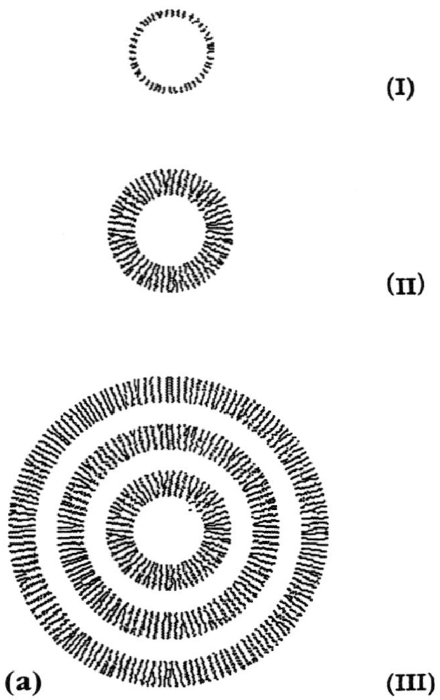

Fig. 6 Computer simulation of the mathematical model with differentiation rule g^* (see the text for an explanation). (a) Initial distribution with equal phases gives rise to concentric ring pattern (diachronous pattern). (I) after 10 time steps, (II) after 30 time steps, (III) after 108 time steps. (b) Phase discontinuity in the initial condition leads to a spiral pattern. (I) after 10 time steps, (II) after 30 time steps, (III) after 101 time steps. Parameters: $|B_1| = 40$, $k_0 = 20$, $\delta = 5$, $\varepsilon = 1$.

ing vegetative hyphae choose one out of two alternatives at every time step; they either activate or inhibit their environment or, alternatively, they differentiate to aerial hyphae/spores. The activation/inhibition (simulated by increasing/decreasing m_α values) influences differentiation, but not mycelial growth and branching. Differentiation is triggered when the inhibition supersedes a critical threshold. The assumption is based on the observation that differentiation can be induced by various stress signals. In particular, it is supposed that induction of differentiation in this case is not dependent on temporal signals released by some endogenous oscillator (e.g., circadian rhythm). Note that the use of the terms *activating* and *inhibiting* influences may be somewhat misleading since differentiation is not induced by highly activating but in contrast by highly inhibitory influ-

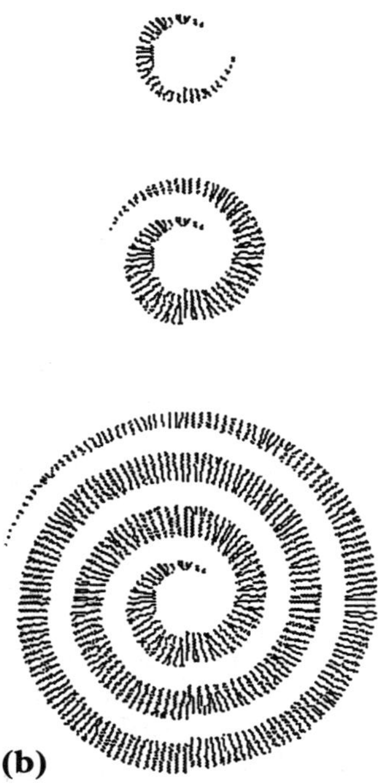

(I)

(II)

(b) (III)

ences. The only use of the terms is to express antagonistic effects. Thus it might be better to talk about +/− influences instead. In the language of the proposed automaton model the hypothesis can be formulated as

$$g(\zeta^t(s(t)) \circ \varphi^r_\alpha(N_{t+1})) = p_\alpha(t+1) := 1_{\{m_\alpha(t)<0\}}$$

Thus a cell α differentiates [i.e., becomes $p_\alpha(t+1) = 1$] if $m_\alpha(t) < 0$ (threshold value: 0). A cell α with phenomenological state $p_\alpha = 0$ (i.e., vegetative hypha) causes a shift of the m-distribution in the surrounding of α in the following way:

Within a distance r_1 of α there is an "activating" influence. Correspondingly, the m_β values of all cells β in the neighborhood $\varphi^{r_1}_\alpha(N_t)$ increase:

$$m_\beta = m_\beta + k_1 \quad \text{for all} \quad \beta \in \varphi^{r_1}_\alpha(N_t), \quad k_1 \in \mathbb{R}^+_0$$

Within a distance r_2 of α there is an "inhibiting" influence. Correspondingly, the m_β values of all cells β in the neighborhood $\varphi^{r_2}_\alpha(N_t)$ decrease:

$$m_\beta = m_\beta - k_2 \quad \text{for all} \quad \beta \in \varphi_\alpha^{r_2}(N_t), \quad k_2 \in \mathbb{R}_0^+.$$

Hence the overall shift of the morphogen value of a particular cell α is a biased sum ($r = \max\{r_1, r_2\}$):*

$$\begin{aligned} m_\alpha(t+1) &:= h(\zeta^t(s(t)) \circ \varphi_\alpha^r(N_{t+1})) \\ &= m_\alpha(t) \\ &\quad + k_1 \cdot \#\{\gamma \in \varphi_\alpha^{r_1}(N_t) \quad \text{and} \quad p_\gamma(t+1) = 1\} \\ &\quad - k_2 \cdot \#\{\gamma \in \varphi_\alpha^{r_2}(N_t) \quad \text{and} \quad p_\gamma(t+1) = 1\} \end{aligned}$$

g and h are independent of the phase information. Thus the rule for phase transition (f) need not be considered (and defined) in this case.

Besides radial zonation patterns (Fig. 3), the model also exhibits concentric ring and spiral patterns. Concentric rings emerge if the differentiation rules are chosen as

$$\begin{aligned} g^*(\zeta^t(s(t)) \circ \varphi_\alpha^r(N_{t+1})) &:= 1_{\{\phi_0 \le \phi_\alpha(t) \le \phi_1\}} \\ f(\zeta^t(s(t)) \circ \varphi_\alpha^r(N_{t+1})) &:= (\varphi_\alpha(t) + 1)(\text{mod } \nu) \end{aligned}$$

where the "temporal signals" ϕ_0 and ϕ_1 mark the beginning and end of the differentiation phase that is superposed on the system by the circadian rhythm. The phases in the initial cells have to be equal (Fig. 6a). A phase discontinuity in the initial condition can lead to spiral patterns (Fig. 6b).

IV. DISCUSSION

Apparently, the band strain of *Neurospora* does not exhibit spiral patterns of spore distribution as they are known from other fungi. This might be due to the existence of diffusible "phasing substances" diminishing initial phase discontinuities (Deutsch, 1991). In testing the proposed hypotheses for differentiation, the molecular analysis of pattern formation of the ascomycete *Neurospora crassa* appears to be a promising project. *Neurospora* is especially suitable for evaluating model predictions and experimental results because more than 2000 auxotrophic and genetic mutants exist. In particular, the molecular processes, which cause the choice between differentiation alternatives g and g^*, leading to synchronous (radial zonation) and diachronous (circular rings) patterns, respectively, have to be analyzed. The structures visible as macroscopic patterns (aerial hyphae/conidia) do not influence the pattern formation process themselves. They are merely secondary structures ("hands"). Thus concen-

---

*In the following, $\#\{a, b, c, d, \ldots\}$ defines the number of elements of a set $M = \{a, b, c, d, \ldots\}$ (i.e., $\#M := |M|$).

tric ring formation may be interpreted as hands of the endogenous temporal oscillation, while radial zonations reflect hands of mutual mycelial interactions of stress processes. Experimentally, the choice can be influenced by appropriate shifts of the medium components. This bifurcation corresponds to a shift in the boundary conditions leading to differentiation alternatives g and g^*, respectively. Growing hyphae cause inhibitory signals in their surrounding, for example, by consumption of nutrients or diminishing the O_2 level. During growth various external protein levels are changed that may also serve as inhibitory signals. Antagonistic signals may be exerted by decreases or increases of external Ca^{2+} or cAMP levels. But so far, nothing is known about synergetic effects of these agents.

The proposed cellular automaton model shows features of self-organized systems: a microscopic feedback mechanism gives rise to ordered structures as radial zonation patterns, concentric rings, and spirals. These are the main macroscopic patterns expressed by filamentous fungi. The deterministic feedback loop consists of a branching and a differentiation mechanism. The initial conditions (i.e., the location of the cells constituting the inoculum ring) are randomly chosen. Only cells at appropriate distances from each other contribute to the radial zonation pattern. Thus the pattern is determined to a high degree by morphogenetic processes occurring immediately after the inoculation. Order, expressed by radial zonations, emerges out of an irregular initial condition (Fig. 3d). In particular, short-range communication is sufficient to account for macroscopic pattern formation. It is interesting to note that this model, despite its nonlinear character, does not include autocatalytic steps. A theoretical analysis of cellular automata based on a sequence of base spaces promises to be not only of theoretical interest; it has been shown here to apply to a large class of biological systems, such as radially growing organisms.

ACKNOWLEDGMENTS

I wish to thank Andreas W. M. Dress (University of Bielefeld) for critically reading the manuscript and many stimulating discussions. This work was supported by Stiftung Volkswagenwerk, Hannover.

REFERENCES

Cellular Automata

Deutsch, A. (1991). Ph.D. thesis, University of Bremen.

Deutsch, A., Dress, A. W. M., and Rensing, L. (1991). In *Nonlinear Wave Processes in Excitable Media*, (A. V. Holden, M. Markus, and H. G. Othmer, eds.), Plenum Press, New York.

Dress, A. W. M., Gerhardt, M., and Schuster, H. (1988). In *From Chemical to Biological Organization*, (M. Markus, S. C. Müller, and G. Nicolis, eds.), Springer-Verlag, Berlin.

Neumann, J. V. (1966). *Theory of Self-Reproducing Automata*, University of Illinois Press, Urbana, Ill.

Wolfram, S. (1986). *Theory and Applications of Cellular Automata*, Advanced Series on Complex Systems, Vol. 1, World Scientific, Singapore.

Neurospora crassa

Mahadevan, P. R., and Mahadkar, U. R. (1970). *J. Bacteriol.*, *101*: 941.

Reissig, J. L., and Kinney, S. G. (1983). *J. Bacteriol.*, *154*: 1397.

Schmid, J., and Harold, F. M. (1988). *J. Gen. Microbiol.*, *134*: 2623.

Scott, W. A., and Solomon, B. (1975). *J. Bacteriol.*, *122*: 454.

Models of Biological Pattern Formation

Bartnicki-Garcia, S., Hergert, F., and Gierz, G. (1989). *Protoplasma*, *153*: 46.

Briére, C., and Buis, R. (1985). *Math. Biosci.*, *74*: 219.

Edelstein, L., and Ermentrout, B. (1989). *SIAM J. Appl. Math.*, *49*: 1136.

Goldwasser, L., Maini, P. K., and Murray, J. D. (1989). *J. Theor. Biol.*, *137*: 271.

Hyver, C., Jerebzoff, S., and Nguyen, V. H. (1979). *Chronobiologia*, *6*: 213.

Meinhardt, H. (1984). *Models of Biological Pattern Formation*, Academic Press New York.

Murray, J. D. (1981). *J. Theor. Biol.*, *88*: 161.

Winfree, A. T. (1973). *J. Theor. Biol.*, *38*: 363.

Young, D. A. (1984). *Math. Biosci.*, *72*: 51.

Index

M